Softwar: A Novel Theory on Power Projection and the National Strategic Significance of Bitcoin

by

Jason P. Lowery

M.S. Astronautical Engineering, Air Force Institute of Technology (2017)
B.S. Mechanical Engineering, Baylor University (2010)

Submitted to the System Design and Management Program
in Partial Fulfillment of the Requirements for the Degree of

Master of Science in Engineering and Management

at the

Massachusetts Institute of Technology

February 2023

Authored by: Jason P. Lowery
System Design and Management Program
January 20, 2023

Approved by: Joan Rubin
Executive Director, System Design & Management Program

Accepted by: Warren Seering
Weber-Shaughness Professor of Mechanical Engineering

This page was intentionally left blank.

Softwar: A Novel Theory on Power Projection and the National Strategic Significance of Bitcoin

by

Jason P. Lowery

Submitted to the System Design and Management Program February 2023
in Partial Fulfillment of the Requirements for the Degree of
Master of Science in Engineering and Management

ABSTRACT

Current analysis of Bitcoin's underlying proof-of-work technology is almost exclusively based on financial, monetary, or economic theory. Recycling the same theoretical frameworks when performing hypothesis-deductive analysis of Bitcoin has the potential to create systemic-level analytical bias which could negatively impact public policy making efforts and could even pose a threat to US national security.

This thesis introduces a novel theoretical framework for analyzing the potential national strategic impact of Bitcoin as an electro-cyber security technology rather than a peer-to-peer cash system. The goal of this thesis is to give the research community a different frame of reference they can utilize to generate hypotheses and deductively analyze the potential risks and rewards of proof-of-work technologies as something other than strictly monetary technology. The author asserts it would be beneficial for researchers to explore alternative functionality of proof-of-work technologies to eliminate potential blind spots, provide a more well-rounded understanding of the risks and rewards of proof-of-work protocols like Bitcoin, and positively contribute to the development of more informed public policy in support of the March 2022 US Presidential Executive Order on Ensuring the Responsible Development of Digital Assets and the May 2022 US Presidential Executive Order on Improving the Nation's Cybersecurity.

Utilizing a grounded theory methodology, the author combines different concepts from diverse fields of knowledge (e.g. biology, psychology, anthropology, political science, computer science, systems security, and modern military strategic theory) to formulate a novel framework called "Power Projection Theory." Based on the core concepts of Power Projection Theory, the author inductively reasons that proof-of-work technologies like Bitcoin could not only function as monetary technology, but could also (and perhaps more importantly) function as a new form of electro-cyber power projection technology which could empower nations to secure their most precious bits of information (including but not limited to financial bits of information) against belligerent actors by giving them the ability to impose severe physical costs on other nations in, from, and through cyberspace. The author calls this novel power projection tactic "softwar" and explores its potential impact on national strategic security in the 21st century. Like most grounded theory research efforts, the primary deliverable of this thesis is a novel theory rather than deductive analysis of a hypothesis derived from existing theory.

Thesis Supervisor: Joan Rubin
Executive Director, System Design & Management Program

Acknowledgements

First, I would like to thank General C.Q. Brown for giving me the courage and top-cover to devote myself to this research topic, and to the Department of the Air Force for entrusting me with this assignment.

I would also like to thank everyone in my personal life who tolerated the amount of time and mental effort I committed to this. You gave me the continual support and confidence I needed to follow through. This endeavor turned out to be more challenging than I expected, in ways that I never would have expected, and your support (plus your sacrifice) played an essential role.

I would like to thank everyone who helped me formulate this grounded theory. When I first indicated my interest in this topic, I could not have imagined the amount of feedback I would receive. I feel very thankful for everyone who took the time to listen to me and challenge my reasoning. Your discourse helped me question my biases and pushed me to formulate the structure of the theory as it exists today. I owe the success of this research effort to you. Thank you for your sincere feedback. I look forward to future discussion and debate.

It's impossible for me to list everyone who I wish to thank by name, but a few specific people I want to give thanks to are as follows: Adam Back, Greg Foss, Robert Breedlove, Jeff Booth, Preston Pysh, Level39, Michael Saylor, Natalie Smolenski, Asher Hopp, Mike Alfred, Dennis Porter, Jason Williams, Luke Gromen, Jim O'Flaherty, SusieB, Max and Stacie Keiser, Peter McCormack, Anthony, Joe, and John Pompliano, Natalie Brunell, Ben Prentice, Joe Burnett, Matthew Pines, Erin Malone, Brandon Quittem, Brian Harrington, Tomer Strolight, George Peacock, Nathan Perry, Cory Swan, Dylan LeClair, Tuur Demeester, MarbellaHODL, Bobby von Hodlwitz, Ck_SNARKs, Mike Hobart, Wealth Theory, Alex Gladstein, Joseph Aguirre, Ghazaleh Victoria, Tarun Chattoraj, Eric Hart, Samson Mow, Phil Dubois, Jimmy Song, Nik Bhatia, Kelly Lannan, Dan Held, Poens, w_s_bitcoin, Mark Moss, and many, many others. It has truly been a pleasure engaging with you.

I would also like to thank Rebecca for the custom drawings she provided.

Biography

Major Jason "Spook" Lowery is a Department of Defense (DoD) sponsored US National Defense Fellow, Department of the Air Force Fellow, MIT System Design and Management Fellow, astronautical engineer, and active-duty field grade officer in the US Space Force (USSF). Prior to attending MIT, Jason served as the director of operations for the USSF Second Space Launch Squadron. Before that, he was a founding member of the cadre of officers who stood up USSF, serving as the deputy chief of the commander's action group for USSF Space Operations Command and US Space Command (USSPACECOM) Combined Force Space Component Command (CFSCC). Jason transferred into USSF from the US Air Force (USAF), where he served as an all-source intelligence analyst and subject matter expert in electronic warfare, blast and ballistics effects, and space weapon system design. Jason has a decade of experience serving as a technical advisor for US senior officials, to include the Office of the President of the United States (OPOTUS), Office of the Secretary of Defense (OSECDEF), and the Office of the Director of National Intelligence (ODNI). He has a master's degree in astronautical engineering from Air Force Institute of Technology, Ohio and a bachelor's degree in mechanical engineering from Baylor University, Texas.

Table of Contents

List of Figures

List of Tables

Acronyms

ABP	Abstract-Power-Based
BCR_A	Benefit-to-Cost Ratio of Attack
B_A	Benefit of Attack
C_A	Cost of Attack
CCCH	Congested, Contested, Competitive, & Hostile
CFSCC	Combined Force Space Component Command
DINO	Decentralized in Name Only
DoD	Department of Defense
EO	Executive Order
OPOTUS	Office of the President of the United States
OSECDEF	Office of the Secretary of Defense
MIT	Massachusetts Institute of Technology
PPB	Physical-Power-Based
SDM	System Design & Management
US	United States
USAF	United States Air Force
USSF	United States Space Force
USSPACECOM	United States Space Command

Executive Summary

Figure 1 shows five different ways that machinery can be used to impose severe physical costs on others in, from, and through five different domains. The image at the bottom shows the specialized machinery that is currently being used to keep special bits of information called "Bitcoin" secure against belligerent actors. This image illustrates the bottom line of this thesis, which is that Bitcoin isn't strictly a monetary protocol. Instead, Bitcoin appears to be emerging as a cyber power projection tactic for the digital age. While most software can only *logically* constrain computers, Bitcoin can *physically* constrain computers and impose severe physical costs (as measured in watts) on belligerent actors in, from, and through cyberspace. Bitcoin's global adoption could therefore represent a revolutionary approach to cyber security and could dramatically reshape how modern society secures their most valuable digital resources.

Figure 1: Five Ways to Impose Severe Physical Costs on Attackers in Five Different Domains
[1, 2, 3, 4, 5]

Bitcoin could represent a strategically vital national security technology for the digital age. However, the American public may not understand why Bitcoin has the potential to be so strategically important because they don't appear to understand the complexity of (1) the computer theory behind the design concept called "proof-of-work," (2) modern power projection tactics, (3) the function of militaries, or (4) the profession of warfighting. If the theories presented in this thesis prove to be valid, then the American public's lack of understanding about these core concepts could jeopardize US national strategic security.

The future of US national strategic security hinges upon cyber security, and Bitcoin has demonstrated that "proof-of-work" functions as a new type of cyber security system. Nations appear to be waking up to the potentially substantial strategic benefits of Bitcoin and learning that it could be in their best strategic interest to adopt it (hence Russian's recent 180-degree pivot to supporting Bitcoin). Another cold war could be kicking off, except instead of a space race, it could be a *cyber* space race. As is often the case with the emergence of any new power projection technology, speed of adoption may be critical.

If the US does not consider stockpiling strategic Bitcoin reserves, or at the very least encouraging Bitcoin adoption, the author believes the US could forfeit a strategically vital power projection technology lead to one of its greatest competitors and set itself back in global power dominance. The current approach that US leaders are taking to analyze the potential risks and benefits of proof-of-work technologies like Bitcoin could therefore represent a threat to US national security. It is particularly concerning that US policymakers have arbitrarily chosen to categorize Bitcoin as "cryptocurrency" and tacitly allow institutions with conflicts of interest to claim to be experts in proof-of-work technology. These institutions could use their misperceived expertise to influence public policy making efforts for their own benefit, compromising US national strategic security in the process.

Computer scientists have been researching proof-of-work protocols for over 30 years – that's more than twice as long as Bitcoin has existed. Since the beginning of this research endeavor, it was hypothesized that proof-of-work protocols could serve as a new type of cyber security system that could empower people to keep computer resources (namely their most valuable bits of information) secure against hacking and exploitation simply by imposing severe physical costs (in the form of computer power) on belligerent actors trying to access or interfere with that information. In other words, computer scientists rediscovered what military officers have known about physical security for thousands of years: to stop or deter bad guys from doing bad things, make it too physically expensive for them to do those bad things.

While academia theorized via formal academic channels about how proof-of-work could work, software engineers and "doers" like Adam Back, Hal Finney, and Satoshi Nakamoto designed, built, and deployed several operational prototypes via informal, non-academic channels. Today, Bitcoin has emerged as by far the most globally-adopted proof-of-work cyber security system to date. Bitcoin is so physically powerful in comparison to other open-source proof-of-work protocols that a popular mantra has emerged, initiated by technologist Michael Saylor (MIT '87): "There is no second best." [6]

But what could Bitcoin possibly have to do with warfare? To understand this connection, one must recall the primary function of militaries. Sovereign nations have a fiduciary responsibility to their people to protect and defend access to international thoroughfares (e.g. land, sea, air, space) to preserve freedom of action and the ability to exchange goods with other nations. When a nation intentionally degrades another nation's freedom of action or ability to exchange goods across these thoroughfares, that activity is often considered to be an act of war. Militaries exist explicitly to protect and defend people's access to these thoroughfares. The way militaries accomplish this is by imposing severe physical costs on those who try to deny access to these thoroughfares or impede a population's ability to exchange goods across them.

Military branches are categorized based on the thoroughfare they assure access to and preserve freedom of action in. Armies assure access to land. Naval forces assure access to the sea. Air forces assure access to the sky. Space forces assure access to space. Regardless of the domain to which access is secured, each service effectively works the same way: preserve the nation's ability to utilize each thoroughfare by imposing severe physical costs on anyone who impedes or denies access to it. Physical power is used to stop and deter belligerent activity in, from, and through these thoroughfares. The more physically powerful, motivated, and aggressive a military is, the better it usually performs. The more a military service can utilize technology to project power in clever ways, the more effective it is at its primary value-delivered function.

One of the most strategically important thoroughfares of the 21st century is colloquially known as "cyberspace." It is of vital national strategic interest for every nation to preserve their ability to exchange a precious resource across this thoroughfare: valuable bits of information. Just like they already do for land, sea, air, and space, sovereign nations have both a right and a fiduciary responsibility to their people to protect and defend their access to this international thoroughfare called cyberspace. If a nation were to intentionally degrade another's freedom of action or ability to exchange valuable bits of information in cyberspace, that activity would likely be interpreted as an act of war, as it would in any other domain.

Until Bitcoin, nations have not had an effective way to physically secure their ability to freely exchange bits of information across cyberspace without resorting to kinetic (i.e. lethal) power. This is because they have not had access to technology which enables them to impose severe physical costs on belligerent actors in, from, and through cyberspace. This appears to have changed with the discovery of open-source proof-of-work technologies like Bitcoin – a complex system which empowers people to physically restrain belligerent actors. This technology works, and adoption has already scaled to the nation-state level.

Thanks to proof-of-work protocols like Bitcoin, nations can now utilize special machinery to impose severe physical restrictions on other nations in, from, and through cyberspace in a completely non-destructive and non-lethal manner. This capability has the potential to transform cyber security by enabling computer networks to run computer programs which don't give a specific group of users special or unimpeachable permissions over the computer network and entrusts them not to exploit those permissions. With the ability to impose severe physical costs on users through cyberspace, zero-trust computer networks (and a new type of internet) can now be designed where users can have their special permissions physically revoked if they abuse or exploit them. The first computer network to prove this design concept appears to be the network of computers utilizing Bitcoin. *Bitcoin is proof that proof-of-work works.*

At its core, Bitcoin is a computer network that transfers bits of information between computers using a zero-trust physical security design. As previously mentioned, bits of information can represent any type of information, including but not limited to financial information that might be used to support international payments and financial settlements. It makes perfect sense that a proof-of-work computer network's first use case would be to physically secure the exchange of vital financial bits of information, but that is clearly not the only use case. This technology could have far wider-reaching applications, as there are many other types of precious information that society would want to physically secure in the information age. To that end, Bitcoin could represent the dawn of an entirely new form of military-grade, electro-cyber information security capability – a protocol that people and nations could utilize to raise cyber forces and defend their freedom of action in, from, and through cyberspace. The bottom line is that Bitcoin could represent a "softwar" or electro-cyber defense protocol, not merely a peer-to-peer electronic cash system. The author believes proof-of-work technology could change the future of national strategic security and international power dynamics in ways that we have barely started to understand.

Chapter 1: Introduction

"We cannot abolish war by outlawing it. We cannot end it by disarming the strong. War can be stopped, not by making the strong weak but by making every nation, weak or strong, able to defend itself. If no country can be attacked successfully, there can be no purpose in war."
Nikola Tesla [7]

1.1 Inspiration

Einstein theorized that mass is swappable with energy. Assuming he's right, this would imply that nations could one day learn how to swap some of their mass-based (i.e. kinetic) defense systems with energy-based (i.e. non-kinetic) defense systems for applications related to physical security and national defense. Modern militaries already utilize both cyber and electronic defense systems, but perhaps there is some other type of defense technology that could combine electric and cyber defense systems together into an electro-cyber form of defense technology. If true, then perhaps one day society will learn how to utilize this special type of technology as a "soft" form of warfare to resolve international policy disputes, establish dominance hierarchies, defend property, rebalance power structures, or even mitigate threats associated with "hard" warfighting, such as nuclear escalation.

Electro-cyber warfighting is not a new idea; it's at least 123 years old. In 1900, Nikola Tesla hypothesized that society would eventually develop such destructive kinetic power that humanity would face a dilemma and be compelled out of existential necessity to fight their wars using human-out-of-the-loop "energy delivery" competitions. He believed humans would eventually invent intelligent machines that would engage in electric power competitions to settle humanity's disputes, while humans observe from afar. [8]

Other titans of the American industrial revolution had complementary ideas about using electricity to mitigate the threat of war. In 1921, Henry Ford (while reportedly standing with Tesla's rival, Thomas Edison) claimed society could eliminate one of the root causes of warfighting by learning how to create an electric form of currency that bankers couldn't control. [9]

Both Tesla and Ford saw potential in the idea of using electricity to either eliminate a root cause of warfare or eliminate a root cause of warfare's associated destruction and losses. However, neither were successful at building the technology required to test or validate their hypotheses. This could have been because both theories predated the invention of "intelligent machines" a.k.a. general-purpose, stored-program computers. Both Tesla and Ford's theories predate the popular theoretical framework we call "computer science" and the development of the abstraction we call "software."

This thesis was inspired by the following question: what if Tesla and Ford were both right, and they were both describing the same technology? What if Ford's theory is valid, and it is indeed feasible to mitigate a root cause of warfare by converting electricity into monetary and financial information? What if Tesla's theory is valid, and the future of warfare does indeed involve intelligent machines competing against each other in human-out-of-the-loop energy competitions? Would this technology not reduce casualties associated with traditional kinetic warfighting? If it did, would this technology not be worth every watt?

Assuming Tesla's theories were valid, then what might "soft" warfighting technology look like? How might this technology impact or re-shape agrarian society's social hierarchies and power structures after spending well over 10,000 years predominantly fighting "hard" or kinetic wars? If Tesla's "intelligent machines" are in fact computers, then wouldn't their power competition be dictated by a computer program? Maybe humanity's "soft" and futuristic form of electro-cyber warfare would take the form of an open source "softwar" computer protocol. And because nothing like it has ever been seen before, maybe nobody would recognize it. This concept is illustrated in Figure 2.

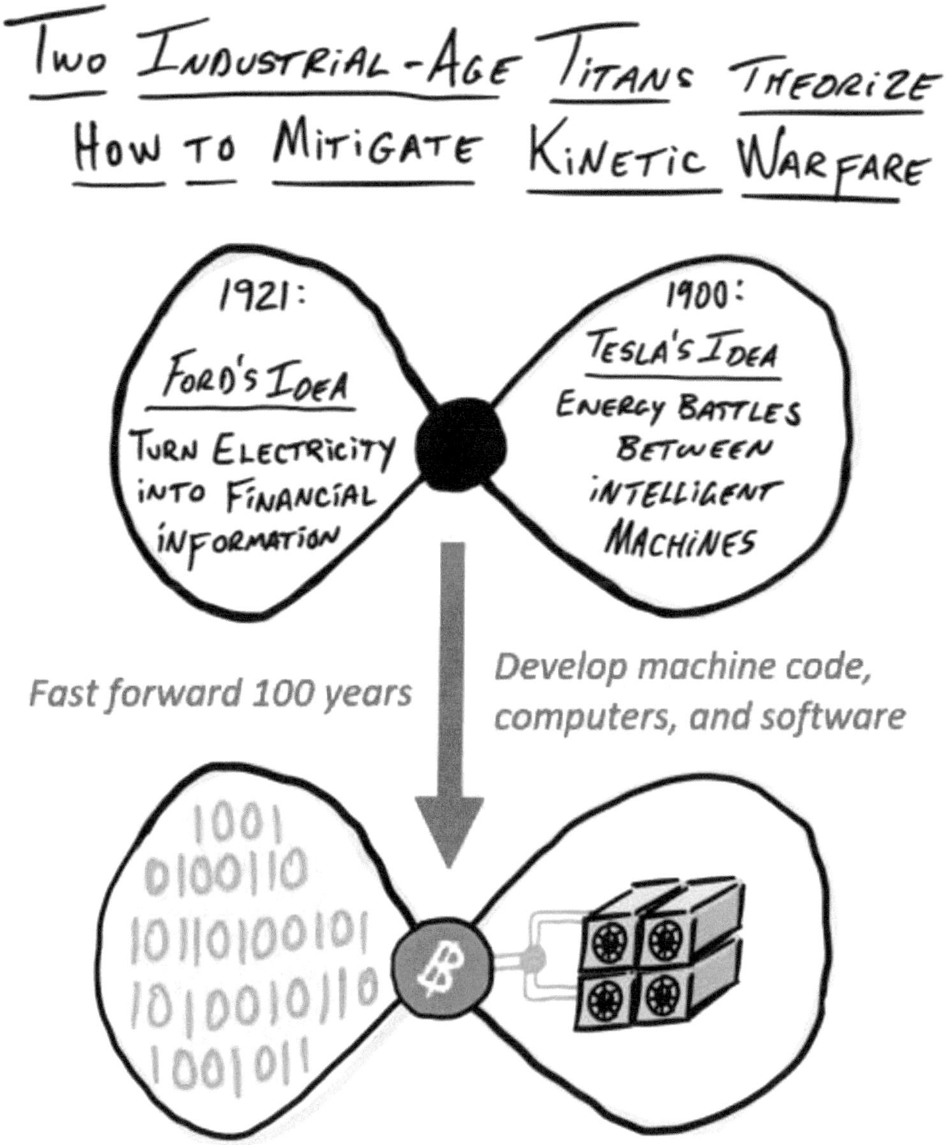

Figure 2: Tesla & Ford's Theories Could Manifest as an Open-Source Computer Protocol

A "softwar" protocol could theoretically utilize society's internationally-dispersed, global electric power grid and existing internet infrastructure to empower computers to impose severe, physically prohibitive costs on other computers in, from, and through cyberspace. It could combine Tesla's and Ford's ideas together and serve as both a "softwar" protocol *and* a monetary network. There's no logical reason to believe it couldn't serve both functions simultaneously, considering how the development and expansion of all technologies need financing – especially defense industrial complexes.

Here's an even more compelling idea: Maybe "softwar" technology already exists and nations are already starting to adopt it. Maybe this new form of power projection technology is already demonstrating how it can empower every nation, weak or strong, to physically secure their interests like never before, thus fulfilling Tesla's prediction. Perhaps this electro-cyber warfighting technology is hiding in plain sight, but people don't recognize it yet because they are mistaking it for a peer-to-peer electronic cash system. Finally, perhaps all it will take for society to recognize that they're entering a completely new and transformational paradigm of non-lethal warfighting is simply a different point of view. To that end, the author presents this thesis.

1.2 Justification

"Airplanes are interesting toys, but of no military value."
General Ferdinand Foch, Supreme Allied Commander of WW1 [10]

1.2.1 If a Softwar Protocol were Invented, is it Safe to Assume we would Recognize its Military Value?

Having suffered through two world wars and fallen into the brink of strategic nuclear annihilation within the past century, it's easy to look skeptically back at Tesla and Ford's theories and think pessimistically about them. But maybe they were truly onto something. Their design concept could have been right, but the computer science needed to implement their ideas simply wasn't developed yet. If that's true, then it would be worthwhile to revisit these design concepts and investigate them further now that society is a century older and more technologically mature (especially with respect to computer technology).

The justification for this research can be explained with a thought experiment. For the sake of argument, let's assume that (1) "soft" warfighting is possible, and that (2) soft wars would be fought using some kind of "softwar" protocol. Is it reasonable to believe that society would recognize the strategic importance or functionality of this technology when it was first discovered? The author asserts there is no reason to believe that a non-kinetic, immaterial, or disembodied form of "soft" warfighting technology would look anything remotely like ordinary warfighting technology. It seems possible – perhaps likely – that this technology would not be recognizable as warfighting technology because it would look and behave nothing like the technologies we normally associate with warfighting.

It's hard to feel confident that society would be able to recognize the functionality of this type of technology if it emerged, considering how many times in recorded history that previous empires failed to recognize strategically vital warfighting technologies when they first emerged – technologies which seem obvious in hindsight (of course airplanes have military value).

In the 9th century, the Chinese alchemists who invented black powder thought it was medicine. It took them centuries to realize that black powder had substantial potential as a new type of warfighting technology for a new type of warfare. For some reason, people during this time weren't inclined to investigate the national strategic security applications of charcoal, sulfur, and saltpeter mixtures. Why? Perhaps it was because no one had ever thought to use black, powdery mixtures to project physical power and impose severe physical costs on adversaries. That changed in the 13th century when iron foundry engineers started inventing complementary technologies to utilize black powder for its capacity to produce lots of power and impose severe physical costs on adversaries. [11]

In the 1450s, Emperor Constantine XI refused to support the adoption of cannons after Orban the iron foundry engineer invented them and offered to build them to defend Constantinople against the neighboring Ottoman Empire. Emperor Constantine was killed a year later during the *cannon* siege of Constantinople. In the 1520s, China burnt down their asymmetrically dominant naval fleet shortly before the discovery of the Americas and the emergence of European naval dominance throughout the age of sail – a mistake which has taken more than six centuries to correct. In the 1860s, the British Royal Navy passionately denounced and refused to adopt self-propelled torpedoes when English engineer Robert Whitehead invented them. [12, 13]

In the 1920s, US Army General Billy Mitchell was demoted and court-martialed for insubordination after lambasting US Army, Navy, and Congressional leaders for their incompetence. General Mitchell accused his superiors of "near-treasonous" incompetence because they refused to accept the validity of emerging theories that airplanes would become equally as strategically vital as battleships and other major military programs at the time. He died 9 years before these theories were conclusively validated by the Japanese attack of Pearl Harbor. He was posthumously restored the rank of general and awarded a Congressional gold medal a year after the conclusion of World War Two. Today, General Mitchell is celebrated as a maverick and widely recognized as the founding father of the US Air Force. [14]

There is no shortage of other examples to demonstrate society's notorious inability to recognize the vital strategic importance of emerging power projection technologies after they're first discovered. Incidentally, there's also no shortage of examples to illustrate how emerging power projection technologies make or break empires. Yet somehow, despite how existentially important it is for empires to recognize and master emerging power projection technologies, their leaders keep forgetting this basic lesson of history and allowing their empires to crumble.

5,000 years of written testimony indicate that failing to recognize the strategic importance of emerging warfighting technologies is the rule, not the exception. Time and time again, empires rise and fall because they keep allowing themselves to be surprised by the emergence of game-changing power projection technologies. Even more absurdly, the people in charge of these empires keep acting like they have an option to refuse or ignore new warfighting technologies after they emerge – as if the cat can be put back in the box, as if they live in an isolated bubble completely separated from the rest of the world, as if their empire is the only empire which gets to decide how they're going to use this technology. Why do rulers keep allowing their empires to be disrupted by new power projection technologies? Why do empires keep forfeiting important technological leads over to their adversaries? There are several explanations.

One simple explanation is that society keeps repeatedly making the same mistake of believing that the next war will look like the last war. To be more specific, people keep making the same mistake of expecting next century's strategically vital warfighting technologies to look like last century's strategically vital warfighting technologies. Faulty assumptions, expectations, and mental models can account for this blind spot. People aren't checking their assumptions, so they aren't aware of how presumptuous they're being.

It's clearly difficult for empires to recognize strategically vital applications of new technologies, even when that technology is placed right in front of their faces. Empires often don't adopt vital new technology even after it's adopted by competing empires. Empires don't take new technologies seriously even when their own military officers literally scream at them to take it seriously before the next major disruption of the existing power dominance hierarchy begins. Instead, they discredit them or discharge them. One would think that the rulers of these empires would learn from the mistakes of their predecessors, but history shows they keep making the same mistakes.

With this history lesson fresh in our minds, let's ask ourselves these questions: is it reasonable to believe that society would recognize a "softwar" protocol if it were invented? Moreover, what would be the potential risks, rewards, and national strategic security implications of a "softwar" protocol were invented, but one empire doesn't adopt it while neighboring empires do? These questions highlight the justification for this thesis. It is of vital national strategic importance to keep an ear to the ground and an eye out for emerging power projection technologies, because failing to recognize them, take them seriously, and adopt them could have dire consequences. Like everything in nature, empires rise and fall based on their ability to adapt to new power projection tactics, techniques, and technologies. Incidentally, this is the author's job as a US National Defense Fellow at MIT – to keep an ear to the ground and an eye out for emerging power projection technologies.

1.2.2 Even if Society Recognized Softwar, is it Safe to Assume they would Adopt it Before it's Too Late?

"[The Manhattan Project] is the biggest fool thing we have ever done.
The bomb will never go off, and I speak as an expert in explosives."
Admiral William Leahy, chief military advisor to President Truman, 1945 [15]

Expanding upon this thought experiment even further, let's assume that (1) "soft" warfighting is possible, that (2) future wars could be partially fought using some kind of "softwar" protocol, and (3) US National Defense Fellows have successfully identified a key-enabling technology for this type of warfighting and are actively working to raise awareness about it – to the point of dedicating more than a year of research to developing a theory about "softwar" to inform the public.

Is it reasonable to believe that society would accept it and adopt it soon enough? Perhaps some people would, but how long would it take for enough people to reach consensus that this technology has vital national strategic significance and should be adopted and mastered – even stockpiled – as soon as possible? Would society reach consensus before their adversaries reached consensus? Would they adopt and master this new, strategically vital power projection technology before their adversaries adopted and mastered it? As has always been the case with the emergence of new, strategically vital power projection technologies, timing is everything.

This thought experiment highlights a national strategic security dilemma. If a new technology does have vital national strategic security implications, this would imply that barriers slowing society down from reaching consensus about the strategic importance of that technology would represent a national strategic security hazard. Like all examples of game-changing power projection technologies to emerge in the past, success depends upon speed of adoption. Thus, anything that degrades speed of adoption would also degrade security.

It's not sufficient for society to *eventually* recognize the vital strategic importance of vital new power projection technology – they must come to consensus about it, adopt it, and master it before their adversaries do. They must not wait around and let their competitors teach them how strategically important this new technology is. As one of many famous examples, the people of Constantinople had less than a year to reach consensus that cannons were of vital strategic importance which must be adopted despite their cost. This was apparently not enough time, so the people of Constantinople allowed Sultan Mehmed II to teach them how important this new technology was the hard way: by example.

During the early 1900s, the US government had a few decades to reach consensus that airplanes, nuclear energy, and rocketry would have game-changing security applications, and it worked out well for them – but not without some hiccups along the way. As those precious few decades ticked along, there were many barriers (e.g. court-marshalling General Mitchell, or bad advice from people like Admiral Leahy) which slowed down public consensus.

In the 1930s, Albert Einstein concluded that nuclear energy likely wouldn't be obtainable in his lifetime. A few years later, Einstein pleaded with the US government to take the national strategic security implications of atomic energy more seriously. By 1939, the situation had become so severe that Einstein – a world-famous pacifist – urged US President Franklin Roosevelt to race to develop atomic warheads before Germany. Many forget that Einstein's famous letter to Roosevelt doubled as a "mea culpa" letter after he famously discredited the potential for nuclear energy and famously advocated against war. [16]

The barriers which slowed the American public from reaching consensus about the strategic implications of these emerging technologies had the potential to seriously jeopardize US national strategic security. Consider how many people would not have been killed, or how much time, effort, and resources would not have been wasted, if military leaders and civil policy makers had accepted Billy Mitchell's ideas sooner and postured the US to be a leader in aerial warfighting in the mid-1930s? Consider what would have happened if Germany – the first to develop jet aircraft and ballistic missiles, had also developed atomic warheads first, months before the US did? The point is, the US dodged several strategic bullets throughout the 20th century. Is it reasonable to assume the US will keep successfully dodging bullets when the next strategically vital power projection technologies emerge throughout the 21st century?

1.2.3 Four National Strategic Security Hazards

"Protect your heads with shields in combat and battle. Keep your right hand, armed with the sword, extended in front of you at all times. Your helmets, breastplates and suits of armor are fully sufficient together with your other weapons and will prove very effective in battle. Our enemies have none and use no such weapons. You are protected inside these walls..."

Purported Final Speech of Emperor Constantine XI during the 1453 cannon siege of Constantinople [12]

These thought experiments are designed to beg a question: what barriers slow the adoption of emerging power projection technologies that have vital strategic security importance? The reason why this question is so important to beg is because every answer to it represents a national strategic security hazard. So what, specifically, are the barriers that slow adoption? The author proposes four answers: lack of general knowledge about the profession of warfighting, pacifism, analytical bias, and cognitive dissonance.

One thing slowing society from adopting emerging power projection technologies is a general lack of knowledge about the profession of warfighting. Some people simply don't have enough experience or understanding with the basics of physical security to make the connection and encourage rapid adoption. This makes sense considering how less than 2% of people actively participate in this profession. Warfare is a niche field of expertise that almost everyone in society outsources to people like the author.

Another barrier is pacifism. Some people are perfectly capable of understanding the potential strategic implications of a new technology, but they have a hard time accepting it because of moral, ethical, or ideological objections. Pacifism slowed the development of both air forces and space forces and was especially prevalent amongst civil scientists during the first decade of nuclear warhead development. At one point, President Truman called Oppenheimer a "cry-baby scientist" and forbid him from visiting the White House because of how much Oppenheimer struggled to emotionally reconcile the development and use of nuclear warheads based on moral, ethical, and ideological objections. [17]

Air flight, space flight, and nuclear energy were technological milestones that many in society wanted to preserve for strictly peaceful purposes. These people were perfectly aware of the fact that air flight, space flight, and nuclear energy technologies could be used for physical power projection and warfare, they just objected to it based on ideological reasons, and discredited the people who talked about using it for security purposes as being "war mongers." Outspoken pacifists like Einstein famously overcame these objections and eventually encouraged the development of nuclear warheads, recognizing the simple fact that **nobody has the option of outlawing their adversaries from utilizing these technologies against them.**

Whether it's due to lack of knowledge about warfighting or pacifism, the core challenge associated with getting the public to quickly adopt strategically vital power projection technologies appears to be a biproduct of self-domestication. Like any other kind of animal, humans are vulnerable to becoming too docile and domesticated. This can cause them to misunderstand the importance of emerging power projection technologies. This may sound impolite, but it's a legitimate assertion backed by no shortage of scientific evidence and written testimony that will be discussed throughout this thesis.

Docility and self-domestication are reoccurring security problems that are pertinent to subject matter concerning national strategic security and any emerging physical power projection tactic, technique, or technology. It's possible for human populations to spend too much time separated from nature to understand their own nature. In their comfort, complacency, or perhaps even hubris, they forget how strategically important it is to remain at the top of the power projection curve, so of course they will struggle to understand how new power projection technology functions and why it is so important for them to adopt it as soon as possible.

Some have argued that expecting a domestic society to see the functionality of emerging power projection technology (i.e. weapons technology) is like expecting a golden retriever to understand the functionality of a wolf collar. This concept is illustrated in Figure 3. In their domesticated state, golden retrievers don't know what they are and where they come from, so naturally they aren't going to understand what happens when they encounter the natural, undomesticated version of themselves. Retrievers don't know that their aversion to physical conflict makes them extraordinarily vulnerable and a ripe target of opportunity for predators. So they aren't going to understand how their wolf collar technology works and why it's so important for them to use it in the presence of wolves.

Figure 3: A Domesticated Wolf Wearing a Wolf Collar

There are other scenarios where people slow adoption even when they have plenty of knowledge about warfighting and no ideological objections to it. These are usually people who are perfectly capable of understanding the potential strategic security implications of new technologies, but they nevertheless still forfeit technological leads to their adversaries. These scenarios illustrate a third and fourth barrier slowing society from adopting strategically vital new power projection technology.

A third barrier preventing society from adopting strategically vital new power projection technology is analytical bias. Sometimes, people aren't aware of how biased their analysis of a given technology is because they don't recognize their assumption that the first intended use case of a given technology is the most important or even the most relevant use case. For example, when alchemists first started to theorize about the medicinal risks and benefits of black powder, they were inadvertently biased because they only analyzed the first intended use case. They weren't aware of the assumptions they were making – namely the assumption that black powder was strictly a form of medicine that couldn't be useful for several other applications.

Why did alchemists make so many assumptions about black powder? Perhaps it was just because they intended to build medicine, so they named it medicine and only evaluated it as medicine. This created a barrier to national adoption of black powder that existed for as long as nobody thought to use a different theoretical framework to analyze black powder as something *other* than medicine. While this is somewhat of an oversimplification of the issue, it illustrates the point that this same phenomenon could also be a barrier slowing down US adoption of what could become critical proof-of-work cyber security technologies like Bitcoin. We need to recognize the assumption we're making that the most important or even the most relevant use case of Bitcoin is its first popular use case (internet money). There is no shortage of examples of emerging technologies where it's not reasonable to assume that the first intended use case will be the technology's primary use case.

A fourth barrier is cognitive dissonance. Sometimes, people can see the existentially important national strategic security implications of emerging technologies, but they struggle to accept and reconcile what they see because it contradicts their preconceptions. Faulty preconceptions can be caused by phenomena already discussed, like a lack of warfighting expertise, ideological objections, or analytical bias, but they can also happen due to fear, shock, or even pride. In plain terms, **change is scary, and it's easier on the emotions to ignore or discredit the threat because we don't like the way it feels to be threatened** (especially if we've become too accustomed to being the top dog).

Esteemed gunnery specialists of the British Royal Navy bitterly opposed adoption of Whitehead's self-propelled torpedoes because of how hard it was for them to reconcile how effective this technology would be at subverting and countervailing the combined strength of their Navy – the world's most powerful military force at the time. It's hard to dedicate one's career to the mastery of one form of warfighting, only for it to become obsolete by the emergence of a new power projection technology. As history has shown time and time again, **all it takes is one engineer to subvert the authority of an entire military institution and undermine its combined expertise**. As Edwyn Gray notes about Whitehead, *"this relatively unknown English engineer exerted more influence over the tactics of naval warfare and the design and development of warships than all the world's top Admirals and naval architects put together."* [13]

Cognitive dissonance due to fear, shock, or pride could also explain why the US rejected General Mitchell's theories about how vulnerable US Naval warships would be against aerial bombardment. General Mitchell famously led a demonstration where a captured German warship was sunken by an airplane-launched torpedo to demonstrate how easy it would be for adversaries to employ the same power projection tactics, techniques, and technologies against US battleships. Nevertheless, officials rejected his assertions about the emerging strategic importance of air power and continued to invest heavily in the development of battleships throughout the 1930s – battleships which now lie at the bottom of the Pacific Ocean.

Accepting the national strategic implications of disruptive new power projection technologies and then pivoting to adopt them as soon as possible is naturally going to be hard in bureaucratic systems like the US government that have been intentionally designed *not* to change quickly. This challenge is compounded by the complexities of novel technologies with a steep learning curve. But other times, such as in the case of the emergence of the Whitehead torpedo, lack of adoption could be attributed to hubris, cognitive dissonance, and sunken cost fallacies.

Being able to recognize the strength and disruptive potential of new power projection technologies requires one to have the courage to accept and recognize one's own weaknesses and vulnerabilities to those technologies. Accepting the emergence of disruptive new power projection technology requires a population to accept the fact that existing defense systems (which they probably paid a lot of money for) aren't going to secure them as well as they thought they would. For example, investing substantially in kinetic warfighting technologies may not provide as much security as expected and could lead to a substantial amount of sunken costs if other nations learn how to utilize non-kinetic or "soft" warfighting technologies via cyberspace that can bypass kinetic strength altogether. If this type of situation were to happen, it would be the fiduciary responsibility of those responsible for sinking so much money into increasingly irrelevant kinetic warfighting technologies to accept these sunken costs and maneuver accordingly.

Moreover, shock and even denial are common responses to the sudden existential dread faced by a person or a population when they realize they are losing a vitally important technology lead to their adversaries because of assumptions they didn't realize they were making – assumptions like expecting the next war to look like the last war or expecting digital-age warfighting technologies to look the same as non-digital-age warfighting technologies. It is hard to reconcile the idea that one's baked-in assumptions about the future of warfare could irrevocably harm one's country, but this is the responsibility of all leaders, especially military officers. Empires rise and fall based on the baked-in assumptions guiding the decisions of those entrusted with the responsibility of national security. And warfare is a path-dependent phenomenon that is highly unforgiving to people who make miscalculations because they don't take the time to seriously question these assumptions.

The point is that no empire is safe from technological disruption. It's strategically essential for populations not to allow fear, shock, hubris, complacency, or sunken cost fallacies slow their adoption of important new power projection technologies when they emerge. Speed of adoption has always been critical, especially when factoring in how severe and highly path-dependent the consequences can be if vital new power projection technologies aren't adopted quickly. Military leaders must hold none of their expertise in existing power projection tactics, techniques, and technologies too sacred, because winning strategies can change as quickly and as often as the technological environment changes, and there's no doubt that our technological environment is changing rapidly in the digital age, perhaps more rapidly than in any other time in the history of the profession of warfighting.

1.2.4 Raising Awareness and Educating the General Public about new Power Projection Technology

"Educate and inform the whole mass of the people...
They are the only sure reliance for the preservation of our liberty."
Thomas Jefferson [18]

For whatever reasons society might struggle to accept the national strategic security implications of emerging power projection technologies, the answer to overcoming these barriers seems to be the same: raising awareness and educating the public. If a population is too domesticated (i.e. too passive or separated from the business of physical confrontation) to understand the dynamics of physical power projection, or they feel inclined to reject a theory based on ideological reasons, this barrier can be mitigated by raising more awareness and educating the public. If the public is suffering from systemic-level analytical bias or they're struggling to reconcile cognitive dissonance associated with a potentially disruptive power projection technology, this barrier can be mitigated by raising more awareness and educating the public.

To that end, the primary justification for this research is to provide more information about the possible national strategic security implications of an emerging power projection technology called Bitcoin. A simple definition of Bitcoin is that it's the world's most widely-adopted open-source proof-of-work computer protocol to date. Proof-of-work is a new type of computing protocol which enables users to keep cyber resources (i.e. software and the corresponding bits of information managed by that software) secure against attacks not just by encoding logical constraints, but by imposing severe physical costs. Whereas most computer systems only use encoded logical constraints to keep themselves secure against systemic exploitation (i.e. hacking), proof-of-work systems like Bitcoin use real-world physical power (i.e. watts) to keep cyber resources *physically* secure against attack by imposing severe physical costs (as measured in watts) on any would-be attackers. Based on a theoretical framework developed and presented in this thesis called "Power Projection Theory," the author hypothesizes that Bitcoin is not strictly a monetary technology, but the world's first globally-adopted "softwar" protocol that could transform the nature of power projection in the digital age and possibly even represent a vital national strategic priority for US citizens to adopt as quickly as possible.

1.3 Background

"War is merely the continuation of policy with other means."
Carl von Clausewitz [19]

1.3.1 Modern Warfare 101

The author recognizes that many readers do not have expertise or training in modern military theory. Therefore, before proceeding into a detailed theoretical discussion about the potential national strategic security implications of proof-of-work technologies like Bitcoin, it might be beneficial to establish a common understanding of the profession of warfighting and briefly elaborate on the "softwar" neologism. For this, the author turns to one of the most respected modern military theorists, General Carl von Clausewitz.

In the early 1800s, General Clausewitz examined the nature of war and defined it as a trinity with three distinct characteristics. First, war is comprised of the same "*blind natural forces*" of "*primordial violence*" observed in nature. Second, war contains "*the play of chance and probability*" rewarding "*creative spirits*." Third, war is an instrument of national policy used to resolve political disputes. [19]

Clausewitz's explanation of warfare is noteworthy for several reasons. The first is because it acknowledges warfare as a *primordial* phenomenon – something that has existed since the beginning of life on Earth. If we were to combine this observation with what we now understand about biology, we could note that physical power competitions like warfare predate humans by billions of years. Physical power competitions play an essential role in evolution as well as the establishment of natural dominance hierarchies, including but not limited to human dominance hierarchies. Physical power competitions can be observed in every corner of life at every scale, helping organisms of all kinds solve the existentially important problem of establishing a pecking order over Earth's limited resources. [19]

The second reason why Clausewitz' definition of war is noteworthy is because it acknowledges how the *natural forces* levied during war are intrinsically *blind*. In other words, physical power competitions are completely unbiased, indiscriminate, and exogenous to people's belief systems. Based on what we now understand about physics, we note that the "natural forces" to which Clausewitz refers are forces displacing masses to generate physical power, a.k.a. watts. These watts are most often generated kinetically, through Newtons of force displacing kilograms of mass. We can independently and empirically validate from our own personal experiences that watts are indeed blind. They don't appear to show favoritism or have any discernable prejudices. They have no known capacity for misrepresentation, and all things appear to be subordinate to them – including and especially the people at the top of existing dominance hierarchies who enjoy high rank, wealth, and social status. [19]

Third, Clausewitz's trinity acknowledges war as a game of probability, the outcome of which favors *creative spirits*. War could therefore be described as not just an indiscriminate physical power competition, but a probabilistic one – a.k.a. a lottery. Moreover, winning the power lottery is not merely about finding ways to amass larger quantities of physical power; it's about combining people's intelligence to utilize physical power in the most creative and innovative ways possible. Why? As an instrument for resolving international policy disputes – which include but are not limited to international *monetary* policy disputes. [19]

It was from this point of view that Clausewitz made the aphorism for which he is famous: "*War is merely the continuation of policy with other means.*" [19] What Clausewitz meant by this statement is that even though war is inherently destructive, nations do not go to war merely for the sake of demonstrating their capacity for destruction. Instead, nations seek to utilize the *blind natural forces* as an alternative means to resolve international policy disputes. In other words, **war is a mechanism for nations to settle policy disputes using physical power (a.k.a. watts) rather than a court of law, because the former is perfectly indiscriminate, while the latter is not**.

1.3.2 War vs. Law

But why would nations prefer something as lethal and destructive as warfare to resolve international policy disputes when they have a far more energy-efficient option of peaceful adjudication through a court? The answer is quite simple: because they don't trust, respect, or sympathize with the court. To borrow a concept from an anonymous software engineer, the root problem with peace is all the trust that's required to make it work. [20]

It's not easy to get a large population to come to consensus about what "right" means, much less what the "right" ruling is, or the "right" rule of law is. It's even more difficult to expect large populations of people to trust their lawmakers not to abuse the abstract power and control authority given to them by the existing rule of law. It's perhaps even more difficult to trust people both inside and outside a given nation to sympathize with a nation's laws. History is full of breaches of the enormous amount of trust that's required to make law-based societies function properly. The incontrovertible truth of the matter is that law-based societies break down because they are systemically vulnerable to corruption and invasion.

Rule of law is a highly energy-efficient cooperation protocol, especially for the purpose of establishing human dominance hierarchies (i.e. pecking order) and achieving consensus on the legitimate state of ownership and chain of custody of limited resources. Rule of law is particularly well-suited for functions like property dispute adjudication. But law-based-societies aren't perfect; they come with substantial tradeoffs. Like all rulesets, laws are inherently inegalitarian. They create a ruling class and a ruled class. Laws are also trust-based. They only function properly when the ruling class can be trusted not to exploit the ruled class, when people can be trusted to follow the rules, and when neighboring societies can be trusted to sympathize with their neighbor's rules. In other words, law-based societies are predisposed to systemic exploitation and abuse because they rely too much on trusting people who do not deserve to be trusted. **Humans are apex predators; trusting an apex predator not to attack or exploit a belief system is not a good security strategy, regardless of how energy-efficient and non-destructive it looks.**

The intent of law is noble, but for reasons that are exhaustively explored in this thesis, the inegalitarian and trust-based nature of law-based social structures make them systemically insecure, hence every corrupt or oppressive government to have ever existed. Law-based societies are prone to reaching a hazardous state over time, leading to substantial losses for their populations. Perhaps the ruling class finds a reason to systemically exploit the law, creating a state of oppression. Perhaps the ruled class finds a reason to stop following the law, creating anarchy. Perhaps a neighboring nation finds a reason to be unsympathetic to their neighbor's laws, creating an invasion. 5,000 years of written testimony about law-based societies makes one thing very clear: give them enough time, and they will inevitably become dysfunctional.

When law-based societies break down, war follows. Like law, war has its own tradeoffs. War is highly energy-intensive and destructive, but it's also egalitarian. Physical power makes no distinction between the ruling and ruled class; a king suffers the same from a sword through the heart as a peasant (across history, kings and other high-ranking people especially have a habit of losing their heads after losing wars). War is zero-trust; it doesn't require trust to function properly. War is also unsympathetic to people's belief systems, thus completely impartial to them. Therefore, physical power competitions work the same regardless of what people believe and whether people are sympathetic to it.

Here we can see that law and war are remarkably complementary to each other in the sense that they represent almost perfectly opposite approaches to achieving the same end state. Together, they form an interdependent system with opposing tradeoffs. Law is an energy-efficient and non-destructive way for society to settle disputes and establish a dominance hierarchy, but it requires people to adopt common belief systems that are trust-based and highly inegalitarian, making them demonstrably vulnerable to systemic exploitation and abuse. On the other hand, war is an energy-intensive and destructive way for society to settle disputes and establish dominance hierarchies, but war also doesn't require people to adopt a common belief system, making it practically invulnerable to systemic exploitation and abuse. These are the tradeoffs that a population must weigh when settling policy disputes.

Understanding the systemic differences and tradeoffs between war and law helps explain why the application of physical power is so effective at restoring law and order when rules-based societies inevitably become dysfunctional. Rulers who exploit rules of law can be compelled to stop by using physical power, in what's often called *revolution*. Participants who can't be trusted to follow the rules can be compelled to follow them using physical power, in what's often called *enforcement*. Outsiders who don't sympathize with the rules can be compelled to sympathize with them using physical power, in what's often called *national strategic defense*.

Whenever laws become dysfunctional, the cure appears to be the same for each affliction: keep projecting increasing amounts of physical power in increasingly clever ways until symptoms improve. As energy-efficient and peaceful as it would be to remedy the exploitation, abuse, and neglect of laws using courts of law, it clearly doesn't work in all cases – else there would be no war in the first place. A great deal of evidence suggests that signing policy is a far less effective way to fix dysfunctional societies than applying brute-force physical power (see the 1938 Munich Agreement for one of countless examples from history).

1.3.3 War has Major Benefits for Society No Matter How Much Society Hates to Admit it

"I wish there were a war! Then we could prove that we're worth more than anyone bargained for."
Alexander Hamilton, *Hamilton* the Musical [21]

What makes physical power so useful and effective as a means for securing property and settling policy disputes? One explanation could be what Clausewitz already observed; physical power is completely blind. Physical power doesn't see people's feelings, much less allow itself to be influenced by them. This makes physical power impartial and virtually immune to politicking. Physical power has no apparent capacity for favoritism, discrimination, or hidden agendas. It can't be manipulated or corrupted. It doesn't appear to have any systemically exploitable attack surface whatsoever because it's exogenous to people's belief systems.

For this reason, the court of power functions as a true meritocracy. It gives people the freedom to challenge and change policy regardless of rank and social status. This makes it serve as a reliable court of appeals for people who are being systemically exploited by an existing dominance hierarchy. Those who feel wronged by their laws can (and often do) turn to the supreme court of brute-force physical power to give them a judge that is mercilessly impartial. Physical power's rulings are quick and decisive. The basis for its judgement is known equally by everyone, and its verdict is very easy to audit – making it an easy way to achieve consensus.

Another explanation for why physical power is so useful is because it is virtually unlimited and relatively easy to access. There are hard limits to the amount of rank, votes, and social status that a person can obtain within their chosen belief system, and these imaginary forms of power are fickle, nepotistic, and inegalitarian. It takes a lot of time and effort to ascend existing rank-based dominance hierarchies. High-ranking positions are often unavailable across multiple generations. It's often far easier and more achievable to simply change the existing dominance hierarchy using physical power than to climb through the imaginary ranks of the existing one.

Physical power is very different than rank. There is virtually no limit to the amount of physical power that people can summon to shape, enforce, and secure the policies they value. Physical power is accessible via one's own ingenuity and merit. People also tend to respect physical power more because of how self-evident it is. Physical power is proof of its own merit; it doesn't need anyone to believe in it to know its worthiness. This is in stark contrast to rank and social status, which are both part of abstract, artificial, and inegalitarian belief systems which are incontrovertibly vulnerable to exploitation and abuse from those who have the most rank and social status.

Therefore, as much as people hate to admit it, there are major benefits to war, which would explain why societies wage it so frequently. Moreover, from a sociotechnical perspective, war has proven efficacy. It has clearly played a major part in the formation of high-functioning agrarian societies. Virtually every nation today was forged through war. National borders are sculpted by war. The development of state-of-the-art technology is often accelerated by war. The most warring societies on Earth have consistently had the largest economies. [22]

Therefore, despite how unpopular it is to talk about the benefits of war, it could be useful to at least take the time to understand what those benefits are, so that we can understand why wars keep happening. Endeavoring to understand the benefits of war could help us learn how to design systems that minimize our need for those benefits. Alternatively, endeavoring to understand the benefits of war could help us gain insights about ways to wage it better, perhaps in "softer" ways that are far less destructive. In the author's opinion, understanding the merit of war is key to understanding the merit of proof-of-work technologies like Bitcoin.

1.3.4 Like any other Profession, the Profession of Warfighting could also be Disrupted by Software

Computing machines are more than two millennia old. For thousands of years, instructing computers how to operate involved mechanical, kinetic activity (i.e. forces displacing masses). Even after the invention of fully-electric general-purpose computers, programmers were still required to spend days pulling levers, turning dials, flipping switches, pressing buttons, and plugging wires to operate them.

Eventually, as technology matured, scientists, engineers, and mathematicians came up with a revolutionary concept: converting kinetic computer programming operations into digitizable states and storing them in electronic computer memory. By digitally converting kinetic computer operations into electrically-actuated states, special-purpose computing machines could be repurposed and reprogrammed instantaneously without the lengthy and expensive process of having to redesign, rebuild, remanufacture, or manually reprogram them. Machines that were either physically impossible to build or impractical to operate suddenly became very feasible, and modern agrarian society was forever changed.

Stored-program general-purpose computing was such a profound invention that it caused society to split the way it perceives computer programming into two separate concepts. Manufactured elements (i.e. wares) of computing machines that are material and that require forces displacing masses were conceptualized as "hard" (i.e. mass-based) wares. The other manufactured elements of computing machines began to be conceptualized as "soft" (i.e. massless) wares.

Since "soft" computing "wares" first emerged, society has raced to convert all sorts of material activities into immaterial activities performed by general-purpose, stored-program computers. The past 80 years have been uniquely characterized by the dematerialization of special-purpose machines (e.g. printing presses, typewriters, phones, calculators, televisions) across every industry, as computer scientists and engineers have continued to hone their skills and perfect their craft. Society seems to be constantly reeling from the disruption of "soft" computer "wares," somehow always surprised by what can be converted into the massless, disembodied form of a computer program despite how routine the disruption has become. No profession appears to be uninterruptable by the technology we now call "software."

With these concepts in mind, the author would like to challenge the reader with the following question: Is it reasonable to expect the profession of warfighting to be any exception to this trend of technological disrpution? Is it reasonable to expect that the profession of warfighting won't be, at least partially, dematerialized and disrupted by software now that there is close to a century of examples of this technology dematerializing and disrupting virtually every other profession? Why *wouldn't* we expect the profession of warfighting to be substantially disrupted by software if so many other activities have already been?

1.3.5 What would Soft War Look Like?

For the sake of argument, let's assume warfighting will be disrupted by software just as all professions have been disrupted by software. Let's assume software could enable the dematerialization of special-purpose war machines just as they have dematerialized countless other special-purpose machines over the past several decades. What might that look like?

Using Clausewitz's definition of war, consider what it would take to use general-purpose computers and software to engineer a massless, immaterial, or disembodied form of the physical power competition which human societies utilize as their "other means" for settling global policy disputes. In other words, let's consider what kind of technology might cause society to split the way it perceives the profession of warfighting into two separate mental concepts: "hard" warfighting machines and "soft" warfighting machines.

To engage in a soft form of warfighting, people would need to figure out a way to take the mass out of global-scale physical power competitions. One way to do this would be to project physical power electronically via charges passing across resistors, rather than kinetically via forces displacing masses. But simply being able to wield and project electronic power in, from, and through cyberspace is not enough to satisfy the criteria for war according to Clausewitz's definition because it's missing two other ingredients.

Soft war machines would also need to incorporate "a *play of chance and probability that rewards creative spirits*." Therefore, in addition to inventing a mechanism for projecting physical power electronically, there would also need to be some sort of probabilistic protocol for people to compete against each other. There would need to be a clear winner of this competition that doesn't require a court or a judge to declare it. To that end, it seems feasible that people could design a common protocol which establishes internationally agreed-upon standards for wielding electronic power and competing against each other in a zero-trust and egalitarian manner.

The final ingredient for soft war according to Clausewitz's definition is for nations to simply start using this electronic power projection technology to settle their policy disputes. Soft war machines would need to serve their nations as a continuation of policy with other means, just as Clausewitz aptly described war. Nations would need to start using it to physically secure the international policies they value.

1.3.6 Soft War would have Strategic Benefits for Rebalancing Power Structures

What exactly would be the benefits of creating a soft form of warfighting? To answer this, let's consider the previously mentioned benefits of warfighting and then reflect upon what would happen if it became more massless, disembodied, or immaterial. A soft form of warfighting would give people access to the supreme court of physical power, and that court would likely be just as indiscriminate and impartial in electronic form as it already is in kinetic form. For this reason, soft warfighting protocols could be ideal for small countries wielding small amounts of kinetic power (i.e. small militaries) seeking to settle policy disputes with larger countries wielding large amounts of kinetic power (i.e. big militaries).

Soft warfighting could also clearly have utility for nuclear superpowers seeking to settle major policy disputes with other nuclear superpowers, because it would enable them to battle each other without the threat of mutually assured destruction. An international monetary policy dispute, for example, represents the type of strategic-level international policy dispute that could be settled using a soft form of warfighting. This kind of dispute would be a prime candidate for nuclear peers seeking to settle a major political conflict in an energy-efficient way that is far less likely to escalate than traditional warfare would. Of course, an international monetary policy dispute wouldn't be the only type of policy dispute that could be resolved using a soft form of warfighting, but it does seem to be an obvious first use case (especially considering how the US is openly denying another nuclear power's access to a financial computing network, such as what the US is currently doing to Russia via sanctions).

If society were to invent a soft form of international warfighting, the tradeoff between law and war would likely remain the same. War would still represent a more energy-intensive way to settle disputes and establish a dominance hierarchy over limited resources than law would be. But a soft form of warfare would only burn watts electronically, not kinetically. It would therefore have profoundly different emergent behavior – a major one being its non-destructive side effects. By definition, there would be no kinetic forces or masses involved in a disembodied or immaterial form of soft warfighting, so there would likely be no practical threat of physical injury. Thus, a soft form of warfighting would represent a non-lethal form of warfighting – making it a potentially game-changing and revolutionary way for nations to establish, enforce, and secure international policy.

Of course, soft warfighting machinery would not completely replace the need for hard warfighting machinery. This makes sense considering how we already know that software doesn't completely replace the need for hardware. We should expect soft warfighting machinery to continue to rely on hard warfighting machinery just like software continues to rely on hardware. As long as people continue to value material things with mass, they will need a mass-based method of kinetic warfare to keep that mass secure. But it just so happens to be the case that much of what people value is as disembodied, immaterial, and massless as the soft war machines which could one day be used to physically secure them. For example, money doesn't require mass. Money's predominate form is already the disembodied, immaterial, and massless form of software. As another example, common belief systems like social contracts or international policies never had mass in the first place. Things like constitutions and rules of law have always been disembodied and immaterial, so who says they can't be physically secured against systemic exploitation and abuse in a disembodied and immaterial way?

Just as software dramatically changed society's understanding of how to build and operate machines, soft war could dramatically change society's understanding of how to build and operate social systems, particularly with respect to the way societies agree on policies, physically enforce them, and physically secure them against systemic exploitation and abuse. Similar to how the amount of hardware needed to compute things substantially decreased following the invention of software, the amount of hard warfighting machinery needed to secure our policies and our belief systems using physical power could decrease following the invention of soft warfighting machinery. Considering how software made it feasible to build what were previously considered to be impossible or impractical computing systems, soft warfighting could make it feasible to build what were previously considered to be impossible or impractical defense systems. Bizarre and counterintuitive solutions for national security (solutions like non-lethal and non-destructive world warfare) would theoretically be feasible if a soft form of warfare were discovered and utilized by nation states. And it could dramatically transform society's ethical calculus in the process.

1.3.7 Soft War could Change a Society's Ethical, Moral, or Ideological Calculus

Systems security is a trans-scientific phenomenon. It involves difficult, unquantifiable questions about system design, morals, and ethics. These types of trans-scientific questions would likely become a major topic of conversation if society were to discover a soft form of warfighting – that is, a new way of settling policy disputes in a zero-trust and permissionless way that is physically incapable of causing injury.

Would it be moral or ethical to design legal systems using pen and parchment which are (1) insecure against systemic exploitation and abuse, and (2) must be enforced and secured using lethal kinetic power, if society discovered a non-lethal and non-destructive alternative to warfare? Imagine if society were to discover a way to write down policies using C++ instead of parchment, then enforce and secure those policies using physically harmless electric power. A discovery like that could change society's perception about the moral value of traditional laws *and* warfare simultaneously.

Would it be ethical to prohibit people from securing property and enforcing policies they freely choose to value in a non-lethal way? Some people might argue that it's not ethical to prohibit people from securing their property and policy at all, even if it does lead to serious injury. For American readers, consider what the authors of the 2nd amendment of the US Constitution would think about government officials advocating for public policies which prohibit citizens from physically securing their property. This becomes a relevant topic of conversation when writing public policy about Bitcoin, if the theories presented in this thesis are valid that Bitcoin represents a physical security system rather than strictly a monetary system.

Attempting to outlaw electric forms of power projection technology like Bitcoin could eventually be seen as a double-standard. Why would it be acceptable for citizens to utilize kinetic (i.e. lethal) forms of physical power to secure the property and policy they value, but not electro-cyber (i.e. non-lethal) forms of physical security? The intent of the second amendment was to make it more difficult for leaders of the US government to infringe upon their citizen's right to *physically* secure what they value. Should it matter if the technology used to physically secure property utilizes electric power rather than kinetic power? And so what if that technology uses a substantial amount of power? As will be discussed in this thesis, using a substantial amount of electric power to impose a substantial amount of physical costs is the primary-value-delivered function of proof-of-work protocols and a key missing ingredient to cyber security.

Civilizations have been projecting physical power to secure the property and policies they value since policies first emerged more than five thousand years ago. There is no evidence to suggest that society has ever found, or ever will find, an energy-free way of doing it. Physical security requires the expenditure of watts. Watts are expended to impose a severe physical cost on attackers, thus stopping or deterring the attack. Therefore, the fact that some policy makers are discouraging the use of proof-of-work cyber security technologies like Bitcoin because of a belief that these technologies expend too many watts could be an affront to the intent of the 2nd amendment and the founding philosophy of the US. [23]

Watts are watts regardless of whether they're generated kinetically or electronically. Rulesets are rulesets regardless of whether they're written on parchment or programmed into a computer. Private property is private property regardless of what form it takes. There may not be a lot of room for moral ambiguity when it comes to the right to defense; Americans already have a right to bear arms to physically secure their access to the property and policies they freely choose to value, including and especially against their own government (this was the express intent of the amendment according to several founding fathers). It's also not difficult to make the argument that it would be morally, ethically, and ideologically preferable to everyone involved if physical security of property and policy could be achieved non-lethally and non-destructively, as it could be if physical costs were imposed electronically.

Now consider the international implications of this technology, rather than just the domestic implications. By converting kinetic warfighting or physical security operations into digital-electric form, written rulesets (e.g. laws) that are inherently vulnerable to systemic exploitation can be secured using (non-lethal) electric power rather than (lethal) kinetic power. International policies (e.g. monetary policy) could be written in C++ and secured using (non-lethal) electronic power rather than being written on parchment and secured using (lethal) kinetic power.

An entirely new defense industrial complex could be built around the concept of using electro-cyber forms of physical power projection to impose severe physical costs on others in, from, and through cyberspace. Cyber forces could devote themselves to the task of eliminating the need for bloody, kinetic physical conflict using soft forms of warfighting wherever applicable. How could someone consider this to be an immoral or unethical pursuit, even if it does require a great deal of energy (which is what people implicitly assert when they argue that proof-of-work technologies like Bitcoin are "bad" because of their energy expenditure)? Aside from a fear of technological disruption, why would a nation want to prohibit the use of non-lethal warfighting technology (which is what they could be doing when they consider banning the use of proof-of-work technologies)?

Note the baked-in assumptions of the previous question – that a nation even has the option of prohibiting anyone but their own population from benefiting from strategically important physical power projection tactics, techniques, and technologies. If proof-of-work does indeed represent strategically important power projection technology for the digital age, then banning it for ideological reasons (namely that it's too energy intensive) would be banning one's own population from benefiting from a strategically important technology that other nations could adopt for their own strategic benefit. It would be akin to denuclearization, a.k.a. banning nuclear weapons for ideological reasons (incidentally, nukes are also considered to be too energy intensive). Those in favor of nuclear disarmament are often criticized for demonstrating a severe lack of understanding about the complexities of global strategic power competition and the dynamics of strategic deterrence. Consider, for example, the fact that Ukraine used to be the world's third-largest strategic nuclear superpower behind the US and Russia. Ukraine once had a strategic nuclear arsenal that housed thousands of nuclear warheads, but Ukraine surrendered them to Russia in exchange for a guarantee that Russia would never invade them. As of this writing, we are one year into a Russian military occupation of Ukraine. [24]

1.3.8 Soft War could be Worth Every Watt

Now consider the potential emergent effects of soft war on humanity. Imagine if this new type of war machine accelerated the development of faster computers and more abundant energy infrastructure. Imagine if the economies of scale for that electric power infrastructure could be shared worldwide with everyone who participated. Meanwhile, this would theoretically be possible while preserving a non-lethal option for preserving zero-trust and permissionless control over valuable resources, like international property (e.g. money) and international policy (e.g. monetary policy).

If soft war had these positive side effects, nations might become eager to go to war rather than avoid it. Society could appeal to soft war as their first line of defense rather than what has traditionally been their last line of defense. Why risk the demonstrable systemic security threats and disfunction of corrupt judges and biased courts to settle property or policy disputes when non-lethal softwar is an option? Systemic security hazards (e.g. oppressive ruling classes) could be stopped before they escalated into widescale losses, as is often the case for citizens who wait too long to go to war.

With soft war, citizens would be more empowered (in the literal meaning of the word) to physically secure themselves against a well-known vulnerability of rules-based society: their own untrustworthy nature. An unthinkable amount of human exertion and sacrifice could be replaced by an electricity bill, and there would be virtually no limit to the amount of electric power citizens could summon to secure the bits of information they value, no matter what it's used for. People could build machines to harness the power of the sun to do what previously required humans to kill each other at unnatural scale. If any of these theories about soft war are valid, then how could it *not* be worth every watt? Even if we ignore the strategic imperative of adopting new power projection technologies when they emerge, there is clearly a moral or ethical imperative to adopt non-lethal physical security tactics that don't result in bloodshed, and proof-of-power protocols like Bitcoin represent non-lethal physical security tactics.

1.4 Objective

"If you don't believe me or don't get it, I don't have the time to try to convince you, sorry."
Satoshi Nakamoto [20]

1.4.1 Responding to Two Presidential Executive Orders

The primary objective of this thesis is to develop and present a new theoretical framework for researching the risks and potential benefits of Bitcoin. This research objective supports President Biden's March 2022 executive order (EO) on Ensuring Responsible Development of Digital Assets and May 2022 EO on Improving the Nation's Cybersecurity. Per White House press release, this EO represents *"the first whole-of-government approach to addressing the risks and harnessing the potential benefits of digital assets and their underlying technology."* [25] The scope of this research effort is limited to the underlying technology of Bitcoin, colloquially known as proof-of-work.

1.4.2 Why aren't Proof-of-Work Protocols Recognized as Security Protocols Anymore?

Another objective of this thesis is to steer the academic community back towards evaluating proof-of-work technology as a cyber security technology rather than strictly a monetary technology, because it seems like people are becoming distracted by the "shiny penny" and devoting a disproportionate amount of time towards addressing the financial use cases for this technology at the expense of devoting more time to understanding general use cases.

Prior to the release of Bitcoin, academic consensus was that proof-of-work protocols were cyber security protocols that could be used to stop common types of cyber attacks like denial-of-service attacks or sybil attacks. Computer scientists discussed how proof-of-work protocols could be used as a foundation for achieving consensus on decentralized and permissionless networks. But after the release of operational proof-of-work protocols, the primary topic of academic conversation changed from cyber security to money. [26, 27, 28]

Today, amidst the buzz around Bitcoin's functional utility as a monetary payment system, few research papers are investigating Bitcoin's utility as a proof-of-work cyber security system that could be used to secure other software systems and computer networks from systemic exploitation and abuse. The author finds this missing piece of academic research noteworthy because, for all intents and purposes, Bitcoin appears to validate the theories regarding proof-of-work that first emerged 30 years ago. Most of these theories had almost nothing to do with finance, money, and economics, so why has money become the primary topic of conversation?

Bitcoin is incontrovertible proof that "proof-of-work" works as a security system, so why isn't it being evaluated as a security system? Bitcoin is a recursively valuable technology that uses proof-of-work to keep its own bits of information secure against systemic exploitation. Bitcoin therefore proves its own merit as a cyber security system, not exclusively a monetary system. Furthermore, the fact that Bitcoin demonstrates that proof-of-work works as a cyber security protocol makes it intrinsically valuable, which alone could explain why it gained and maintained monetary value. People have many reasons to want to keep their bits of information secure against cyber attacks, including and especially their financial bits of information but not strictly limited to financial information. It makes perfect sense that a system designed to physically secure bits of information would double as an ideal monetary system, but that wouldn't be its only possible use case, nor potentially its primary use case in the future, as financial bits of information are just one of countless types of information that people would want to keep secure.

Bitcoin could therefore represent something far more strategically valuable than just a new financial system architecture. Once we have figured out how to keep *financial* bits of information physically secure against attack, that means we have figured out how to keep *all* bits of information physically secure against attack. This would imply that Bitcoin could represent a special new type of computing architecture – a novel way for computers to send bits of information back and forth across cyberspace in a zero-trust and physically secure way that isn't vulnerable to systemic exploitation and abuse like existing computer networks connected to cyberspace are. Bitcoin could have far more national strategic security implications than just a monetary system, and we could be overlooking it for no other reason than the fact that people aren't questioning their assumptions. We could be like the alchemists of the past, looking at this new black powder concoction and assuming it's medicine for no other reason than the fact that its creator intended for its first use case to be medicine, so they called it medicine.

1.4.3 Hypothesis-Deductive Approaches to Researching New Technologies are Highly Presumptuous

A third objective of this thesis is to produce a new theoretical framework for analyzing general use cases for proof-of-work technologies like Bitcoin. Current approaches to analyzing the risks and benefits of proof-of-work technologies are most often centered on theoretical frameworks related to financial, monetary, or economic theory. Because analysis of this technology has been centered around these same recycled theoretical frameworks, it is possible that academia, industry, and government are creating systemic-level analytical bias when evaluating the risks and potential benefits of this technology. This is problematic considering how the express purpose of the aforementioned presidential executive orders is to establish responsible public policy regarding this emerging technology. It should go without saying that it's not possible to ensure the responsible development of public policies about Bitcoin and proof-of-work cyber security technology if the analysis used to shape that public policy is analytically biased.

Academia, industry, and government could be introducing analytical bias into their research because of the presumptions that must be made when performing hypothesis-deductive research. Researchers are tacitly making assumptions about proof-of-work technologies like Bitcoin when they use hypothesis-deductive approaches to analyzing its risks and potential rewards. They are presuming this technology is strictly a candidate form of monetary or financial technology for essentially no other reason than the fact that an anonymous programmer arbitrarily called it a peer-to-peer electronic cash system, and that financial use cases happen to be among the first operational use cases for proof-of-work protocols. Based off these presumptions, researchers are almost exclusively using financial, monetary, or economic theoretical frameworks to derive their hypotheses regarding the risks and rewards of this technology. **This thesis represents one of few exceptions where the author doesn't automatically assume that Bitcoin is strictly monetary technology just because money is its first intended use case**. [29]

The problem with using the same presumption for every research effort is that it leads to systemic-level bias. The presumption could be wrong. Or at the very least, it could be incomplete. Bitcoin could be useful as more than just peer-to-peer electronic cash. An easy way to illustrate that this presumption exists is to observe how much time, effort, money, and talent have been committed to analyzing Bitcoin based on the idea that proof-of-work technology would only be useful as a monetary technology – that it doesn't have other functionality aside from money which would justify the use of a different theoretical framework to form a hypothesis about its risks and potential benefits.

The hypothesis-deductive approach to research requires a theoretical framework from which to derive a hypothesis to deductively analyze. A researcher must choose a theoretical framework to use to analyze Bitcoin before they start analyzing it, which means they must presume Bitcoin is a certain type of technology before they analyze it. But Bitcoin is a novel technology which may or may not have functionality outside the artificial boundaries of the theoretical frameworks chosen to design or analyze it. **It's not reasonable to expect new technology to fit perfectly into existing theoretical frameworks of analysis.**

One contributing factor to this problem is that it is unclear what other fields of knowledge or theoretical frameworks would be appropriate to derive hypotheses and perform deductive analysis of proof-of-work technologies. Exploring the potential risks and benefits of this technology as something other than a peer-to-peer electronic cash system is largely uncharted territory. This technology is unique and still quite new, so alternative functions and use cases are either unknown or speculative, making it unclear what other theoretical frameworks to apply, assuming any exist.

Therefore, the choice of theoretical framework itself, from which all hypothesis-deductive approaches to analyzing the risks and benefits of Bitcoin are derived, is highly subjective and vulnerable to bias. The capacity for bias is because researchers are making the same presumption that proof-of-work technologies like Bitcoin only function as monetary technologies when they form their hypotheses about Bitcoin's risks and benefits. **The author has yet to find a formally-published research paper which even acknowledges this presumption.** Economists aren't indicating that they understand how tacitly subjective and biased it is to label this technology as strictly monetary technology. This observation alone is a red flag; there's clearly a risk of analytical bias because researchers aren't even acknowledging the presumption they keep making and their own capacity for bias.

This issue also presents a research dilemma: it's difficult to perform alternative hypothesis-deductive analysis of new technology when it's unclear what alternative theoretical frameworks (if any exist) to use to derive a hypothesis to analyze in the first place. This dilemma suggests that a missing ingredient for research related to proof-of-work technologies like Bitcoin is the exploration or the development of different theoretical frameworks from which to generate hypotheses and deductively analyze it. For the sake of developing informed public policy on proof-of-work technologies like Bitcoin, it would perhaps be beneficial to generate a different theory about proof-of-work technology to guide analysis – something that doesn't regurgitate the same presumptions predominating current research. This is the objective of this thesis.

1.4.4 Computer Science 101: All Computer Program Specifications are Abstract, Subjective, and Arbitrary

The potential for analytical bias regarding Bitcoin is especially noteworthy considering how proof-of-work was first and foremost described by computer scientists as a cyber security protocol for fifteen years preceding the release of Bitcoin. Computer scientists have been researching proof-of-work concepts since the early 1990s, but this is less common knowledge, perhaps because when the design concept was first introduced, it wasn't called proof-of-work. Like all design specifications, "proof-of-work" was an arbitrary name that emerged several years after the design concept was first introduced in academic literature.

Speaking of arbitrary names and design specifications, the first formally published paper to introduce the term "proof-of-work" originally called it "bread pudding." [28] Other papers called it a pricing function, client puzzle, or a stamp. [26, 27] Like all software specifications, the names assigned to programs which implement proof-of-work designs (to include the name "proof-of-work" itself) are arbitrarily-derived metaphors based on the personal whim of the inventor.

A foundational principle of computer theory is that software is an abstraction, therefore all software specifications use semantically ambiguous and arbitrary descriptions. Software engineers arbitrarily and subjectively choose how to describe their software based on what information is important to share about their design. Some engineers (like Satoshi) choose names based on the software's intended use case. Others choose names to emphasize design concepts. In either case, the name is arbitrary. In fact, even the term "software" itself is arbitrary. [30]

The names and descriptions that software engineers use to describe the design and functionality of their computer programs are not intended to be technically accurate (it's impossible to produce a technically precise description of an abstract concept like software). The names used are intended to make it easier for people to understand a program's *intended* use case and desired complex emergent behavior, which may not be the program's primary use case in the future. This is something that's critical to understand when analyzing software-intensive systems like Bitcoin.

1.4.5 Bitcoin's "Coins" Have only ever Been a Metaphor

The names we give to our computing systems are metaphors; these names are not meant to be taken literally. At the risk of insulting the intelligence of the reader, the computer system which stores our emails and cat pictures is not literally a "cloud." Similarly, the computer system used to sell personal and preferential information of billions of people to advertisers is not literally a "face book." Moreover, any object described using any type of object-oriented software design specification is not an actual object – these descriptions are abstractions used to make it easier to understand the desired functionality and behavior of our software.

In 2008, a pseudonymous software engineer named Satoshi Nakamoto decided to describe a variation the first reusable proof-of-work system developed by Hal Finney as a "coin" rather than to continue to call it a "proof." Instead of calling it a reusable proof-of-work protocol that utilized a decentralized server architecture rather than a trusted server architecture, this pseudonymous engineer named it "Bitcoin" and asserted that it could be used as a peer-to-peer electronic cash system. This pseudonymous engineer was famously short with their description of this technology. Nobody seems to know who the engineer was or where they worked (although they clearly had subject matter expertise in NSA cryptography). The specification they wrote was informally published, and it's only 8 pages long. This pseudonymous engineer did not elaborate much about the design in follow-on conversations, and they famously disappeared just 2 years after first announcing the project. Nothing was formally published or peer-reviewed.

The following point should be made explicitly clear: what academia and industry discuss about Bitcoin – including and especially what has been formally published about this technology – is what *other* people who *didn't* design it have to say about it based off (1) one of many potential use cases for proof-of-work technologies, and (2) a metaphorical design specification produced by a pseudonymous entity that orphaned the project. Everything that has been written about Bitcoin through formal channels was written by people speculating about someone else's metaphorical design concepts, developing their own theories about it, connecting their own dots based on the same minimal public information. Consequently, **there's no expert or authority on general-purpose use cases of proof-of-work technologies like Bitcoin. There are only people with expertise on *singular* use cases of proof-of-work technologies like Bitcoin.**

An overwhelming majority of the professional and academic analysis surrounding Bitcoin is centered around a presumption that the *only* use case for proof-of-work technology is to serve as a peer-to-peer electronic cash system, for apparently no other reason than the fact that peer-to-peer payments were the first operationally successful use case for this technology made by the pseudonymous engineer who developed it. The public appears to be ignoring the principles of computer theory and interpreting Bitcoin's name and design specifications literally, not metaphorically, despite the fact that "coin" was not even the first name or theorized use case of proof-of-work "bread pudding" protocols.

People are not only adopting the habit of assuming the only possible use case for proof-of-work technology is financial; they're also adopting the habit of acting like Bitcoin's "coins" are *only* coins, even though it's incontrovertibly true that all object-oriented software design specifications are abstract. In other words, it's incontrovertibly true that Bitcoin's "coins" don't exist – "coins" are a completely imaginary concept. Like anything abstract, Bitcoin's "coins" could just as easily be abstracted as anything else the imagination is capable of conceiving – hence why proof-of-work technologies were called something else for more than a decade before Nakamoto published the Bitcoin design specification.

Yet people keep acting like Bitcoin is strictly a monetary protocol. Moreover, people with economic or financial expertise keep masquerading as experts in proof-of-work technologies like Bitcoin for practically no other reason than the fact that this technology was arbitrarily called a "coin" and has miscellaneous operational use cases in finance. Internal combustion engines are useful for cutting down trees with chainsaws, but that doesn't make lumberjacks experts in internal combustion engine design. So why are financiers acting like the experts in proof-of-work technologies like Bitcoin, and why are we even listening to them? The most they can claim to be experts in is how to use proof-of-work technologies for singular financial use cases.

Theoretically speaking, anything – to include an arbitrarily-named software abstraction – can be monetized. Monetary value itself is an abstract concept, so of course something abstract can have monetary value. Moreover, bits of information transferred and stored via computers can represent any kind of information, so of course it can represent monetary information. It's not the fact that people have assigned monetary value to proof-of-work protocols like Bitcoin that the author finds noteworthy; it's the fact that people aren't acknowledging how the term "coin" is just as much of an arbitrary name for proof-of-work protocols and their underlying bits of information as the name "stamp" or "bread pudding." For some reason, much of current academic research doesn't acknowledge this basic principle of computer science. This would explain why researchers keep recycling the same theoretical frameworks when analyzing Bitcoin. This would also explain how academic consensus about the primary value-delivered function of proof-of-work protocols changed following the operational success of Bitcoin. But why has academic consensus about proof-of-work changed if underlying theories in computer science haven't? Add all these questions together, and the issue seems clear: we need to consider taking a completely different approach to analyzing proof-of-work technology.

1.5 Thesis Structure

1.5.1 Overall Structure of the Thesis

This thesis is designed to serve as an open letter to the Office of the President (OPOTUS), Office of the Secretary of Defense (OSECDEF), Office of the Joint Chiefs of Staff (OJCS), the National Security Council (NSC), and the American public. In response to the March 2022 OPOTUS EO on Ensuring Responsible Development of Digital Assets and the May 2022 EO on Improving the Nation's Cybersecurity, this thesis provides an argument for why accommodative and supportive Bitcoin policy could be a national strategic imperative. It also provides an argument for why the Department of Defense (DoD) should consider accumulating a strategic stockpile of Bitcoin. These arguments are made from the author's perspective as an active-duty US officer and national defense fellow assigned to MIT to research the national strategic impact of emerging technology.

This thesis also explores the significance of Bitcoin's proof-of-work protocol from a broader perspective of computer theory. It presents an argument for why Bitcoin deserves to be treated as something wholly different than so-called "blockchain" or "cryptocurrency" technology. It explains why Bitcoin is both physically and systemically different than other "cryptocurrency" protocols and is therefore inappropriate to categorize as the same type of technology despite having similar semantic descriptions.

This thesis introduces a new theoretical framework for analyzing Bitcoin called Power Projection Theory and presents its core concepts in serial fashion, starting with foundational concepts and gradually building upon each other to arrive at a novel description of Bitcoin in later chapters. If the reader skips ahead, they might miss important context needed to fully understand each concept in detail. However, each chapter uses the constant comparative method of grounded theory, repeating the core concepts presented in each section. Therefore, the reader will likely be able to grasp the core concepts of the theory no matter how far they skip ahead. The reader is encouraged to read each chapter sequentially and to revisit previous chapters as needed to understand core concepts.

Each chapter is structured as a collection of essays or what grounded theorists call "memos" which capture the core concepts of the theory and assemble them together according to similar conceptual categories and themes. The title of each section states the core concept presented in each memo. Because the author utilized the interpretivist approach to grounded theory, there is no separate chapter providing a literature review. Instead, the literature review is spread out across the thesis and is used to illustrate core theoretical concepts in a structured manner to provide additional conceptual density.

If readers find themselves wondering what a given topic has to do with Bitcoin, that's 100% intentional. The author deliberately searched for anecdotes and diverse information that seem as unrelated as Bitcoin as possible to add conceptual density to the theory and make it richer (a strategy encouraged by those who created grounded theory). Each memo, no matter how unrelated they seem to be, is linked together by core conceptual categories. An underlying "theme" plays out and eventually culminates in a new and unique specification of Bitcoin as a power projection tactic rather than a monetary technology. The intent of this approach is to capture the complexity of this technology, address its wide range of sociotechnical implications, and inspire researchers from across multiple different fields of research to derive new hypotheses to analyze in future research endeavors from their respective fields of knowledge.

1.5.2 Chapters 1 & 2 Set up Power Projection Theory

Chapter 1 gives the inspiration and justification for conducting this research. The author explains what inspired him to create a new theory about Bitcoin and why he felt like it was his fiduciary responsibility as a military officer and US national defense fellow to do so. The reader is provided with background information about military strategy and the author's profession of warfighting. The reader is also given some introductory thoughts about the "softwar" neologism. The purpose of this chapter is to present the argument for why a different theoretical framework is needed to analyze the strategic implications of proof-of-work technologies like Bitcoin.

Chapter 2 provides a breakdown of the methodology used to formulate this theory. The author discusses why he chose to use grounded theory and provides an overview of the methodology and analytical techniques used. The author concludes this chapter by highlighting the advantages and disadvantages of the grounded theory methodology and the lessons he learned along the way.

1.5.3 Chapters 3 Explains the Basic Principles of Physical Power Projection

Chapters 3-5 represent the main deliverable of this thesis: a novel theory called Power Projection Theory. The theory is divided into three parts, where each part corresponds to a separate chapter and represents a separate core conceptual category. Chapter 3 provides the foundational theoretical concepts of Power Projection Theory. The author utilizes different fields of knowledge and theoretical frameworks (namely biology and military strategy) to explain why physical power projection is essential for survival and prosperity in the wild, and how it's used to establish dominance hierarchies (a.k.a. pecking order).

This chapter explores theoretical concepts associated with property ownership and physical security, using examples from nature to illustrate how organisms develop increasingly clever power projection tactics to settle disputes, determine control over resources, and achieve consensus on the legitimate state of ownership and chain of custody of property. The purpose of this chapter is to introduce the core theoretical concepts needed to understand the complex sociotechnical relationships between physical power, physical security, and property ownership. The core theoretical concepts presented in this chapter frame the discussion presented in follow-on chapters. The primary takeaway from this chapter is a detailed understanding of why antlers are such a profound power projection technology, because of how they enable intraspecies physical power competition while minimizing intraspecies injury.

1.5.4 Chapter 4 Explains the Basic Principles of Abstract Power Projection (and Why it's Dysfunctional)

Chapter 4 provides a deep-dive into how and why humans use different power projection tactics than animals. The author provides a deep-dive on different power projection tactics, techniques, and technologies employed by modern agrarian society. This chapter can be viewed as having two parts. Part 1 focuses on how humans create and use abstract power as a basis for settling disputes, controlling resources, and establishing pecking order. Part 2 focuses on explaining why sapiens inevitably revert back to using physical power as the basis for settling disputes, controlling, resources, and establishing pecking order the same way other animals in the wild do. In other words, Chapter 4 provides a theory about why humans try but never succeed at escaping from war. The chapter concludes with a discussion about how a strategic nuclear stalemate may place human society in a highly vulnerable position, which could be alleviated by a "soft" or non-kinetic form of warfighting. This sets up the reader for understanding the potential sociotechnical and national strategic implications of Bitcoin.

The purpose of Chapter 4 is to highlight the complex sociotechnical tradeoffs and implications associated with different power projection tactics, techniques, and technologies employed by human societies. This chapter rigorously explores moral, ethical, ideological, and design decisions that people make when they use both abstract and physical power projection tactics. Across a long series of memos, the author summarizes the emergent behavior associated with these different types of power projection tactics using a constant comparative method. The point of this lengthy discussion on human power projection is two-fold. First, it illustrates how the subject of national strategic security is a complex, sociotechnical, trans-scientific phenomenon that involves frustratingly unquantifiable questions related to ethics and design. Second, it highlights the glaring vulnerabilities and systemic security flaws of our existing systems of governance and resource control. The core concepts discussed in this chapter are highly relevant to follow-on discussions about Bitcoin because the systemic security hazards identified in this chapter represent cyber security hazards too. This discussion is also designed to present the background needed to understand the "so what" of Bitcoin because it alludes to how substantial Bitcoin's impact could be on the organization of future human societies. The primary takeaway from this chapter is that humans need to find their own version of antlers that would enable intraspecies physical power competition while minimizing intraspecies injury.

1.5.5 Chapter 5 Explains how Computers Change Power Projection Tactics, both Abstract and Physical

Chapter 5 takes the core concepts presented in chapters 3 and 4 and uses them to present a novel explanation about why Bitcoin could be a groundbreaking new type of physical power projection technology rather than merely a monetary technology. This chapter begins with a deep dive into computer science and the challenges associated with software systems security. The author utilizes core concepts presented in the previous chapter to point out how the emergence of modern computing has empowered software engineers to create abstract power and use it to give themselves asymmetric and unimpeachable control over one of 21st century society's most precious resources: bits of information.

The first half of Chapter 5 illustrates how the current architecture of the internet makes society highly vulnerable to massive-scale systemic exploitation and abuse from computer programmers. It argues that society is going to invent new types of physical power projection technologies to secure themselves against exploitation and abuse via cyberspace. After highlighting this vulnerability, the author proceeds into a multi-part explanation of how proof-of-work technologies like Bitcoin could be used to mitigate these emerging threats by empowering people to impose severe physical costs on belligerent actors in, from, and through cyberspace.

The second half of Chapter 5 presents novel theories about proof-of-work protocols. Here, the author performs a deep-dive about why he believes Bitcoin may represent the discovery and utilization of completely new type of state mechanism called the "planetary state mechanism." He argues that Bitcoin could represent humanity's adoption of global-scale planetary computer that has been intentionally reversed-optimized to be as expensive as possible to operate, giving it irreproducible emergent properties that would be physically impossible for ordinary computers to replicate.

The purpose of Chapter 5 is to present a completely different perspective about Bitcoin using a completely different theoretical framework that has little to nothing to do with money, finance, or economics, but everything to do with computer science and national strategic security. Using Power Project Theory, the author highlights how technologies like Bitcoin could have sociotechnical implications which exceed our current understanding of this technology, not just with respect to computer science and cybersecurity, but also with respect to national strategic security as a whole. The author concludes this chapter with an argument that Bitcoin could represent a new way of warfighting called "softwar" that could forever change international power dynamics, and even mitigate a strategic-level stalemate between nuclear superpowers. The primary takeaway from this chapter is that Bitcoin could represent the discovery of what the author describes as "human antlers."

1.5.6 Chapter 6 Discusses Key Takeaways from Power Projection Theory

Chapter 6 is the closing chapter of the thesis. The author enumerates several new hypotheses about Bitcoin which he derived from Power Projection Theory and encourages the research community to consider analyzing them. The author gives some advice about next steps for researchers and offers some brief advice to US policy makers. The author concludes the thesis with some closing thoughts about the potential historical, strategic, and ethical implications of proof-of-work technology.

Chapter 2: Methodology

"Two roads diverged in a wood, and I – I took the one less traveled by, and that has made all the difference."
Robert Frost [31]

2.1 Four Reasons for Grounded Theory

"The growth of the internet will slow drastically, as the flaw in Metcalfe's Law ... becomes apparent: most people have nothing to say to each other! By 2005 or so, it will become clear that the Internet's impact on the economy has been no greater than the fax machine's."
Paul Krugman [32]

2.1.1 Grounded Theories are used to Inform Public Policy

This thesis utilizes the grounded theory research methodology. Grounded theory has been a popular research methodology in the social sciences for the better half of a century. It is commonly used for developing novel theories that are grounded in, and inductively derived from, systemically-analyzed data. This methodology is commonly used for research related to public policy. As co-creator of grounded theory Anselm Strauss once explained, grounded theory "*can be relevant and possibly influential either to the understanding of policy makers, or to their direct action.*" [33]

The author chose the grounded theory methodology for four primary reasons. The first has already been mentioned: it is one of the most popular forms of qualitative research used for shaping public policy. Second, it is flexible enough to accommodate analysis from multiple different theoretical frameworks. Third, because it accommodates disciplined qualitative analysis, and qualitative analysis is necessary for addressing questions related to public policy making and national strategic security. And fourth, because of the author's personal desire to challenge himself with a different type of research methodology.

Expanding on these four reasons further, an explicit goal of this thesis is to help policy makers, military senior leaders, and the public become more aware of the social, technical, and national strategic implications of Bitcoin following two Presidential EOs released by the White House within the same 3-month period. The first was President Biden's March 2022 EO on Ensuring Responsible Development of Digital Assets. The second was President Biden's May 2022 EO on Improving the Nation's Cybersecurity. Grounded theory seemed like a natural fit because it enables the author to address both EOs simultaneously to help inform White House staffers and public policy makers.

2.1.2 Analysis of Emerging Technology Should Not be Performed under Singular Theoretical Frameworks

The second reason why the author selected grounded theory is because the methodology's flexibility is necessary to explore the full range of implications of an emerging technology from multiple different perspectives. The author felt it was necessary to use a research methodology that could accommodate a wide range of data analysis across a broad range of subject matter. When this research endeavor started, it was clear to the author that Bitcoin could be analyzed using many different theories, but it wasn't clear what subject area was most appropriate for conceptualizing Bitcoin's sociotechnical implications.

To expand upon the argument presented in the introduction, some theoretical frameworks for analyzing Bitcoin are clearly more popular within academia, but that doesn't necessarily mean they're more appropriate. This point of view became a core theme of this thesis because one of the primary hypotheses to emerge from data analysis was that academia could be inappropriately categorizing Bitcoin as strictly a monetary technology, thereby creating a blind spot which ignores other categorizations that expose potentially more significant national strategic security implications. In short, current research efforts are not addressing two basic principles of computer science: (1) object-oriented software design is an arbitrarily-derived abstraction, (2) the semantics used in all software design specifications are also arbitrary-derived, making bit "coin" an arbitrary-derived name for a software abstraction.

The problem with using a traditional scientific model for researching Bitcoin is that a theoretical framework of analysis must be presumptuously chosen upfront, and then one or more hypotheses must be derived from within the boundary of that theoretical framework prior to the collection of data to assess the validity of the hypotheses. By researching Bitcoin using a traditional scientific model, academics are compartmentalizing this technology into one theoretical category from which they analyze it, which almost always ends up being one of the same three frameworks: financial, monetary, or economic theory. To illustrate this point, the reader is invited to perform their own literature review on Bitcoin to find a paper that *doesn't* analyze it using one of these same frameworks.

Shoehorning Bitcoin as strictly a monetary technology and then relying on the expertise of economists to influence public policy could be a major problem. There could be very high stakes game theory at play for the US if proof-of-work protocols like Bitcoin represent more than just a candidate form of internet money. If everyone researching Bitcoin is complicit in making the same tacit assumption that Bitcoin is strictly a monetary technology prior to analyzing it, then that is going to create a pool of skewed and biased research that gives public policy makers a massive blind spot. This could be devastating to US national security interests and global power dominance in the 21st century if Bitcoin does indeed represent more than just peer-to-peer electronic cash.

Bitcoin is multidisciplinary technology, and multidisciplinary technology should not be compartmentalized under single theoretical frameworks or fields of knowledge like economics. As Harvard professor Orlando Patterson argued in 2015, overreliance on economic "pseudo-science" is an emerging problem within academia in general, not just as it relates to research on emerging technologies like Bitcoin. "*Have we given economists too much authority based on mistaken views about their scientific reputation among established scientists and the public*?" [34]

Thanks to the grounded theory research methodology, the author didn't have to choose a single theoretical framework like economics from which to analyze Bitcoin. Grounded theory gave the author the flexibility needed to analyze Bitcoin from multiple different domains of knowledge and "chase the rabbit" down a complex, multidisciplinary rabbit hole that involved concepts and categories from multiple different scientific and engineering disciplines. After collecting data using different theoretical frameworks, the author was free to use inductive and deductive reasoning to develop several unique, counterintuitive, but informed hypotheses, which could be followed by more targeted data collection techniques for validation. This methodology allows the author to present the academic community with a theoretical framework they can use to do their own analysis, develop their own ideas and hypotheses, and to think for themselves using their own area of expertise.

2.1.3 National Security is a Trans-Scientific Question Involving Unquantifiable Variables

The third reason why the author selected grounded theory is because questions regarding public safety or systemic security are fundamentally trans-scientific questions that incorporate frustratingly unquantifiable phenomena like security design decisions and ethical considerations. It's impossible to objectively quantify what "good security design" is, just like it's impossible to quantify what the "socially right thing to do" is. These are fundamentally trans-scientific questions that demand rigorous qualitative, not quantitative, analysis. The grounded theory methodology is famous for its ability to support flexible yet structured qualitative analysis of precisely these sorts of trans-scientific social questions. This thesis explores social and security implications of an emerging technology, so it seems appropriate to utilize the methodology that was explicitly designed to address social questions like public safety and security.

2.1.4 Grounded Theory is More Challenging, More Fun, and Perhaps more Likely to be Seen

Finally, in the interest of being fully transparent to the reader, the fourth and last reason why the author selected a grounded theory methodology is because of selfish, personal motivations. In the author's subjective opinion, grounded theory looked like it would be more fun, more challenging, and produce more interesting deliverables than the traditional scientific method. Many have argued that grounded theories offer more conceptually dense and intellectually satisfying results.

This thesis represents another DoD-sponsored research endeavor and a second chance for the author to produce another thesis. The author saw this as an opportunity to challenge himself with a different research methodology with which he had no previous experience. Additionally, the author thought that using a grounded theory methodology might produce something more people would want to read. Public research that nobody reads is poorly performing public research; it's frankly a waste of dollars, not to mention a waste of a researcher's time, talent, and expertise, to write something that nobody reads.

2.2 Overview

"Reality simply consists of different points of view."
Margaret Atwood [35]

2.2.1 The Goal of Grounded Theory is to Generate a New Theory, not to Analyze a Hypothesis

The goal of grounded theory (the methodology) is to produce a grounded theory (the deliverable). The central idea behind this methodology is to evolve a novel theory through an iterative and continuous interplay between data collection and analysis. This research approach directly contrasts with the hypothesis-deductive approach of the traditional scientific method because it includes and emphasizes both inductive and deductive reasoning aimed at developing a new theory, rather than strictly deductive reasoning aimed at testing a hypothesis derived from an existing theory. [36, 33]

In the traditional hypothesis-deductive approach, a hypothesis is derived from a previously existing theory and then proven or disproven based on analysis of data. Grounded theory uses a different process where data collection, analysis, and theory development happen simultaneously and iteratively during research until the researcher eventually arrives at theoretical saturation – the point where additional data stops providing additional theoretical insight. Then, after the novel theory has been developed, it is possible to formulate and test hypotheses using the traditional hypothesis-deductive approach. [36]

Grounded theory methodologies are useful in situations where there are not formally defined or existing theoretical frameworks from which to formulate a hypothesis in the first place. The author asserts that this is precisely the case for emerging proof-of-work technologies like Bitcoin. Existing theoretical frameworks don't paint a complete picture of the technology. This could be creating an undetected bias in academic analysis because nobody is asking questions like, "should we assume that this is only monetary technology and not something more?" To remedy this, grounded theory can be used to develop a novel theory that researchers can use as a new starting point to formulate new hypotheses. The primary deliverable of this grounded theory research endeavor is therefore a new theory, not another hypothesis-deductive analysis using the same recycled presumptions. [33]

The goal of any grounded theory effort is to discover emerging patterns in analyzed data and to develop a novel, generalized theory. Researchers who use grounded theory do so with the intent to provide a usable explanation for an existing phenomenon. As grounded theory researcher Kailah Sebastian notes, "*Rather than relying on past analyses or assumptions to highlight the right answers to the wrong questions, grounded theory pushes researchers to be enthusiastic and driven towards finding the right answers to the right questions*." [37] The challenge of a grounded theory researcher is to find the right theoretical framework from which to ask the right questions.

In their book which first introduced grounded theory, Glaser and Strauss assert that conventional research methods pressure people into verifying theories rather than attempting to generate new ones. They argue there is no reason to believe that verifying theories should have primacy over generating new ones – that both types of research are equally valid and can be equally as beneficial, especially in fields of research that could be asking the wrong questions because they aren't using an appropriate theoretical framework, or fields of research that go stale because they keep asking the same questions over again. [36]

"... generating theory goes hand in hand with verifying it; but many sociologists have been diverted from this truism in their zeal to test existing theories... Surely no conflict between verifying and generating theory is logically necessary during the course of any given research... however, undoubtedly there exists a conflict concerning primacy of purpose, reflecting the opposition between a desire to generate theory and a trained need to verify it. Since verification has a primacy... the desire to generate theory often becomes secondary, if not totally lost, in specific researches." [36]

2.2.2 The Interpretivist Approach to Grounded Theory

Over the course of the past fifty years, several different approaches to grounded theory have emerged, the most common being classical, interpretivist, and constructivist approaches. This thesis utilizes the interpretivist approach championed by the co-creator of the grounded theory methodology, Anselm Strauss. The interpretivist approach to grounded theory differs from other approaches in four primary ways. The first difference is that it allows for the researcher to be engaged with the data and to actively make their own interpretations of it, rather than striving to be as distant and detached from data analysis as possible and restricting oneself exclusively to other people's interpretations of the data. [37, 36]

The second difference with the interpretivist approach to grounded theory is that it encourages the use of prior knowledge to influence the interpretation of data. The interpretivist style of grounded theory allows for researchers to leverage prior knowledge on a given subject to strengthen the overall research using their own insights on relevant issues. Strauss encourages researchers to be aware of the influences of their prior knowledge so they do not negatively impact their research focus, data collection, and categorization efforts, but he insists that prior disciplinary or professional knowledge is highly valuable and should be incorporated into the inquiry. [33]

The ability to utilize prior knowledge is the primary reason why the author chose the interpretivist approach to grounded theory, as it allows for the author to incorporate his own professional experience into the interpretation of collected data. The author is a field grade officer in the military and a weapons system developmental engineer with subject matter expertise in blast and ballistics engineering, electronic warfare, satellite system design, and software design, who was part of the founding team of officers who stood up US Space Command and US Space Force. The author has more than a decade of experience with physical security, systems safety, and software development missions for the DoD, which can be useful for detecting patterns and developing insights about emerging proof-of-work technologies that appear to be closely related to cyber security.

Critics of the interpretivist approach to grounded theory argue that incorporating one's own experiences into the interpretation of data can bias the results. "*If you're trained to be a hammer*," the critique goes, "*then you're more likely to see a nail*." Strauss argues that this phenomenon is a primary advantage to the interpretivist approach because analogical thinking is a core part of all theory development. In other words, the hammer's inclination to see a nail is a good thing because it can uncover new insights and patterns that were previously undetected by those who don't have the same disciplinary or professional knowledge as the hammer. In a crowd full of screwdrivers who only see a screw, it can be useful for a hammer to enter the scene to explain why it sees a nail – especially if the phenomenon being examined is indeed a nail that nobody recognizes yet because they aren't thinking like a hammer thinks.

The author has a niche field of expertise. If we assume that his interpretation of data related to the subject of Bitcoin is biased, then we must also admit that this interpretation can be no more biased than the interpretation of a financial or economic theorist on data related to the subject of Bitcoin. It could therefore be useful for people to listen to a different interpretation, considering how all interpretations are biased and few people analyzing Bitcoin have the same background as the author. If the goal of existing EOs is to investigate the national strategic security implications of technologies like Bitcoin for the sake of developing informed public policy, it could be useful to consider the interpretation of the US National Defense Fellow – the person whose full-time job is to investigate the national strategic security implications of new technologies.

The cross-pollination of insights derived from a researcher's field of expertise is something to encourage, Strauss argues, not something to condemn. Prior disciplinary or professional knowledge greatly enhances the development of the theory because it empowers analogical thinking and makes the researcher more attentive to relevant matters from different bodies of knowledge that would otherwise go unnoticed by other people without similar expertise. Researchers endeavoring to perform interpretive grounded theory *"carry into their research the sensitizing possibilities of their training, reading, and research experience, as well as explicit theories [from outside disciplines] that might be useful if played against systemically gathered data, in conjunction with theories emerging from analysis of these data."* [33, 37]

A third major difference with the interpretivist approach to grounded theory is that it allows for the researcher to have a vaguely-established research question prior to data collection, rather than not being allowed to have any form of pre-set research question prior to data collection, no matter how loose or vague. This was another reason why the author favored the interpretivist approach, as the author already had a vaguely-established research question prior to starting the research: what are the national strategic security implications of Bitcoin?

The fourth major difference with the interpretivist approach to grounded theory is related to literature reviews. The interpretivist approach allows the researcher to incorporate the literature review into the data collection process so that it can be used for data analysis, to make data comparisons, to stimulate new observations, and to confirm or explain certain results. Therefore, rather than completing a literature review prior to or after data collection, a researcher using interpretivist grounded theory can incorporate or blend the literature review into the theory itself. The reader should note that this thesis does not have a separate or distinct chapter for the literature review like a traditional thesis would have (particularly the ones which utilize the traditional scientific method). Instead, the literature review of this thesis is "woven" into the development of the theory across multiple chapters. [33, 37]

3.2.3 Common Pitfalls to Avoid when Developing Grounded Theories

One of the most commonly-cited mistakes of researchers using the grounded theory methodology is that they become too self-restrictive. Corbin and Strauss emphasize that *"qualitative research is not meant to have a lot of structure or rigid approach to analysis. It is an interpretive, very dynamic, free-flowing process, and unless researchers understand the basics of what they are trying to do, they lose these aspects of analysis. Their research becomes superficial and fails to provide the novel insights into human behavior that give qualitative research its dynamic edge."* [38]

Another common problem associated with this methodology is that researchers often struggle to find the right way to summarize or explain complex ideas using abstraction and inductive reasoning. A core goal of an interpretivist approach to grounded theory is finding the right way to break down and better understand complex social phenomenon using systems thinking techniques like abstraction, but without oversimplifying the issues at hand. Strauss describes this challenge as follows:

"The world of social phenomena is bafflingly complex... How to unravel some of that complexity, to order it, not to be dismayed or defeated by it? How not to avoid the complexity nor distort interpretation of it by oversimplifying it out of existence? This is of course, an old problem: Abstraction inevitably simplifies, yet to comprehend deeply, to order, some degree of abstraction is necessary. How to keep a balance between distortion and conceptualization?" [38]

To mitigate the challenging of having to connect the dots between multiple different fields of knowledge, some researchers will deliberately avoid using diverse theoretical frameworks when performing their grounded theory analysis, and instead choose to stick to familiar areas of expertise. To put it in simpler terms, researchers are sometimes tempted to avoid chasing the rabbit down the rabbit hole if it starts to appear like it's leading them too far away from their comfort zone. Strauss warns against this behavior because it can cause grounded theories to become too categorically narrow. He argues that one of the primary values of the interpretivist approach to grounded theory is that it allows for the exploration of a given phenomenon in very diverse and perhaps even counterintuitive fields of knowledge. He asserts that researchers should seek to utilize a wide array of concepts from different fields of knowledge to showcase interesting patterns and produce conceptually dense theories. The more diverse the subject matter within a given theory, the more intellectually satisfying the results can be. In his words:

"Part of the risk is that users don't understand important aspects of the methodology, yet claim to be using it in their research. For instance, they discover a basic process but fail to develop it conceptually, because they overlook or do not understand that variation gives a grounded theory analysis conceptual richness." [33]

Without incorporating varied data from separate fields of knowledge, Strauss warns that researchers can end up writing sterile theories that don't provide novel, unique, or interesting insights. Strauss asserts that being overly restrictive with inductive research and not utilizing the grounded theory methodology's strength of flexibility is a missed opportunity that, quite frankly, produces boring results. It's ok for researchers to analyze previously-developed theories that were restrictive with their approach, but he urges researchers not to allow previously-developed theories to restrict the development of their own theory. *"Thoughtful reaction against restrictive prior theories and theoretical models can be salutary,"* Strauss explains, *"but too rigid a conception of induction can lead to sterile or boring studies."* [33]

To avoid these pitfalls, the author deliberately incorporated concepts from highly diverse fields of knowledge. Most of the core concepts developed in this theory have nothing to do with finance, money, or economics. Instead, the core categories of this theory are derived from fields like biology, neuroscience, computer science, and systems security. They incorporate concepts related to natural selection, dominance hierarchies, human metacognition, political science, military theory, and even strategic nuclear policy. This theory is intentionally different than any other approach to analyzing Bitcoin because the author believes that a different approach to analyzing Bitcoin is what's needed.

2.3 Process

2.3.1 Four Phases of Grounded Theory Development

This section summarizes the interpretivist approach to grounded theory that was used for this thesis, as outlined by Corbin and Strauss in their book, *Basics of Qualitative Research: Techniques and Procedures for Developing Grounded Theory.* The methodology consists of four general phases illustrated in Figure 4 below. The first phase is data collection. Once enough raw data have been collected for analysis, the second phase is data coding. Once enough data have been coded and categorized, the next phase is theoretical sampling. These three phases repeat in a cyclical, continuous pattern until reaching a point called theoretical saturation, at which point the final theory is assembled. [38]

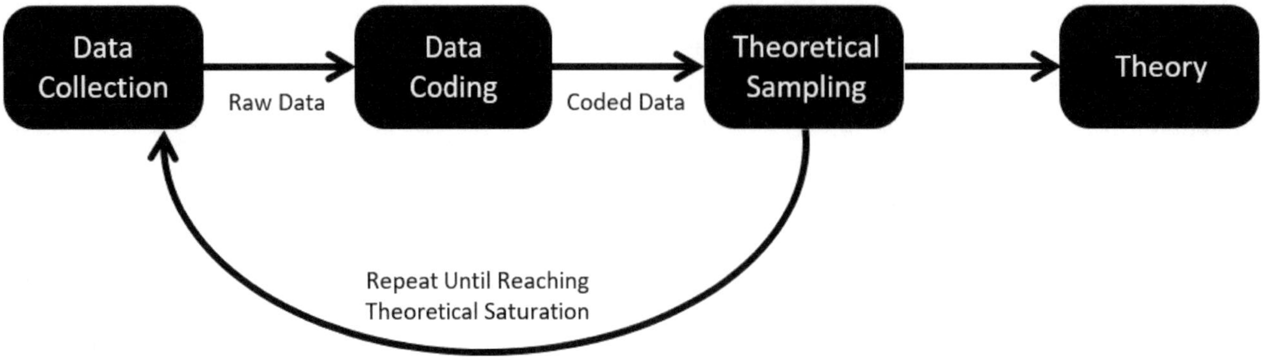

Figure 4: Four Phases of Grounded Theory Development [33]

2.3.2 Phase 1: Data Collection

The grounded theory approach relies heavily on constant comparative analysis of collected data. Using the constant comparative method, a researcher continuously moves back and forth between data collection and data analysis in an iterative manner, asking a series of questions designed to encourage inductive reasoning and lead to the development of a new theory regarding some phenomenon. The continuous, generative questioning of data (more colloquially known as "pulling threads") leads the researcher through multiple iterations of data collection, data coding, and theoretical sampling. This helps the researcher identify what data to be collected and analyzed. [38]

One of the biggest advantages of the grounded theory methodology is that it allows for a researcher to comb through highly diverse and unconventional sources of data, including but not limited to videos, documents, drawings, diaries, group meetings, memoirs, news articles, opinion pieces, historical documents, biographies, books, journals, technical papers, non-technical papers, and studies. Grounded theory researchers can use one or several of these sources in combination with each other depending upon what they're investigating. [38]

Data diversity is especially helpful when researching a field of technology as novel as Bitcoin, because much of the latest and most informed subject matter related to this technology comes from informal sources. The Bitcoin white paper, for example, was published via a private mailing list, rather than through academic journals. Likewise, the first operational proof-of-work software was circulating amongst a largely anonymous online community of "cypher punks" for years before the idea of "proof-of-work" was first discussed in formal academic literature. Moreover, this subject matter is still new, controversial, divisive, and has yet to arrive at academic, professional, or legal consensus surrounding it, making it virtually impossible to define what constitutes an "informed" source of information. [38, 39]

The primary source of data collected was technical literature and non-technical literature. Technical literature consisted of scientific research papers, research reports, theoretical papers, philosophical papers, and other sources of information characteristic of professional and disciplinary writing. These primary data were mostly sourced from the author's academic studies and research (the author was enrolled in MIT's system design and management curriculum and took several graduate elective classes like systems security and software engineering to support this research endeavor). Non-technical literature was also used as supplemental data. These data included books, letters, biographies, diaries, reports, videos, memoirs, news articles, catalogues, memos (scientific or otherwise), and a variety of other materials. [38]

2.3.3 Phase 2: Data Coding

Data coding is a process where a researcher engages in a process of quantitative microanalysis, interpretation, and conceptual abstraction by assigning concepts (a.k.a. codes) to singular incidences of data. Concepts are words or phrases used by the analyst to stand for the interpreted meaning of a given incidence of data. After enough data have been coded, a grounded theory researcher engages in a process of conceptual ordering where concepts are organized into discrete categories according to their properties (i.e. characteristics that define, give specificity, and differentiate one concept from another) and dimensions (i.e. the range over which a conceptual property can vary). [40, 38]

Quantitative microanalysis of data involves careful consideration and interpretation of meaning. Every concept represents a researchers' own subjective understanding of the meaning implicit in the words and actions of participants. To arrive at meaning, an analyst will brainstorm, make constant comparisons, try out multiple different ideas, eliminate some interpretations in favor of others, and expand upon others before finally arriving at a final interpretation. This is designed to be a productive process which can generate multiple meanings of the same event, object, or experience. The goal is to open minds to new points of view and to illuminate other people's experiences through the context of different fields of knowledge. This cross-pollination of different ideas and careful consideration of interpretation gives people a way to explain things which might not otherwise be easy to recognize or understand within a given theoretical framework – particularly the ones that are considered more popular or conventional. [38]

As Corbin explains, quantitative microanalysis of data's interpreted meaning is useful because it enables people to think differently about things and uncover new, unconventional insights which might otherwise go undetected. Novel theories often arrive at conclusions that go against conventional wisdom because researchers were careful observers of detail that kept an open and exploratory mindset about what they observed. [38] Corbin cites an explanation of this phenomenon by social economist William Beveridge:

"New knowledge very often has its origins in some quite unexpected observation or chance occurrence arising during an investigation... Interpreting the clue and realizing its significance required knowledge without fixed ideas, imagination, scientific taste and a habit of contemplating all unexplained observations... In reading of scientific discoveries, one is sometimes struck by the simple and apparently easy observations which have given rise to great and far-reaching discoveries making scientists famous. But in retrospect, we see the discovery with its significance established. Originally, the discovery usually has no intrinsic significance; the discoverer gives it significance by relating it to other knowledge, and perhaps by using it to derive further knowledge." [38]

The basic-level concepts generated from data coding create the foundation of a grounded theory. Concepts are organized in varying levels of abstraction into categories based on their themes, properties, and dimensions. Categories provide the framework or skeleton of a grounded theory which gives it greater explanatory power. Categories themselves can be further organized into higher levels of abstraction according to their properties and dimensions to create what are known as core categories. Core categories form the backbone of a theory; they represent what a researcher has determined to be the main theme of the data. Core categories are comprised of broad, holistic, and abstract concepts. When a grounded theory is finalized, it is usually ordered, assembled, and presented according to its core categories. [38]

Figure 5 provides an illustration of how a grounded theory is constructed. Researchers engage in data collection and data coding to develop basic-level concepts, then use inductive reasoning to generate more generalized and abstract categories. Core categories serve as the highest-level abstractions of the theory. When a researcher presents the core categories of their grounded theory to an audience, the word "grounded" serves as a reminder that each core category is grounded to basic-level concepts that were developed after quantitative microanalysis and interpretation of coded data.

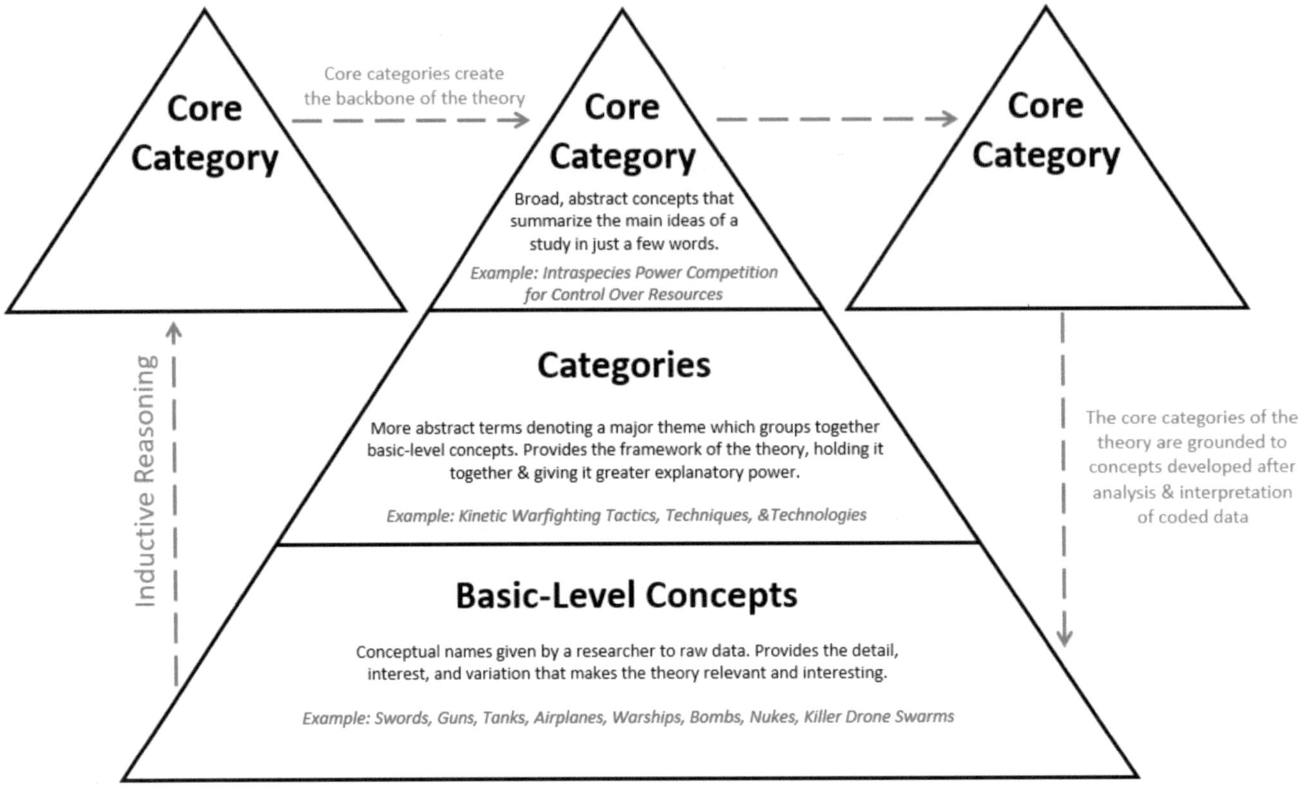

Figure 5: General Construction of a Grounded Theory
[33]

2.3.4 Phase 3: Theoretical Sampling

The goal of generating a new theoretical framework is to create a foundation for explaining phenomena and for providing concepts and hypotheses for subsequent research. As Corbin and Strauss explain, *"at the heart of theorizing lies the interplay between researcher and data out of which concepts are identified, developed in terms of their properties and dimensions, and integrated around a core category through statements denoting the relationships between them all."* [38]

Theories can range from substantive, middle-range, or formal depending on how specific, broad, and dense they are. For this thesis, the author endeavored to create a formal theory. Formal theories are the broadest and most dense kind of theory, used to understand a wider range of social concerns or problems. Constructing a formal theory requires an idea to be explored fully and considered from multiple different angles or perspectives. To aide in this process, researchers utilize analytical tools like diagrams (visual devices that depict relationships between analytical concepts) and memos (written records of analysis). Diagrams and memos represent more than just repositories of analysis, but a form of analysis in and of itself where a researcher can form a dialogue with their data to move their analysis further. An example of a diagram generated during the author's data coding effort is shown in Figure 6. [38]

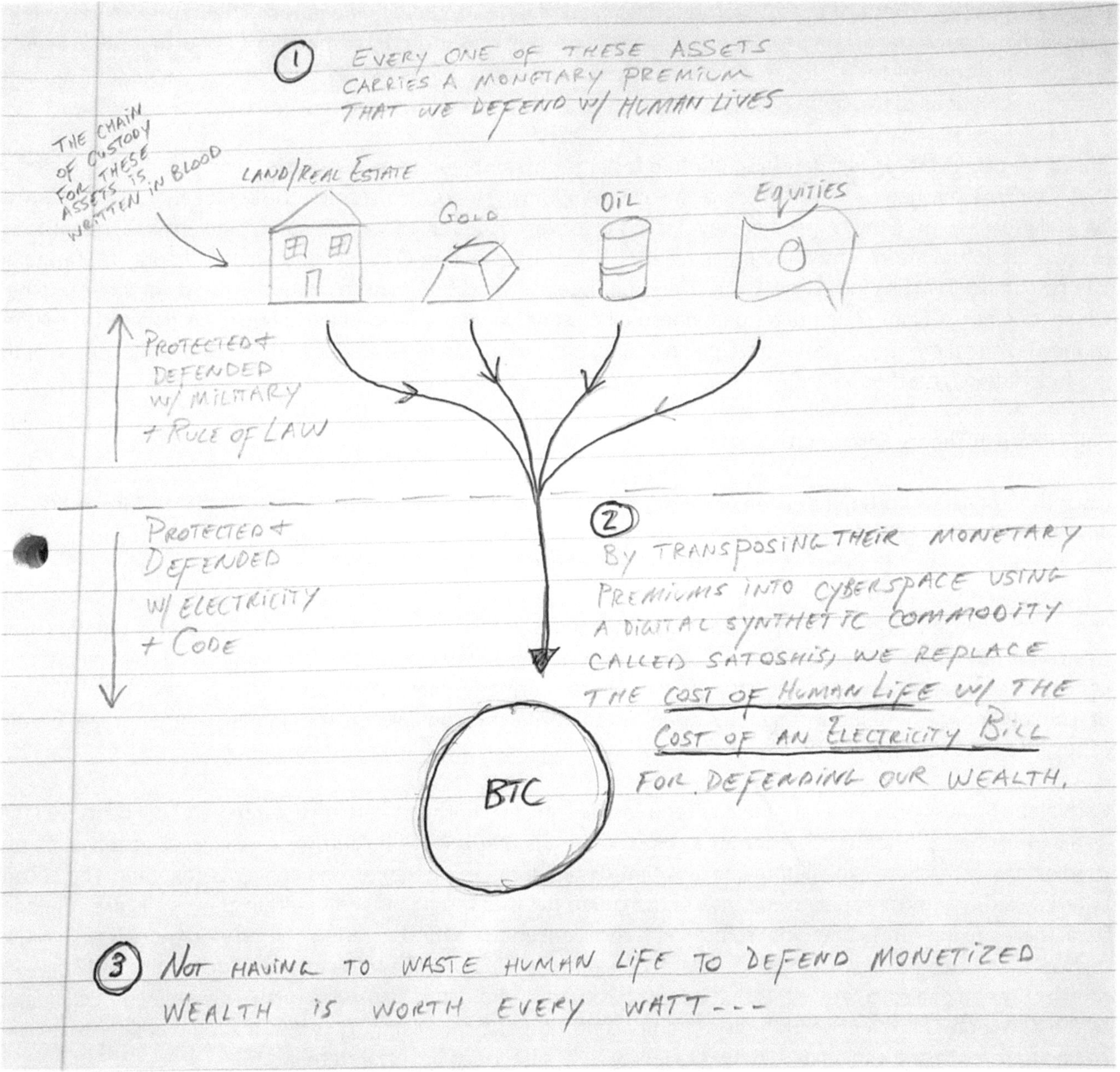

Figure 6: Conceptual Diagram Generated During Data Coding

When performing data analysis, Corbin and Strauss explain that researchers must constantly interact with data by examining it, making comparisons, asking questions, coming up with new concepts to stand for meaning, and suggesting possible relationships between different concepts. These activities create a dialogue in the mind of a researcher that can be captured in diagrams and memos, allowing the researcher to brainstorm, and let loose with their thoughts. In the beginning, memos and diagrams are rudimentary representations of thought. But as research progresses, they grow in complexity, density, clarity, and accuracy, and serve as a useful tool for keeping track of the complex and cumulative thought processes which go into detailed qualitative analysis. Memos and diagrams provide functional utility because they serve as a method for opening data exploration, identifying or developing the properties and dimensions of concepts, asking questions, exploring relationships, and developing a theory's overall story line. [38]

As the properties and dimensions of different concepts and categories become more developed, grounded theory researchers transition into a more targeted approach to data collection and coding known as theoretical sampling. Theoretical sampling is a method of data collection that is based on concepts derived from previously collected and coded data (as opposed to early-phase data collection which was not collected based on concepts). In other words, theoretical sampling is a method of data collection that enables a researcher to "follow-up" or "close the loop" on specific concepts that are interpreted from previous data coding. The purpose of theoretical sampling is to collect additional data from people, places, and events that maximize the researcher's opportunities to develop concepts in terms of their properties and dimensions, identify relationships between concepts, and uncover different variations of the same concept. [38]

During theoretical sampling, data is scrutinized for tensions, ambiguities, contradictions, and conflicting codes, as these suggest the need for further data collection and analysis to help resolve the dissonance. This creates a cyclical process where the researcher stays locked in a loop of constant data collection, data analysis, quantitative microanalysis and interpretation of meaning, memo-writing, diagram-drawing, and further theoretical sampling. The researcher remains in this cyclical process until they reach theoretical saturation – the point where no new concepts emerge from coded data, and where all existing concepts have been fully explored in terms of their properties and dimensional variation. Upon reaching saturation, the researcher can move to the final phase of the grounded theory methodology. [38]

2.3.5 Phase 4: Theory Formation

Once theoretical saturation has been achieved, the final phase of a grounded theory effort is the integration and writeup of the theory itself, a process where formulated categories are linked together via core categories to form the overall theme of the theory. Integration is essential for creating a holistic view of underlying concepts, as concepts alone don't make a theory. Categories must be linked together and filled with conceptual detail to construct a dense and explanatory theory that represents more than just the sum of different categories. Core categories often have the greatest explanatory power because they expose the common thread relating different concepts together in new and interesting ways. If chosen correctly, core categories create the "mind blown" effect of a novel theory where interesting facts or enlightening information link together in new ways to create a sense of surprise or excitement. [38]

Corbin and Strauss provide a list of five criteria for identifying whether a particular category of concepts qualifies as a core category. The first criterion of a core category is that it must be sufficiently abstract so that it can be used as an overarching explanatory concept that ties underlying categories and concepts together. The second criterion is that it must appear frequently in the coded data, to the point where within almost all cases of coded data, there are indicators that point to the same core concept. The third criterion of a core category is that it must be logical and consistent with the coded data. Concepts should not have to be forced under a core category. The fourth criterion of a core category is that it must be abstract enough to be used in further research that could lead to the development of a general theory. The final criterion of a core category is that it should appear to grow in conceptual depth and explanatory power as lower-level categories are related to it. [38]

Figure 7 provides a breakdown of the core categories identified by the author. These core concepts were identified based on the most commonly reoccurring concepts coded after achieving theoretical saturation. The central core category of this grounded theory is "power projection." Every other core category presented in this theory is centered around power projection. The theory begins with an exploration of power projection tactics in nature and explores sub-categories of basic-level concepts related to power-based resource control, principles of survivorship, and inter/intra species power competitions. These concepts create the foundational understanding needed to explore the next core category of human power projection tactics. From here, the theory dives deep into sub-categories of concepts related to abstract power projection tactics and physical power projection tactics employed by modern agrarian human societies. These concepts create the foundational understanding needed to explore the final core category of power projection tactics in cyberspace. From here, the theory dives deep into a sub-category of concepts related to abstract power projection tactics in, from, and through cyberspace. This lays the groundwork for understanding Bitcoin not as a monetary technology, but as a potentially new form of software-instantiated physical power projection technology, which the author encapsulates with the neologism "softwar."

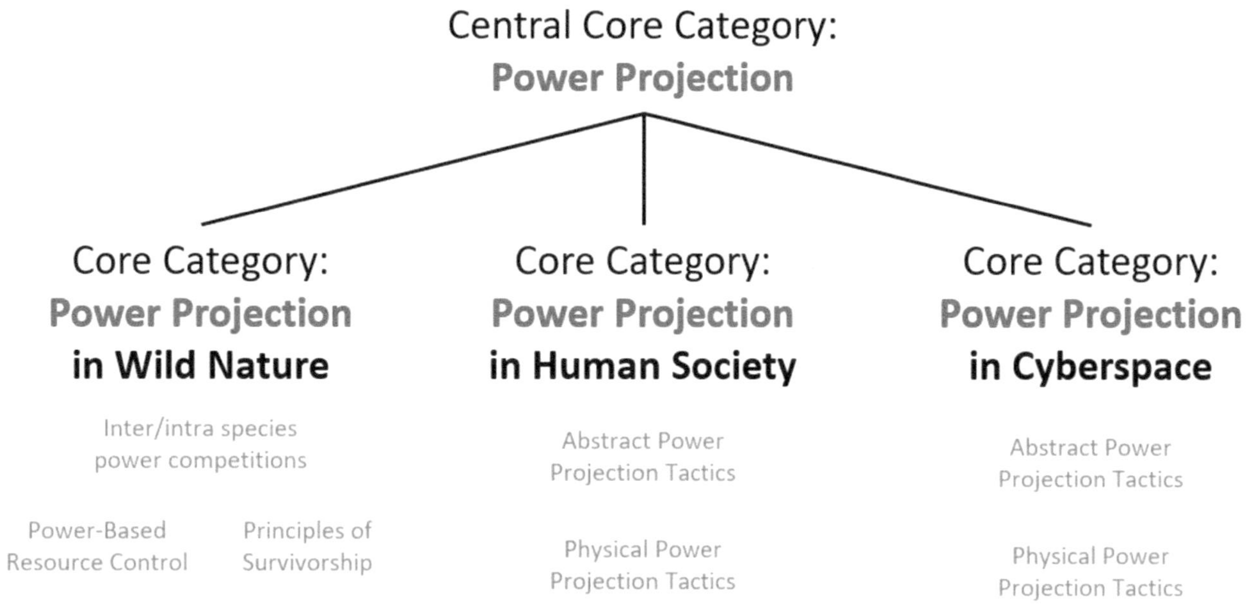

Figure 7: Core Categories Chosen for this Research Effort

With these core categories identified, the author assembled the most relevant conceptual memos under each core category to form the final integrated theory. The theory itself is simply a collection of conceptual memos written by the author throughout the duration of the data collection and analysis effort, which expand on the basic-level concepts that were interpreted during coding. It should be noted that the final integrated theory only includes categories and concepts that were most relevant to the core categories, which were not known prior to data collection and analysis. In other words, what the reader sees as the final deliverable of this research endeavor represents only a fraction of the concepts explored throughout the duration of quantitative microanalysis. The author's job was to effectively discover all the "dead end" ideas or clues in pursuit of finding a new "common thread" linking different concepts together in a previously undetected way. The novel theory presented to the public as the final deliverable represents a small tip of a much larger iceberg of concepts analyzed throughout the formation of the theory.

By linking all these diverse concepts together under the same core categories centered on power projection, the reader is (hopefully) able to gain a newfound appreciation for the potential sociotechnical and national strategic implications of proof-of-work protocols like Bitcoin that expands beyond the boundaries of the current theoretical frameworks that are being used to analyze this technology. The theory incorporates multiple different fields of knowledge together in novel (and hopefully interesting) way that highlights how this technology could have broader implications than what is currently being addressed using singular frameworks like financial, monetary, or economic theory.

2.4 Disadvantages & Advantages

"Experience without theory is blind, but theory without experience is mere intellectual play."
Immanuel Kant [41]

2.4.1 Four Commonly-Cited Disadvantages of Grounded Theory

Many formal studies and papers have discussed the advantages and disadvantages of grounded theory. What follows is a summary of those which stood out to the author based off his experience completing this thesis, starting with four disadvantages, and concluding with three advantages.

As previously discussed, the most cited disadvantage of grounded theory is that interpretations and findings are vulnerable to intrusion of perspectives, biases, and assumptions. There are strategies that can be used to highlight and mitigate these intrusions, but it certainly appears to be a valid criticism. However, it is important for the reader to understand that subjectivity of interpretation is often not considered to be a bad thing in qualitative research like it is with quantitative research, because different interpretations lead to the formation of new knowledge. People like hearing diverse and unique perspectives on issues that are important to them, and subjectivity of interpretation is precisely what provides these unique perspectives.

A second commonly-cited disadvantage of grounded theory is that it doesn't provide objective results. This appears to be another form of general discontent with qualitative research. Corbin and Strauss argue that it is not possible for qualitative research to have objective results and assert that researchers should instead aim for sensitivity rather than objectivity. In their words:

"Data collection and analysis have traditionally called for objectivity. Today it is acknowledged that objectivity as it is traditionally defined in research can't be applied to qualitative research. The reason is that qualitative researchers interface with participants and the data. They bring with them their perspectives, training, knowledge, assumptions, and biases, which in turn influence how they interact with participants and interpret data. Instead of objectivity, qualitative researchers aim for sensitivity, or the ability to carefully listen and respect both participants and the data they provide." [38]

A third commonly-cited disadvantage of grounded theory is that the presentation of research findings is not straightforward. [42] This is perhaps a reflection of how difficult it can be to categorize concepts and their interrelationships – which is something that the author certainly struggled with. Finally, a fourth commonly-cited disadvantage of grounded theory is that it's time consuming and difficult to conduct. To this criticism, the author would wholeheartedly agree.

2.4.2 Three Commonly-Cited Advantages of Grounded Theory

One of the most commonly-cited advantages of grounded theory is that that it's helpful for developing new understandings of complex phenomena that cannot be explained using existing theories or paradigms. Qualitative research in general is good for exploring areas that have not yet been thoroughly researched because they are flexible and allow for researchers to search and discover relevant variables that can later be tested through quantitative forms of research. Theoretical explanations can be developed that reach beyond the known, or beyond what humanity is currently capable of measuring, offering new insights into a variety of experiences and phenomena that should be explored in the future. It is not uncommon for a theory to be developed in one person's lifetime, only for it to be confirmed several lifetimes later using quantitative forms of research when the right measuring tools or techniques eventually become available. [42, 38]

This is especially true for theories related to computer science, as the first theories related to general-purpose computing – not to mention the first published computer program – preceded operational general-purpose computers by more than a century. In other words, the first computer programs were nothing but theories. What we now call computer science was founded by theorists and is still dominated by theorists. Sometimes the only option is for people to theorize until the means or resources to perform more rigorous quantitative analysis becomes available. And it should go without saying that every quantitative analysis of an underlying hypothesis requires a theoretical framework from which to derive a hypothesis in the first place. Without dreamers coming up with theories, there would be no hypotheses to validate using quantitative analysis.

A second commonly-cited advantage of grounded theory is that it creates a systematic and rigorous process for data collection and analysis, enabling researchers to study phenomena with a great level of depth. The author found this structure to be especially useful since he didn't have previous experience using qualitative research methodologies. [42, 38]

A third advantage of grounded theory qualitative research is also good for taking a holistic and comprehensive approach to the study of phenomena because they can incorporate multiple different theoretical frameworks. General concepts can be identified, and theoretical explanations can be developed that reach beyond what is currently known. This helps people give new meaning to what they encounter in their lives and perhaps make more sense of it, giving individuals and groups the ability to make sensible plans of action for managing problems (hence the methodology's popularity for public policy making). [38]

2.5 Lessons Learned

"...without the making of theories I am convinced there would be no observation."
Charles Darwin [43]

Overall, the author was satisfied by the grounded theory methodology, particularly the interpretivist approach used for this research effort. The experience turned out to be far more intellectually demanding (thus satisfying) than expected. It was nice to have the flexibility to dive into diverse fields of knowledge in pursuit of underlying clues or concepts that could resolve some conflict, ambiguity, or dissonance that emerged in the analyzed data. It was extremely exciting to discover a concept that linked two completely different fields of knowledge together in unexpected ways.

Being able to use a wide variety of diverse and unconventional data sources proved to be a critical-enabling factor for this research effort, particularly during the theoretical sampling phase. It's hard to imagine that a detailed analysis of Bitcoin could be done without incorporating unconventional data sources, as it is still very new, and formal literature on the subject is quite scarce.

For anyone considering grounded theory in the future, the author offers three lessons learned. The first lesson learned about this methodology is that it can take the researcher far outside of their academic background and create a steep cognitive switching penalty. The author spent most of his time digging through technical literature that has nothing to do with his academic background. This is both a time consuming and mentally exhausting process because it requires the researcher to essentially teach themselves the basic principles of multiple different fields of knowledge to establish a general understanding of each field and be able to relate similar concepts from different fields together under the same theoretical categories.

The author found it easy to dive deep into very narrow fields of knowledge to increase depth of understanding for each topic, but much harder to increase breadth of knowledge by diving into multiple narrow fields. There was a notable cognitive switching penalty when performing research because of having to switch attention between different fields of knowledge and their associated contexts (e.g. switching from papers about computer science, to biology, to anthropology, etc.). The upside, of course, is that doing deep dives into diverse topics is intellectually satisfying – far more so than any other research the author has done in both his professional and academic capacity.

A second lesson learned about this methodology is that it's surprisingly frightening. Throughout most of the data collection and analysis process, the researcher does not know what the final theory is going to look like, how it's going to be perceived, or if it's even valid. Because much of the interpreted qualitative data is subjective, there is no satisfactory feeling of "being right" like there is with the objective analysis of highly quantitative data sets. Additionally, unlike the traditional hypothesis-deductive approach where a researcher can choose a research question and formulate a hypothesis in such a way that is virtually guaranteed to have interesting results regardless of whether their hypothesis is validated or invalidated, a grounded theory approach is far more open-ended, unstructured, and uncertain. The researcher doesn't get the comfort of feeling as though their research effort is going to lead to the formulation of a meaningful and interesting theory until the very end of the analytical effort, long after most of the work has been done and all the pieces of the theory are integrated together.

It takes high risk tolerance and comfort with uncertainty to put so much effort into something without having a clear idea about what the end state will look like. Along the way, grounded theory can be disheartening because the analytical process requires a lot of experimentation with different interpretations and categories, creating many conceptual dead ends. Additionally, the author never felt a sensation of being done with the analysis like he did using the traditional scientific method on previous research efforts. Even after achieving theoretical saturation, there always seemed to be more concepts, properties, or dimensions to uncover. It was impossible for the author to feel like he had reached the so-called "bottom of the rabbit hole."

A third lesson learned about this methodology is that it is far more time consuming than expected, despite being explicitly warned that grounded theory is often harder and more time consuming. Coding data, drawing diagrams, and writing memos is extremely time-intensive, especially when it's related to technical literature surrounding complex topics with lots of semantic ambiguity and jargon, like computer science. It takes a lot of time "digging" to discover the right data needed for theoretical sampling. Most of the author's time was spent learning what interpretations *not* to make, what categories *not* to use, and what *not* to include in the theory.

It was a challenge to sort through thousands of pages of technical and non-technical literature, interpret them, and link them together via conceptual relationships. It was even harder to identify what categories to use, and it could be very disheartening when an exciting candidate category turned out to be a dead end, causing the author to have to return to square one, or even rework or redo major parts of the theory. Moreover, the struggle to find the "right" categorizations is undetectable by the reader because all they see at the end of the effort are the precious few categorizations which survived a rigorous down-selection process. This creates a "tip of the iceberg" phenomenon where the final grounded theory represents a tiny sample of the concepts collected and categories explored throughout the development of the theory.

As a final thought about grounded theory, it's worth echoing what Corbin and Strauss summarized as the characteristics of people who are attracted to qualitative research. This methodology is for people with a humanistic bent who are curious about people and how they behave. It is for those who are creative and imaginative but have a strong sense of logic. It's for those who are detail oriented, who can recognize variation as well as regularity in the data they analyze. To succeed at grounded theory, a researcher needs to be willing to take risks and live with ambiguity. [38]

Chapter 3: Power Projection Tactics in Nature

"The most important causes of change are not to be found in political manifestos or in the pronouncements of dead economists, but in the hidden factors that alter the boundaries where power is exercised... subtle changes in climate, topography, microbes, and technology..."

The Sovereign individual [44]

3.1 Introducing Power Projection Theory

Many human power projection tactics can be understood by simply observing what happens in nature. Spend enough time studying how organisms behave, and it becomes clear that there is a causally inferable relationship between two phenomena: (1) the physical power projection capabilities of living organisms, and (2) the amount of freedom, prosperity, and resource abundance they enjoy. Nature appears to disproportionately favor its strongest and most intelligent power projectors. Why is that? Nature doesn't necessarily have to behave like this; there are other characteristics aside from strength and intelligence that animals could asymmetrically reward by placing them higher in the pecking order. What's so special about physical strength and intelligence? Why do so many species focus their attention on rewarding their strongest and most intelligent members? Why do animals even need to be picky about who they feed and breed in the first place?

Power Projection Theory lays the groundwork for understanding why physical strength and intelligence is so intrinsically valuable in the wild, and why it's often used as the basis for settling disputes, managing limited resources, and establishing intraspecies dominance hierarchies (a.k.a. pecking order). Nature has an incontrovertible bias towards its strongest and most intelligent organisms. Animals which master their capacity and inclination to project physical power in increasingly clever ways tend to prosper better in the wild than animals which don't. In other words, the strong and the aggressive often survive. The weak and the docile often don't. There must be an explanation for this – an explanation which could provide some insight into why humans project power, settle disputes, and manage resources the way they do. This explanation could help shed light on why emerging power projection tactics, techniques, and technologies like Bitcoin are so remarkable.

To understand the sociotechnical and national strategic significance of Bitcoin, it is first necessary to develop a first principles understanding of the primary value-delivered function of physical power projection. A mental model is needed of how physical power projection works, why it works, and what its primary value-delivered function is for the organisms which master it. To that end, the author begins a grounded theory about Bitcoin with a theory about power projection.

This chapter explores the concept of property ownership and retraces the evolutionary steps that life took to become increasingly more prosperous. The reader is guided through examples of power projection tactics in nature. A series of anecdotes explore complex relationships between life, power, property, and prosperity. Throughout this chapter, the word "power" is used strictly in a physical context to describe energy (joules) transferred per unit of time (joules/second) to form a phenomenon called watts. From both a systemic and psychological perspective, physical power (a.k.a. watts) serves many useful functions in nature and society. One of the most useful yet underappreciated functions of physical power in nature is providing living creatures with a basis for settling their disputes, managing their resources, and establishing a pecking order in a zero-trust, egalitarian, and permissionless way.

The author explores how physical power provides life with what it needs to undertake the existentially imperative task of gaining, maintaining, and sharing access to limited physical resources using dominance hierarchies. An assertion is made that physical power is the primary means through which all living creatures – including and especially sapiens – achieve consensus on the legitimate state of ownership and chain of custody of precious resources. This assertion forms the basis of an argument that without the presence of physical power, most animals cannot achieve consensus about who owns what property. This simple observation feeds into later discussions about power dynamics in modern agrarian society.

After explaining the link between physical power and resource ownership, the author discusses how organisms in nature adopt different physical power projection strategies to increase their capacity to capture resources while simultaneously defending themselves against predators. The author introduces a concept called *"primordial economics"* to explain the dynamics of naturally occurring phenomena like predation. A novel technique called *"bowtie notation"* is introduced to offer a simple explanation for why animals organize the way they do. These concepts are used throughout the remainder of the theory to explain why humans project power and how new technologies like Bitcoin could affect this behavior.

3.2 Physical Power & Resource Ownership

"Veni, vidi, vici."
Julius Caesar [45]

3.2.1 Proof-of-Power is Proof-of-Ownership

Imagine precious resources (e.g. land, water, food, gold) residing on Earth many billions of years ago, long before our planet was inhabited by life. Assume there are no living organisms capable of exerting physical power to secure access to these resources. Would it make sense to claim they're *owned*? Would it make sense to claim that these resources qualify as something's *property*? If the reader answered yes, then how could the owner be identified if there are no living things securing access to these resources? Resource must have an owner to qualify as being owned or to qualify as being something's property.

If the reader answered no, then we have just established that the phenomenon we call property ownership is not strictly an abstract idea. Ownership has a physical signature in shared objective reality that is somehow related to physical power projection. The real-world physical power projected by animals to gain and maintain access to resources is somehow related to the phenomenon we call ownership.

Since the dawn of life on Earth, organisms have evolved increasingly more creative ways to project physical power to settle property disputes, secure control authority over resources, achieve consensus on the state of ownership and chain of custody of property, and establish dominance hierarchies (a.k.a. pecking order). The control authority over Earth's natural resources that many plants and animals enjoy today appears to be the byproduct of energy exerted over time (joules/sec). This would imply that **property ownership has a physical signature that can be denominated in watts.**

Watts signal ownership. Organisms determine what they own based *not* on what they say, but on what they do – how they project their watts. When an organism decides to stop owning a resource, they stop spending the watts needed to gain and maintain their access to it. Perhaps an organism stops spending watts because their priorities changed; perhaps it simply doesn't value the resource anymore. Either way, **when an organism stops using watts to secure their access to a resource, their perceived ownership of that resource disappears.** Ownership of the discarded resource then passes onto the next able-bodied organism capable of and willing to spend the watts necessary to gain and maintain access to it.

Watts appear to be the only part of ownership that is based in that place we call objective physical reality. Physical power appears to be the only means through which organisms (99.9% of which are incapable of abstract thought) can achieve consensus on the legitimate state of ownership and chain of custody of physical property. Most organisms are not capable of abstract human constructs regarding ownership. There are no "laws" or "property rights" in nature like there are in human society. There is nothing to which people agree to determine who owns what. And even if there were, there's little to compel wild animals to be sympathetic to abstract human constructs about property rights and legal ownership.

It is incontrovertibly true that all organisms rely heavily on physical power to achieve inter and intra species consensus on the ownership status of limited resources. Even for sapiens, Earth's master abstract thinkers, physical power is still the primary means through which they settle territory disputes and resolve conflicting abstract beliefs about property rights. They write rules of law to define property rights, but then they use physical power to solve disputes about what the "legitimate" rule of law is, or what the "right" property rights should be. While there are many examples in everyday life where law successfully settles human intraspecies property disputes, what people often overlook is the long history of physical disputes that were used to instantiate those laws (in other words, our property rights exist because of the wars fought to establish those property rights).

Nature's way of sorting out property ownership can therefore be conceptualized as a proof-of-power protocol. The physical power exerted to own a resource is self-evident by the fact that a resource is perceivable as owned in the first place. Power appears to be the only non-abstract characteristic about the phenomenon of property ownership that can be seen, detected, or measured, thus independently validated. If an organism detects ownership of a resource, it's probably because power is being projected by another living thing to signal their ownership of that resource.

The proof-of-power protocol is easy to overlook for people who subscribe to the power projection capacity of others to gain and maintain access to limited resources. Most people living in modern society do not participate in the proof-of-power protocol like wild animals do, so it's easy for them to lose sight of the fact that people are constantly projecting physical power to settle property ownership disputes and establish intraspecies dominance hierarchies. Nevertheless, the proof-of-power protocol is always running. The physical power bill is always being paid whether we pay it ourselves or outsource it to others. [22]

The property rights we enjoy today exist because people were willing to project lots of physical power to claim and maintain those rights. Without the expenditure of watts by living things to secure access to precious resources, living creatures are unable to perceive a resource as being something's property in the first place. Without physical power, resources are either perceived to be unclaimed (therefore not property), or resource ownership is purely an abstract construct that manifests as a belief system – belief systems which can be ignored, exploited, or considered illegitimate.

3.2.2 Signaling Ownership by Showing One's Capacity & Inclination to Impose Severe Physical Costs

To illustrate how physical power is used to signal property ownership, consider a scenario where the reader attempts to take freshly hunted meat (a precious physical resource) from a wolf. The wolf would likely signal her ownership of that resource by projecting physical power. She would accomplish this by displaying her capacity and willingness to impose severe physical costs on the reader for trying to deny her access to the meat. This proof-of-power display would probably look something like Figure 8.

Figure 8: Organism Signaling Ownership of a Resource using the Proof-of-Power Protocol

The wolf's capacity and willingness to impose severe physical costs on the reader to secure her access to the meat is displayed via her snarl, and it would likely leave a clear impression on the reader. Two things should be noted about this display. The first is that her power projection capacity is physically quantifiable. With the right combination of sensors, we could measure her capacity to project power in watts. The second thing to note about this display is the fact that those watts are the only independently verifiable and objective signal of ownership based in physical reality. Her ownership of the meat manifests itself through the power she projects to secure her access to the meat. That snarl serves as her certificate of ownership. In other words, her proof-of-power is her proof-of-ownership.

Now imagine what would happen if the wolf were docile. Imagine if you picked up the meat and the wolf did not snarl and threaten to bite you. She projects no physical power and signals no willingness to impose physically prohibitive costs on you to prevent you from accessing the precious resource. In that scenario, you and neighboring organisms would likely perceive that she is either being friendly and *sharing* her property with you, or she does not believe the meat is *her* meat in the first place. **Without her physical projection of power, there is no physical signature from which we can perceive or denominate her ownership** of the meat, so it is not clear if she owns it at all.

This scenario illustrates how closely physical power projection is metacognitively linked to the concept of property ownership. Physical power and aggression are signals of resource ownership. Organisms rely on other organisms to signal property ownership by projecting power. Without physical power projection, it's hard for organisms to detect ownership unless they have the capacity to think abstractly like humans do and communicate via common language. But talk is cheap; abstract constructs of ownership are extremely weak signals of ownership that are often ignored unless they're backed by physical power.

The proof-of-power protocol for property ownership is energy-intensive and prone to causing injury, but it has many positive tradeoffs. The main benefits of the proof-of-power ownership protocol is that it's a zero-trust, egalitarian, and permissionless protocol. Proof-of-power is zero-trust because it doesn't require trust to function properly. It works the same regardless of whether organisms are trustworthy and sympathetic to our beliefs or not. Proof-of-power is egalitarian because all organisms are equally subordinate to watts. Proof-of-power is also permissionless; the wolf doesn't need to ask for permission from the animal it hunts to take its meat – her physical power gives her the freedom to do what she wants.

Another major benefit of the proof-of-power ownership protocol is that it's exogenous to belief systems. Ownership of the meat passed from the prey to which it originally belonged to the wolf who hunted it down for no other reason than because the wolf projected physical power to gain and maintain access to the meat. She doesn't own the meat because she *believes* she should own the meat. Beliefs don't put dinner on the table; physical power does. The wolf's continual projection of physical power is why she continues to own the meat. If she were to stop displaying proof-of-power to signal proof-of-ownership of the meat, then she would likely lose her access to the meat regardless of what she *believes* she owns.

Now imagine if you picked up the meat, the wolf snarled at you, and you doubled down and snarled back at her. You and the wolf would produce two conflicting signals of ownership because you're both projecting power. In this situation, it wouldn't be clear to neighboring organisms who truly owns the meat. To resolve this property dispute and achieve consensus on the legitimate state of ownership and chain of custody of the meat, more physical power would need to be applied to the situation.

It would not be possible to file a lawsuit against the wolf to challenge her custody of the meat. It would not be possible to engage in diplomatic talks with the wolf to draft an agreement about what the proper abstract construct of ownership should be. These options would not be on the table because they require the wolf to be sympathetic to the reader's beliefs about property rights, and she isn't physiologically capable of that. She doesn't have the biological circuitry nor the brain power to understand the reader's abstract explanation about why the reader believes they are somehow the proper owner of the meat; much less does she have the inclination to be sympathetic to the reader's belief system when she can simply use her superior physical power to shred the reader to pieces and have even more meat for herself.

In lieu of the option for peaceful adjudication, the reader would have to settle the property dispute by entering a probabilistic physical power competition to determine the meat's "legitimate" owner. Some call these probabilistic physical power competitions *battles*. The winner of the battle would become the newly recognized owner of the meat. Why? For no other reason than because the newly declared owner won a probabilistic physical power competition. **Since physical power is the only part of the phenomenon of property ownership that appears to be based in shared objective physical reality, physical power competitions are the only non-imaginary way for organisms with no belief systems or conflicting belief systems to resolve disputes about property ownership.**

3.3 Life's War against Entropy

"Thought itself is a limited lifetime phenomenon in the cosmos... the relentless rise in entropy ensures that any cogitating being that happens to still be able to persist in this unusual realm of particles will ultimately burn up in the entropic waste generated by its own process of thinking. So the process of thought itself in the far future will generate too much heat for that being to be able to release that heat to the environment and to avoid burning up in its own waste."
Brian Green [46]

3.2.1 To Live is to Project Power

"Big Things have small beginnings."
David 8, the Android [47]

A key observation from nature is that resource ownership for all living organisms seems to be fundamentally linked to physical power projection. The proof-of-power protocol is primordial. It has existed since abiogenesis, the dawn of life. It is half a million times older than sapiens and their belief systems about resource ownership and property rights. Proof-of-power exists in every corner of life, at every scale. Everywhere you look, you can see that resources are owned insofar as organisms have the capacity and inclination to project physical power to gain and maintain access to those resources. This begs a question: how did the proof-of-power property ownership protocol begin, and how does it work?

Among the first resources that would likely qualify as being owned by life were mineral-rich deposits of nutrients captured from deep-sea hydrothermal vents shortly after the formation of oceans around four billion years ago. Life's first major power projection technology wasn't sharp teeth like what we saw with the example of the wolf in the previous section. Instead, it was a pressurized membrane – little more than a bubble. A pressurized membrane is a wall or thin mass stretched across a volume that exerts force to displace surrounding mass, as illustrated in Figure 9. When external forces from the environment contact a membrane, the membrane exploits Newton's third law to passively project opposing forces back at the environment to displace the mass of the surrounding volume. [48]

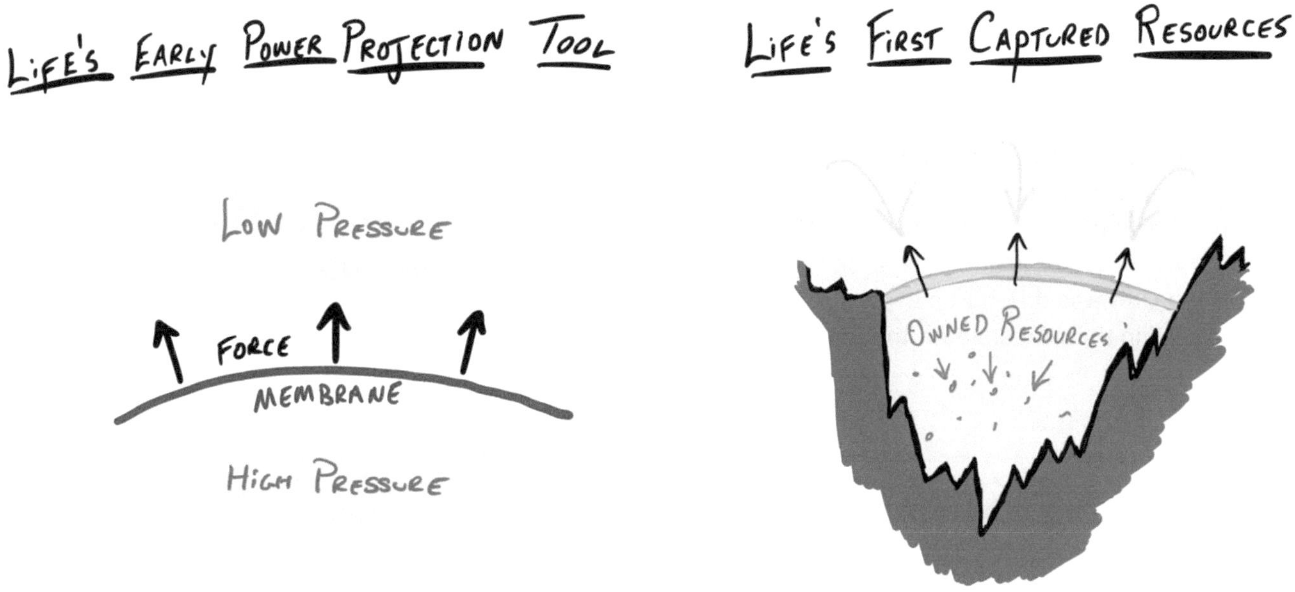

Figure 9: Illustration of One of Life's Most Dominant Power Projection Tactics

Using pressurized membranes to capture resources and survive in the wild would probably qualify as life's earliest and most successful power projection tactic, technique, and technology to date. Pressurized membranes enabled life to exert physical power to capture nutrient-rich volume from their surrounding environment. Despite consisting of nothing more than thin films stretched across microscopic gaps in rocks, these early lifeforms were nevertheless global superpowers. They stood as iron citadels capable of projecting infinitely more power than the lifeless void which existed before them. [48]

The tiny fraction of watts exerted by these microscopic bubbles were anything but insignificant; they were monuments of defiance against what could be described as life's mortal enemy: the cold and unsympathetic entropy of the Universe. If we ignore the technicality that these microorganic structures emerged billions of years before the evolution of sight, then we could describe the emergence of pressurized membranes as life's first "Veni, Vidi, Vici" moment. At this early stage, membranes were only capable of passively exerting equal and opposite forces upon the surrounding environment. But this passive power projection strategy didn't make membranes a "defense-only" power projection tactic. The nutrient-rich incubatory volume occupied by these pressurized membranes was captured the same way Caesar captured Rome: by force. [48]

As discussed in the previous section, physical power is how all living organisms achieve consensus on the "legitimate" state of ownership and chain of custody of resources. "Legitimate" is put in quotes to serve as a reminder that "legitimate" resource ownership is an abstract construct invented by sapiens to assist with the peaceful adjudication of intraspecies property disputes. In other words, nature doesn't care about what sapiens think "legitimate" resource ownership means. In fact, nature does not appear to care about any abstract sapient construct. Nature could care less about people's property rights, or rules encoded into property law. Nature only appears to recognize proof-of-power. The first living organisms didn't have the capacity to *think*, much less *believe* that the nutrient-rich volume they captured was "legitimately" theirs or not – they simply took it by force, the same way all animals (including and especially sapiens, as much as they hate to admit it) gain and maintain access to their precious resources.

Early life forms "owned" deep-sea hydrothermal nutrients for the same reason a wolf "owns" meat: because they had the capacity and inclination to project physical power to successfully capture and secure their access to it. Since these first little organisms emerged, life's pressurize membrane power projection tactic has evolved and taken many different complex forms over the past four billion years, but the function has not changed. From microscopic bubbles, to armor, to castle walls, to militarized national borders, all pressurized membranes work the same way: they passively project physical power to gain and maintain access to precious resources. These resources are captured by force. Period.

Abiogenesis reminds us that **living is an act of projecting physical power to capture physical resources**. Life physically captures the oxygen it breathes by force. Life physically captures the food it eats by force. Life physically captures the volume it occupies by force. What life needs to survive is "owned" for no other reason than the fact that life has the capacity and inclination to project power to capture it. A quick glance into the night sky reminds us that the Universe does not owe us our lives; we have what we have because we *take* it using physical power. As discussed in the next chapter, the rest of what we believe about resource ownership is abstract.

3.2.2 To Live is to Convert Chaos into Structure

"In any fight, it is the guy who is willing to die who is going to win that inch."
Tony D'Amato, *Any Given Sunday* [49]

The emergent behavior of life is something remarkable. By projecting lots of physical power to capture and secure access to resources, life is miraculously able to turn the inexorable chaos of the Universe into something more structured. It then leverages that structure to exert more physical power to capture more resources and convert those resources into even more structure. Life owes its existence and prosperity to this process. Few things are as aligned with the fundamental nature of all living things than this physical power projection process through which organisms secure access to resources and then use those resources to build additional structures, for no other discernable reason than to simply improve its ability to countervail entropy and survive a little longer.

Having defied entropy and established its first beachhead of nutrient-rich territory, life's first pressurized membranes were fully equipped for battle. Fighting inch over inch for more nutrient-rich volume, pressurized membranes expanded in size and strength until they created enough structure to where they no longer needed the structural support of rocks. Using clever power projection tactics like closed-loop pressurization control, life was able to construct fully self-contained membrane bubble fortresses such as the one shown in Figure 10, capable of floating to unexplored, nutrient-rich heights.

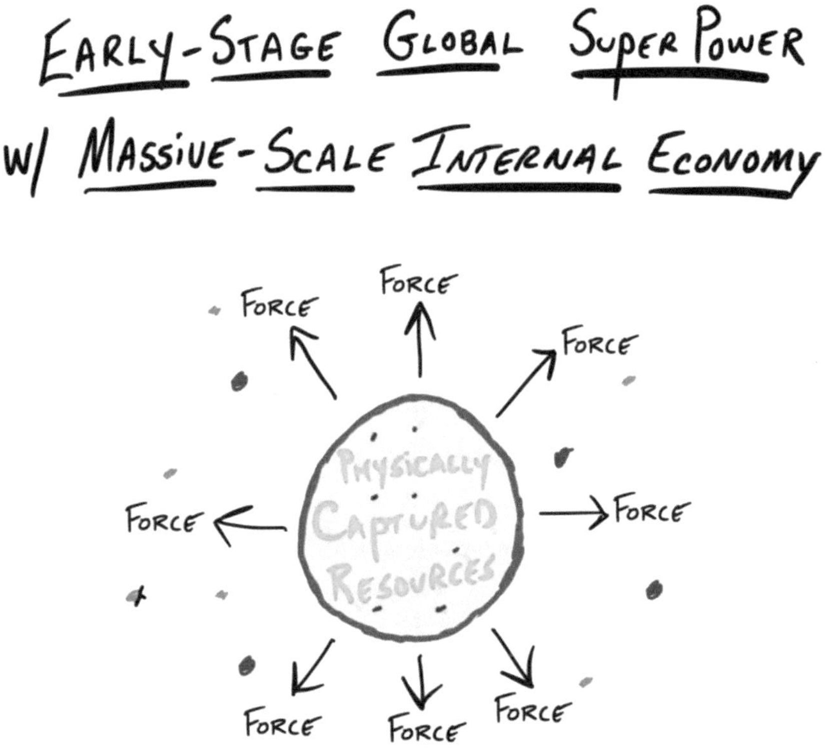

Figure 10: Illustration of an Early-Stage Global Superpower

Under the protection of their pressurized membranes, these new global superpowers were able to form highly complex internal microorganic economies. Subcellular molecules self-assembled into increasingly more specialized workforces, trading various microorganic goods and services and becoming ever more efficient, productive, and resource abundant. Through this special combination of robust membrane power projection and high-functioning internal economy, life was able to follow a multistep biochemical path towards ever-increasing structure until it managed to build complex, massive-scale economies we now call singled-celled bacteria. [48]

3.4 Primordial Economics

"It is not the most intellectual of the species that survives; it is not the strongest... the species that survives is the one best able to adapt and adjust to the changing environment in which it finds itself."
Leon C. Megginson [50]

3.4.1 Benefit-to-Cost to Ratio of Attack

There is estimated to be more bacteria on Earth today than there are stars in the Universe. Suffice it to say, Earth's nutrient-rich volume has become significantly more congested than it was 4 billion years ago. As our oceans began to fill to the brim with bacteria, organisms began to face a new challenge: resource scarcity. It was in response to resource scarcity that life appears to have discovered one of its most primordial economic equations: the benefit-to-cost ratio of attack (**BCR_A**). [51, 52]

Every organism could be described as a nutrient-rich bounty of precious resources. Inside every organism are the building blocks necessary to create other organisms. For this reason, most organisms represent an attractive target of opportunity for other organisms to do what we have established that life does demonstrably well: capture with force. Consequently, a weak, docile, or ineffectual nutrient-abundant organism is essentially a floating gift basket of vital resources for neighboring life forms to devour. This is because of the primordial economic dynamics shown in Figure 11.

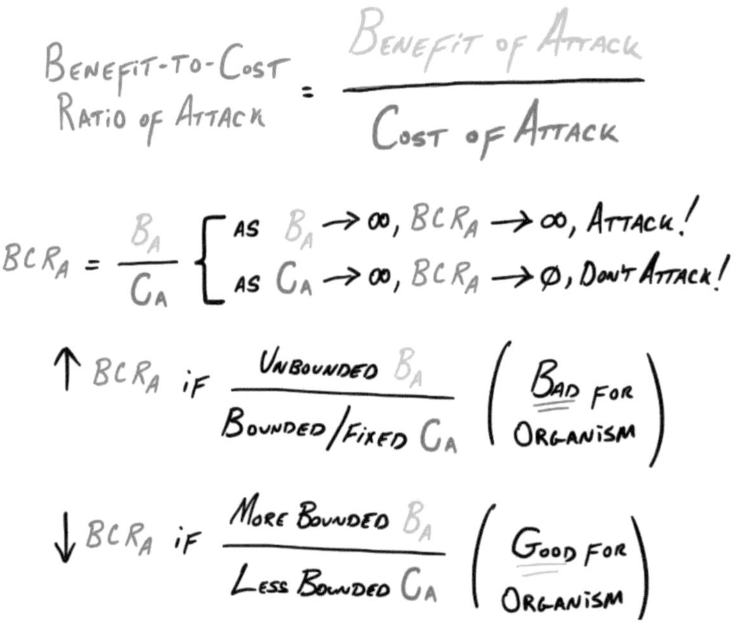

Benefit-to-Cost Ratio of Attack = Benefit of Attack / Cost of Attack
BCRA = BA / CA
as BA → ∞, BCRA → ∞, Attack!
as CA → ∞, BCRA → Ø, Don't Attack!
↑ BCRA if Unbounded BA / Bounded/Fixed CA (Bad for Organism)
↓ BCRA if More Bounded BA / Less Bounded CA (Good for Organism)

Figure 11: The Benefit-to-Cost Ratio of Attack (BCR_A) a.k.a. Primordial Economics

Every organism can be attacked; therefore, every organism has a **BCR_A**. An organism's **BCR_A** is a simple fraction determined by two variables: the benefit of attacking it (B_A) and the cost of attacking it (**C_A**). B_A is a function of how resource abundant an organism is. Organisms with lots of precious resources have high B_A. Organisms with less precious resources have lower B_A. On the flip side of the equation, **C_A** is a function of how capable and willing an organism is at imposing severe physical costs on attackers. Organisms capable of and willing to impose severe physical costs on neighboring organisms have high **C_A**. Organisms that are not capable of or willing to impose severe physical costs on neighboring organisms have low **C_A**.

Higher **BCR_A** organisms are more vulnerable to attack than lower **BCR_A** organisms because they offer a higher return on investment for hungry neighbors to devour. Organisms therefore have an existential imperative to lower their **BCR_A** as much as they can afford to do so by increasing their capacity and inclination to impose severe physical costs on neighboring organisms. An organism can't just devote all their time and energy towards increasing their resource abundance and expect to prosper for long, because doing so would cause their BCR_A to climb and jeopardize their chances of long-term survival.

To survive, organisms must manage both sides of their BCR_A equation. To prevent their BCR_A from climbing to hazardous levels, organisms must either shrink their numerator or grow their denominator. They must either decrease their resource abundance to decrease their B_A or grow their C_A by increasing their capacity and inclination to impose physical cost on attackers. Decreasing resource abundance is not an ideal solution for organisms seeking to grow, so increasing C_A (i.e. increasing the denominator) is a preferable option. If organisms choose to grow the denominator, they must grow C_A at an equal or higher rate than the rate at which their B_A increases, or else BCR_A will climb.

3.4.2 Lower Benefit-to-Cost Ratio of Attack means higher Prosperity Margin

A simple way to visualize the primordial economic dynamics of survival is shown in Figure 12. To survive, an organism must keep their BCR_A level lower than a hazardous BCR_A level that will motivate neighboring life to attack them. The space in between the organism's BCR_A level and the hazardous BCR_A level can be called the *prosperity margin*. This margin indicates how much an organism can afford to increase its BCR_A before it risks being attacked.

Primordial economic dynamics seem simple and straightforward, but there's a catch. There's not really any way for organisms to know how large their prosperity region is because they don't know exactly what level of BCR_A would qualify as being hazardous. How hazardous a BCR_A level is depends almost entirely on factors outside of an organism's sight and control. This is because it depends on external circumstances within the environment. If neighboring organisms (i.e. potential attackers) choose to grow their C_A to lower their BCR_A, then the hazardous BCR_A for that environment drops and the organisms which don't lower their own BCR_A lose prosperity margin. Thus, the same organism with the same BCR_A could have two completely different property margins based exclusively on the conditions of the environment the organism can neither see nor control. This phenomenon is illustrated in Figure 12.

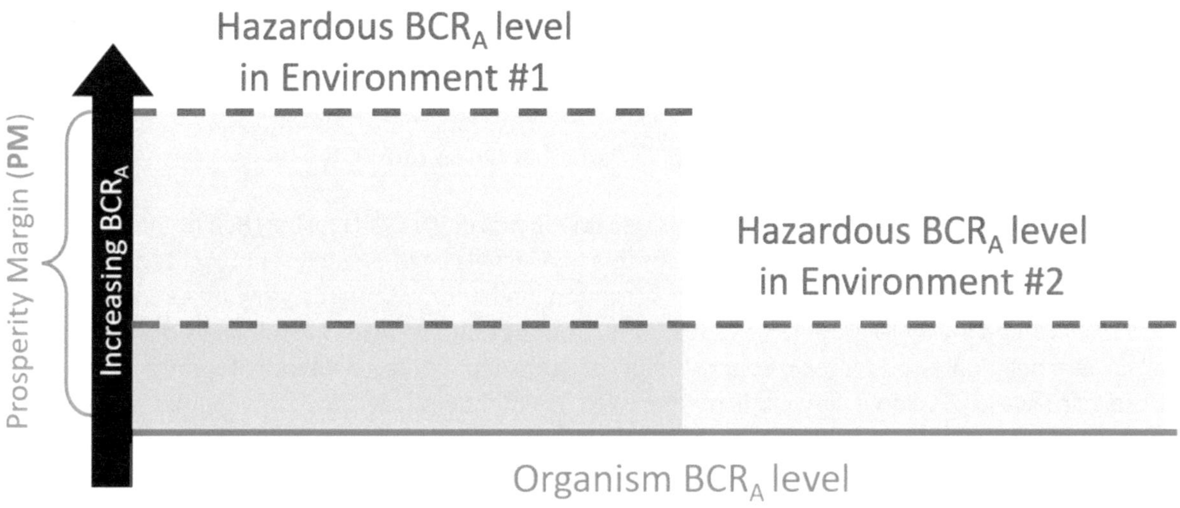

Figure 12: Prosperity Margin Changes among Different Environments

Organisms operating in empty neighborhoods have intrinsically higher prosperity margin than organisms operating in environments filled with neighboring life. With that surplus of margin, they can devote more time and energy towards boosting their resource abundance (thus increasing B_A) and increasing their BCR_A without having to focus much attention on growing C_A. They simply don't have to worry about their BCR_A as much because there's nothing around to attack them (hence animals like manatees).

Environments tend to change, however – sometimes quickly. They become *congested*; they fill up with a lot of other organisms. When environments become congested, they become *contested*; organisms increasingly oppose one another's attempts to access the same limited resources. As environments become more contested, they become more *competitive*; organisms seek to gain an advantage over each other. While all of this is happening, environments remain intrinsically *hostile*; entropy is a constant, looming threat. And if entropy doesn't attempt to kill an organism, a hungry neighbor will undoubtedly try to devour it. Add these factors together and we get the type of environment all organisms live in today: congested, contested, competitive, and hostile (CCCH) environments.

Organisms can try to move to different environments that naturally afford higher prosperity margin, but wherever life goes, other life inevitably follows, making the new environment CCCH as well. Consequently, finding a non-CCCH environment is not really an option. Survival therefore becomes a task in learning how to adapt to the local environment by learning how to throttle down **BCR_A** and buy oneself as much prosperity margin as possible. Different organisms have different successes at this task. Figure 13 illustrates how organisms which succeed at lowering their **BCR_A** level enjoy more prosperity margin.

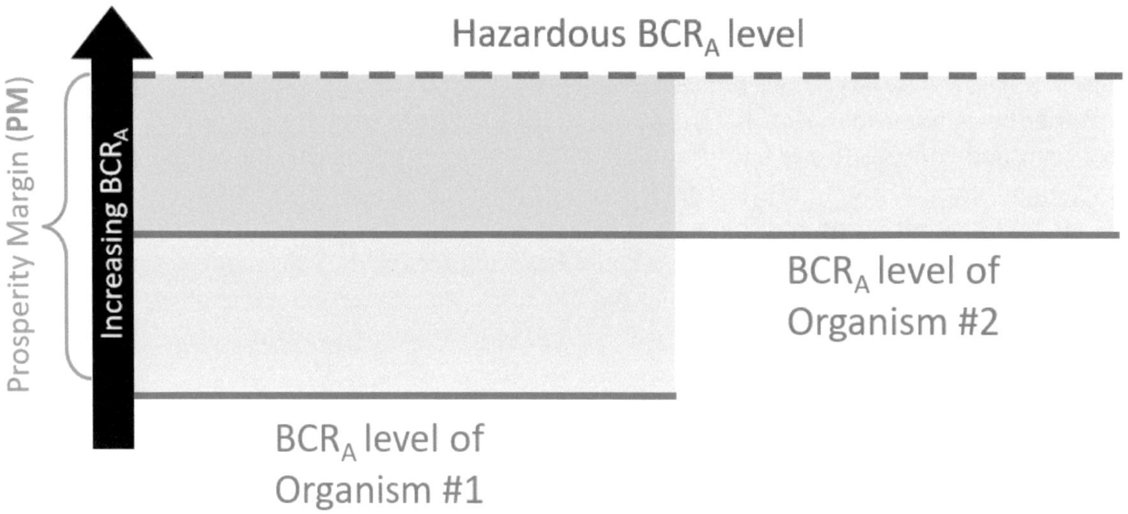

Figure 13: Prosperity Margin Changes amongst Different Organisms

A hazardous state arises when local environments become increasingly CCCH faster than organisms can adapt to them. When these changes occur, previously acceptable **BCR_A** levels become unacceptably hazardous. It becomes more of an existential imperative to devote time, attention, and energy towards growing **C_A** to lower **BCR_A**. If an organism doesn't find a way to grow their **C_A** fast enough, they compromise their chances of survival by making themselves the neighborhood target of opportunity for surrounding life to devour. Most organisms learn this lesson the hard way, but some are intelligent enough to adapt and develop new power projection tactics, techniques, and technologies to continuously lower their **BCR_A**.

3.5 Innovate or Die

"What doesn't kill you makes you stronger."
Friedrich Nietzsche [53]

3.5.1 The Rise of Predation

Whenever we study nature, we should remind ourselves that the behaviors we observe in nature are incontrovertibly winning strategies for survival. This is something to keep at the forefront of our minds when we observe that killing and fratricide are some of the most common, routine, and predictable behaviors in nature. Life appears to be well-versed in primordial economics, devoted to the task of devouring those who don't devote their time and attention to lowering their BCR_A.

Early forms of predation used dual-use power projection tactics which capitalized on pressurized membranes, like *phagocytosis*, or cell eating. Forming cavities in their membranes, organisms figured out how to use rudimentary mouth-like structures to capture resources by engulfing particulate matter. This capability is extremely useful because it's multifunctional; mouths capture resources *and* impose severe physical costs on attackers. In order words, mouths influence B_A and C_A simultaneously, giving an organism more control over their BCR_A. This explains why mouth-like structures have become such a popular power projection tactic employed by multiple different species. [54]

Phagocytosis illustrates yet another vital function of physical power. Not only do organisms use physical power to achieve consensus on the state of ownership and chain of custody of resources, but they also use physical power to regulate their BCR_A levels. Porous membranes and mouths demonstrate how some power projection tactics can influence both sides of the BCR_A equation, making them highly desirable.

Other power projection tactics only affect one side of the BCR_A equation. For example, if you take a pressurized membrane but remove its ability to subsume particulate matter, you get armor plating. Armor plating is useful for growing C_A by imposing higher prohibitive physical costs on attackers via Newton's 3rd law, but its inability to subsume particulate matter makes it not particularly useful at capturing resources to increase B_A. Nevertheless, armor plating is still a winning power projection tactic often seen in nature because of how it helps organisms with the existential imperative of lowering their BCR_A and buying themselves as much prosperity margin as possible to keep themselves secure against neighboring life.

Some other examples of important dual-use power projection tactics that emerged during the early days of predation were evolutions like *pili* (hair) and *flagella* (tails). These innovations allowed life to swim around and capture resources to increase B_A, while simultaneously allowing them to impose physically prohibitive cost on attackers by outrunning them or by using them as whips to break apart their neighbors' membranes. Mixing these technologies with phagocytosis proved to be an especially powerful combination, leading to the emergence of what we now call *predators*.

A predator is a proactive primordial economist. Predators are BCR_A bargain shoppers who hunt down the best BCR_A deals within their local environment. Armed with dual-use power projection technologies like whips, tails, and mouths, life's early predators mastered the art of BCR_A bargain shopping by swimming around and eating resource-abundant organisms with the highest BCR_A levels. As oceans gave rise to more of these BCR_A bargain shoppers, neighborhoods became increasingly more CCCH. Organisms which developed the most effective dual-use power projection tactics for their neighborhood became what we call *apex predators*. Organisms which couldn't buy themselves enough prosperity margin to adapt to their new environment were promptly devoured by these apex predators, as illustrated in Figure 14 below.

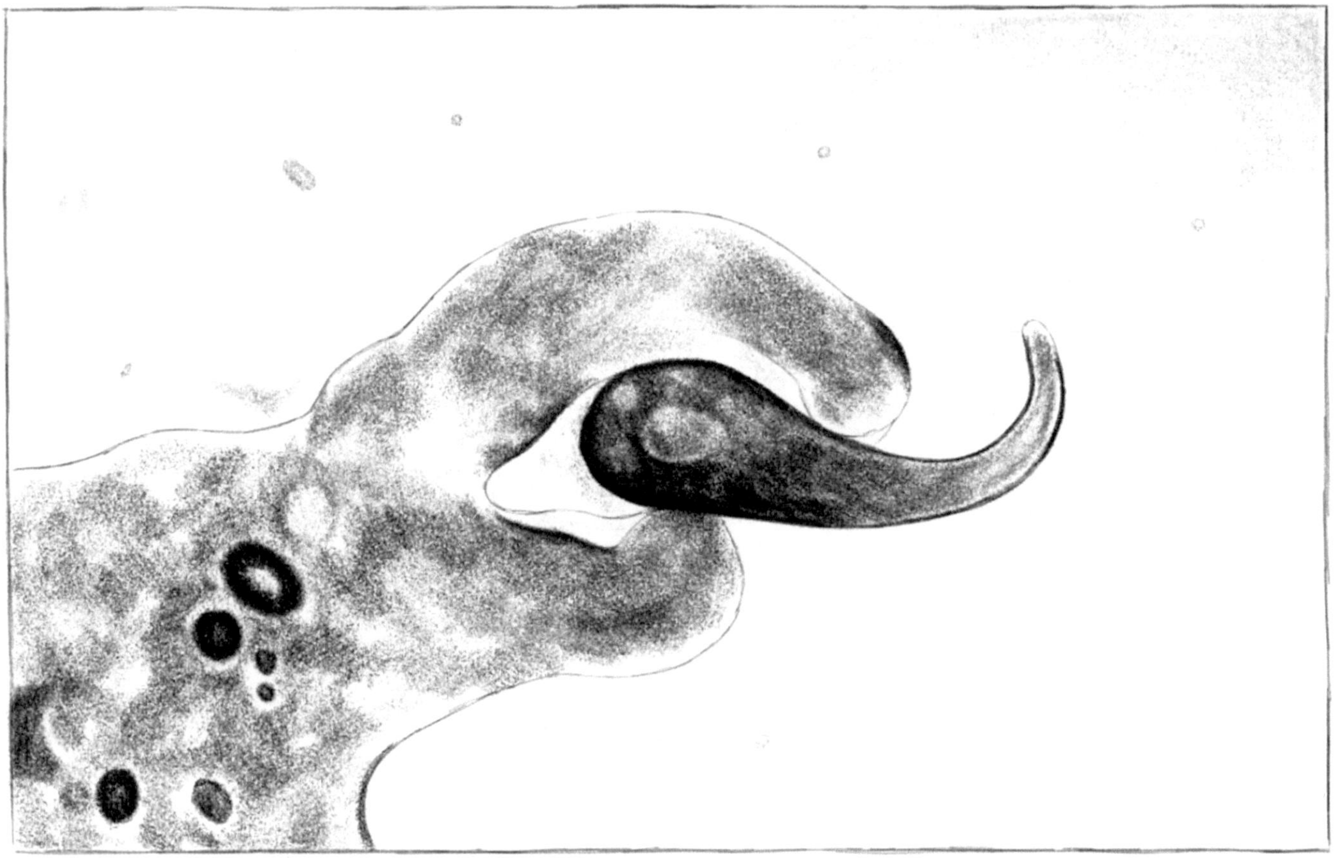

Figure 14: Apex predator using Phagocytosis to Devour a Neighboring Organism with High BCR_A

Some might say predation is a negative phenomenon because of how murderous and fratricidal it appears to be. This assertion is based purely on human ideology. From a systemic perspective, predation has benefits for life. In sufficient moderation, predation acts like a filter that weeds out life's most unfit and unadaptable organisms. By passing organisms through this filter, life revectors Earth's precious limited resources away from its worst prosperity-margin-growers towards its best prosperity-margin-growers, consequently buying more prosperity margin for life as a whole. In other words, **the stronger and more adaptable organisms become at surviving against predation, the more capable life itself becomes at surviving against entropy**.

It takes a stoic mindset to recognize and appreciate the complex emergent benefits of predation. In a Universe without entropy or resource scarcity, there might not be a lot to gain by filtering out organisms that are not optimized for their environment. Alas, that is not the Universe we live in. Entropy and resource scarcity are very much at play, which means there is a lot for life to gain by filtering out its unfit members and revectoring limited resources towards its fittest members who are most capable of surviving against entropy.

Without predation, lifeforms might operate on something like a "first come, first serve" or "finders keepers" basis of resource management. A lack of predation would mean that organisms automatically gain monopolies on the nutrient-abundant territory they discover because they are uncontested. Regardless of how strong, resourceful, or adaptable they are, the first to arrive at a resource would automatically be allowed to have monopoly control of that resource by virtue of their being unchallenged. Without the competitive stress of predation, these organisms would have far fewer external motivators to become stronger, more resourceful, and more adaptable. In other words, without predation, there would be nothing but unimpeachable, centralized, monopoly control over precious resources.

Many business professionals have made similar arguments that monopolies aren't good for consumers. They argue that competition is holistically beneficial for consumers because it compels organizations to innovate and build better products. If we accept this argument as valid, then it stands to reason that predation is a positive phenomenon because it prevents environmental resource control monopolies from forming. Predation doubles as an induced competition for resources; a way to force organisms to earn their seat at the table. The result? Better products (i.e. fitter organisms more capable of survival against entropy) for the consumer (life itself).

Without predation, the rate of environmental change would be comparatively slow. Organisms would only have to adapt to Earth's elemental changes, and the sudden onset trauma of rapid elemental changes are relatively rare. Species can live for millions of generations unaffected by asteroids, supercontinent breakups, landmass adjustments, ice ages, glacial events, volcanic activity, and major changes in the chemistry of the atmosphere. Without predation to keep them busy during Earth's elemental downtime, organisms would not need to be as quick to adapt, making them far less capable of rising to the challenge of surviving entropy's next attempt to kill it. [48]

Predation kicks the rate of existentially threatening environmental change into high gear. Survival no longer depends on adapting to Earth's comparatively slow elemental changes. Instead, it depends on outpacing the threat posed by other lifeforms. The eat-or-be-eaten dynamic of predator-prey relationships gives rise to a self-reinforcing feedback loop where the continuous discovery of increasingly more effective and lethal power projection tactics, techniques, and technologies begets the need for increasingly more effective power projection tactics, techniques, and technologies.

More predation leads to a more CCCH environment with faster-falling hazardous **BCR_A** levels. In response to this, organisms must figure out how to make their own **BCR_A** levels fall faster, which ends up making the environment even more CCCH, and the dynamo continues. Living prosperously becomes a task devoted to making new discoveries that will help each organism survive a rigorous natural selection process. The emergent effect of this dynamic is that life becomes faster, stronger, more adaptable, more intelligent, and better at surviving against the cold and unsympathetic cruelty of entropy.

3.5.2 We Owe our Lives to the Ecological Arms Race caused by Predation

Endothermy serves as a great example of the complex emergent benefits of predation. Around 250 million years ago, life above the surface of the Earth was very CCCH due to the Permian-Triassic mass extinction. Organisms couldn't afford to spend a lot of time out in the open because of harsh conditions caused by predation and entropy. This presented a challenge for organisms seeking to regulate their body temperature. Not spending a lot of time on the surface means not being able to capture an important resource from the sun: heat. Shrew-like organisms like the one shown in Figure 15 were able to overcome this challenge thanks to biological mutations which allowed their bodies to actively produce their own heat by metabolizing fats and sugars in their food. This power projection tactic is called endothermic body heating, or warm blood. [48]

Figure 15: A Warm-Blooded Organism that Sparked an Ecological Arms Race

The reader should note that when predation and entropy aren't factored into our calculus, warm blood looks like an extremely inefficient use of energy and a rather unhelpful power projection tactic. Why pay for something you can get for free? As paleontologist Mike Benton explains, *"endotherms have to eat much more than cold-blooded animals just to fuel their inner temperature control."* [55] Metabolizing food to heat the body is not nearly as energy efficient as receiving heat passively from the sun.

Why would an organism volunteer to compete over scarce watts and then burn those watts just to heat themselves when they have the more energy-efficient option of receiving heat passively for free? The answer is because they live in a CCCH environment filled with predators and entropy. If organisms don't learn how to warm their bodies underground where it's safe, they must go above the surface during the daytime where predators eagerly wait to devour them. With warm blood, organisms can keep themselves warm and safe underground during the heat of the day when their natural predators are out heating themselves. Endothermic weasels can therefore save their resource-capturing activities for the night, after the sun goes down and their ectothermic predators are less likely to see and devour them.

Because the sun is effectively a free fuel supply of exogenously available heat, cold-blooded animals don't have to compete over sunlight like they do for their food. But for cold-blooded predators whose food supply suddenly turns endothermic, these predators have an existential imperative to become endothermic too, or else they risk starvation. This phenomenon leads to what has been called an ecological arms race, where both predator and prey adopt the same adaptations and engage in a cat-and-mouse game where they try to out-evolve each other. In game theory, these are called strategic Schelling points. Thus, predation creates a game-theoretic dynamic where predator and prey adopt the same Schelling points. This leads to complex emergent behavior which benefits both predator and prey, as both become increasingly fitter and more adapted to their local CCCH environment.

As the University of Bristol explains, "In ecology, arms races occur when predators and prey have to compete with each other, and where there may be an escalation of adaptions." [55] Endothermy sparked an intense ecological arms race. Shrew-like prey and their reptilian predators both developed warm blood. After many years of innovate-or-die predatory dynamics, these creatures both developed distinctively upright bone structures which allowed them to move faster. They both developed better eyesight and more advanced brain circuitry. Consequently, both animal classes found themselves much more capable of survival when the next major elemental change happened on Earth.

Cold-blooded ectothermic body heating is indeed a more energy-efficient design – except when the sun stops shining and the world's free fuel supply of heat suddenly disappears. 66 million years ago, Earth's biggest, strongest, and most energy-efficient organisms learned the hard way that survival is not strictly about optimizing energy efficiency; it's also about adapting to a harsh environment. To be more specific, survival is about not freezing to death when entropy throws a 7.5-mile-wide asteroid at Earth sixteen times faster than a bullet, creating such a large debris cloud that direct sunlight didn't reach the surface of the Earth for years. Can the reader guess what power projection tactics are useful in *that* environment? Self-warming blood and the full suite of improved speed, eyesight, intelligence, and other capabilities developed during the ecological arms race between endothermic predators and prey.

Deprived of the power projection tactics, techniques, and technologies enjoyed by the animals which had participated in a highly competitive ecological arms race, many dinosaurs died en masse. The resulting food supply chain disruptions led to mass starvation and eventually mass extinction for ~80% of life on Earth. Meanwhile, the smaller, faster, smarter organisms with endothermic body heating that had been locked into a highly competitive arms race found themselves much better equipped to adapt to their new environment. With 80% of their compatriots gone, these animals were free to feast on what the rest left behind. These special animals are still thriving today; we call them birds and mammals.

3.5.3 Lighting a Fire Under Life's Hind Quarters as an Extrinsic Motivator to Get them to Adapt Faster

"It is the knowledge that I'm going to die that creates the focus that I bring to being alive. The urgency of accomplishment, the need to express love now, not later. If we live forever, why even get out of bed in the morning? Because you always have tomorrow. That's not the kind of life I want to lead... I fear living a life where I could have accomplished something and didn't."

Neil deGrasse Tyson [56]

Innovation is a tricky thing. As Clay Christensen taught us, the best innovations often don't appear to be better than the status quo. They often have worse performance characteristics. They often look highly inefficient and wasteful, and they routinely don't satisfy an existing need. This leads to the infamous innovator's dilemma, where searching for an innovative strategy is predestined to look like a poor economic decision because it means burning through resources searching for a solution to a problem nobody recognizes yet. [57]

For these reasons, the life of an innovator is often characterized by condescension, mockery, and underappreciation. Nevertheless, entropy demands that all lifeforms innovate or die. Life must find increasingly clever power projection tactics, techniques, and technologies to grow its prosperity margin and continue to survive. When entropy inevitably strikes again and the next major elemental change occurs, innovators often have the last laugh. The mighty dinosaurs fall, and the shrews inherit the Earth.

Now that we have reminded ourselves that we owe our existence to an ecological arms race, let's revisit the topic of predation and ask ourselves how *should* life compel its organisms to overcome the innovator's dilemma and maximize its chances of survival against entropy? How do we motivate organisms to pursue innovative power projection tactics that are predestined to look inefficient and wasteful? How do we compel ourselves to find the best survival tactics, techniques, and technologies if we can't know what they are *a priori*?

The answer doesn't appear to be intrinsic motivation because that's not what we observe in nature. What we observe in nature is a whole bunch of organisms constantly trying to devour each other, and then narrowly surviving extinction as a direct result of the clever power projection tactics developed to avoid being devoured. Organisms didn't develop warm blood and superior speed, eyesight, and intelligence to survive against a meteor. They did it to survive against each other, and those tactics, techniques, and technologies just happened to make them more capable of surviving a meteor. This would imply that life's approach to solving the innovator's dilemma is *extrinsic* motivation via predation.

From a systemic perspective, life effectively lights a fire under its organisms' hind quarters and tells them to innovate or die; figure out better power projection tactics to grow prosperity margin so we can survive in this Universe longer, or else be devoured by those who are willing to step up to the task. Like Olympians training under the oxygen-depravation stress of high altitudes, life seems to have figured out how to deliberately stress itself and spur innovation using predation. This process breaks up local resource monopolies and filters out the ecologically unfit and un-innovative, revectoring precious limited resources to the organisms which are stronger, more intelligent, and more adaptable.

Why would life want to do this? Because an organism that is incapable of innovating is an organism that is incapable of adapting to the environment. Organisms which can't adapt to the environment are destined to die to entropy anyways, so there's little for life to lose by cutting their losses, killing off their weak and unadaptable organisms early, and revectoring those resources to better survivors. On the surface, this seems like a cold and unsympathetic strategy. But it's not as cold and unsympathetic as the Universe hovering above our heads, seemingly determined to kill us. Moreover, four billion years' worth of data suggests predation is an incontrovertibly winning strategy for survival, hence its ubiquity in nature.

3.6 The Survivor's Dilemma

"Do not go gentle into that good night... rage, rage against the dying of the light."
Dylan Thomas [58]

3.6.1 Survival is not a Birthright; it must be Earned

Survival and prosperity do not appear to be birthrights. Nothing in our observations of the Universe indicates that life on Earth has an inherent right to live or to keep living. This would imply that survival is an act of earning life's seat at the table by countervailing the formidable entropy of the Universe. The key to accomplishing this daunting task is for life to innovate and develop increasingly clever power projection tactics, techniques, and technologies. The power we project must be used to capture the precious resources we need to survive, because the Universe does not appear to be inclined to part with them otherwise. The power we project must also be used to continually secure our access to those resources, because predators and entropy seems to always want to take them.

To innovate as quickly as possible, life's emergent behavior is to compel its organisms to keep searching for better power projection tactics under threat of predation. As birds and mammals demonstrated, the competition for better power projection tactics is daunting, but the strategy clearly works. While predators and prey compete against each other in ecological arms races, their discoveries double as a means to countervail entropy. The better organisms get at projecting power between and amongst themselves, the better life itself gets at earning its place between and amongst the stars.

Nature gives us abundant supporting evidence to indicate that the more organisms battle with each other, the more they develop better power projection tactics which help them lower their BCR_A and grow their prosperity margin. On a planetary scale, this helps life vector its precious resources to its best survivors. Food, energy, and territory flow to the fittest – those which prove their capacity to countervail entropy. It's as if life uses predation as a testing environment to try out and incubate different power projection tactics in a controlled environment, so that when the next meteor hits, it's better prepared.

Unfortunately, nature's process of compelling life to overcome the innovator's dilemma seems to have led to yet another dilemma, one that has driven sapiens to the brink of self-destruction via nuclear annihilation (more on this in the next chapter). The author calls this dilemma the *survivor's dilemma*.

3.6.2 Organisms Have No Way of Knowing How Secure is Secure Enough, Creating a Survivor's Dilemma

As previously discussed, organisms must ensure their BCR_A levels stay below the environment's hazardous BCR_A level to survive and prosper. Every local environment has a BCR_A level which qualifies as hazardous, serving as a threshold where an organism is virtually guaranteed to be attacked. This level changes depending on the environment and organisms can increase their chances of survival if they adjust their own BCR_A so that it stays below the environment's hazardous BCR_A level. The farther an organism's BCR_A level is below the hazardous BCR_A level, the better. These dynamics are illustrated in Figure 16

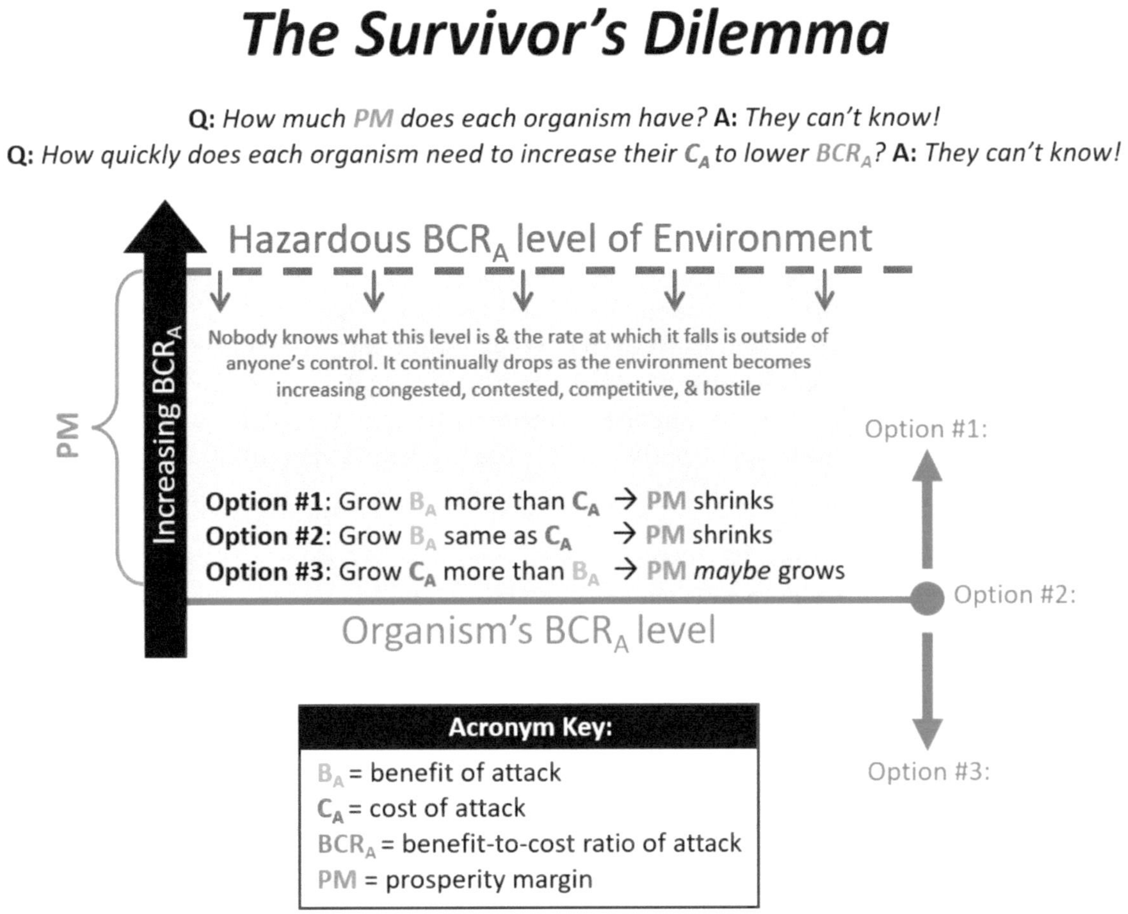

Figure 16: An Illustration of The Survivor's Dilemma

The margin between the organism's BCR_A level and the environment's hazardous BCR_A level (i.e. the BCR_A level at which an organism is likely to be attacked within a given predatory environment) is the organism's prosperity margin. An organism is safe to increase its BCR_A within the prosperity margin without reaching a hazardous state. Keeping one's BCR_A below the hazardous BCR_A level seems straightforward, but it's challenging because organisms don't know what the hazardous BCR_A level is. That level is a moving target; it changes depending on conditions outside of what organisms can see or control. They can only guess how far they can afford to raise their BCR_A before they put themselves in a hazardous state. To make matters even more complicated, the hazardous BCR_A level tends to drop as the environment becomes increasingly more predatory and CCCH, and the rate at which it drops is also unknown. This means **organisms don't know how much prosperity margin they have, nor how quickly it's shrinking.**

3.6.3 Organisms have Three Options for Solving the Survivor's Dilemma

"I don't have to outrun the bear; I just have to outrun you."
Proverb, Origin Unknown

This lack of critical information creates a dilemma for organisms seeking to survive and grow in a CCCH environment. The author calls this dilemma *the survivor's dilemma*. The dilemma happens because **growing resource abundance causes an organism's B_A to increase, which in turn causes its BCR_A to increase**. In response to this challenge, organisms have three strategic options.

If an organism does nothing to counterbalance the effect of increasing B_A, then its BCR_A will climb, and its prosperity margin will shrink. We can call this "option #1." If an organism perfectly counterbalances the effect of increasing B_A with an equivalent amount of C_A, then its BCR_A will stay fixed, but its prosperity margin will still shrink because the increasingly CCCH nature of the environment naturally causes the environment's hazardous BCR_A level to fall. We can call this "option #2." The last remaining option for an organism seeking resource abundance growth is to counterbalance the effect of its growing B_A with a greater amount of C_A to ensure its BCR_A continuously shrinks. We can call this "option #3." Note how even option #3 doesn't necessarily prevent prosperity margin from shrinking. The organism must execute option #3 in such a way that the organism's BCR_A falls faster than the rate at which the environment's hazardous BCR_A level falls to prevent prosperity margin from shrinking.

Survival is therefore much like the proverb about outrunning the bear, where the bear is the local environment's hazardous BCR_A level, and it's completely invisible to the organisms trying to outrun it. Fortunately, one doesn't need to outrun the invisible bear, they just need to outrun their neighbors who are also trying to outrun the invisible bear. In this scenario, it's clear that option #3 is the most strategically optimal because it's the only option which actually runs *away* from the direction of the invisible bear; option #2 stands still and option #1 runs *towards* the invisible bear.

It seems straightforward, but option #3 is deceptively difficult because **organisms don't know how much C_A they need to survive**. The organism needs to grow enough C_A to ensure its BCR_A drops quicker than the hazardous BCR_A level, but the organism knows practically nothing about that level. It doesn't know what level qualifies as hazardous, nor the rate at which that level is falling as the environment becomes more CCCH. The organism's prosperity margin could even drop to zero while it executes option #3, simply because it isn't aggressive enough.

The reader is now invited to place themselves in the shoes of an organism faced with the task of survival (you are technically already in these shoes, whether you accept that or not). You have a power projection budget of X watts. What do you do with those watts? Do you put those X watts towards growing resource abundance (thus increasing B_A) or towards growing your ability to impose severe physical costs on your murderous and fratricidal neighbors (thus increasing C_A)? The precise amount of B_A or C_A you need to grow is impossible for you to know because it depends on factors outside of what you can see and control, factors like what hungry, envious neighbors are choosing to do with their watts.

3.6.3 The Strategic Schelling Point is to Continuously Decrease the Benefit-to-Cost Ratio of Attack

The survivor's dilemma creates a game theoretic situation where you can't trust your neighbor, you don't know what BCR_A level qualifies as hazardous, and you don't know how quickly the hazardous BCR_A level is chasing you. This means you don't know how much prosperity margin you have or how quickly it's dwindling. You know you should try to outrun your neighbors from the invisible bear, but you don't know how quickly you need to run because you can't see the invisible bear.

In this situation, the optimal strategy is to simply run as fast as you can afford to run – to invest your watts into keeping your BCR_A falling as quickly as you can afford for it to fall. This will minimize the probability of causing your prosperity margin (i.e. the distance between you and the invisible bear) to close while still giving you the opportunity to grow your resource abundance. It's like two-pedal driving where C_A is the throttle pedal and B_A is the brake pedal. To survive, you have to keep a heavy foot on the C_A pedal at the same time you press the B_A pedal to outrun the neighbors, and you have to constantly manage this.

The survivor's dilemma represents the same fundamental challenge as national strategic security: there's no way to know how much security is *enough* security. A nation can only guess how much security they need based on the intelligence they can collect about their opponent's power projection capabilities, but the only way for a nation to know for sure that they haven't dedicated enough resources towards security is the hard way. This is the same dilemma that all organisms face, no matter what kind of organism they are and no matter what they think about primordial economics. The organisms which survive are the ones that adopt the Schelling point of lowering BCR_A as consistently and as affordable as possible. In other words, **the organisms which survive are the ones who learn to continuously maximize their capacity and inclination to impose severe physical costs on their neighbors**.

In this environment, a significant premium is placed on dual-use power projection tactics, techniques, and technologies. Tactics which only serve to increase B_A are strategically compromising; it's more favorable to develop power project tactics which allow an organism to spend their watts capturing resources *and* imposing physical costs on neighbors to ensure their BCR_A level continues to fall as quickly as possible. Unfortunately, when multiple organisms adopt this same Schelling point, it makes the local environment more CCCH because everyone dedicates a disproportionate number of their watts towards finding increasingly more clever ways to impose severe physical costs on each other. These dynamics cause the hazardous BCR_A to fall even faster, making the invisible bear pick up more speed.

3.6.4 The Survivor's Dilemma Explains why Nature's Top Survivors are so Powerful and Mean-Looking

The emergent effect of these dynamics explains why wild animals look and act the way they do. Ever notice how nature's survivors at the top of the food chain look so consistently *tough*? The survivor's dilemma offers an explanation: top-performing organisms are pressing the C_A pedal harder than the B_A pedal. They focus on making sure they stay well-equipped with all the latest and greatest dual-use power projection tactics that enable them to grow B_A *and* grow C_A simultaneously with heavier emphasis on growing C_A. Their teeth are *sharpened*; their nails are *sharpened*. Why? To impose severe physical costs on their neighbors by puncturing them. Life's top survivors are frequently covered in equipment which empowers them to pinch, puncture, and bludgeon. They have thick suits of armor backed by rigid skeletons, often with big muscles hanging from those skeletons. They're what some people might call "scary," or aggressive, or even repugnant due to their capacity and inclination to impose severe physical costs on their neighbors.

Have you ever stopped to consider *why* top-performers in nature look the way they do? Why aren't nature's top-performers consistently fat, soft, and docile? Once we understand primordial economic game theory, it makes perfect sense why Earth's top performers keep converging on the same characteristics despite being separated by vast quantities of time and distance. We can't allow ourselves to overlook the fact that what we observe in nature is incontrovertibly what survives in nature. The fact that life's top survivors keep converging into the same lean, mean, fighting machines is probably not a coincidence; to believe otherwise is to be guilty of survivorship bias (more on this later). Nature is clearly telling us something. It seems to be telling us that emphasis on C_A

matters, and it matters quite a lot. It tells us that **organisms who burn watts to increase C_A are organisms which survive**. It tells us that if we want to prosper and grow, we need to become sharper, both physically and intellectually.

3.7 Chasing Infinite Prosperity

"My life, old sport, my life has got to be like this. It's got to keep going up."
Jay Gatsby, *The Great Gatsby* [59]

3.7.1 Illustrating the Survivor's Dilemma using Bowtie Notation

Another way to illustrate primordial economics is by using what the author calls *bowtie notation*. An organism's BCR_A can be represented by the knot in the center of a bowtie, where each side of the tie represents B_A and C_A, as shown in Figure 17. This notation is useful for learning how to think like a predator by visualizing how "appetizing" organisms are for neighboring life to devour, a practice known as *adversarial thinking*. Adversarial thinking is useful for improving security because it helps one recognize one's vulnerabilities. Bowtie notation will be used throughout the following chapters to illustrate multiple survival strategies.

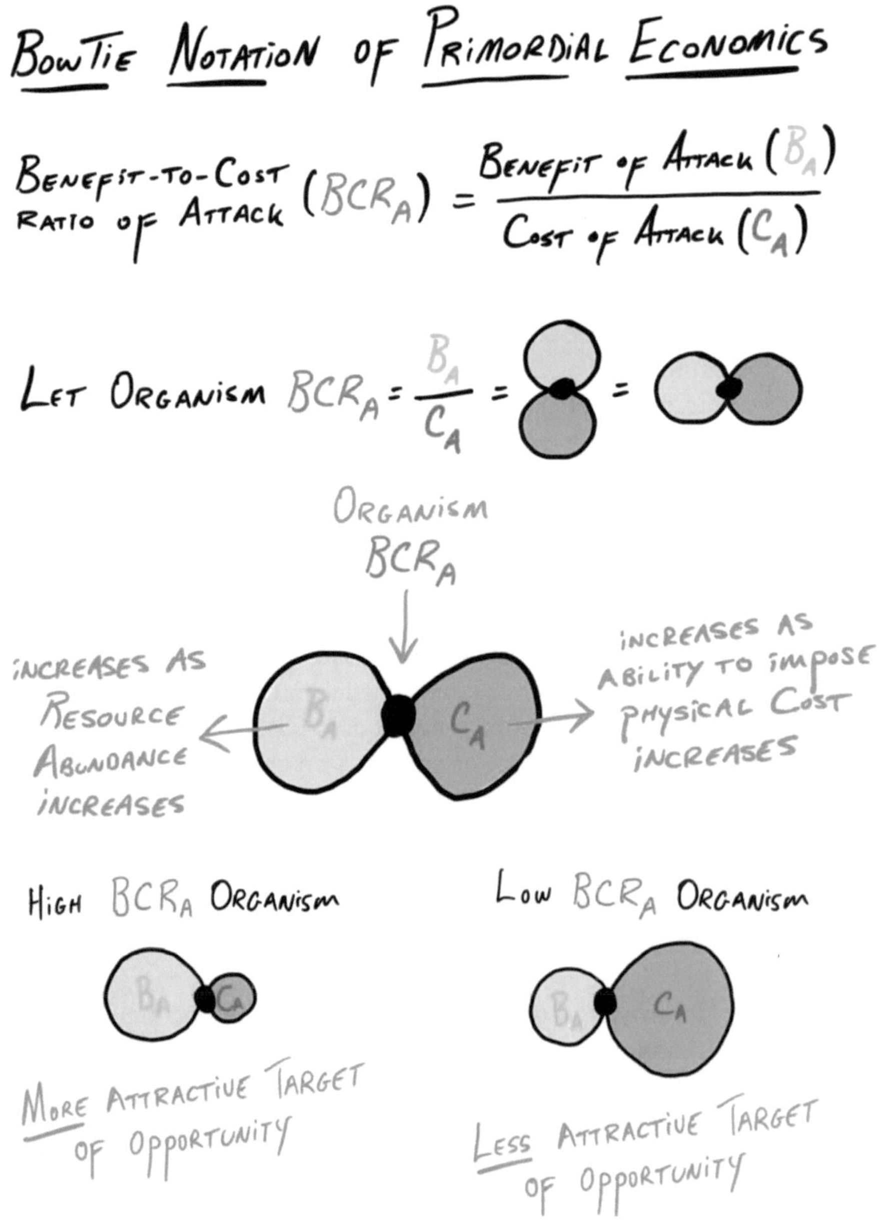

Figure 17: Bowtie Notation of Primordial Economics

As discussed in the previous section, a core challenge associated with living in a CCCH environment is growing one's resource abundance (thus increasing B_A) without making oneself an attractive target of opportunity for predators to devour (i.e. without increasing BCR_A or shrinking prosperity margin). To overcome this challenge, organisms have three options for responding to the survivor's dilemma outlined in the previous section. We can revisit these options and gain further insight using bowtie notation, as shown in Figure 18.

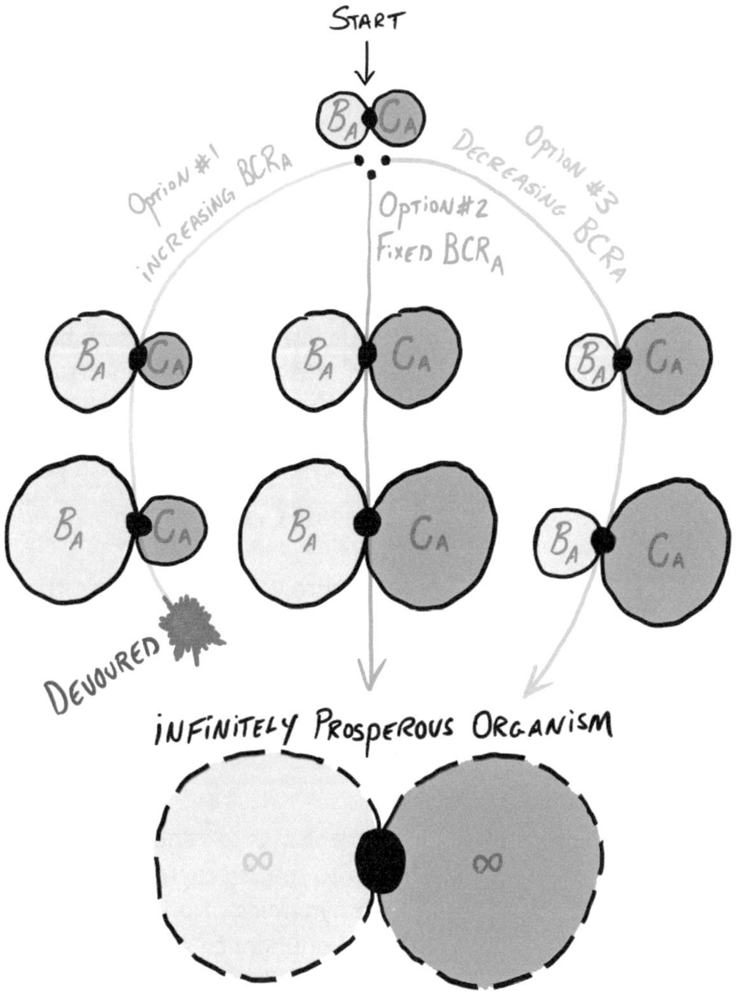

Figure 18: Bowtie Illustration of Three Power Projection Strategies for Pursuing Infinite Prosperity

Option #1 (shown on left side of Figure 18) represents the strategy where organisms use their available watts to grow their resource abundance at a faster rate than their capacity to impose physically prohibitive cost on attackers (i.e. grow B_A faster than C_A). This is illustrated as a lopsided bowtie in Figure 18 where the quantity associated with B_A is clearly larger than the quantity associated with C_A. From the organism's point of view, the upside to deploying this strategy is energy-efficiency; it can grow more resources with its budget of watts than either of the two other strategies, leading to more rapidly expanding resource abundance. However, from a predator's point of view, organisms which try "option #1" represent a target of opportunity because there is more to be gained from attacking the organism than there is to lose. The downside to "option #1" is therefore that it shrinks an organism's prosperity margin and makes it increasingly more likely to be devoured.

Option #2 (shown in middle of Figure 18) represents a power projection tactic where organisms grow their resource abundance and capacity to impose severe physical costs on attackers at an equal rate (i.e. grow B_A at same rate as C_A). This is illustrated as an even-sided bowtie which grows over time. The upside to this strategy is the ability to increase resource abundance without causing one's BCR_A to increase. The downside of this strategy is that fixed BCR_A levels become increasingly more hazardous over time as environments become increasingly more CCCH, causing prosperity margin to naturally shrink (the reader is invited to turn back to Figure 12 for an illustration of this phenomenon).

Option #3 (shown on right side of Figure 18) represents a power projection tactic where organisms increase their capacity and inclination to impose severe physical costs on attackers at a faster rate than they grow resource abundance (i.e. grow C_A faster than B_A). This is illustrated as a lopsided bowtie where the quantity associated with C_A is clearly larger than the quantity associated with B_A. From the organism's point of view, the upside to deploying this tactic is decreasing BCR_A and possibly increasing prosperity margin. The downside is *perceived* energy-inefficiency and a potentially slower rate of resource abundance growth. The reader should note that in reality, this downside is not inefficient because the energy is being spent for a clear reason: security. From a predator's point of view, an organism using this strategy represents an undesirable target. **The fact that predators choose *not* to devour this target is proof that the energy expended to increase the organism's C_A was not wasted energy; it was worth every watt**.

Of these three options, option #3 has the highest probability of long-term survival because it minimizes the organism's BCR_A and results in the largest prosperity margin. Option #3 accounts for the unescapable reality that Earth is a dynamic CCCH environment filled with murderous, fratricidal, and cannibalistic predators determined to devour high BCR_A organisms. To achieve long-term prosperity, option #1 and #2 require non-CCCH environments or trust that predators will not be motivated to attack them. Option #3 assumes there's no such thing as a non-CCCH environment.

3.7.2 An Infinitely Prosperous Organism is One that can Increase C_A ad Infinitum

Note how all three power projection options shown in Figure 18 point towards the same desired end state, something the author calls the *infinitely prosperous organism*. We can define an infinitely prosperous organism as an organism that is capable of growing its prosperity margin ad infinitum. Mathematically, this is only possible if the organism can grow its C_A ad infinitum, thus decreasing its BCR_A ad infinitum and potentially increasing prosperity margin ad infinitum. With an infinitely growing prosperity margin, an organism can grow its resource abundance without the threat of being devoured.

The purpose of conceptualizing an infinitely prosperous organism is to generate the following insight: **the key to survival, resource abundance, freedom, and prosperity is to maximize your ability to impose severe physical costs on neighbors**. This insight helps us understand the dynamics of primordial economics and the resulting emergent behavior of life at increasing scales. The resource abundance enjoyed by all organisms, organizations, and civilizations alike are simply the byproducts of life aspiring to become an infinitely prosperous organism. **We owe our survival, resource abundance, freedom, and prosperity to our physical power projectors who burn watts to increase our C_A and decrease our BCR_A.**

Through this lens, a macroscopic topic as complex as national strategic security can be simplified down to a simple illustration. All of the power projection tactics employed by sapiens are merely a higher-scale version of the same type of power projection tactics that first emerged during abiogenesis. The lessons of survival are exactly the same, whether we're talking about bacteria, nation states, or anything in between.

3.8 Sticking Together

"Our need will be the real creator."
Plato [60]

3.8.1 Hitting a Bounded Prosperity Trap

Once we understand the strategic dynamics of primordial economics, we can further appreciate the behavior we observe in both nature and society. Organisms of all shapes and sizes appear to be devoted to the task of solving the survivor's dilemma and becoming increasingly prosperous by mastering their ability to project power. However, the path to infinite prosperity is not straightforward. Life has no way of knowing what combination of power projection tactics it needs, nor in what sequence to develop them. This leads to self-induced reversions, unexpected side effects, and enumerable setbacks.

Lacking the ability to predict the future or comprehend the complex emergent properties of its environment, life appears to favor trial and error; it simply rolls the dice repeatedly until it lands on something that works. Like fighting fire with fire, life adapts to its randomly changing environment by randomly changing itself (a.k.a. evolving). No matter how much prosperity margin it enjoys, life doesn't seem to be inclined to stop searching for better power projection functionality. The threat of predation and an increasingly CCCH environment ensures organisms stay motivated to keep building and testing new features to help them continually minimize their BCR_A.

With the survivor's dilemma fresh in our minds, let's turn back to the example of our bacterial predators. For nearly 2 billion years, life existed as a murderous and fratricidal soup of single-celled organisms. Under the stress of their environment, bacteria invented different power projection tactics until they eventually stumbled upon a way to mitigate one of life's biggest exogenous threats: the relentless bombardment of radiation from the Sun.

"*As a means of defense*," biologist Henry Gee writes, "[bacteria] *evolved pigments to absorb these harmful rays. Once their energy had been absorbed, it could be put to work… cyanobacteria used it to drive chemical reactions. Some of these fused carbon, hydrogen, and oxygen atoms together to create sugars and starch. This is the process we call photosynthesis. Harm had become harvest*." [48]

Earth is the only known planet where fire exists. Astronomers have not discovered another planet with enough oxygen in its atmosphere to catch fire. If they did, this would be a revolutionary discovery, because it would represent a tell-tale sign of life on another planet. The ability to effectively feed on solar energy and poop out oxygen was no doubt a major evolutionary step for life on Earth, but like many new innovations, photosynthesis did not initially look like a success story. The oxygen exhaust produced by photosynthesis was devastating to the local environment because of its tendency to catch fire. For bacteria born into a world without oxygen, getting covered with oxygen was like getting covered with napalm. The discovery of photosynthesis literally backfired on life, causing "*the first of many mass extinctions in Earth's history, as generation upon generation of living things were burned alive*." [48]

Existing as nothing but a murderous, fratricidal, and burning soup of single-celled organisms, the state of life around two billion years ago could be described as a *bounded prosperity trap*. **The author defines a bounded prosperity trap as a situation where the inability to sufficiently grow C_A causes an organism to be unable to grow their prosperity margin any further, and they become trapped within either a fixed or shrinking margin of prosperity**.

Having a bounded prosperity margin means an organism can no longer grow its resource abundance without automatically causing its prosperity margin to shrink to the point of being devoured by the local CCCH environment. **Bounded prosperity traps show that when organisms hit a ceiling on their ability to project power, it translates directly into a degradation in prosperity. The ability to countervail entropy is severely degraded or halted altogether, and progress plateaus**.

Life plateaued at the single-cellular level as it struggled to overcome its CCCH environment. It hit a barrier on its ability to scale its power projection capacity to lower **BCR_A** and countervail entropy. Fortunately, as Plato observed, life's needs are its creator (the modern form of this expression is "*necessity is the mother of invention*"). Through continuous iteration, life managed to find the power projection tactics it needed to overcome this plateau and escape its bounded prosperity trap. Among those innovations was another one of the most effective power projection tactics ever discovered: cooperation.

3.8.2 Projecting More Power by Summing it Together

Before diving into a technical discussion about cooperation, it's important to note that two billion years ago, organisms had neither eyes nor brains. They had no capacity to see or understand what they were doing at any conceptually meaningful level because, as far as we can tell, both sight and foresight require multiple cells to form a brain. This means early cooperation was an unconscious phenomenon; bacteria weren't consciously aware of what they were doing or what impact it would have.

Organisms didn't decide to stick together because of their desire for a better future. As best as we can tell, bacteria are incapable of understanding abstract concepts like *the future*. Cells didn't wake up and decide to turn off their predatory nature and begin cooperating because they suddenly felt bad about billions of years of murder and fratricide. They also didn't start cooperating because they believed teamwork and interdependence could lead to a greater good for all single-celled kind. On the contrary, the reason why cells first started sticking together was because *they were literally stuck together*.

Early cooperation appears to have taken two primary forms: colonization and clustering. Colonization occurs naturally when there is limited volume available for life to occupy. When a group of individual organisms occupy the same space (e.g. the surface area of a rock), they inadvertently form a colony. As each organism acts in their own self-interest to defend their individual access to their space, they mutually reinforce each other at a wholistic level, forming a single cohesive colony which sums their collective power together to impose physical costs on external organisms seeking access to that same space. [48]

This phenomenon explains why **one of the most common forms of attack in nature is a colonization attack. Individual organisms simply act in their own self-interest to capture a small piece of territory for themselves and defend their access to it. They don't need to be intentionally working together or conscious of what's happening to execute this strategy, they just need to have aligned self-interests.** At larger scales, a colonization attack is sometimes called *invasion*. An invasive species is one that colonizes a given territory, and colonization can happen either intentionally or unintentionally.

The undisputed masters of colonization attacks on Earth are the serial invaders we now call *plants*. Powered by photosynthesis, plants have abundant watts available to devote themselves to mastering colonization attacks. This power projection tactic has worked quite well for them, as flora now represents 80% of Earth's biomass, dwarfing bacteria's 15%. Remaining organisms (to include all animals, reptiles, birds, fishes, etc.) only represent a measly 5% of Earth's biomass. [61]

The second example of cellular cooperation is clustering. Mutations in small bacterial cells called *archaeon* (which several scientists argue was spurred by the sudden onset trauma of photosynthesis and the resulting great oxidation extinction event) empowered them to form Velcro-like tendrils that physically capture neighboring cells by sticking to them and entrapping them under a common membrane. It's important to emphasize that **archaeon did not bargain with neighboring cells, engage in diplomacy with them, or draft treaties. They captured their neighbors by force, the same way all living creatures capture all physical resources.**

Archaeon became dependent on neighboring cells for nutrients, so they learned how to entrap their neighbors with sticky tendrils to secure access to those nutrients. The fact that the relationships they formed were symbiotic doesn't negate the fact that this evolution succeeded because one organism developed a power projection technology to physically overpower and entrap the other organism. At larger scales, clustering organisms via entrapment and forcing them to work together is given a different name: *conquering.* As we will be discussed in the next chapter, the undisputed masters of clustering attacks on Earth are the serial conquerors we now call *sapiens*. [48]

3.8.3 Cooperation is First and Foremost a Physical Power Projection Tactic

As previously discussed, pressurized membranes like cellular walls are highly effective power projection tools capable of exerting physical power to capture nutrient-rich volumes of space from the surrounding environment. Entrapped under the protection of a common pressurized membrane formed by archaeon tendrils, semi-symbiotic cells experienced a step-function increase in their ability to impose severe physical costs on their neighbors, and thus enjoyed a step-function increase in their prosperity margin. This enabled them to form booming, interdependent economies that produced vast amounts of resource abundance in virtually the exact same way as what had occurred two billion years prior, when subcellular particles found themselves in a similar situation under the protection of cellular membranes.

These colonies of clustered cells grew increasingly more interdependent and reliant on each other for vital nutrients, materials, and gene swapping. They were able to form highly complex structures, self-assembling into increasingly more specialized workforces, trading various organic goods and services and becoming ever more efficient, productive, and resource abundant. Through this special combination of robust membrane power projection, security, and high-functioning internal economy, life was able to follow a multistep biological path towards ever-increasing structure until it managed to self-assemble into complex, massive-scale economies we now call multicellular life. [48]

The takeaway: **Cooperation is a physical power projection tactic that emerged unconsciously, useful first and foremost for its ability to help organisms survive longer by summing their C_A together to decrease their BCR_A and buy them more prosperity margin**. Multicellular life and its remarkable levels of interdependence can be described as the byproduct of organisms simply having to occupy the same space or by inventing power projection tactics to physically capture and entrap their neighbors. By entrapping neighboring organisms for their nutrients, archaeon inadvertently discovered how to tap into their neighbor's physical power projection capacity to buy themselves more prosperity margin for comparatively little effort.

The emergence of cooperation is quite remarkable when you stop to think about it under the lens of primordial economics. It is such a deceptively simple and effective power projection strategy that it's hard to believe it took nearly two billion years for life to begin mastering it at the cellular level (although clearly life mastered cooperation at the subcellular level much earlier). To increase C_A, simply sum it together. To increase your BCR_A more than you could ever do alone, tap into your neighbors' exogenous supply of physical power. This strategy doesn't even require thinking; brainless organisms cooperate at a systemic level by merely by occupying the same volume (colonization) or by being involuntarily and/or unconsciously enveloped by the same membrane (clustering). Either way, these organisms achieved something remarkable: they increased in their ability to impose physically prohibitive costs on attackers with little additional expenditure of watts, and they escaped a bounded prosperity trap.

Like much of our biological history, the emergence of cooperation is an ironic story and quite relevant to our own lives and personal experiences. By constantly fighting against each other and then literally having a fire lit under them, single-celled life escaped their bounded prosperity trap by evolving new cooperation tactics. The corresponding drop in BCR_A gave cooperative cells abundant prosperity margin to work with. Not only were they empowered to develop thicker, stronger membranes that could survive against oxygen napalm, multicellular organisms were also able to grow their resource abundance to new and unexplored heights. By simply sticking together, organisms started solving the survivor's dilemma using option #3 without even being conscious of it.

3.8.3 "Cooperate or Die" Survival Dynamics

"There's always a bigger fish."
Jedi Master Qui-Gon Jinn, *Star Wars* [62]

Just like how the emergence of phagocytosis (i.e. cell eating) represented a dual-use power projection tactic which sparked predation and the "innovate or die" survival dynamic observed in nature, so too did the emergence of cooperation. Sticking together is a dual-use power projection tactic which influences both sides of the **BCR_A** equation. Cooperation can be used to grow resource abundance, or it can be used to increase capacity to impose physically prohibitive costs on neighbors. Cooperation therefore introduces its own cooperate-or-die Schelling point where the emergence of cooperation begets the need for more cooperation.

We have established that the survivor's dilemma gives organisms a strategic imperative to grow **C_A** as fast possible. They can grow **C_A** individually by discovering innovative new ways to impose higher physically prohibitive costs on attackers all on their own, but this requires them to spend their own watts. Alternatively, organisms can grow **C_A** without having to spend their own watts, simply by learning how to cooperate with neighbors and sum their **C_A** together as if they were a single, cohesive organism. In both cases, organisms execute the option #3 strategy where they grow **C_A** first to buy themselves enough prosperity margin to grow **B_A** without the threat of raising their **BCR_A** to hazardous level, enabling them to increase resource abundance *and* survive in a CCCH environment filled with predators and entropy. These dynamics can be illustrated in bowtie notation, as shown in Figure 19 below.

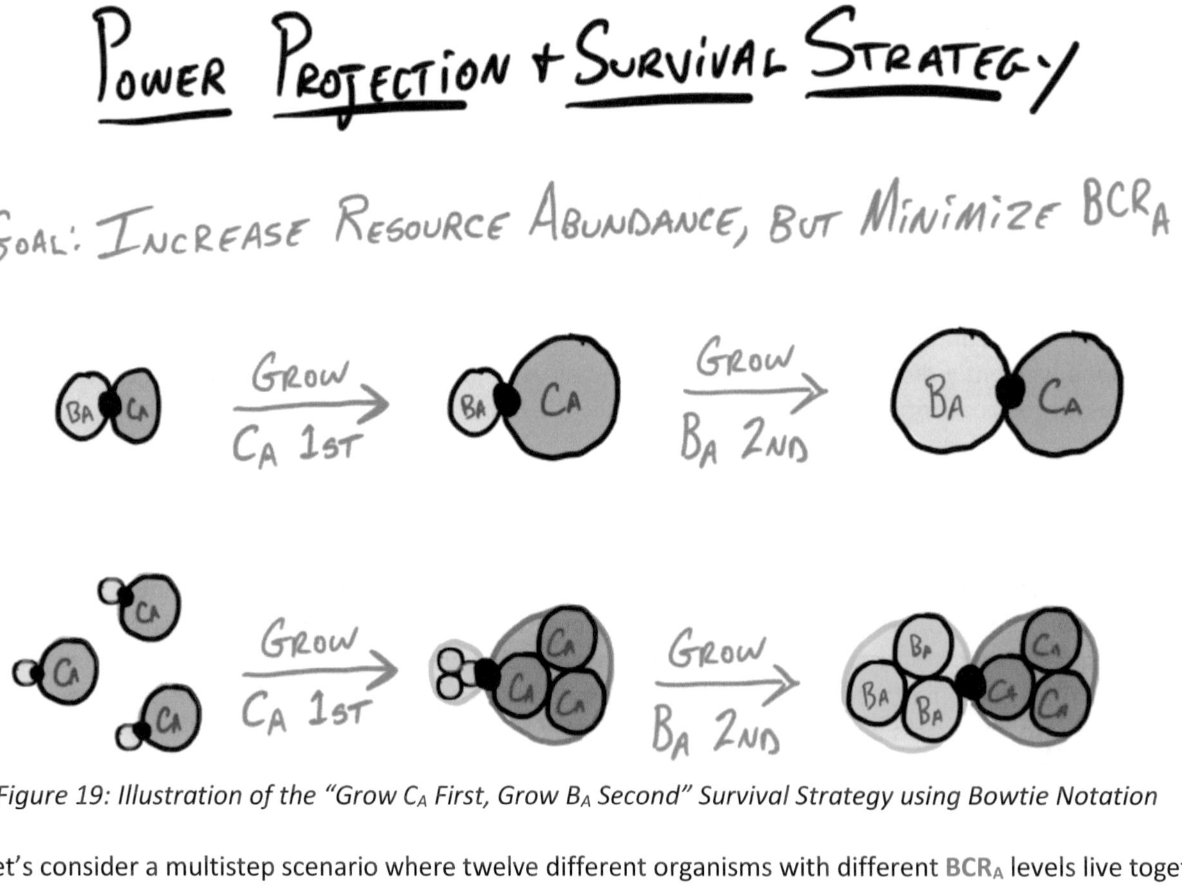

Figure 19: Illustration of the "Grow C_A First, Grow B_A Second" Survival Strategy using Bowtie Notation

Now let's consider a multistep scenario where twelve different organisms with different **BCR_A** levels live together within a highly CCCH environment. Let's call this scenario the "Bigger Fish" scenario and illustrate it in bowtie notation using Figure 20.

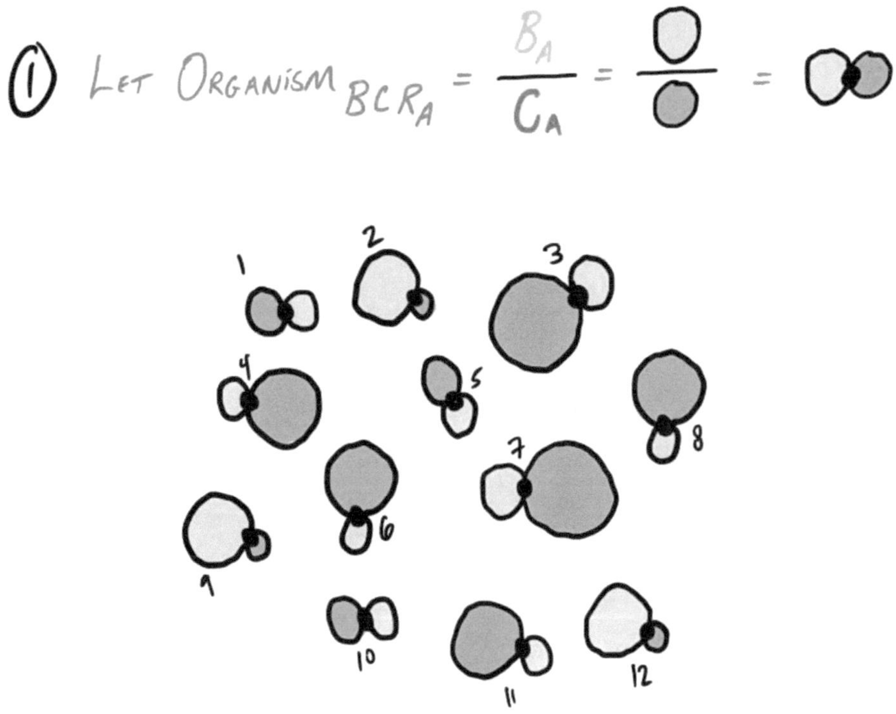

Figure 20: Step 1 of "Bigger Fish" Scenario

We have established that organisms with high **BCR_A** levels are likely to be devoured by hungry neighbors. This means we can expect organisms #2, #9, and #12 will not survive in this CCCH environment, so we can go ahead and cross them out in Step 2 shown in Figure 21.

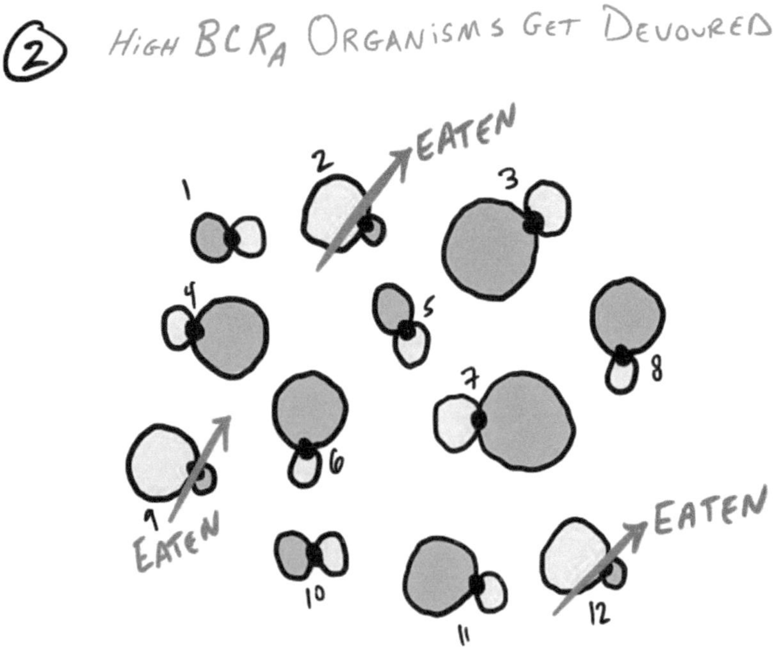

Figure 21: Step 2 of "Bigger Fish" Scenario

Now let's say organisms #1, #5, and #6 start cooperating to form a multicellular organism named Alpha. At the same time, organisms #7, #8, and #11 are compelled to start cooperating to form a multicellular organism named Bravo. This is illustrated in Figure 22.

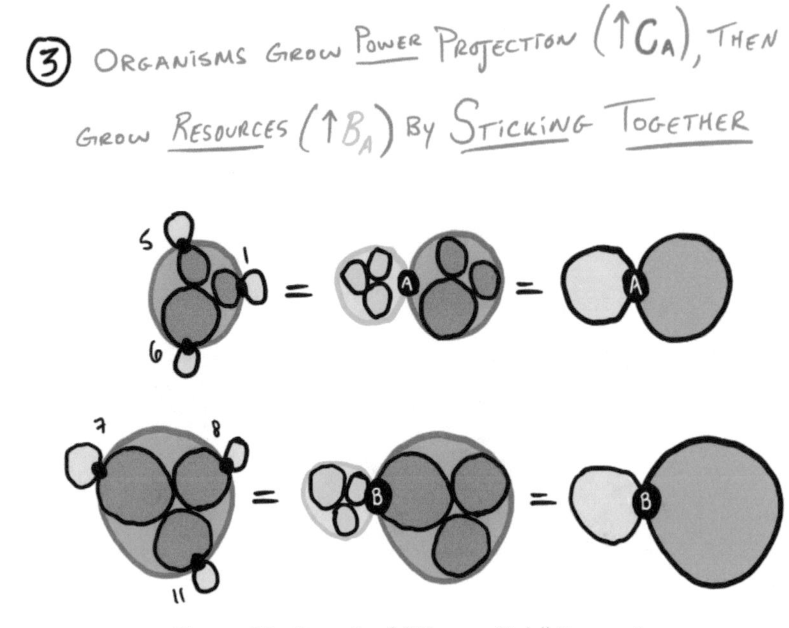

Figure 22: Step 3 of "Bigger Fish" Scenario

We now have a situation where the environment has changed to become substantially more hazardous for the remaining three organisms which didn't get the memo on survival and have done nothing to increase their C_A by learning how to cooperate with each other. The emergence of cooperation has driven the hazardous BCR_A level for this environment down substantially since multicellular organisms now have much larger C_A together than they did individually. Therefore, they have more ability to project power to capture resources than their non coopering neighbors. Now organisms like #10 are in much greater danger even though their BCR_A didn't change. Organism #10 is now the most attractive target of opportunity in this environment, therefore the most likely to get devoured. This is illustrated in Figure 23.

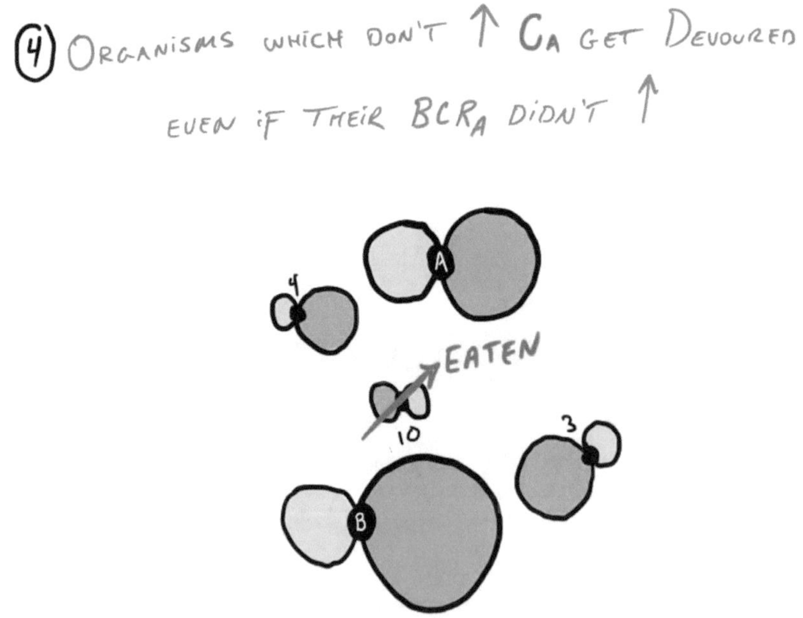

Figure 23: Step 4 of "Bigger Fish" Scenario

The "Bigger Fish" scenario illustrates how keeping one's BCR_A fixed is not enough for survival when operating in increasingly CCCH environments. There are major existential and strategic benefits to growing C_A that shouldn't be discredited, even if the organism has no intent to actually use them. The emergence of cooperation only accelerates the benefit of growing C_A. The top survivors in this scenario were the ones who learned to cooperate. Now take this dynamic and scale it up from 12 organisms to more bacteria on Earth than there are stars in the

Universe. At this scale, the benefits of cooperation become unfathomably complex and blossom into the world of multicellular organisms we see around us today.

Cooperation started its own ecological arms race by giving rise to a self-reinforcing feedback loop where the discovery of increasingly more effective cooperation tactics begets the need for increasingly more effective cooperation tactics. If your murderous and fratricidal neighbors figure out how to cooperate at scale, then you better figure out an equal or better form of cooperation, or else you could become their dinner. It might not be feasible for you to grow your C_A by 100X to fend off an army of 100 cooperating cells on your own. But it *is* feasible for you to raise your own army of 100 cells. Cooperation is therefore a strategic imperative because of the survivor's dilemma. [22]

What does this cooperate-or-die dynamic do for life at the systemic level? The same thing predation does for life at the systemic level. The emergence of cooperation and its adoption by predators makes life exceedingly more compelled to innovate better cooperation tactics to countervail threats and to buy more prosperity margin. This in turn makes life better equipped to survive its environment and to countervail entropy. Organisms which emerge from the fray as champions of survival in a world full of multicellular predatory armies are the most fit *and* the most cooperative.

3.9 Pack Animals

"Wherever we go, this family is our fortress."
Jake Sully, *Avatar* [63]

3.9.1. Keeping up with the Joneses: New Power Projection Tactics Create New Schelling Points

Microscopic life conquered Earth at a pace that would likely make Genghis Khan blush. As it grew, life split across different evolutionary branches and developed unique internal economies with clever approaches to regulating activities like growth, metabolism, reproduction, and gene carrying. Some of the more advanced cells formed dedicated command and control centers called *nuclei*. Armed with highly advanced command and control centers backed by highly productive internal economies, eukaryotic life dramatically increased the rate at which it innovated, developing new power-projection tactics at a significantly more rapid pace. The combination of nuclei and multicellular cooperation proved to be especially powerful, giving rise to the formation of large multicellular superpowers called *plants* and *animals*. [48]

Primordial economics accounts for innumerable power projection tactics employed by plants, animals, and other multicellular eukaryotic organisms. A tiger's stripes, for example, help it capture resources more effectively by closing the animal's attack distance on prey. A zebra's stripes help it impose more physically prohibitive cost on attackers by degrading their attacker's visual targeting capability. Camouflage is therefore a dual-use power projection technology which enables animals to affect both sides of their BCR_A equation. It's therefore not surprising that many of life's top survivors utilize camouflage.

Teeth, armor, skeletons, digestion systems, eyes, claws, brains, feathers, and countless other innovations can be conceptualized as successful power projection tactics too. These features enabled life to fulfill its primordial economic duty of capturing more resources and imposing higher physical costs on attackers. Each one bought their host more prosperity margin and contributed to a global-scale ecological arms race. Once skeletons emerged, skeletons became a new biological Schelling point. If your murderous and hungry neighbors have vertebrae, then it's in your survival interest to have vertebrae too. If your murderous and hungry neighbors have eyeballs, then it's in your survival interest to have eyeballs too.

Whenever a new power projection tactic is successful, it often becomes a focal point. Absent the ability to trust your neighbor or leave the neighborhood, it's in your best interest to ensure your power projection capabilities match or exceed your neighbors'. If your neighbors can use their fancy new power projection technology to grow their C_A faster than you can, that means you could become the organism with the highest BCR_A in the neighborhood, which means you become the most attractive target of opportunity for your neighbors to devour. The survivor's dilemma creates a "keeping up with joneses" effect where it becomes existentially necessary to stay on par with the power projection capacity of your neighbors.

3.9.2 Same Functions, Different Forms

Despite their different appearances, many power projection tactics developed by multicellular eukaryotic life over the past two billion years have been variations of the same well-worked themes. As life grew to massive multicellular scales, primordial economics didn't change, therefore the dynamics of survival didn't change. Innovate or die, cooperate or die, lower BCR_A as fast as affordable, avoid bounded prosperity traps, keep changing and searching for better power projection tactics and strive to become an infinitely prosperous organism. These survival strategies have proven themselves to be effective over vast timeframes filled with many elemental shifts and entropic curveballs.

We can look back at our biological history and observe that **many of life's most successful evolutions were significant because of how they helped life execute these proven power projection strategies more effectively or on larger scales**. Many of them appear to be functional repeats of the same dual-use power projection tactics discovered billions of years ago by our pioneering, microscopic ancestors. The wolf's snarling mouth is a more complex and higher-scale functional repeat of phagocytosis. Other power projection tactics like pressurized membranes have too many functional repeats to count. We give them different names like *bark* or *epidermis* depending on the species they protect, but their function is practically identical.

In systems engineering, it is common for the form of a system to change over time, but its function to remain the same. The function of a typing system, for example, has remained consistently the same regardless of whether its form changes from a mechanical typewriter to a touch screen. Systems thinking can be useful for making insightful observations about the phenomenon we observe in everyday life. By recognizing the difference between the form of a system and the function of a system and focusing one's attention on functional similarities, it's often possible to see similarities between things which wouldn't otherwise appear to be related.

Using systems thinking, we can conceptualize how a microscopically thin soap bubble of organic molecules fighting for nutrient-rich volume around a deep-sea hydrothermal vent four billion years ago are not all that functionally different from a nation state. Neither of these two vastly different-looking things are all that functionally different from a plant or animal fighting over food and territory, either. Why? Because they are different forms of the same system (life) performing the same function. All lifeforms are cut from the same cloth and fighting for survival effectively the same way, including and especially humans.

But what specifically is the function of life? This is impossible to know. Perhaps it is simply to countervail the entropy of the Universe. If we accept this, then consider what sub-functions are necessary to countervail entropy. To countervail entropy, life must learn how to survive and prosper. If we accept this, then consider what sub-functions are necessary to survive and prosper. To survive and prosper, life must grow its C_A to lower its BCR_A as much as possible and buy itself as much prosperity margin as it can afford using scarce natural resources. This will allow it to grow and thrive in a reality that is incontrovertibly congested, contested, competitive, and hostile, teeming with predators. If, for the sake of argument, we accept these assertions, then we can't overlook the fact that **survivorship, prosperity, and the ability to countervail entropy all depend on the exact same activity: burning watts to grow C_A as much as possible.**

With these concepts in mind, the reader is invited to use systems thinking to re-examine what they observe across both nature and society and look for functional similarities. There are many insights to be gained from this approach. The reader is invited to use systems thinking to study nature and society and answer the following question: "How does this behavior help increase C_A?" Our answers to this question can be quite insightful, especially when it comes to understanding the value of new power projection technologies like Bitcoin.

3.9.3 Organizations have the Same Power Projection Dynamics as Organisms

The strategic imperative to grow C_A by tapping into the exogenous power supply of one's neighbors explains why life's top survivors became so inclined to cooperate. The phenomenon of cooperation has emerged at every biological scale: subcellular, cellular, multicellular, multi organism, and even multi species. When multiple organisms tap into each other's power projection supply, they form a single cohesive unit called an *organization*. An organization is a larger form of an organism, no different than how multiple subcellular microorganic compounds organize to form a single cell, or how multiple cells organize to form a single multicellular organism. We can therefore account for an organization's **BCR_A** in the same way that we account for an organism's **BCR_A**. We can also represent it the same way in bowtie notation, as shown in Figure 24.

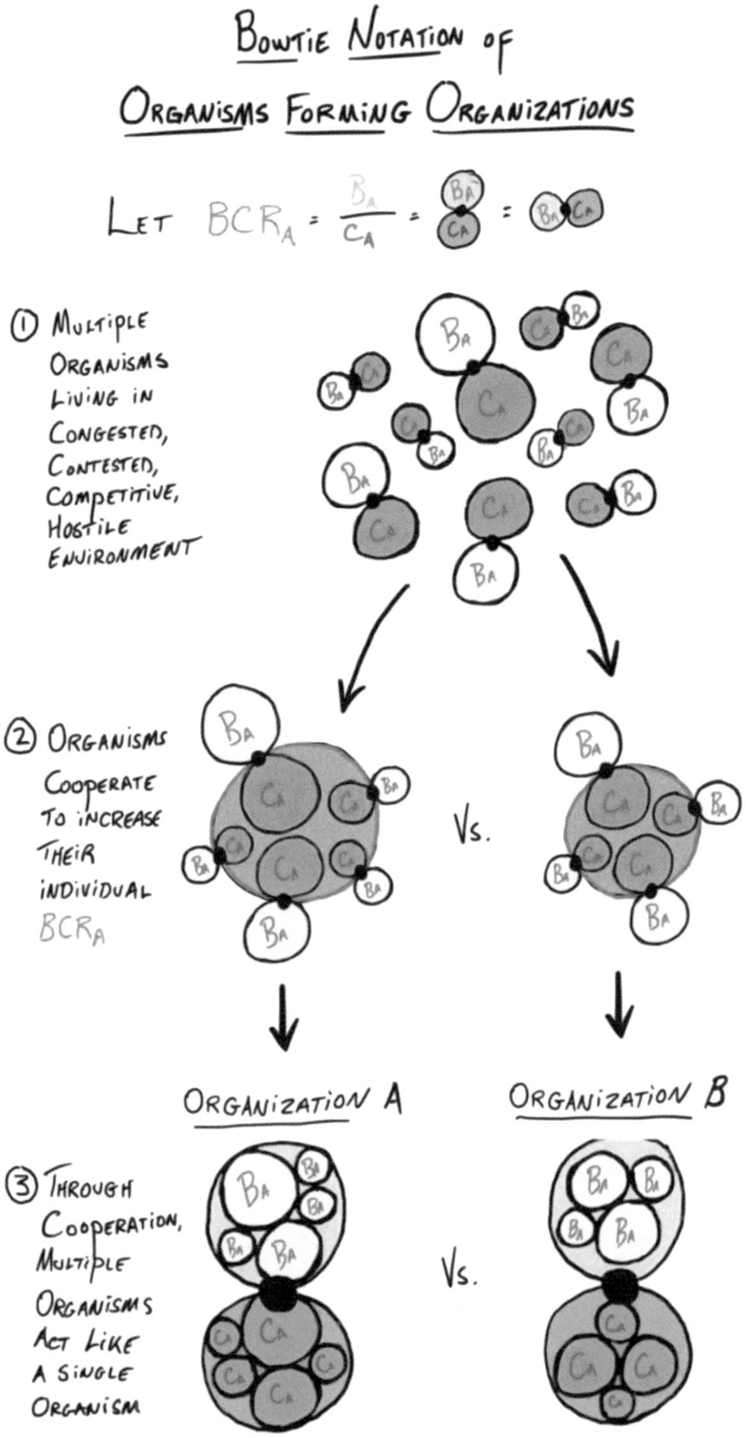

Figure 24: Bowtie Notation of Organisms Forming Organizations

Given what we know about "cooperate-or-die" survival dynamics and the existential imperative to increase C_A, it's no surprise that many of life's most successful plants and animals are instinctively programmed to cooperate. Plants used cooperation tactics to become the most massive form of life on Earth. Animals can also cooperate at an impressive scale, and in seemingly more dynamic ways with arguably more complex emergent behavior. Animals which are especially attuned to cooperating on a large scale are called *pack animals*.

Pack animals clearly understand that an effective way to impose severe physical costs on attackers is to leverage the power of their pack. One wildebeest may not be able to project enough power to prevent a menacing predator like a lion from attacking it, but one wildebeest backed by the exogenous power supply of 99 other wildebeests certainly can. By working together as a pack, each animal gains access to significantly more power projection capacity to impose more physical costs on attackers. Every animal operating within a pack enjoys a step-function increase in C_A, a substantial reduction in their individual BCR_A, and an increase in their prosperity margin at virtually no individual cost to themselves. In many ways, cooperation is a survivor's life hack.

Highly cooperative packs function as single organisms devoted to the task of growing their collective C_A, minimizing their BCR_A, and buying as much prosperity margin as they can to safely grow resource abundance and increase B_A. Pack animals are connoisseurs of the survivor's dilemma option #3 strategy shown in Figure 18, demonstrably capable of achieving high levels of prosperity in comparison to non-cooperative organisms.

When highly cohesive packs function as singular organisms, they form yet another variation of a pressurized, porous membrane capable of capturing nutrient-rich volumes by force. By sticking together, packs repeat the exact same evolutionary process as subcellular and cellular organisms. They grow increasingly more interdependent and reliant on each other for vital nutrients, materials, and gene swapping. They form highly complex internal economies, self-assembling into increasingly more specialized workforces, trading various organic goods and services and becoming ever more efficient, productive, and resource abundant. Through a special combination of robust defense and high-functioning internal economy, packs follow a multistep biological path towards ever-increasing structure until they self-assemble into the complex, massive-scale economies we give names like *flocks*, *flights*, *herds*, *mobs*, *gaggles*, *tribes*, *cities*, *city-states*, and *nation states*.

As pack animals become increasingly more interdependent, they specialize in performing different functions. Because of the strategic importance of increasing C_A, many animal packs have physiologically specialized workforces devoted to the task of projecting power to impose physically prohibitive costs on attackers. How they specialize varies between species (it's usually via sexual dimorphism, either the male or female evolves the strength/tools to be the primary power projector), but a ubiquitous trend is to dedicate some portion of the workforce to doing battle. This trend is quite noteworthy.

Despite its benefits, cooperation isn't easy. Pack animals have their own individual brains. Individual members of every pack have their own individual needs and priorities independent of the pack's collective priorities**. To cooperate at a large scale, pack animals must learn how to negotiate between their individual needs and the needs of the pack as a whole. Navigating this becomes especially tricky when it comes to feeding and breeding. Compromises must be made between the individual's needs and the pack's needs regarding resource control and ownership. Packs must adopt heuristics for determining the state of ownership and chain of custody of the pack's collective resources between and among pack members**. This is called establishing a dominance hierarchy, or more colloquially known as a "pecking order."

3.10 Domestication is Dangerous

"You ever plow a field, Summer?
To plant the kewaa or sorghum or whatever the hell it is you eat? You kill everything on the ground and under it. You kill every snake, every frog, every mouse, mole, vole, worm, quail – you kill them all. So I guess the only real question is, how cute does an animal have to be before you care if it dies to feed you?
John Dutton, *Yellowstone* [64]

3.10.1 Disclaimer #1 – This Subject Hits Close to Home

This section lays the conceptual bedrock for understanding complex social behavior in pack animals, how they establish dominance hierarchies, settle intraspecies disputes, establish control authority over limited resources, and achieve consensus on the legitimate state of ownership and chain of custody of property.

A few disclaimers before entering a discussion about animal pack behavior dominance hierarchies and the strategic implications of different pecking order heuristics. Sapiens are pack animals, and like many other packs, the most brutal and painful struggles between sapiens are centered around the establishment of pecking order. The following chapter has a more thorough discussion about the power projection tactics that sapiens use to establish pecking order, but it's helpful to acknowledge the sensitivities in this section upfront so we understand how emotionally charged this topic can be. Our emotions and our ideologies will most likely affect how we react to conversations about pecking order strategies. In the author's experience doing research for this grounded theory, the topic of pecking order often made people feel uncomfortable or upset, likely because of how closely these concepts are linked to our own personal experiences. This part of the conversation simply hits close to home.

For the sake of formulating a cohesive and conceptually insightful argument about the complex emergent behavior of new power projection technologies, the reader is asked to indulge the author temporarily and to stay cognizant of the broader context of what they're reading. This is a thesis about Bitcoin – a technology that strikes directly at the heart of topics like power projection and resource control. Pecking order is just another name for a resource control protocol, and it will be useful to develop a conceptually dense understanding of naturally-occurring resource control protocols before entering into a discussion about Bitcoin.

What follows does not reflect the author's personal ideology about what resource control protocols should or shouldn't be used within human organizations. This is a logical discussion about mutually observable behavior in nature, designed to help the reader draw out insights that help us better understand the emergent behavior of Bitcoin. What follows is uncomfortable to talk about, but critically important to understand.

3.10.2 Disclaimer #2 – Don't Forget About Survivorship Bias

Survivorship bias is a logical error which causes people to be susceptible to making incorrect conclusions because they discount information they can't see because it didn't survive some selection process. Survivorship bias can affect a discussion about pecking order heuristics by causing people to discount heuristics they can't see merely because it didn't survive the *natural* selection process. On the flip side, survivorship bias can also make people prone to discounting information they *can* see, because they lose sight of the fact that the reason why they see what they see is because what they see is what survives.

Survivorship bias is why the author has reminded the reader multiple times that what we see in nature today is incontrovertibly proven to survive nature. This is a phenomenon we can use to our advantage when it comes to gaining insights about different heuristics related to survival. **<u>If you want to know what pecking order heuristics are best for survival, simply observe nature</u>.** Nature has spent the past four billion years separating the wheat from the chaff and has already figured out the difference between heuristics that survive and heuristics that don't. This is something to keep at the forefront of our minds throughout the remainder of this thesis.

Survivorship bias is important to understand when talking about pecking order heuristics employed by different pack animals. There is a wide range of different pecking order heuristics available to wild animals. For example, "finders keepers" or "first-come-first-served" are popular pecking order heuristics used by sapiens (hence why sapiens stand in lines so often, a uniquely human behavior). "Family first," "oldest first," and "youngest first" are other popular pecking order heuristics. For pecking order heuristics which don't involve high levels of intelligence, intentionality, or theory of mind, it's very likely that over the past several hundred million years, animal packs of all shapes and sizes operating in all types of environments have experimented with practically all of them.

There is an extremely low probability that we can come up with a non-abstract pecking order heuristic which hasn't been tried many times before over the past several hundred million years by innumerable animal packs whose survivorship depended on finding the most strategically optimal pecking order heuristic. Therefore, if we have a good idea for a pecking order heuristic that we don't see in nature, it's most likely *not* because that heuristic hasn't been tried. It's more likely because that heuristic is demonstrably incapable of surviving, therefore we don't see it in the wild.

3.10.3 Disclaimer #3 – Remember that Nature is a Sociopath

"Look at the eyes of that [jaguar]. Nature has created, in those kinds of eyes, the perfect vision of terror.
If you looked into those eyes, there's no forgiveness. There're no emotions.
There's just ferocity and aggression and death."
Joe Rogan [65]

Appreciating nature requires recognizing up front that most organisms are sociopaths. Nature has no apparent capacity to see, understand, or care about sapient theology, philosophy, or ideology. Moral "good" is a highly subjective and abstract construct which exists in an ontologically different category than the study of nature (this is further explained in the following chapter). Therefore, to better understand nature, it's useful to ignore human belief systems about right or wrong, fair or unfair, moral or immoral. These topics are almost irrelevant to the subject of natural selection and survival.

Things which sapiens consider to be theologically, philosophically, or ideologically repugnant and reprehensible are often routine in nature. For example, suicide, murder, and cannibalism are so common in nature that they're often considered to be unremarkable behavior. Aging, for example, is an evolved form of suicide; cells appear to have *learned* the habit of deliberately destroying themselves after a select duration of time as a tactic of evolution. Why would nature do something so inefficient and wasteful? Because the world is filled with predators and entropy, which means the environment is constantly changing. Organisms with long lives can't change their genetic features as quickly as organisms with short lives and are therefore less adaptable to their environment, thus less likely to survive.

Aging is a way for organisms to counteract this strategic problem; it is a highly effective way to get a species to become more adaptable by forcing them to cycle through more evolutionary genetic features faster and test them faster in a live production environment. Shorter lifespans mean shorter mean time to deploy new genetic features that will help the species discover what it needs to prosper. The lifespans we see in different species today represent the optimal mean-time-to-deployment-speed of new genetic features for that given species. For sapiens, it's around 70 years.

Life also has no problem killing itself with genetic malfunctions caused by its evolutionary prototyping strategy. Nature doesn't have a pre-production environment or a test net to work out its bugs; it deploys new features directly into a live production environment and simply accepts the consequences. An organism with a crippling genetic mutation is a prototype for the future version of that species that clearly doesn't function properly in an operational environment. Combined with aging, these malfunctions enable life to fail fast, fail often, and fail forward. The downside of this process is crippling genetic malfunctions at the individual organism level. The upside is the long-term survival of the species as the species remains adaptable to a changing environment. This process is how life gets to endothermy in time to build a suite of self-heating organisms which can survive a meteoric winter.

As we have already established, killing is routine in nature for many of these same reasons. Life appears to take advantage of predation as an external motivator to accelerate the pace of innovation, weed out the weak and unadaptable, and revector resources to the more qualified survivors.

We tend to want to overlook these uncomfortable parts of nature. We don't upvote these moments to the top of our social media feeds. Instead, at the top of our social media feeds we get cute, adorable moments. We watch documentaries with carefully selected depictions of nature designed to thrill or inspire us. The epic music plays in the background and the distinguished voice with an English accent makes an insightful remark. From our high tower of prosperity, this false depiction of nature becomes a generation's primary source of information about the real world, and it creates a beauty complex. We only see a carefully-edited and thematically airbrushed version of nature; a corporately censored version designed to keep our attention.

Missing from the top of our social media feeds is the scene where the mother squirrel eats her own babies alive to make it through January. Disney skipped over the part where Mufasa murdered every cub in the pride and then raped their mothers after killing the previous lion. That part doesn't fit the inspiring storyline they're trying to feed to their audience, so they skip over that part and start with the birth of Simba. Then, somehow, we're just supposed to accept the fact that the animals which have been hunted by lions their entire life are inclined to celebrate the birth of yet another lion.

The truth is that nature is not nearly as pleasant as what we see on our TV screens. Survival is an ugly business. It always has been, and it probably always will be. This ugly part of nature is less entertaining or inspiring to us, so we don't see it as often. The unfortunate side effect of this behavior at scale is that it distorts our perception of reality and inhibits us from understanding primordial economic dynamics, the survivor's dilemma, and the existential importance of physical power projection.

For the sake of gaining deeper insight into the potential sociotechnical impact of new technologies like Bitcoin, the reader is invited to recalibrate their understanding of nature. This means allowing ourselves to feel uncomfortable for a short period of time so we can better understand the dynamics of physical power projection and how it relates to security and survivorship. There's a very clear (but ideologically repugnant) reason why lions kill their own cubs. Nature is giving us a lesson about survival, and it could be beneficial for us to pay attention if we want to survive against our predators, too.

3.10.4 Correlation Doesn't Imply Causation, but Randomized A/B Experimentation Does

It is possible to prove that changing an animal pack's pecking order strategy (how they choose who to feed and breed) to reward different behavior than strength and aggression is systemically hazardous. However, that would require the design of a series of very large, very long, and probably unethical experiments. Fortunately, we don't need to do perform these experiments, because we have already done them on other animals for tens of thousands of years. The domestication of animals is incontrovertible proof that changing an animal pack's pecking order strategy to reward different behavior than strength and aggression is systemically hazardous to them.

Discovering the root causes of social phenomena is difficult. It requires rigorous measurement and design of randomized experiments to control observable and unobservable factors while simultaneously isolating the relationships we want to examine. Randomized experimentation is critical for ensuring that observable and unobservable factors outside of the relationships we want to examine don't account for the differences in emergent behavior. With enough randomized experimentation data, it is possible to analyze the true causal effect of changing specific variables between a treatment and control group. In other words, it's possible to determine with high confidence that factor Y *causes* effect Z, rather than merely *correlates* to it. [66]

With the ability to gain causal inference via randomized experimentation in mind, consider animal dominance hierarchies and what effect pecking order has on an animal pack's security and prosperity. If one were to ask the question, "how would a different pecking order heuristic, where pack animals *don't* reward their most physically powerful and aggressive members with feeding and breeding rights, change an animal pack's capacity to survive and prosper in the wild," one would have to find a way to generate enough randomized experimentation data to causally infer a relationship between these two variables. It is simply not possible to determine causal relationships between a species' pecking order strategy and capacity for survival without randomized experiments.

If scientists wanted to investigate whether making a group of animals less inclined to impose severe physical costs on neighboring animals has a direct impact on their safety, security, and survival, they would have to run randomized experiments on dozens of different pack animal species where they control for the same variable each time. They would need to find a way to interfere with an animal population's pecking order instincts to prevent them from feeding and breeding their most physically powerful and aggressive members. Then they would need to measure changes in emergent behavior by comparing each population to a control group of animals which didn't have their pecking order altered.

This experiment would have both practical and ethical challenges. Scientists who want to examine how interfering with an animal's natural instincts impacts their safety would have to design randomized experiments that would endanger large populations of animals. They would have to change an animal pack's natural inclination to feed and breed their powerful and aggressive members against their will. They would have to force them to breed in ways they wouldn't naturally choose to breed, and place them in hazardous environments surrounded by predators, and then measure how well they survive. Scientists would have to repeat these experiments enough times with enough animal species across multiple environments and time periods to create a sufficiently randomized data set from which they can causally infer that changing an animal pack's pecking order so that they're less physically powerful and aggressive does indeed cause them to be less secure against predators, thus less likely to survive. Scientific rigor would make it necessary to send large populations of animal species to their demise to generate enough data to causally infer this sort of relationship.

In these types of situations where experimentation is not feasible due to practical or ethical concerns, scientists can take an alternative approach. They can look for serendipitous sources of random variation in existing data sets. There are ways to analyze data ex post facto that statistically mimic randomized experimentation well enough to infer causal relationships between different variables with sufficient confidence (e.g. propensity score matching, instrumental variable analysis). All one needs to do is find the right data set on which to perform this type of analysis. In other words, a scientist who wants to look for a causally inferable relationship between variables like systemic security and physical aggression wouldn't have to design unethical experiments which endanger animals. They could search for sufficiently randomized data sets which already exist, and study those instead.

Fortunately (or unfortunately, depending on how you think about it), there is already a plethora of serendipitous sources of randomized experimentation on large populations of animals, across many different regions and timeframes, from which it's possible to causally infer a relationship between an animal population's capacity and inclination to be physically aggressive, and their capacity to survive. Humans have already adopted the habit of interfering with the pecking order of pack animals to prevent them from feeding and breeding their most physically powerful and aggressive members, and then placing them in hazardous environments where they are highly vulnerable to predation. Humans have been slaughtering dozens of different types of animal species across diverse environments for more than ten thousand years. We slaughter and devour billions of these animals. These experiments have become so ubiquitous and routine that many people don't even notice them anymore.

From these experiments, sapiens have already created a data set from which it's trivial to causally infer a relationship between security and lack of physical aggression. Over the course of tens of thousands of years, we have created many A/B testing experiments which demonstrate quite clearly what happens to the safety, security, and survival of animals when they become less inclined to impose severe physical costs on their neighbors.

3.10.5 An Honest Description of the Systemic and Sociotechnical Implications of Domestication

The difference between a Siberian wolf and a dachshund is the difference between pecking order heuristics. Different feeding and breeding heuristics result in clear differences between each animal's capacity and inclination to project physical power and impose severe physical costs on neighbors. One pecking order strategy produces something optimized for independence and survival in the wild (the wolf), the other produces something optimized to serve its master (the dog). This example alone sums up (1) why pecking order matters, and (2) why it's existentially imperative for the freedom and prosperity of animals not to stop feeding and breeding the strongest, most intelligent, and most aggressive members of their pack. Considering how the word "wiener" is slang for weak and ineffectual, the term "wiener dog" is highly appropriate. Dogs – dachshunds especially – are weak and ineffectual wolves.

As sapiens have shown multiple times across thousands of years of randomized experimentation in multiple different environments with at least forty different species of animals of multiple different classes (mammals, birds, and even fishes), slight adjustments to the way animal packs feed and breed the strongest and most physically aggressive members of their pack lead to substantial differences in their ability to keep themselves secure against predators. The domestication of animals offers a large data set with conclusive evidence to show that there is a direct, causally inferable relationship between an animal's capacity and inclination to be physically aggressive, and their capacity to survive, prosper, and live freely. The less an animal is inclined to impose severe physical costs on their neighbors, the easier they are to be systemically exploited and led straight to slaughter.

Changing pecking order heuristics to something other than "feed and breed the most physically powerful and aggressive members of the pack" has had an incontrovertible impact on the safety, security, and survival of dozens of animal species. This implies that **pecking order strategies – how animals instinctively choose to establish control authority over their resources – fundamentally represents a power projection tactic which directly affects their capacity for survival**. The survivor's dilemma (i.e. the strategic imperative to increase C_A) therefore applies to pecking order heuristics just the same as it applies to other power projection tactics. The "better" pecking order heuristics are the ones which maximize an animal pack's ability to decrease BCR_A by imposing severe, physically prohibitive costs on attackers. In other words, animal packs don't reward their most powerful and aggressive members with feeding and breeding rights because it's the "right" thing to do – they do it because all the animals that didn't do it didn't survive.

Establishing the right pecking order strategy represents an existential imperative for pack animals. It is no less vital for the survival of a pack of hyenas to establish an advantageous pecking order over a fresh kill than it is for a starving family of humans to ration their bread. Choosing who to feed and breed first is one of the most critical decisions a pack of animals can make, and there is a lot to be learned from observing how nature's top-surviving animal packs make this decision.

Ironically, the animals we most commonly observe in nature today are not *wild* animals. So not only do we get a false representation of nature on our TV screens, but we also get a false representation of nature during our most common interactions with animals. This further distorts people's perception of how ugly the business of survivorship really is. We can recalibrate our distorted perception of reality by identifying the source of the distortions and filtering them out. To better understand the merits of different pecking order strategies, the distortions of reality we need to filter out are the animals we routinely slaughter.

Sapiens have shown it's possible to change a wild animal pack's pecking order heuristics by genetically entrapping and enslaving them. If you entrap a herd of aurochs and then feed and breed the muscular and docile ones, you get a herd of oxen. If you entrap a herd of aurochs and then feed and breed the obese and docile ones, you get a herd of cows. If you entrap a litter of boar and then feed and breed the obese and docile ones, you get a litter of pigs. If you entrap a flock of junglefowl and then feed and breed the obese and docile ones, you get a flock of chickens. These activities produce A/B testing experiments where oxen, cows, pigs, and chickens become the treatment, and aurochs, boar, and junglefowl become the control. To measure how removing the physically powerful and aggressive members of an animal pack affects their ability to survive against predators, simply take inventory of the difference between the bacon on your plate, and the boars you *aren't* eating. From that data it's possible to infer a causal relationship between docility and survival. Across a wide range of randomized variables, the *docile* animals are the ones we keep encaged to eat or to do our manual labor.

Time and time again, for multiple different species, in multiple different experiments with high variability, we have proven that (1) a pack's capacity for survival and prosperity depends upon their pecking order strategy, and (2) the best way to degrade a pack's safety, security, freedom, and independence is to prevent or undermine their capacity and inclination to impose physical costs on their oppressors by preventing them from feeding and breeding their most physically powerful and aggressive members.

3.10.6 To make it to the Top of the Dominance Hierarchy, Domesticate your Peers

Natural selection caused many pack animals to become sexually dimorphic, where one gender is genetically optimized to be physically stronger and more physically aggressive than the other. For some species, the female is genetically optimized to be more physically powerful and aggressive. For others, it's the male. Either way, with few exceptions, natural instincts make animals sexually attracted to physically powerful, intelligent, and assertive members of the pack. These instincts ensure the species genetically self-optimizes itself for survival by passing on the genes of the most physically powerful, intelligent, and assertive members.

In the mammalian class, males often have higher testosterone levels, contributing to sexual dimorphism and making them physically stronger and more aggressive members of the pack. This sexual dimorphism is a design feature that sapiens learned how to exploit. To change the "feed and breed the powerful first" pecking order heuristic employed by mammalian pack animals, sapiens learned how to neuter the strongest and most physically aggressive males to remove their genes from the gene pool. This tactic is given polite-sounding names like *selective breeding,* but what it represents from a sociotechnical (and honest) perspective is a way to force an entire species to become less physically powerful and aggressive through genetic modifications, thus less capable of and inclined to impose severe physical costs on their human oppressors. Domestication is therefore a form of predation – a power projection tactic that dramatically reshaped our world and put sapiens at the top of a global interspecies dominance hierarchy.

Wild mammals which have had their pecking order strategy exploited via domestication are called *livestock*. Wild birds which have had their pecking order strategy exploited via domestication are called *poultry*. Today, the biomass of domesticated livestock is comprised mostly of cattle and pigs and is about 14X higher than the biomass of the rest of the world's non-domesticated wild mammals combined. The biomass of domesticated poultry is about 3X higher than the biomass of the rest of the world's wild birds combined. It's harder to domesticate birds because they can fly away, hence why most poultry are flightless or nearly flightless birds. [67]

A domesticated animal is a wild animal that has had its C_A unnaturally shrunken, as illustrated in bowtie notation in Figure 25. This type of exploitation is possible because many pack animals employ a specialized workforce devoted to the task of being physically powerful and aggressive. By simply identifying that workforce and not allowing them to multiply, it's possible to dramatically reduce an animal pack's overall C_A over time, thus raising their BCR_A, shrinking their prosperity margin, and making them easier to devour.

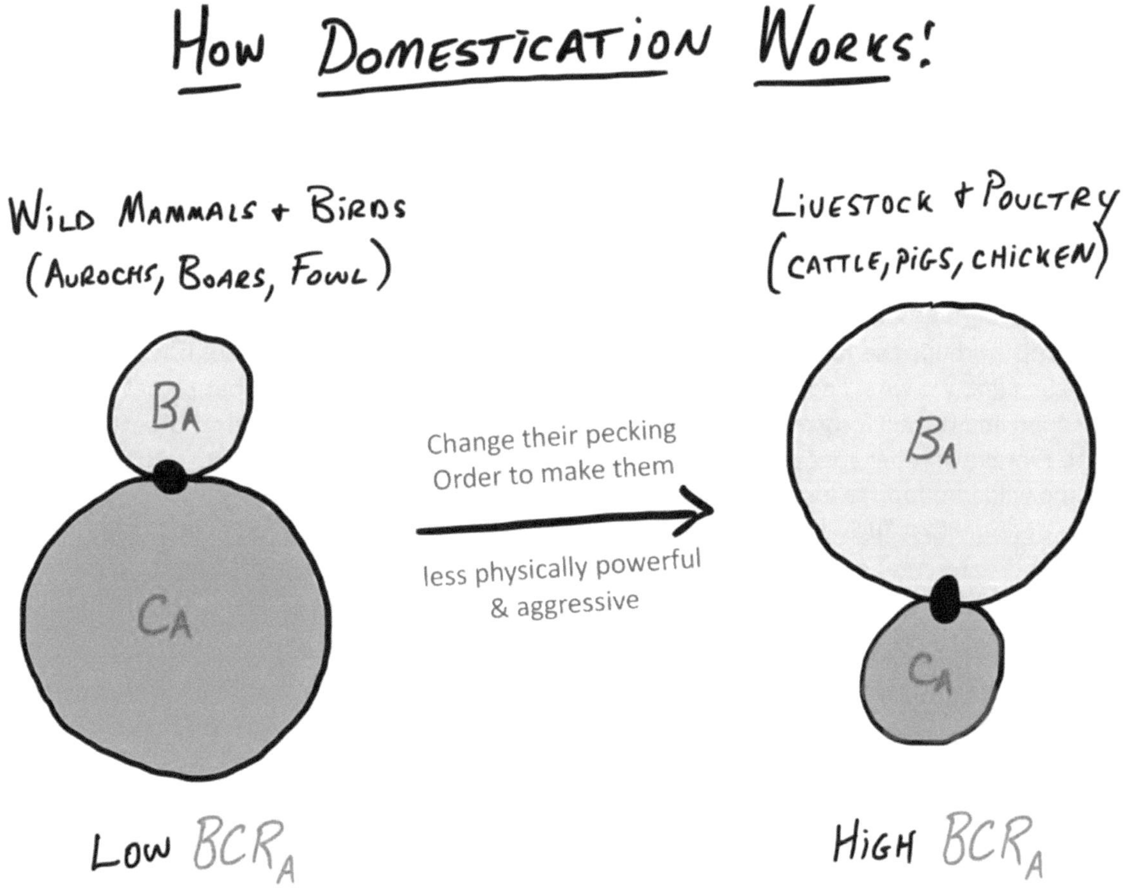

Figure 25: Bowtie Representation of Domestication

The primary value-delivered function of domestication is to make it easier for humans to exploit and devour enslaved animals. The process is centered around reducing an animal's ability and inclination to project power and impose severe physical costs (i.e. raise C_A and lower BCR_A). By simply *not* allowing these animals to instinctively feed, breed, and multiply their strongest and most aggressive power projectors, it becomes much easier to oppress them.

With their physical aggression intentionally bred out of them, oxen will allow themselves to become pack animals, routinely whipped and forced into drawing heavy loads for their masters (namely plows). These plows are used to dredge up nutrients from the soil to aid a process called *irrigation*. Irrigation helps produce more food for other animals which have also been systemically entrapped or forcefully enslaved by sapiens, creating a positively-reinforcing feedback loop of plant and animal exploitation called *agriculture*. An overwhelming majority of all domesticated animals are herbivores for this reason; it's easier to feed herbivores using the fruit of their own slave labor. The ox is whipped to irrigate land to grow grain to feed more oxen and their other domesticated friends.

The genetic entrapment and enslavement of animals via domestication is the practice upon which "civilized" human society was built in the Neolithic age. This is one of many reasons why modern sapiens should think twice before condemning the physically aggressive behavior of a lion or any other wild animal species that have successfully avoided being domesticated by humans. We have a large enough data set to causally infer that it's precisely *because* these animals are physically aggressive that they have not yet been domesticated.

3.10.7 The Dominant Species on Any Given Planet is the Species with Pets

A dog is a wolf which has had its pecking order exploited over the course of 40,000 years to remove its capacity and inclination to impose severe physical costs on humans. Take a pack of wolves, neuter the mean and aggressive ones, breed the docile, subservient, and physically deformed ones, and the end state of that process is a shorter, stubbier, and more co-dependent creature which exists to serve its master. The reason why dogs are "man's best friend" is because they were genetically modified to worship humans by exploiting their pecking order. Meaner and less obedient dogs (i.e. more wolf-like dogs) aren't fed and bred. Nicer and more obedient dogs which literally lick their masters' feet are called "good" dogs.

Domestication is perhaps the most vivid display of interspecies domination possible. If we were to discover a planet with alien life, we would easily be able to identify the dominant species of that planet by finding the one which entrapped and turned forty other species into their pets, slaves, and food supply. Domestication represents the ability to remove another species' physical power altogether rather than fight them – the ability to change a survivor of the wild into food to eat, or a tool to use, or a pet to cuddle. This is an honest and undistorted picture of human predation – that "ugly" part of survivorship that humans don't like to be reminded about.

What is the point of this uncomfortable conversation? To prove a point. **We have conclusive, causally inferable empirical evidence to indicate with a high degree of confidence that substantial impairments to safety, security, and survival are the direct result of *not* being physically powerful and aggressive.** Domestication has created a data set of highly randomized A/B testing experiments across more than three dozen species of animals of multiple classes in multiple environments over tens of thousands of years. It's incontrovertibly true that changing an animal's pecking order strategy to prevent them from giving their resources to their most physically powerful and aggressive members has a direct causal impact on their security. When populations become less capable of or inclined to impose severe physical costs on their attackers or oppressors, they become less safe, less secure, and less free.

3.10.8 If Domesticating Wild Animals is Predatory Behavior, then so is Domesticating Humans

The domestication of animals has proven to be a very effective power projection tactic. In other words, domestication is a highly effective form of predation. This is important for the reader to understand because if domestication represents a systemic security risk to the freedom and prosperity of forty different animal species, then domestication has the potential to threaten sapiens too.

Not only is there a systemic danger of self-domestication, but domestication itself is a form of attack against human society. Remove society's capacity and inclination to impose severe physical costs on other humans, and that will have a direct and measurable effect on their ability to survive and prosper. Human societies therefore have a fiduciary responsibility to not allow themselves to become too self-domesticated. **Societies who are interested in survival and prosperity should not allow themselves to become less capable of and inclined to be physically aggressive to potential attackers or oppressors.**

Honest descriptions of domestication are useful because they demonstrate how poorly designed resource control systems represent a strategic security hazard. Our capacity to prosper depends on the heuristics we adopt to settle our disputes, control our resources, and determine the legitimate state of ownership and chain of custody of our property. Animals are demonstrably susceptible to entrapment and enslavement if they don't adopt resource control strategies (i.e. pecking orders) which minimize their **BCR_A**. It has been proven, time and time again, that if packs don't utilize the optimal pecking order strategy, they become vulnerable to exploitation and abuse.

The domestication of animals represents multiple, repeatable, randomized experiments where pecking order was the isolated variable. Not rewarding physically powerful and aggressive members of the pack with higher levels of control authority over the pack's valuable resources didn't produce higher levels of prosperity for these species; it produced the meat we put on our sandwiches. There is no shortage of empirical data to indicate that once a population stops feeding and breeding their physical power projectors, they start plowing fields, worshiping their masters, and lining up for slaughter.

This has significant implications for sapiens who condemn the use of physical power and physical aggression to establish pecking order over resources because of the energy it uses or the injury it causes (this is further explored in the next chapter). Physical power and aggression clearly have a substantial effect on a population's safety, security, survival, and prosperity. Domesticated animals prove a causal link between docility and enslavement. Therefore, **we should be cautious of people who encourage docility and condemn physical power as the basis for settling disputes, establishing control authority over resources, or achieving consensus on the legitimate state of ownership or chain of custody of property. It's clearly a security hazard and can also be a deliberate attack vector.** This is a critical concept for the reader to understand for discussions in the following two chapters about power projection tactics in both society and cyberspace. This concept is also critical for understanding the sociotechnical implications of Bitcoin.

3.10.9 Check Your Power Projection Privilege

"Studying dogs is more anthropology than zoology... If you want to know how far we've moved from the place we were designed to inhabit, look at modern dogs. The tragic, wheezing ones with bows in their forelocks, and squashed faces and bent legs. Not proper dogs – the ones with faces like wolves..."
Charles Foster [68]

As discussed in the previous section, how a population of animals chooses to divvy up its resources can significantly impact its ability to survive and prosper. If a pack doesn't put the strongest, fastest, leanest, meanest, and most intelligent members of the pack at the top of its pecking order like it is has been instinctively programmed by natural selection to do, it will likely experience different complex emergent behavior related to safety, security, survival, and prosperity. To repeat, this is not military dogma speaking; this is backed by an overwhelmingly large data set created by thousands of years of randomized experiments on animal populations from which causal inference is possible. The reader probably ate a piece of that dataset today.

If animal populations could survive by employing "first come, first served" or "finders keepers" or other resource control protocols, then we would observe them in the wild forming neat and orderly lines to access their food and territory. If wild animal packs could survive using alternative heuristics like "divide food evenly," then we would observe this behavior in the wild. If animal packs could survive by letting their youngest, oldest, sickest, and most injured members have priority over resources, we would likely observe this behavior in the wild. But we don't observe these behaviors in the wild. In fact, we routinely observe the exact opposite behavior. Parents eat their babies or throw them out of the nest for being weak and ineffectual. Lions kill their cubs.

With very few exceptions, wild pack animals abandon their sick, elderly, and injured. The physically powerful and aggressive members of the pack are rewarded with mating rights and control authority over resources, and physical power is used as the basis for achieving consensus on the legitimate state of ownership and chain of custody of resources. **The overwhelming majority of animals adopt physical power-based resource control hierarchies where they settle property disputes using physical power.**

Somehow, different evolutionary paths all converged on practically the same physical power-based resource control and dominance hierarchies. In these systems, power projectors don't wait in line for food or breeding rights; they automatically get first dibs. Instead of being angry at their physical power projectors for these offenses, many pack animals (including sapiens) are sexually attracted to their power projectors – they instinctively want their most physically fit and powerful to have first dibs on food and mating rights to ensure their genes remain in the population's gene pool. **Completely different species separated by major landmasses all independently converged on the same physical-power-based resource control protocols despite dozens of other perfectly viable options for settling disputes, establishing control authority over resources, and achieving consensus on the legitimate state of ownership and chain of custody of their property.** These animals didn't compare notes, but they converged on practically the same "might is right" pecking order heuristic.

Why did this happen? A knee-jerk reaction to this question is to be condescending towards pack animals and label their behavior as savage, as if humans aren't cut from the exact same cloth. As if the foundation of Neolithic sapient civilization *wasn't* based on our mastery of this exact same power projection tactic over animals. We hunted competing animal populations to extinction and then entrapped, enslaved, and devoured the survivors to form what we call "civilized" society (civilization is often considered to be a pejorative term to those who study nature). Sapiens are also instinctively attracted to signs of physical strength, and they constantly celebrate their capacity to project physical power against each other. Humans have no intellectually honest ground on which to stand and accuse animals of being savage, or act like "might is right" pecking order heuristics are beneath them. Sapiens use the same physical-power based resource control protocols whether they like to admit it or not (this is a core concept of the next chapter).

Another common reaction to the question of why pack animals overwhelmingly favor "feed and breed the powerful first" (a.k.a. "might is right") pecking order heuristics is to discredit their intelligence and label this behavior as unsophisticated. An argument is often raised that pack animals simply don't have the mental capacity to employ other, more sophisticated resource control heuristics that are less destructive or harmful. This argument presumes that alternative heuristics that aren't as prone to injury like "first come first served" or "feed and breed the weakest first" are somehow more cognitively complex than "feed and breed the strongest first." This presumption has no rational grounds because it's arguably more difficult to establish and maintain pecking order hierarchies based on power projection capacity.

Physical power-based resource control protocols like "feed and breed the most physically powerful first" or "might is right" require members of the pack to spend a great deal of time and effort constantly asserting themselves and finding opportunities to both improve and display their capacity and inclination to project power. Watching a pack of wild animals establish pecking order is usually an exercise in watching them constantly battle each other. Pack animals (e.g. birds and mammals) commonly snap at each other to assert physical dominance to keep a running tally of who deserves to be fed and bred first and what their position is within the dominance hierarchy. This is an energy-intensive chore which appears to take up more time and energy than alternative pecking order heuristics like "first come first served" or "feed the youngest first" or "feed the oldest first." It's also clearly more dangerous and prone to causing physical injury, creating a disincentive to adopt it. There are simply too many downsides to the "might is right" pecking order heuristic to claim that it's unsophisticated. There must be a reason why this pecking order strategy overcame the disincentive to use it and became the dominant strategy in nature.

A third reaction to the question of why pack animals use "feed and breed the powerful first" to establish control authority over their resources is to argue that weaker members simply don't have the option to do anything but allow physically powerful members to continuously eat first, due to being physically overpowered. This simply isn't valid. Mutiny is clearly an option. Power projectors like Mufasa may be strong, but they also need sleep. There's plenty of opportunities over the course of living together for disenfranchised members of the pack to cut the throat of abusive pack members to stop them from habitually cutting the line and allowing other pack members to starve. So why don't pack members team up and kill their "alpha" power projectors in those moments when they're vulnerable? Why do pack members allow themselves (and their offspring) to starve to death in service to the strongest members of their pack? Weaker members of the pack certainly have the mental capacity to cooperate together and overthrow abusive members, else they wouldn't have the cognitive ability to cooperate and live together as a pack in the first place.

The subject of pecking order and resource control is where it becomes important for sapiens to check themselves on their survivorship bias and their privileges as the world's apex predator. We need to remind ourselves that only the best strategies for safety, security, and survival can survive four billion years of entropic chaos. **Modern sapiens judge and criticize the behavior of wild animals from behind a safety moat filled with the blood of competitors they drove to extinction. Humans burned and speared and slaughtered their way to the comfort of the armchairs from which they pontificate about "right" and "wrong" or "fair" and "unfair" ways to settle disputes, establish control authority over resources, and determine the legitimate state of ownership and chain**

of custody of property. There is currently not a surviving mammal on this planet more predatory and destructive than humans.

3.10.10 Pack Animals Need their Power Projectors to Survive

Now that we have explored these core concepts of power projection, let's return to the example of Mufasa. Mufasa killed the pride's cubs because resources are scarce, and because the world is a cruel and unforgiving place full of predators and entropy. The pride must remain physically powerful and aggressive if they are to survive in this CCCH environment. Their prosperity depends on being able to maximize their C_A as much as they can, lest they risk extinction or perhaps worse, domestication. The pride will not allow themselves to muddy their gene pool or waste their precious time, energy, and food resources raising the offspring of demonstrably weaker and ineffectual lions. So Mufasa follows his natural instincts – instincts developed over millions of years of natural selection – and kills the cubs.

The mothers of the slain cubs will not retaliate against Mufasa because they instinctively understand how vitally important Mufasa's physical power and aggression are for their pack's mutual survival. Physical power and aggression are virtues for safety, security, and survival in the wild. The pack *needs* Mufasa's physical power and aggression to survive as much as they need oxygen. They need to keep Mufasa fed, and they need to breed with him to add his genetics to the gene pool. Why? Because of primordial economics and the survivor's dilemma.

The pride needs to manage resources effectively while lowering the pride's BCR_A as much as possible, and they can accomplish both by killing the offspring of demonstrably weaker parents. It sounds cruel, but what works is what survives. We have very compelling evidence to indicate this strategy is effective because most of us have no idea what a lion tastes like.

In conclusion, the behavior we see in nature should be regarded with solemn respect, not condescension or sanctimony. Nature offers a free lesson in survival from something that has been around far longer than we have, and it's good to listen to the hard-won wisdom of our elders, no matter how uncomfortable it makes us feel. By virtue of its ubiquity, nature's chosen pecking order strategy of "feed and breed the power projectors first" a.k.a. "might is right" deserves our reverence because it has been regression tested over thousands of millennia. Physical power-based dominance hierarchies are demonstrably more secure against predation, exploitation, enslavement, and abuse – capable of passing a very unforgiving natural selection process. This means we should take physical power-based resource control protocols seriously and have some intellectual humility before we condemn them. There's a reason why they survive, and we should seek to understand that reason from a systemic and sociotechnical perspective. Moreover, we should recognize that humans are clearly not "above" these sorts of physical power projection strategies. If anything, humans are the undisputed masters of them.

3.11 Physical Power-Based Resource Control

"Don't hate the playa. Hate the game."
Ice T [69]

3.11.1 There May not be Such a Thing as "Fair" in Nature, but There is Such a Thing as "Fit"

Systemic security is a trans-scientific phenomenon that forces us to confront frustratingly unquantifiable variables like ethics and design considerations. There is almost always a tradeoff between what's "good" and what actually works, both from an ethical perspective and from a design perspective.

One of the most frustratingly trans-scientific questions for any pack animal (to include sapiens) is how to settle property disputes. What is the "right" way to establish control over an animal pack's precious resources and to achieve consensus on the legitimate state of ownership and chain of custody of property? This is fundamentally a question that cannot be answered objectively. However, it is possible to observe nature and independently verify from empirical observation what pecking order heuristics are employed by nature's top survivors. In other words, it's possible to see what "good" resource control designs are by simply choosing to define "good" as "demonstrably capable of survival in a CCCH environment filled with predators and entropy."

Another way of saying the same thing is that regardless of whether people believe that "might is right" is "right," people can't deny that it survives. The ubiquity of "might is right" in nature proves that proof-of-power is a highly effective survival strategy and a time-tested power projection tactic that has proven itself over hundreds of millions of years to be able to keep pack animals systemically secure against predation. Intelligent, physically powerful, and aggressive animals survive and prosper, period.

Humans are incontrovertible proof of this basic fact of life. Therefore, **if we have ideological objections about "might is right," we should also have the intellectual humility to recognize that we have a fiduciary responsibility to the survival of our own species to recognize that these ideological objections are just that: ideological.** As the mantra goes, "Don't hate the player, hate the game." The fact that Mufasa kills the cubs, or that mother goose kicks the runt out of her nest, isn't something to condemn or lament. If there's anything to condemn or lament, it's the fact that we live in a cold, hard, cruel, and unsympathetic world filled with predators and entropy, where it's necessary to kill cubs and abandon runts to survive and prosper. As much as any organism might prefer not to, they all must fight to survive.

Welcome to life on Earth.

The takeaway? There's no such thing as "fair" in nature. "Fair" is a subjective and unquantifiable ideological construct that apparently only humans (the most physically powerful and destructive apex predators on the planet to date) are capable of thinking about. There is, however, such a thing as "fit" in nature. "Fit" is something we can objectively quantify and independently validate through empirical observation, simply by observing what survives. So the primary question to ask is, "what power projection tactics do pack animals employ to be fit for survival?" This question leads us to physical power hierarchies.

3.11.2 Why are Physical-Power-Based Dominance Hierarchies so Fit for Survival?

What makes physical power-based resource control strategies like "feed and breed the powerful first" more capable of survival? Primordial economics provides a simple explanation. Borrowing from the three strategic options for survival outlined in sections 3.6 and 3.7, how a pack chooses to settle disputes, establish control authority over resources, and achieve consensus on the legitimate state of ownership and chain of custody of property can cause an emergent effect of either increasing the pack's B_A more than C_A (option #1), increasing B_A the same as C_A (option #2), or increasing C_A more than B_A (option #3).

The survivor's dilemma indicates that option #3 is the best strategy. This could explain why "feed and breed the power projectors first" represents an optimal strategy for managing internal resources, because it ensures the most powerful and aggressive pack members with the most capacity and inclination to systemically secure the pack by imposing severe physical costs on attackers are well-fed and multiplied. This helps the pack maximize its C_A and minimize its BCR_A to generate enough prosperity margin to grow resource abundance without substantially increasing the threat of being devoured (or domesticated) by neighboring organisms.

Organizations which employ "feed and breed the powerful first" are more likely to have lower BCR_A than organizations which use alternative pecking order heuristics, as illustrated in Figure 26. But when more packs adopt this same pecking order strategy, it drives the environment's hazardous BCR_A level down and compels other packs to adopt the same strategy. This makes "feed and breed the power projectors" yet another power projection Schelling point which organizations are strategically inclined to adopt, hence its ubiquity in both the wild and in human society. If neighboring animal packs are feeding and breeding their power projectors, then your organization must feed and breed your pack's power projectors too, else risk becoming an attractive target of opportunity.

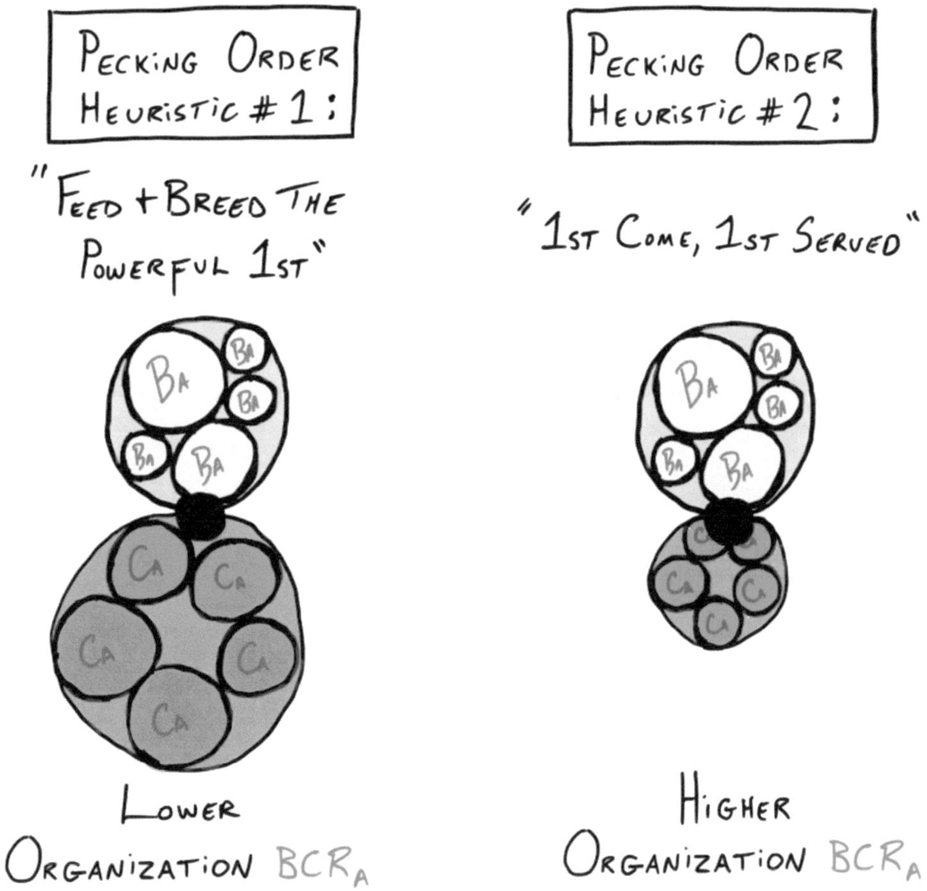

Figure 26: Bowtie Representation of Different Pecking Order Heuristics

3.11.3 Modeling Physical Power-Based Resource Control Protocols

It is possible to model the physical power-based resource control protocol using system theoretic processes. If we treat an animal pack as a system and treat security and survival as complex emergent behavior of that system, then we can conclude that security and survival are complex behaviors which emerge from the structure of individual components within the pack as well as from the interactions between and among those components. By modeling these individual components and their interactions, it is possible to compare them to other resource control models to determine if they might share similar emergent effects (this will be done in the following chapter with a different type of resource control structure developed by agrarian sapiens, known colloquially as governments).

For this thesis, the controlled process we want to examine is the state of ownership and chain of custody of a pack's valuable internal resources. With the controlled process formally defined in Figure 27, the next step in modeling a physical power-based resource control protocol is to model the controllers within the system. By default, every member of an animal pack doubles as a system controller with some amount of control authority over the controlled process. We can call these controllers "Members" and treat them as a component within the system capable of executing certain control actions. In addition to Members, packs have specialized workers genetically optimized to project power. These workers have special control authorities which Members don't have and can therefore be represented as different controllers within the system. We can call these controllers "Power Projectors."

Another controller within an animal pack's physical power-based resource control protocol is one that has control authority over the entire pack: physics. More specifically, physical power (a.k.a. watts). Physics is a naturally occurring control authority to which all other system controllers are subordinate. No animal gets to unsubscribe from the control authority of physical power. This means both Members and Power Projectors operating within an animal pack automatically subscribe to physical power's control authority as an involuntary control action, whether they like it or not. In return, physical power exercises control over the animal pack by giving Power Projectors the resources (physical power) needed to exercise control over the controlled process. Physical power itself can therefore be modeled as a controller which executes the control action of empowering Power Projectors. Together, Physical Power, Power Projectors, and Members represent three controllers within an animal pack's resource control system model. This is shown in Figure 27.

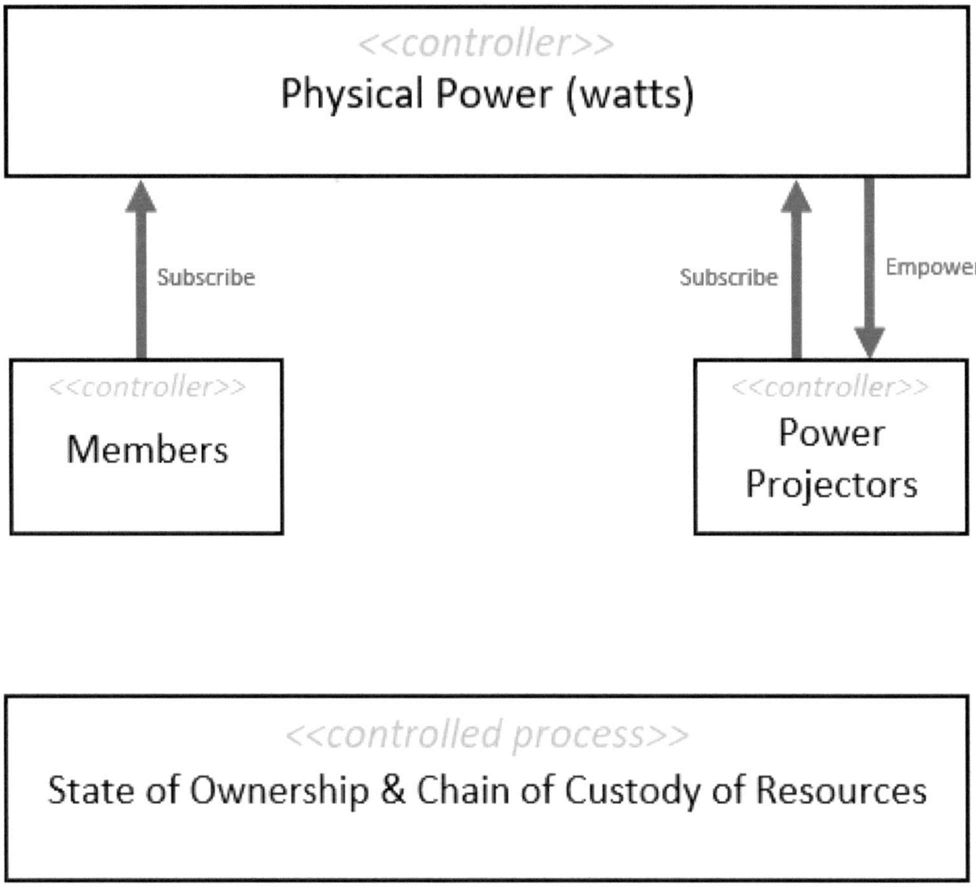

Figure 27: Physical Power-Based Resource Control Protocol used in the Wild (Part 1/4)

As discussed at the beginning of this chapter, resources are only owned insofar as able-bodied organisms are willing and able to project power to gain and maintain access to those resources. We can incorporate this concept into our model as two control actions assigned to Power Projectors which they can exercise over the controlled process: (1) gaining access to resources, and (2) defending access to resources. Both of these control actions are shown in Figure 27.

Additionally, Members don't get to eat unless Power Projectors are willing to use their power to gain and maintain access to the pack's resources (i.e. food). But even if Power Projectors are willing to use their power to gain and maintain access to the pack's resources, Members still don't get to eat unless Power Projectors permit them to eat (in other words, alphas can and often do deny some members of the pack from having access to the pack's internal resources). Pack Members therefore must tacitly request access to the pack's resources, and Power Projectors must tacitly approve those requests. These two additional control actions are also shown in Figure 28.

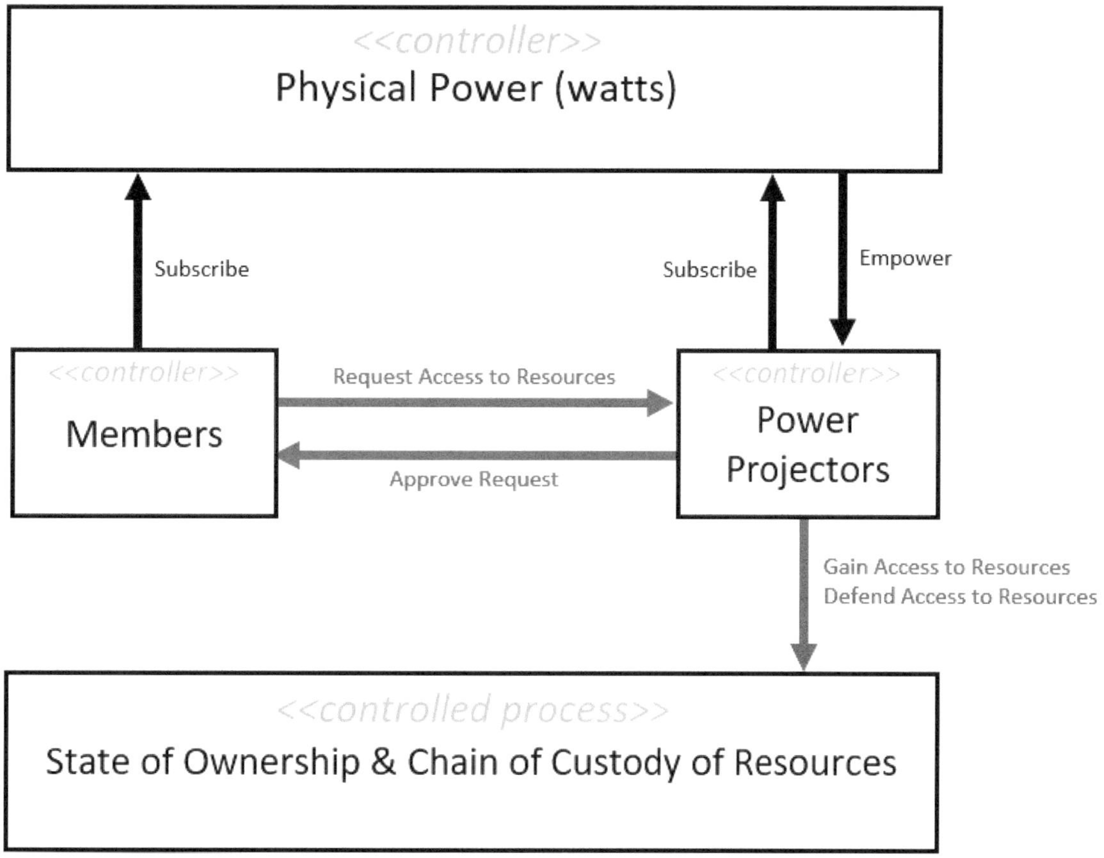

Figure 28: Physical Power-Based Resource Control Protocol used in the Wild (Part 2/4)

Even though Power Projectors have seemingly disproportionate control authority over resources, Members are not without their own form of control authority to provide a counterbalance. Members exercise substantial control over the controlled process by assigning value to the resources. This is a subtle but very important control action which often gets overlooked, that will become important to point out later in a discussion about Bitcoin.

The need for Power Projectors to establish consensus on the state of ownership and chain of custody of resources hinges on the assumption that pack Members actually value those resources in the first place. If Members don't assign value to the resources over which Power Projectors compete for control authority, then the Power Projectors' control authority over those objects is practically useless.

Members therefore have a unique ability to render their Power Projectors' control authority practically useless by simply revoking the value they assign to the resources controlled by Power Projectors. The reason why this control action is often overlooked is because it's rarely exercised by the majority of all species of pack animals. Most resources have existential importance that is virtually impossible for pack Members *not* to value. For example, it's not likely that Members of any species will stop valuing food, water, oxygen, and other physical resources which are essential for their survival. But the value of some resources is flexible and prone to change. For example, the value of lakefront territory can change rapidly if the Sun dries up the lake. Pack Members could care less about their Power Projectors' control authority over lakefront territory if they don't value it anymore.

This control action is a major factor for human packs (a.k.a. polities like governments), because sapiens often compete to exercise control authority over immaterial resources with abstract value (e.g. money). A pack of sapiens could have the most dominant Power Projectors in the world, but their control authority over abstract resources like money can be rendered useless if Members simply decided to stop valuing that abstract resource. With this subtle control action, we can close the loop and complete our physical power-based resource control model in Figure 29.

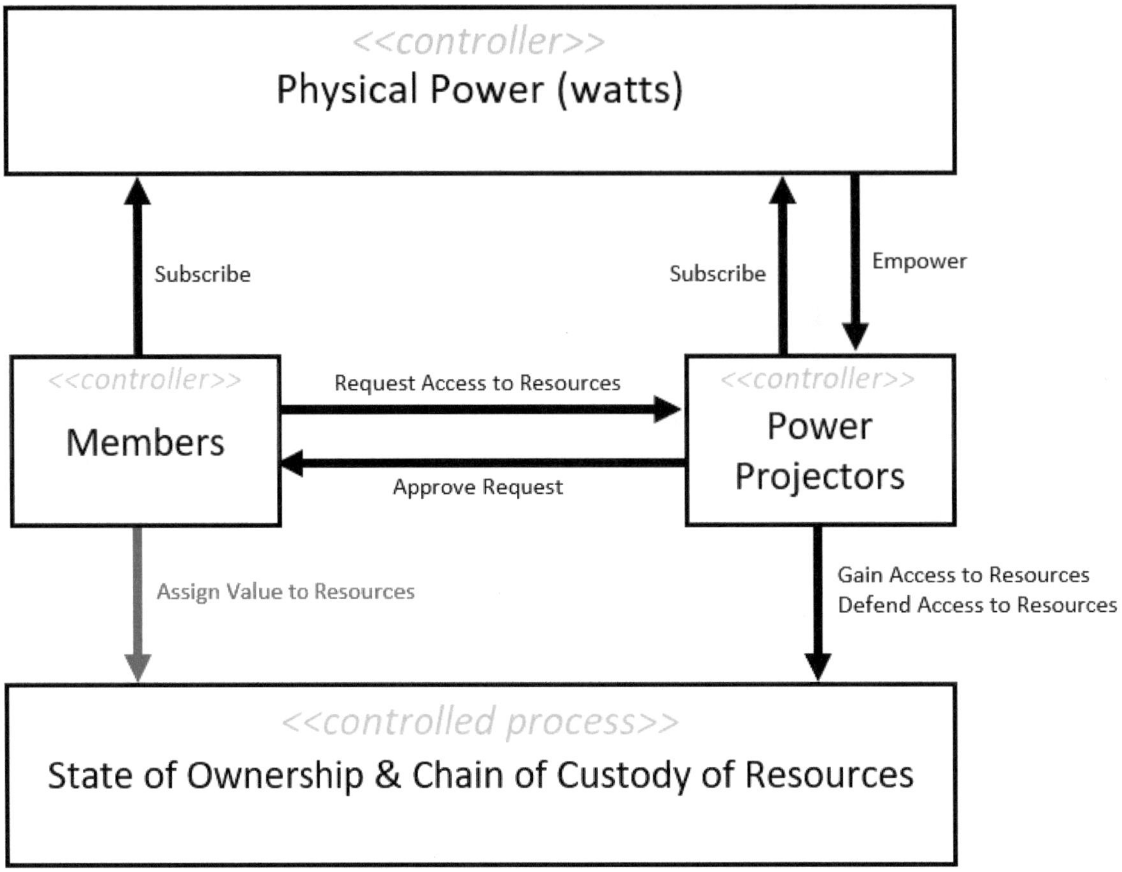

Figure 29: Physical Power-Based Resource Control Protocol used in the Wild (Complete Build)

3.12 The Beauty of Antlers

"There wasn't enlightened, compassionate democracy on the tundra...
Dogs really do eat dogs and, more importantly for the hunter-gatherer world, stags battle stags."
Charles Foster [68]

3.12.1 Physical Power Projection is Necessary, but it Clearly has Drawbacks

Most surviving wild pack animals use physical power to settle disputes, establish control authority over internal resources, and achieve consensus on the legitimate state of ownership and chain of custody of property. Pack animals spend a great deal of time establishing physical power-based dominance hierarchies to manage their resources, constantly seeking to display to their peers their physical strength and aggression, all to showcase their worthiness for the pack's resources because of their capacity and inclination to impose severe physical costs on attackers. For carnivores, this pecking order communication strategy often looks something like that shown in Figure 30.

Figure 30: A Power Projection Strategy Prone to Causing Injury

For animal packs which adopt physical power-based resource control protocols, the strongest and most aggressive power projectors are awarded with food and mating rights, perpetuating a virtuous cycle of systemic security that develops and sustains a well-resourced workforce of power projectors who keep the pack's BCR_A as low as possible given limited resources. The result of this heuristic is safety, security, and survival in a CCCH environment filled with predators and entropy.

An unfortunate side effect for using physical power as the basis for settling disputes, establishing control authority over physical resources, and achieving consensus on the legitimate state of ownership and chain of custody of property is that physical power projection is prone to causing injury. When done kinetically (i.e. using forces to displace masses), displays of physical power can lead to fratricide. And when physical strength and aggression is disproportionately rewarded with food and mating rights, it's easy to see why this protocol can lead to life-threatening injuries.

Nature mitigates this risk by making pack animals instinctively disinclined to fatally injure each other. For example, when wolves infight to establish their pecking order, it is common for one wolf to successfully pin down their opponent. The dominant wolf will press its teeth against the jugular of the opponent, but it will not bite. For existential reasons, the wolf is instinctively disinclined to kill a fellow member of the pack; it needs every member of the pack to hunt for prey, secure the pack against predators, and to survive and prosper for as long as possible. This explains why many animals do not battle to the death over food and mating rights; they battle to the point where they can discern that one is clearly more powerful and aggressive than the other (thus more fit for survival in a CCCH environment).

The reason why the author has dedicated so much time explaining these concepts in such great detail is so the reader can understand that humans have similar natural instincts. It takes a considerable amount of training to get sapiens to stop instinctively pulling their punches when fighting hand-to-hand with other sapiens. This is not because sapiens want to minimize injury to themselves. It's because sapiens have an unconscious instinct to minimize injury to their opponent.

Soldiers must be trained in how to overcome their natural disinclination to cause lethal injury. Hundreds of years ago, when soldiers still used close-ranged rifles where they could see their opponent's faces, it was not uncommon for them to refuse to fire, even under the threat of death. As one famous example, 87% of the rifles recovered from dead soldiers after Gettysburg (the third-bloodiest battle in American history) were fully loaded. Several of those rifles had been double and triple-stuffed with ball and powder, suggesting that soldiers weren't firing shots, and were even faking their shots and reload sequences during battle. It's hard to argue that these soldiers didn't realize their rifles weren't firing because of the substantial amount of sound, recoil, and visible smoke that every rifle produced during this time. [70]

Gettysburg was one of many battles which demonstrated how powerful human instincts are to avoid injuring fellow sapiens. Even when facing a clear and imminent threat to their own life, soldiers won't fire their rifles because their instinct to preserve a fellow human's life is often psychologically stronger than the instinct to preserve their own life. This is part of the reason why militaries require so much training. Militaries must train their members to overcome this instinct when needed, because deliberately killing one's own kind is something instinctively unnatural for some species, to include sapiens. [70]

Nevertheless, wolves still end up killing other wolves. Humans end up killing other humans, and so on. The simple fact of the matter is that physical power-based resource control protocols are destined to cause lethal injury; it's an unfortunate side effect of a demonstrably necessary protocol for survival and prosperity. We know this because the pack animals which survive in the wild are the ones who use this protocol. Is physical injury wasteful? It certainly seems to be. But it also seems to be the case that the risk of physical injury is a price that natural selection demands for survival; a price that Earth's top survivors are willing to pay. [71]

Based on extensive randomized experimentation, we know what happens to wolves and countless other animal populations when their physical power-based resource control hierarchies are unnaturally disrupted. They become docile and domesticated; they start to depend on their masters for safety and security, and they get systemically exploited at extraordinary scales. The emergent effect of interfering with an animal pack's natural instincts to reward their physically powerful and aggressive members is incontrovertible. We see proof of it every day. We eat the proof for breakfast, lunch, and dinner. We play fetch with the proof. We post pictures of the proof on our social media pages.

3.12.2 There are Less Harmful Ways to Establish Interspecies Dominance Hierarchies using Physical Power

Nevertheless, using physical power as the basis for settling disputes, establishing control authority over internal resources, and achieving consensus on the legitimate state of ownership and chain of custody of property doesn't necessarily have to cause so much physical injury. Nature shows us there are different ways to project physical power to establish intraspecies dominance hierarchies. Different animals have adopted different strategies to perform the same function, and we can take note of them. For example, mammalian carnivores and herbivores employ much different power projection strategies for establishing their dominance hierarchies.

Mammalian carnivores hunt for their food, so they are already equipped with the power projection technology needed to settle their internal property disputes and establish internal control authority over resources. They often use the same teeth and claws to take down prey as they do to establish pecking order within their intraspecies dominance hierarchies. Mammalian herbivores, on the other hand, are often equipped with much different-looking power projection technology. Instead of having sharp teeth and claws to bite and cut each other, many herbivores sprout heavy and cumbersome protrusions out of their foreheads, then clash them together to settle disputes and establish pecking order. We call this power projection technology *antlers*.

From a primordial economic perspective, antlers are a fascinating evolution. They can impose as much, if not more, severe physical costs on attackers as a wolf's teeth or claws can. A well-grown pair of antlers are no doubt effective at lowering the **BCR_A** of a stag and a herd of deer. But antlers aren't simple stabbing devices like rhino or triceratops horns; they have bizarre geometries like those shown in Figure 31.

Figure 31: Awkward-Looking & Underappreciated Power Projection Technology

The geometry of antlers makes them prone to entanglement with other antlers, a problem that more simplified and easy-to-grow puncturing devices like horns, teeth, and claws don't have. What could possibly be the benefit of such a bizarre design? Why do stags have such awkwardly-shaped devices growing out of their foreheads when they could grow a simple puncturing device like a rhino horn, and use that to impose severe physical costs on attackers instead? Would that not be just as effective? To answer these questions, simply ask yourself the following question: if you were a stag and you had to go head-to head with your brother stag to settle an intraspecies dispute, would you rather have antlers or a rhino horn? Unless you desire to kill your brother, you should prefer to be equipped with antlers rather than a rhino horn, because your antlers would probably entangle with your brother's antlers without stabbing him, whereas a rhino horn probably wouldn't. Therein lies the subtle eloquence of antler design.

When two pairs of antlers go head-to-head, their entanglement functions as a cushion and creates a safe standoff distance large enough to prevent either side from being severely injured. Thus, the tendency of antlers to entangle is an emergent property of their awkward geometry and a special design feature, not a flaw. **The primary value-delivered function of antlers is to empower stags to defend their packs and establish their intraspecies dominance hierarchy using physical power, but to do it in a way that protects the species against fratricide.**

Entanglement protects members of the same species from impaling each other, yet still enables them to impale other species (notably the predators who don't have antlers) as needed. Consequently, antlers impose far more injury on predators than they do on peers. If you're a member of a herd of deer or similar species, you can butt heads all day long to settle disputes, establish control authority over resources, and determine the legitimate state of ownership and chain of custody of property with far lower probability of causing mortal injury to each other. But if you're not the same species equipped with the same power projection technology (like if you're a predator who intends to bring harm to them), then you can expect a much higher probability of injury.

The design isn't perfect. Puncture wounds still occasionally happen, and it's not uncommon to find two dead stags stuck together because their antlers entangled a little too well. But it's still less prone to mortal injury than rhino-style horns, or predators who cut each other with their claws and fangs. Natural selection is willing to accept the loss of having stags occasionally get mortally interlocked or impaled by accident. Stags which pass on their genes are the strongest stags with the most perfectly awkward antler geometry needed to minimize fratricide while maximizing strength and physical aggression. What survives is what works, and that process led to the fantastically awkward and beautiful antler geometries we see today.

Awkward antler geometry entanglement is a safety feature, not a bug. Antlers allow stags to secure their pack against outside threats (wolves) while also preserving their ability to use physical power as the basis for establishing an intraspecies dominance hierarchy, but do it as safely as possible. This means antlers represent life's discovery of a safer, less-lethal strategy for physical power-based resource control. It's a special type of evolved power projection tactic that retains the strategic benefits of physical strength and aggression but minimizes the harmful side effects.

Survival demands that animal packs must feed and breed their strongest and most aggressive power projectors to increase the pack's overall $\mathbf{C_A}$ and lower $\mathbf{BCR_A}$, but that doesn't mean they have to severely injure each other. Stags don't need to hunt their food, so they have a design trade space which carnivores don't have. With that trade space, stags show us ways to play the power projection game in a way that minimizes the pack's capacity for injuring each other.

Sadly, very few people appreciate the eloquent design of antlers because they don't understand the dynamics of physical power projection. To those who don't factor in primordial economics and the dynamics of security and survival, the unnecessarily large and awkward protrusions growing out of a stag's head may look like bad design – particularly a waste of energy and resources. Clashing antlers together to establish a physical power-based resource control hierarchy seems unnecessary. The unnecessarily large structure of antlers looks like a waste of keratin. Why burn so many calories carrying around the weight of that much extra keratin, and waste more energy clanging them together? What could possibly be the point of such an energy inefficient-looking design?

It would be tragic to condemn antlers for their inefficiency and waste because the intent of the design is quite noble: the prevention of fratricide and the preservation of life. There's nothing inefficient about the preservation of life. In fact, few things are more wasteful, not to mention existentially risky, than allowing the pack's power projectors to routinely commit fratricide. Antlers may look weird and inefficient on the surface (particularly to those who don't understand power projection), but they enable the pack to lower BCR_A and establish pecking order the way natural selection demands, using a strategy which maximizes safety by minimizing the potential for causing mortal injury to a member of the same species. The pack can still feed and breed its strongest and most aggressive power projectors as is necessary for safety, security, and survival. It can still settle disputes in a fair and meritorious way using physical power competition. It can still establish control authority over precious resources using physical power. But it can strive to do all these things without endangering their own species.

A reason why people may not appreciate antler design is because they don't understand the governing dynamics of primordial economics. After spending too much time at the top of the food chain, it's easy to forget that we live in a world filled with predators and entropy. For those who value security, it is an existential imperative to secure oneself by imposing severe, physically prohibitive costs on attackers. That means it's also an existential imperative to ensure the pack's most powerful members get the most resources. But how does the pack identify their most physically powerful members without engaging in physical power competitions? Therein lies the challenge. Animal packs (to include humans) need physical power competitions to identify the members of the pack who are the most deserving of the pack's resources, thus the most qualified to be at the top of the physical power-based dominance hierarchy. But unfortunately, those physical power competitions often lead to fratricide. Antlers fix this.

Without factoring in primordial economics, it's easy to look at antlers and see a flawed, inefficient design, but the reality is quite the opposite. Antlers represent an eloquent display of nature's determination to preserve life by minimizing fratricide, combined with its stoic acknowledgement that physical power-based dominance hierarchies are existentially necessary in a world filled with predators and entropy where the physically powerful survive and prosper. Antlers represent a compromise between two opposing design variables – a way to alleviate the tension between security and safety. Pack animals must maximize their ability to impose physically prohibitive costs on neighbors to improve their security, but they can endeavor to make their physical power competitions as non-lethal as possible.

In summary, pack animals must be able to identify their strongest and most physically aggressive members and award them with control authority over the resources they need to lower the pack's BCR_A, but they don't have to inflict injury and commit fratricide in the process. Stags prove it's possible for pack animals to survive and prosper by projecting power the way natural selection demands, all while minimizing physical injury. In other words, **it's possible to maximize security and safety simultaneously. All it takes is the right kind of bizarre technology, and the right kind of people to recognize the value of the design.**

This concludes the first chapter of Power Projection Theory. In the remaining two chapters, the author will outline why Bitcoin may represent the human equivalent of antlers.

Chapter 4: Power Projection Tactics in Human Society

"Man cannot remake himself without suffering, for he is both the marble and the sculptor."
Alex Carrell [72]

4.1 Introduction

"So long as there are men, there will be wars."
Albert Einstein [73]

Organisms fight and kill each other for their resources. This struggle is real and directly observable. But when evaluating human behavior, there are clear differences between the way sapiens behave and the way other organisms behave, particularly in the way they fight and kill each other for resources. Behaviorally modern sapiens are unique in the animal kingdom in that they fight and kill each other not just for resources, but also for what they chose to believe in. They use their powerful brains to think abstractly, adopt belief systems that other organisms are physiologically incapable of perceiving, and then they physically compete over those belief systems at unrivaled scale.

Ironically, amongst the most commonly adopted belief systems over which humans routinely fight and kill each other is the belief that people *shouldn't* have to fight for their resources – that sapiens and sapiens alone have "natural rights" to their lives, liberties, and properties which other animals don't have, and that humans are special exceptions to primordial phenomena like predation, entropy, and the existential necessity to establish dominance hierarchies using physical power. It is not uncommon for modern humans to believe that the creator of life itself has placed them on some special pedestal above all the other lifeforms on Earth, so they don't have to struggle the same as the rest of the lifeforms "beneath" them to settle their intraspecies property disputes, establish control authority over intraspecies resources, and determine the legitimate state of ownership and chain of custody of their property. In comparison to the rest of life on Earth, human power projection is both unique and bizarre.

In this chapter, the author will break down to their core concepts around human power projection and physical conflict and offer some explanations for why humans behave the way they do – particularly how and why society is routinely compelled to fight and kill each other over their resources and their belief systems. The goal of this discussion is to build some conceptual insights about why humans fight wars, which will develop the conceptual foundation necessary for understanding why Bitcoin might be used not as a monetary technology, but as a soft (a.k.a. non-kinetic) form of warfighting technology that (literally) empowers people to physically compete for their resources and for what they choose to believe in, but in a non-lethal way that doesn't harm others.

Now that we have a foundational understanding about how and why organisms use physical power to settle intraspecies disputes and establish their dominance hierarchies, we can turn our attention to human beings. The remarkable thing to note about behaviorally modern sapiens is that they strive *not* to use physical power as the basis for settling their disputes and establishing their pecking order. Instead, they use their imaginations to conceive of *abstract* sources of power, and then they *attempt* to use these abstract sources of power to settle their disputes and establish their pecking order. These attempts often don't succeed, hence the continual and inevitable reversions back to warfare.

4.1 Introduction

As much as humans wish they could cheat nature and establish their dominance hierarchies using something which transcends physics, they have yet to transcend warfare. They still rely on physical power to enforce, legitimize, or illegitimate abstract power. Ironically, despite how much they strive to avoid conflict, human infighting is the most physically destructive intraspecies competition on the planet. For some reason, no matter how much sapiens try *not* to use physical power to settle their disputes and establish their dominance hierarchies, they always resort back to the same primordial economic behavior as practically every other species in nature. Why is that?

This chapter endeavors to explain why humans attempt to use abstract sources of power to settle their intraspecies disputes, establish their dominance hierarchies, determine who has control authority over their resources, and achieve consensus on the legitimate state and chain of custody of their property. The author begins by guiding the reader through a deep-dive in human metacognition and abstract thinking. Once a baseline understanding in human metacognition has been established, the author explores the differences between abstract and physical power, how abstract-power-based dominance hierarchies work in comparison to the physical-power-based resource control structures discussed in the previous chapters, and then offers some explanations for why they break down and lead humans back to war.

The purpose of this chapter is to establish a thorough understanding of the root causes of warfare in order to develop the "why" behind emerging power projection technologies like Bitcoin. A key assertion of this thesis is that Bitcoin could theoretically serve as an extension of warfare for human society as it enters the digital age. Before introducing the concept of "softwar" and entering into a discussion about the sociotechnical and national strategic security implications of Bitcoin as a warfighting protocol rather than strictly a monetary protocol, it's necessary to understand how and why wars break out in the first place.

A word of warning, though. Warfare is difficult and trans-scientific topic that is highly emotional & politically charged. It's impossible to know why all wars break out, but the author provides some viable explanations based off concepts that emerged from research. The point of doing this is to provide the reader with conceptual insights about why humans can't seem *not* to fight wars. In other words, the reader should leave this chapter with an understanding for why society's ostensibly "peaceful" alternatives to warfare inevitably lead back to warfare. There is something about how humans attempt to settle intraspecies property disputes and establish pecking order that isn't working, that routinely breaks down and becomes dysfunctional.

Humans keep having to return to the primordial economic behavior of fighting and killing each other to establish decentralized, zero-trust, and permissionless control over their valuable resources. Once the reader understands why humans keep resorting back to fighting and killing each other over resources like wild animals do, the author proceeds into a discussion about the dynamics of national strategic security. This leads to an exploration of how and why humans have scaled their physical power projection tactics to the point of mutually assured destruction, and why this situation represents a major systemic security hazard. The chapter concludes with a discussion about how humans would benefit from their own version of antlers which would allow them to settle their disputes and establish their pecking order using non-kinetic (thus non-lethal) physical power projection tactics, techniques, and technologies.

Human behavior could be described as ironic because in their attempts to avoid intraspecies infighting and transcend "uncivilized" physical-power-based methods of resource control, they have become amongst the most physically destructive mammals on the planet. This chapter attempts to provide some explanations for *why* this destruction takes place, so that these concepts can be utilized later to explain the context about why people would be motivated to use Bitcoin as a warfighting technology and how Bitcoin could help people mitigate warfare's destructive effects.

4.2 A Whole New World

"The true sign of intelligence is not knowledge, but imagination."
Albert Einstein [74]

4.2.1 Fire Power equals Computing Power

Brain tissue requires about 20 times more power than muscle tissue. The most energy-consuming part of our brain is the modern part, the neocortex, which is used for higher-level processing and abstract thinking. For shrew-like mammals, the neocortex represents about 10% of total brain volume. The average mammal's brain volume is 40% neocortex. Primates have an above-average neocortex volume of 50%. But even a primate's neocortex is small compared to anatomically modern humans. A staggering 80% of sapient brain volume is neocortex. This substantial difference is shown in Figure 32. [68]

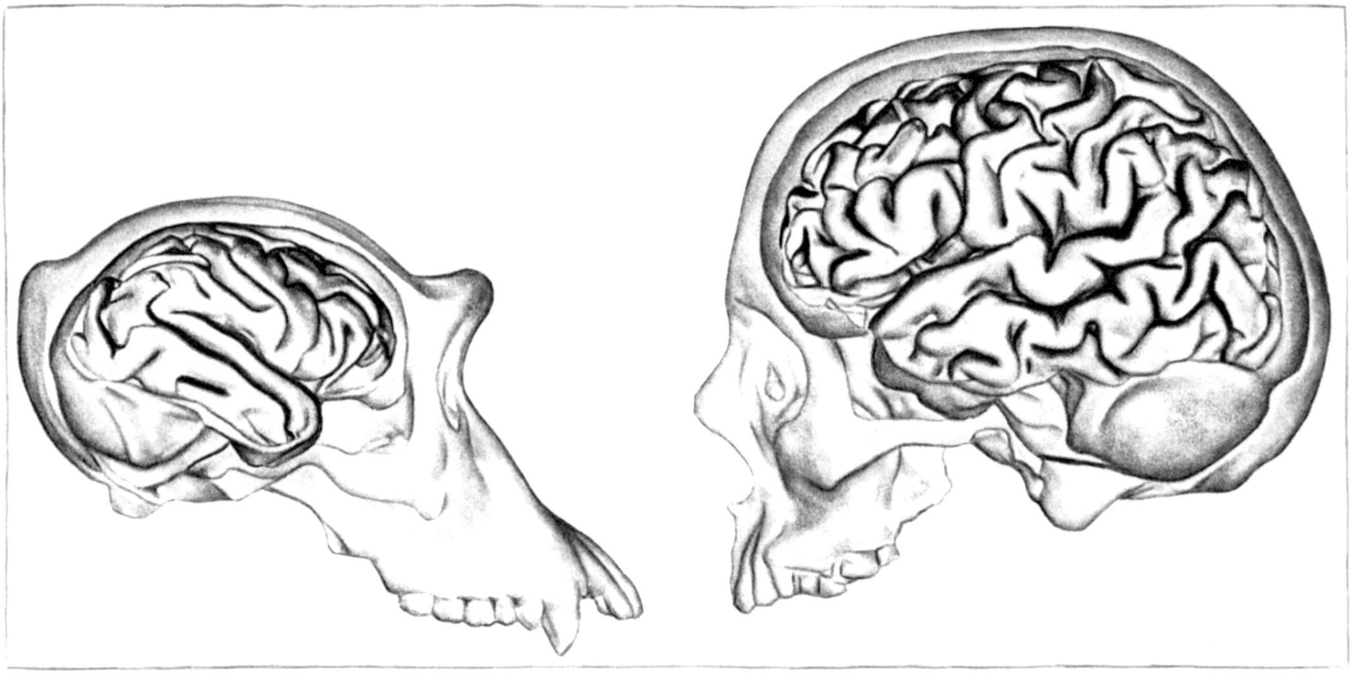

Figure 32: Comparison Between a Sapient Brain and That of its Closest-Surviving Ancestor

Humans can afford to power their large neocortices because of the energy abundance they achieved by learning how to control fire. Humans unlock far more energy per unit of food consumed than other animals thanks to their ability to cook. Just like a staged combustion cycle on a modern rocket can unlock more energy for relatively little penalty in size, weight, and power by using a pre-burner to combust fuel and oxidizer before it reaches the combustion chamber, humans operate the same way. By cooking their food, humans pre-burn their fuel as a form of pre-digestion before it reaches their combustion chamber (stomachs), allowing them to unlock substantially more energy per unit of food with only a minor penalty to size, weight, and power. And just like how more energy allows a rocket to carry more payload on top of it, more energy allows a human to carry more payload on top of it, too.

Humans enjoyed a step-function rise in surplus energy when they took control of fire, which they vectored towards performing the highly energy intensive task of thinking. The brain represents only ~2% of a modern human's body weight, yet it consumes ~20% of its energy. Fortunately, humans became so rich in surplus energy that they were able to habitually overclock their neocortices to the point where it drove dramatic physiological changes, giving rise to the modern sapiens' massive neocortices and oddly bulbous heads. [68]

Learning how to control fire and gain access to exogenous energy caused sapient brains to grow so quickly that they outpaced the growth of their own pelvises and birth canals. Combined with their tendency to walk upright, these sudden anatomical changes cause sapiens to have far more complex and painful childbirths than other primates. Sapient heads are so large that they must be born approximately 50% prematurely with only partially assembled skulls and necks just to fit through their mother's birth canal. For this reason, sapiens are significantly more fragile and helpless at birth compared to other mammals, and they must go through a longer adolescent period to account for having to incubate outside the womb. But as the saying goes, "the juice is worth the squeeze." A sapiens' massive neocortex is the source of the most significant power projection technique observed on Earth. [68]

4.2.2 Using Imagination to Create a Virtual Reality

Anatomically modern sapiens and their absurdly large foreheads shown in Figure 33 emerged from Africa at least 200,000 to 300,000 years ago and expanded into Europe as the ice melted. Despite their physiological similarities, sapiens don't appear to have behaved like the ones today until the Upper Paleolithic era started around 50,000 years ago. This is when the fossil record first starts showing signs of sapiens' having a higher degree of intentionality and theory of mind required for the highly self-consciousness behavior of modern humans, a phenomenon often called *behavioral modernity.* [68]

It was during this timeframe when sapiens started tracing their hands on cave walls and signaling extraordinarily high levels of self-consciousness compared to other animals. They started making distinctions between themselves and their environment, with both objective and abstract qualities. It's unclear what exactly caused human consciousness to spark, but when it did, it appears to have spread quickly. [68] Charles Foster describes the change as follows:

"Something tectonic happened to human consciousness in the Upper Paleolithic – whether by revolution or revelation or evolution. A new type of consciousness emerged out of or in addition to or in substitution for the consciousness that had been there before... for however long it had been gestating, a new type of self-perception and self-understanding had burst. It was manifested in a new symbolic sense... so much better at expressing itself that it looked different in kind or degree from anything that had existed before." [68]

Figure 33: An Anatomically Modern Homosapien with its Characteristically Large Forehead

It should be noted that sapiens are first and foremost hunter-gathering nomads, having spent only the last 5% of their history on Earth doing anything except traveling the world searching for fauna and flora to eat. Their overclocked, overpowered, and oversized neocortices are especially useful for these activities because they help their hosts perform advanced pattern finding. The ability to connect dots between sensory input information enhances sapiens' ability to detect and exploit patterns of behavior in surrounding fauna and flora for improved hunting and gathering.

The advanced dot-connecting and pattern-finding capability of a brain is colloquially known as intelligence. The more an animal can use their brain to connect dots between their sensory inputs or detect valid patterns of behavior within their environment, the more intelligent the animal is perceived to be. Not surprisingly, with 80% of their brain volume comprised of advanced pattern-finding neocortex hardware fueled by excess energy and fire power, modern sapient brains are the most intelligent on Earth.

Sapient brains are so effortlessly good at dot-connecting and pattern-finding that they don't even need physical sensory inputs to detect patterns. It is possible to put a behaviorally modern human in a dark, empty, sound-dead room and their brain will have no trouble envisioning many sights, sounds, and objects. They will connect dots between sensory inputs and detect patterns which don't physically exist. This remarkable capability is known as abstract thinking.

Imagination could be described as the phenomenon which occurs when human neocortices use their abstract thinking skills to form ideas, images, or concepts independently and without physical sensory inputs. This thesis uses the term *imaginary* to describe phenomena detected by human brains which don't physically exist in shared objective reality. Imaginary patterns can occur either due to a false correlation to physical sensory inputs or because the pattern was formed without sensory inputs in the first place (the author acknowledges there are detailed fields of philosophy with far more detailed and varying definitions of imaginary – this is how the author will use the term throughout this thesis).

Because of their ability to think abstractly and find imaginary patterns, sapiens operate in two different realities simultaneously: one in front of their eyes and one behind them, as shown in Figure 34. The author defines the concrete reality in front of a human's eyes as *objective physical reality*, the domain of energy, matter, space, and time which precedes humans and produces our physical sensory inputs. This reality is shared by all humans regardless of whether they can detect or conceive of it. Moreover, physically objective reality exists in and of itself although sapiens technically can't process it objectively without the abstract biases caused by their own brains.

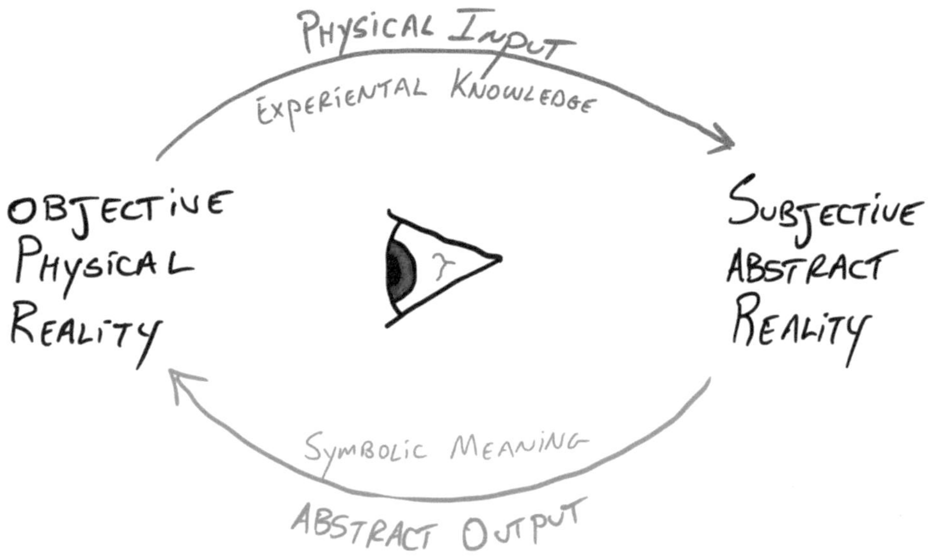

Figure 34: Illustration of the Bi-Directional Nature of Abstract Thinking

The reality behind a human's eyes can be defined as *subjective abstract reality*, a non-physical domain constructed out of the abstract thoughts of sapient neocortices and filled with imaginary patterns like symbols and semantic meaning. This reality can either be exclusive to one human mind, or it can be combined with other human minds. This abstract reality can either be an individual reality, or it can be a shared reality, where the latter occurs when sapiens get other people to see and believe in the same abstract reality together.

No other species on Earth appears to be as capable of perceiving abstract reality as sapiens. It is possible that other animals simply aren't physically capable of it because they don't have enough brain power and neurological circuitry needed to think of abstract reality. For the purposes of this thesis, abstract reality is defined as a new, imaginary world that recently emerged within the last 0.001% of life's total time on Earth. Abstract reality seems unique to humans, the only animal to have survived the evolutionary journey and prospered enough to have the capacity to think of it (the author acknowledges there are detailed fields of philosophy, theology, ideology, phenomenology, and metaphysics with far more detailed and varying definitions of these terms – this is how the author will use the term throughout this thesis).

Juggling these two different realities at the same time is quite an energy-intensive burden for human brains to bear, so to make it more efficient, sapiens show extreme favoritism towards their subjective, abstract, and imaginary reality. Then, they superimpose their abstract and imaginary beliefs onto sensory inputs seen, smelled, tasted, touched, and heard from objective physical reality. Brains appear to have no way of knowing if a detected pattern is anything but abstract when it's first generated, so they rely on their hosts to cross reference their abstract thoughts against physical sensory inputs to determine if the imaginary pattern is a physically "real" pattern (more on this in the next section).

The way human brains think is noteworthy for two reasons. First, it suggests humans experience the world foremost through their imaginations, then determine what is "real" based off what imaginary patterns happen to correspond to matching sensory inputs. Second, it suggests sapiens use abstract thinking in both directions of data processing, for dual purposes. Brains process physical inputs from objective physical reality at the same time they generate abstract beliefs about physical reality for the senses to investigate. These two separate tasks occur simultaneously, all the time.

Human minds produce a mental model of the world which influences the way they think, act, and perceive the information they receive back from their senses. Brains therefore act like a lens through which sapiens understand the world in abstract reality, while simultaneously acting as the mechanism through which they shape the world in objective reality. This bi-directional feedback and dual-use type of abstract thinking, where the brain's imagination influences the processing of its physical sensory data inputs as well as its outputs, appears to be the key enabling skillset required for a phenomenon called *symbolism* (the author acknowledges there are detailed fields of psychology with far more detailed and varying definitions of symbolism – this is how the author will use the term throughout this thesis).

Humans are so skilled at using their habitually overenergized brains to perform bi-directional and dual-use abstract thinking that it happens automatically without being conscious of it. It appears to be extraordinarily difficult for humans to turn off this behavior unless the brain becomes physically damaged or chemically impaired. It's practically impossible for humans *not* to distort their senses with their abstract thoughts or act purely off experiential knowledge (i.e. knowledge gained based exclusively off sensory inputs without biasing those inputs with our own subjective and abstract thoughts).

Ironically, humans can't do what other animals can do effortlessly: experience objective physical reality for what it is, without skewing sensory inputs through a neocortical lens of abstract biases. **Whereas most non-human species can't perceive symbols and abstract meaning in the first place, sapiens can't *not* perceive symbolic patterns and abstract meaning** once a given pattern has been committed to memory. The reader is invited to test this out. Try to look at this page without detecting symbols like letters and words or try to listen to someone produce the audible wave pattern of your name without detecting that abstract concept called your name. Most people find this to be impossible except for people drugged, brain damaged, or experiencing severe memory loss.

Another way to show how difficult it is to *not* think symbolically is to look at Figure 35. The blue and black scribbles drawn above and below the grey dashed line are identical in every way except for their topology. Simply change their topology and, like magic, a bunch of nonsensical and objectively meaningless scribbles suddenly turn into something with rich symbolic meaning even though the only thing that changed was the topology of the scribbles. If you don't see the same objectively meaningless scribbles above and below the grey line, then you are guilty of applying abstract, symbolic meaning to something that is objectively meaningless. Technically speaking, the scribbles drawn above and below the dashed line are equally meaningless, but you can't see it that way because you have committed the topological patterns below the grey dashed line to memory as symbols denoting semantically and syntactically complex abstract meaning. That's how gifted sapient brains are at abstract thinking. You don't even have to try, and you can't turn it off. You are slave to your neocortex, incapable of *not* perceiving symbolic patterns and imaginary meaning which directly interfere with how you process physical sensory inputs and how you perceive the shared objective physical reality we're living in.

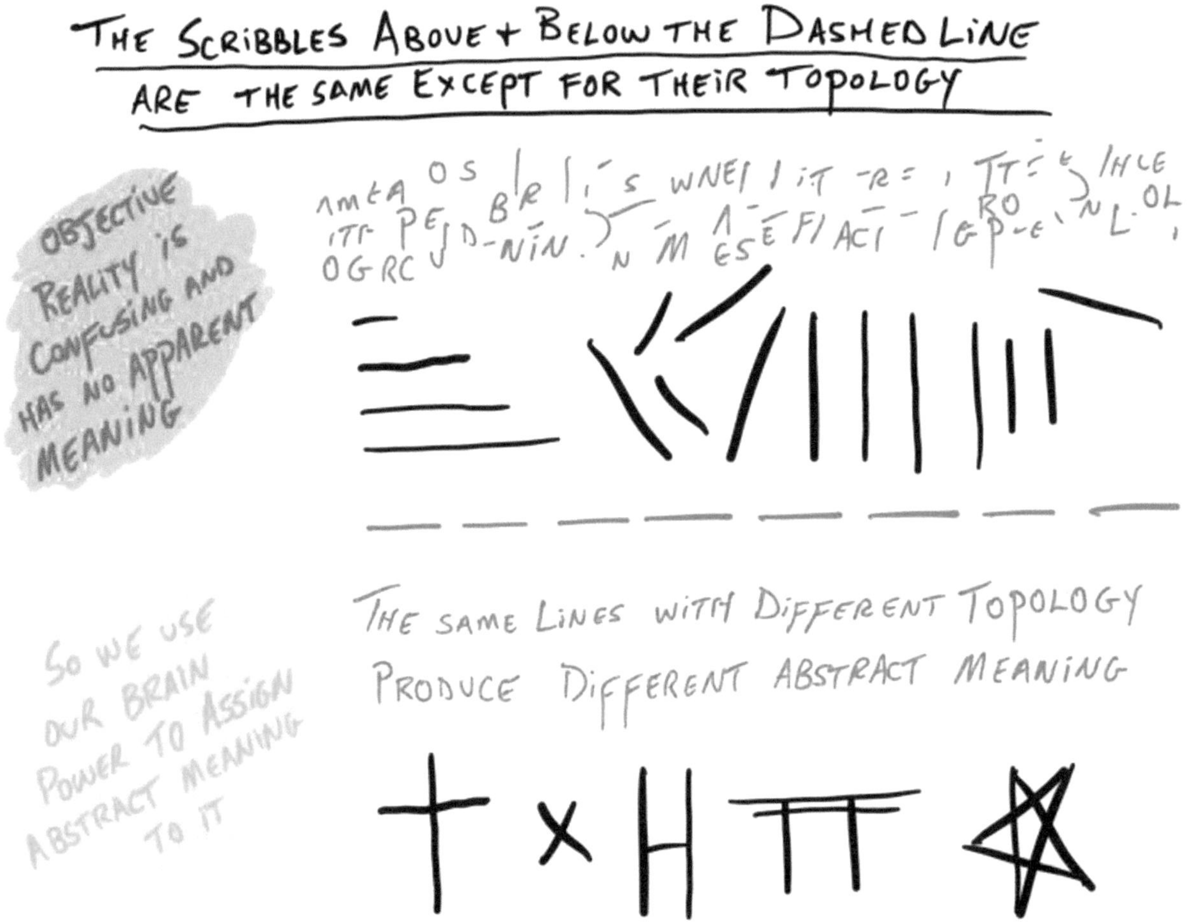

Figure 35: Illustration of How Practically Impossible it is Not to Think Symbolically

4.3 How to Detect if Something is Real

"We suffer more often in imagination than in reality."
Lucius Annaeus Seneca [75]

4.3.1 Sapiens Struggle to Detect What's "Real"

There are dedicated fields of knowledge like psychology, phenomenology, and metaphysics devoted to the subject of understanding what "real" means. It turns out, "real" is a surprisingly difficult thing for sapiens to define because, as demonstrated in the previous example, our objective sensory inputs are tainted by the subjective abstract thoughts and interpretations of our big, fat, overpowered, hyperactive, and overclocked brains. Sapiens are effectively trapped behind a neocortical cage, unable to interpret the world objectively for what it is without skewing it with imaginary meaning and symbolism. This makes it exceptionally difficult to know what "real" is.

Therefore, for the sake of producing a simple argument, this thesis uses the word "real" as a synonym for "physical" and the terms "imaginary" or "abstract" as synonyms for "non-physical." The author acknowledges that what we call "physics" is technically an experiential process mediated by the brain's abstractions and therefore not mutually exclusive to or divisible from abstract thoughts. However, for lack of better words to describe ontologically precedent, exogenous, and distinct phenomena like time, mass, space, and energy which predate sapiens by nearly fourteen billion years – this is how the author will use the term "real" throughout the remainder of this thesis.

Real things, according to the author, are categorized as things which produce their own physical signature in the domain of shared objective physical reality – as best as we currently understand what that means. On the contrary, imaginary things can be categorized as things which aren't real and therefore don't produce their own physical signature in the domain of shared objective physical reality. The author understands that subject matter experts in the fields of knowledge devoted to studying what "real" means don't make this same non-physical versus physical distinction. He apologizes in advance to professional philosophers for an engineer's bias towards physics and asks for temporary indulgence for the sake of illustrating a point.

A human brain's advanced bidirectional and dual-use abstract processing is done automatically and subconsciously, without requiring control inputs from a human host. Imaginary patterns are produced up front, then cross-referenced with sensory inputs received from shared objective reality to determine if a given pattern is real or unreal. This processing logic can be modeled in Figure 36 below.

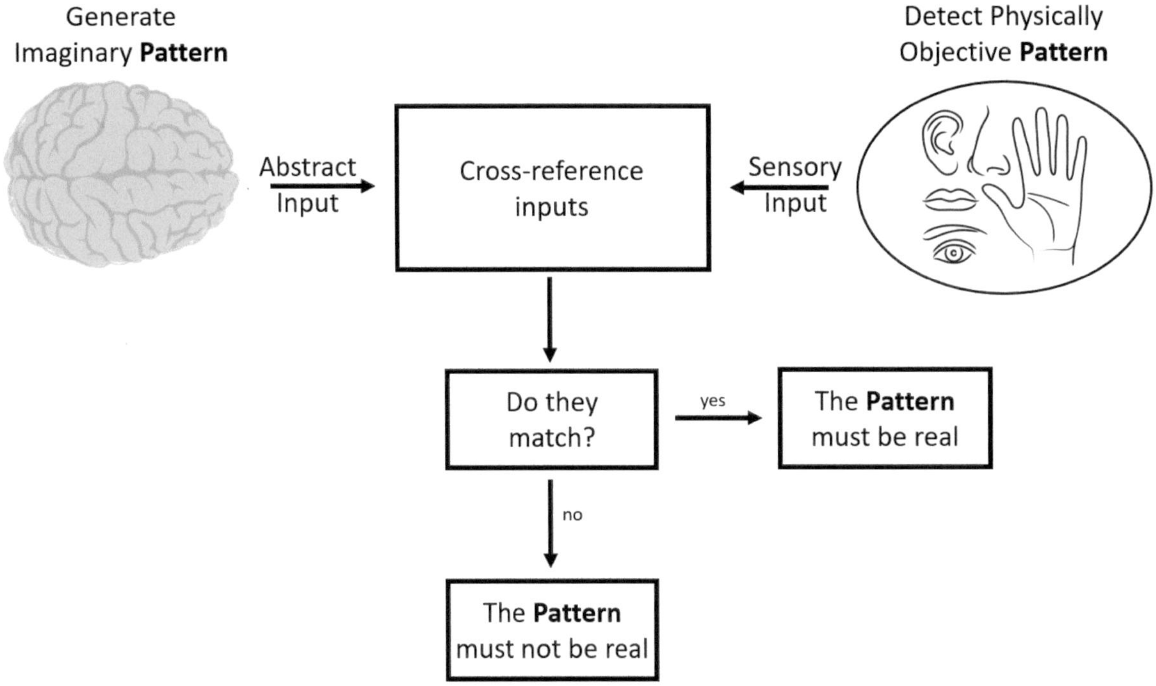

Figure 36: Model of a Realness-Verification Algorithm Performed by Sapient Brains
[76, 77]

Because of the physical constraints associated with receiving sensory inputs (namely the fact that a human can't be everywhere and see, smell, touch, taste, and hear everything all the time), it's far easier for the brain to produce imaginary patterns than it is for the brain to physically verify them by cross-referencing them with the body's sensory inputs. The former only requires imaginary or symbolically-gained knowledge (e.g. knowledge gained from activities like thinking or reading or looking at a computer screen), whereas the latter requires experiential knowledge (e.g. knowledge gained by collecting physical signals through the body's sensory organs).

This is an important distinction to make because it means a strong majority of what humans think they know hasn't been physically verified or cross-referenced by their own sensory organs. Additionally, this algorithm is fully automated and subconscious, so sapiens are often not even aware of the fact they do it. In other words, most of what sapiens "know" about the world is derived completely from symbolism and the imagination, not from what they can directly physically experience through their senses, and they commonly don't realize there's even a difference between these two different types of knowledge. Most people get their information about the world from interpretations presented on TV screens or computer screens (i.e. symbolic knowledge) rather than from actually experiencing the world first-hand (i.e. experiential knowledge). Note how the term "first-hand" implies that information gained or learned directly must directly come from physical sensory organs like hands.

So much of what sapiens experience is imaginary compared to what they can physically validate using sensory inputs, that they simply forget the fact that much of what they believe is physically unverified. This would explain why it's so easy for sapiens to be psychologically manipulated. It's easy to lose sight of the difference between objective reality and abstract reality without devoting a non-trivial amount of brain power towards understanding metacognition. It should therefore come as no surprise that sapient brains produce a lot of false positive beliefs about objective reality which go completely undetected.

The way sapiens think appears to be a highly effective survival tactic. Humans may be trapped behind a cage of symbolism from which they can't escape, but the disadvantages of being stuck behind a prefrontal cortex where no conscious distinction is made between real (a.k.a. physical) and imaginary (i.e. non-physical) things is offset by some major advantages of abstract thinking.

One advantage of abstract thinking is that it helps people with advanced pattern finding and detecting threats. As an example, consider a human walking through the woods. If the eyes detect some sticks, the sapient brain can utilize its advanced pattern finding and abstract thinking skills to produce an imaginary image of a snake. Because the imaginary image of a snake can be cross-referenced to a visual input of sticks gained from physical reality, the brain can produce a false correlation about "realness," as illustrated in Figure 37. Unaware of the fact that the snake isn't real, the brain's host will take quick and decisive maneuvers to avoid what could be a serious threat.

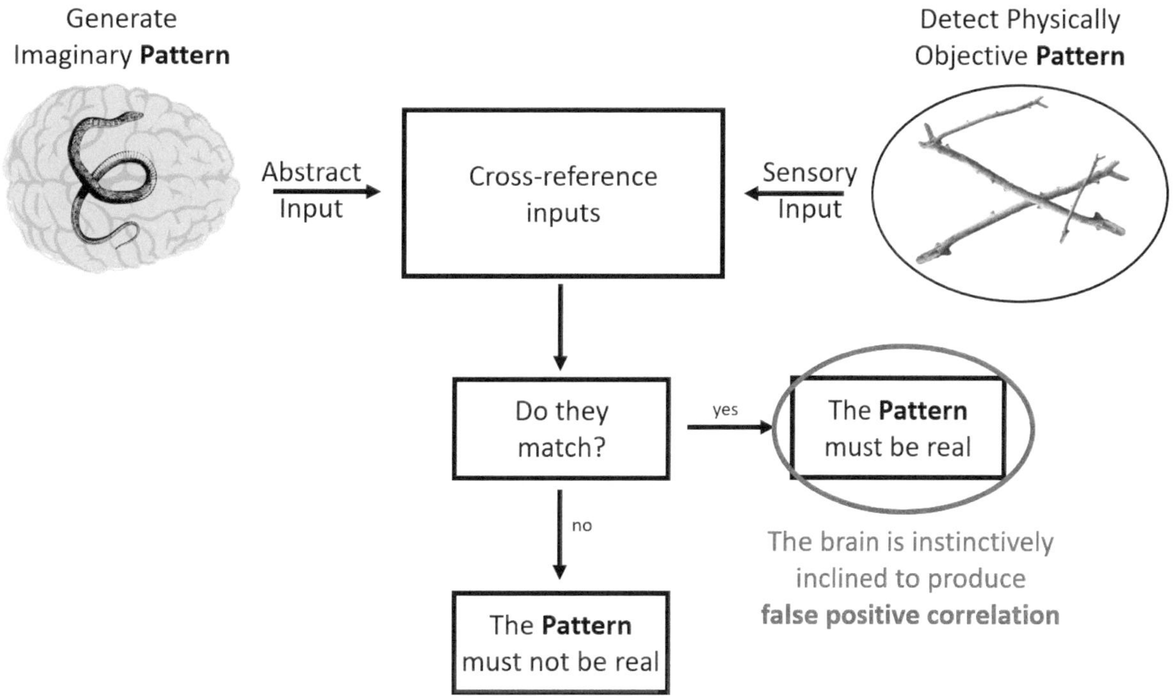

Figure 37: False Positive Correlation Produced by the Brain's Realness-Verification Algorithm [76, 78, 79]

This example illustrates a scenario where the brain's realness-verification algorithm produced a false positive correlation. It turns out, false positive correlations about the realness of imaginary patterns are useful in a dangerous world filled with predators because it results in a tendency to err on the side of caution. It's better for survival to produce false positive beliefs about the realness of a threat than to fail to detect a true threat. Because of this instinctive programming, even when sapiens are metacognitively aware of the fact that they can't verify what they are experiencing is something physically real, they will still weigh it highly or even more important than what their sensory inputs can physically verify. [68]

Likewise, even when sapiens are metacognitively aware of the fact that the source of the sensory input isn't the same as the abstract input, they will still falsely correlate it. This is one of the reasons why it's so easy to scare people using fictional stories like horror movies. Even when we consciously know that the threat isn't real because it's on a movie screen, we can still feel afraid about what we witness for days after watching a scary movie. This phenomenon also explains why large populations are quick to believe that high-ranking people (e.g. monarchs, presidents, etc.) are powerful people. We falsely correlate the abstract or symbolic power exercised by a king to the real or physical power exercised by the king's army (much more on this in the following sections). The bottom line is, it's simply more efficient and easier on the brain to assume imaginary things are real because it saves a substantial amount of energy-intensive thinking power, and because it results in false positive correlations which are beneficial for survival. [68]

Recalling the lesson on survivorship bias from Chapter 4, it is reasonable to believe there were plenty of early brains which weren't as instinctively inclined to favor abstract inputs over sensory inputs or quick to produce false positive beliefs about the realness of imaginary patterns. However, their resulting lack of producing false positive beliefs would have made their hosts more complacent, less cautious, and less responsive to genuine threats. In other words, they were less likely to survive. Therefore, the reason why humans constantly struggle to distinguish between imaginary things versus real things could simply be explained by natural selection. The tendency to believe in imaginary things leads to a higher probability of surviving real threats and consequently passing on one's genes. Fast forward over hundreds of thousands of years, and here we are today, routinely making false positive correlations between real and imaginary things, because that's what survives.

Sapiens have so much thinking power and are so inclined to believe imaginary things are real, that they behave unlike any other animal on Earth. They strongly and passionately believe in things they have never seen, smelled, heard, tasted, or touched. They act, react, and show extreme favoritism towards symbols, and operate either oblivious to or consciously unconcerned with the difference between abstract things and physical things. They will respond to stimuli which exist nowhere except within their imagination, more often and far more passionately than information received by their sensory inputs from physically objective reality. They will ignore their experiential knowledge altogether and act strictly according to symbolic knowledge, often not even aware of the fact that they're doing it. They will stop paying attention to the fact that they're sitting down and doing practically nothing, because they are entranced in an abstract world spoken or written into existence by a stranger using a complex written language (if you are reading this, look around for a second and realize that you're just sitting still, doing nothing but staring at symbols). Sapiens, unlike any other animal, can and will subject themselves to a great deal of struggle, suffering, and personal sacrifice for reasons which don't exist anywhere except within their collective imaginations. They will adopt belief systems and participate in population-scale consensual hallucinations, grateful for the opportunity to labor their entire lives over imaginary things, and even to die for them. This is a defining feature of the human experience.

4.3.2 Proof-of-Power is Proof-of-Real

"...pinch me, because I must be dreaming."
Origin Unknown

We have now established that sapient neocortices are so effortlessly skilled at abstract thinking that sapiens struggle to know what real is and, thanks to natural selection, they constantly make false positive correlations between imaginary things and real things. Because of this, humans need workarounds or protocols they can use to help them distinguish between imaginary things and real things. For situations where sapiens struggle to distinguish between real and imaginary and desire to know if what they are experiencing is real, many subscribe to an adaptation of the realness-verification protocol illustrated in Figure 38. Whenever sapiens wonder or doubt if an object is physically real or not, the commonly accepted protocol is to attempt to manually generate the physical sensory inputs needed to cross reference the brain's abstract input with a physical sensory input. A common way of doing this is by poking/pinching something to generate haptic feedback for touch sensory organs.

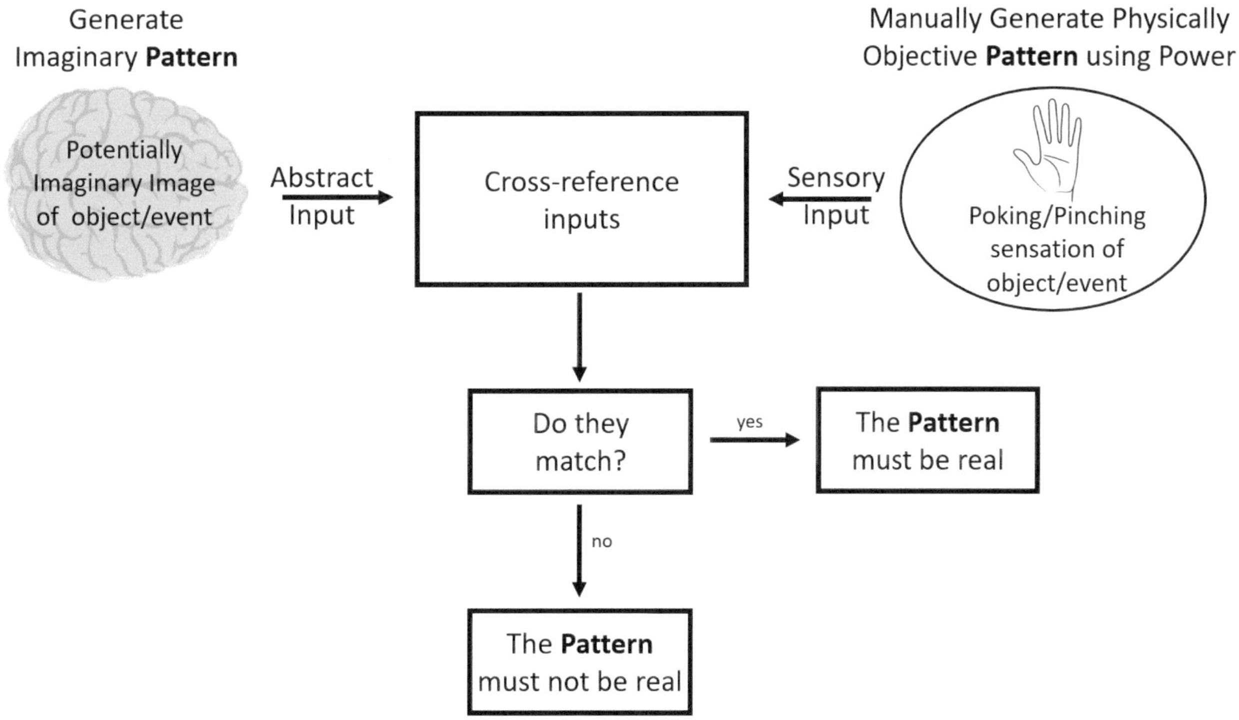

Figure 38: The Poking/Pinching Realness-Verification Protocol
[76, 77]

It's a simple and effective protocol. Not sure if an object is real? Try poking it. Not sure if an experience is real? Try pinching yourself. Pinching is especially useful during special occasions when the neocortex is busy producing very convincing imaginary patterns, but the host is deprived of the physical sensory inputs from objective physical reality it needs to cross-reference inputs and attempt to validate realness. This often happens when a human is sleeping, or when operating in a dream-like environment like cyberspace (hence why haptic feedback systems are popular, such as vibrating video game controllers).

Note how the act of poking or pinching something involves the application of force to displace mass. If the object is real, then a human knows from experience (and from reading Newton) that an equal and opposite force will displace the mass of their hand. Likewise, if the event is real, then the force used to pinch the skin will cause neuroreceptors to detect the displacement. What the brain is doing in these situations is manually generating the sensory inputs needed to cross-reference its abstract thoughts with physical sensory inputs to help make it easier to decide if an object or event is real.

Here the reader should note how this protocol requires physical power to work. What's another name for the displacement of mass with force? *Power*. Therefore, what's another name for the act of using power to *prove* that something is real? *Proof-of-power*. When a human pokes or pinches something, they apply a force to a mass over time and displace it across space. In other words, they're projecting physical power.

Why is projecting power in this way useful? Because power is comprised of the concrete phenomenon of objective physical reality: energy, mass, time, and space. It is impossible to (kinetically) project power without these phenomena, so a proof-of-power protocol like poking or pinching doubles as a one-stop objective-reality-verification-shop for the sapient brain. If a human can generate proof-of-power, then they can take comfort in knowing that the concrete phenomena of shared objective reality (energy, mass, time, space) are present and accounted for within the context of what they're experiencing, helping their mind reach a quicker conclusion about what's real.

The purpose of this section is to illustrate yet another reason why physical power is useful, and to demonstrate yet another application for the proof-of-power protocol. **Not only is proof-of-power necessary for survival and helpful for wild animals seeking to establish dominance hierarchies, it's also essential for helping sapiens – creatures who are trapped behind their abstract thoughts – determine what is and isn't real**. When we factor in sapient metacognition and the processes our brains use to cross reference abstract thoughts with sensory inputs, we can see that proof-of-power doubles as proof-of-physical-reality. Without the ability to detect the presence of physical power using protocols like poking or pinching, it's much harder for people to be confident that what they're experiencing is real or not.

Trapped in an inescapable and imaginary world behind their eyes and continuously spammed by abstract thoughts produced by an overpowered, oversized, and overclocked neocortex, proof-of-power serves as a reliable signal for the human brain to identify what is physically and objectively real in an otherwise imaginary world. Physical power is like the north star for our brains, it helps us navigate across an ocean of imaginary thoughts to get to what's "true." Without being able to manually generate a physical power signal for the purpose of cross-referencing abstract thoughts with sensory inputs, a human cannot verify if what they see and hear physically exists, or if it is just another one of a countless series of abstract beliefs produced by their hyperactive imagination. Imprisoned by our own abstract thinking, physical power is a lifeline. We almost desperately rely on physical power signals to know what's real.

What does this have to do with Bitcoin? The bottom-line up front is that cyberspace is an imaginary reality (virtual reality is, by definition, not physical reality). Cyberspace is nothing but abstract, symbolic meaning applied to the combined state space of all the state mechanisms connected to each other via internet. Operating online is akin to being in a dream state; it's a shared abstract medium through which people in physical reality communicate. But until the invention of proof-of-work protocols like Bitcoin, cyberspace was missing something: an open-source, decentralized proof-of-power protocol. People who operate in, from, and through cyberspace have no way to cross-reference the imaginary things they experience online with a genuine proof-of-power signal that they can use to validate if something is real or not. As will be discussed in the next chapter, Bitcoin appears to fix this by creating a proof-of-power signal to serve as proof-of-real signal.

4.4 Evolution of Abstract Thinking

"Symbolism begets symbolism... once you're locked into this synergy,
it's hard to stop it from building a whole new world."
Charles Foster [68]

4.4.1 Applications of Abstract Thinking

In comparison to the timescale of evolution, human abstract thinking became very advanced very quickly. Humans started using their massive neocortices for advanced pattern-finding to enhance their hunting and gathering activities. The ability to detect and exploit patterns is very useful for learning the behavior of surrounding fauna and flora. Sapiens also benefited from the ability to produce false positive patterns because it made them exercise more caution and increased the probability of detecting true threats. Another benefit to be gained from bi-directional abstract thinking is that it allows the imagination to influence the processing of the brain's sensory inputs, leading to the phenomena called symbolism. Thanks to symbolism, humans can produce and detect meaning from otherwise objectively meaningless sensory input data. [68]

The author has just described three of ten applications of abstract thinking mastered by sapiens over the past 50,000 years: (1) advanced pattern finding, (2) exercising caution, and (3) symbolism. A complete list of ten applications of abstract thinking discussed in this chapter is listed in Table 1 below.

Table 1: Applications of Abstract Thinking

Application	Description
Advanced Pattern Finding	Patching up experiential knowledge gaps about shared objective physical reality using abstract thoughts and hypotheses.
Exercising Caution	Producing false positive imaginary beliefs about threats to improve probability of survival.
Symbolism	Assigning imaginary, abstract meaning to patterns observed in shared objective reality.
Planning & Strategizing	Producing abstract simulations to forecast future scenarios within the safety & comfort of one's own mind.
Semantically & Syntactically Complex Language	Developing semantically & syntactically complex languages as a medium to exchange symbolically meaningful, conceptually dense, mathematically discrete & precise information.
Storytelling	Leveraging abstract thinking, imagination, & high-order language to construct virtual realities to share with other humans for enhanced relationship building, knowledge sharing, entertainment, & vicarious experience.
Solving physically Unverifiable Mysteries	Explaining phenomenon impossible to objectively know or understand via physical sensory inputs, most notably what happens after death.
Developing Abstract Constructs & Belief Systems	Constructing theological, philosophical, and ideological constructs to explain, justify, or shape sapient norms of behavior.
Creating & Wielding Abstract Power	Creating abstract power & control authority over sapiens as a non-energy-intensive & non-injurious surrogate to using physical power as the basis for settling disputes, establishing control authority over resources, and achieving consensus on the legitimate state of ownership and chain of custody of property.
Encoding Abstract Power Hierarchies	Formal codifying (via spoken/written language) belief systems where people with imaginary or reified abstract power are organized into hierarchies.

4.4.2 Abstract Thinking Application #4: Planning & Strategizing

Sapiens learned how to use their imaginations to improve their hunting capabilities by performing a special type of abstract thinking called *planning*. Driven by the need to develop better hunting strategies to take down far more physically powerful pack animal species with high C_A and low BCR_A, sapiens started using their overclocked and oversized brains to do something their prey couldn't: build hypothetical hunting simulations and test out strategies from the safety of their own minds.

Using imagination, humans learned how to render highly complex scenarios about the future that were not coupled to their sensory inputs or directly linked to any concrete experiential knowledge within shared objective physical reality. They could run multiple hunting simulations which never occurred and fill them with people, animals, objects, and environmental conditions that didn't physically exist. They could down-select their best hunting strategies, test them in real-world operational environments, and use the experiential knowledge as feedback to train their mental models and make more realistic hunting simulations. Perhaps even more impressively, sapiens learned how to communicate and share these hunting simulations with their peers, creating shared abstract realities (a.k.a. virtual realities) filled with hunters and prey which don't exist anywhere except within their collective imaginations.

This is the process through which sapiens mastered the primordial economic dynamics of hunting and became the world's peak predator. The primordial economic dynamics of hunting is illustrated in Figure 39. Thanks to their big brains, humans were able to develop strategies which enabled them to increase the BCR_A of their targets by reducing their C_A. Gradually, over time, humans learned that **the hunter's job is not merely to kill a target, but to prep the target for the kill over time by reducing its C_A**. Over thousands of years of planning, strategizing, and testing imaginary hunting scenarios through real-world trial-and-error, humans mastered the art prepping their target for the kill by reducing their C_A using tactics like increasing standoff distance or taking advantage of the topography of the local environment.

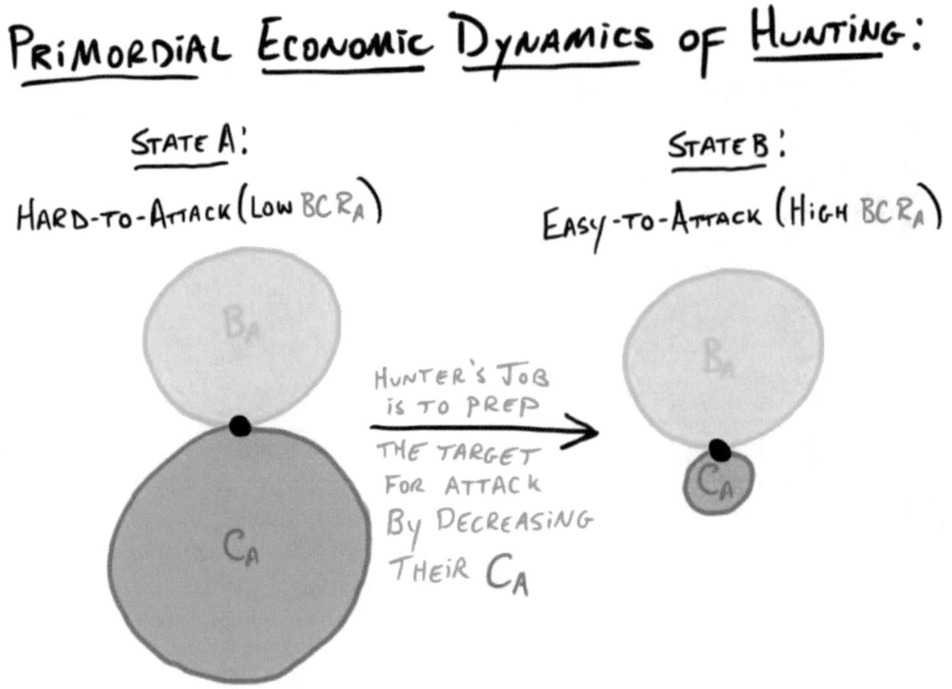

Figure 39: Primordial Economic Dynamics of Hunting

For example, instead of trying to take on a powerful herd of caribou in an open field like a pride of lions would try to take on a herd of wildebeest, humans took advantage of their advanced pattern-finding capabilities to study the migration behavior of caribou, predict what paths they would take, and find the ideal spot along that path where their C_A is lowest, to ambush them. This tactic is illustrated in Figure 40.

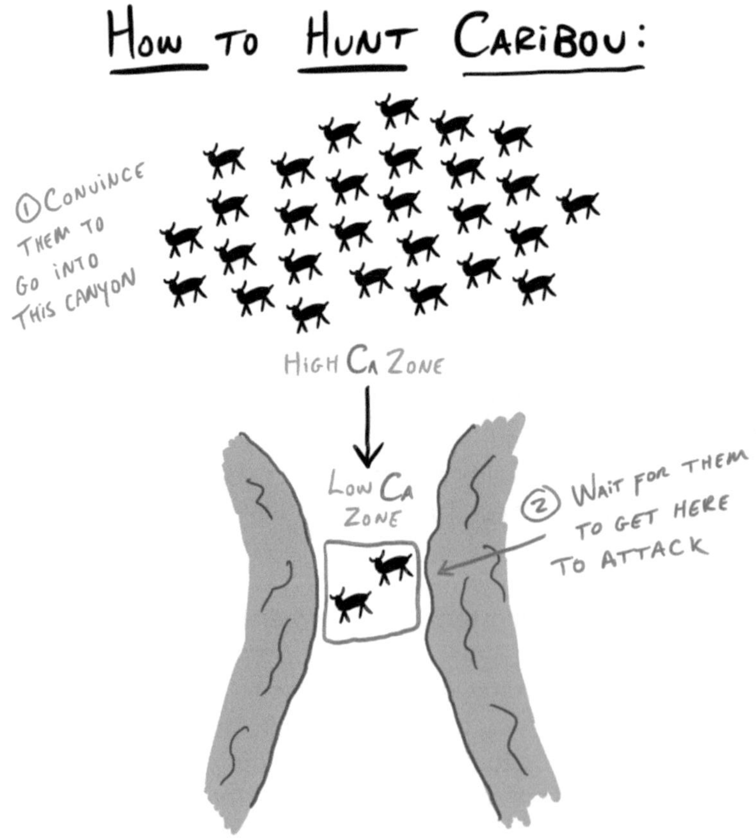

Figure 40: Illustration of a Hunting Strategy Mastered by Humans

A herd of caribou is quite powerful and can impose a lot of physical cost on attackers in an open field (this area is marked in Figure 40 as the "high C_A zone"). But caribou can't impose severe physical costs on attackers as effectively when they're wedged in the bottom of a canyon with spears raining down on them (the area is marked in Figure 40 as the "low C_A zone"). Thus, an effective hunting strategy for increasing the BCR_A of a herd of caribou is to predict their migration patterns and find a spot where their C_A is the lowest due to the topography of the environment, then simply wait for them to get there. To that end, humans can park themselves at the top of a canyon and wait for the caribou to arrive. Once they do, they can throw spears down on them from a safe standoff distance where their capacity to impose severe physical cost on their attackers is rendered useless or ineffective.

While there are many peak predators intelligent enough to employ similar hunting capabilities, no surviving animal appears to be as skilled at hunting as humans are. Hunting strategies like this require lots of abstract thinking because humans must use their imaginations to predict their prey's behaviors and intentions. This type of abstract thinking is what's known as *intentionality* or *theory of mind*, and sapiens are extraordinarily good at it. In fact, one of the most defining characteristics of behaviorally modern sapiens is that they have a remarkably high order of intentionality and theory of mind. Simply put, modern sapiens can predict and understand the thoughts and intentions of *other* brains remarkably well. The more thoughts and intentions one can predict, the higher order of intentionality they are said to have, but the more cognitively demanding it is to make those predictions. Fortunately, because sapiens evolved much better thought processing hardware, they have become much better at predicting the mental states of other creatures and therefore have much higher-orders of intentionality and much higher degrees of theory of mind.

Another way to think about this is that human brains are so large and powerful that they have enough spare thought processing margin leftover to dedicate towards thinking *as* their neighbors. Not surprisingly, this makes the hosts of sapient brains exceptionally talented at hunting despite how weak their bodies are in comparison to the animals they frequently hunt. Sapiens have driven dozens of gigantic species to extinction not by projecting more physical power than them (although the ability to wield fire clearly helped), but by out-thinking them.

A simple way to think about the strategic advantages of high-order intentionality is that it makes sapiens way better mind readers than any other species on Earth. **Because they're better mind readers, that makes sapiens better hunters. What sapiens lack in physical strength, they make up for ten times over with intelligence – specifically their ability to predict what neighboring organisms are going to think and do long before they think and do them**. Animals often find themselves helpless against humans despite being far more physically powerful. This is because fighting a human is like fighting a jedi – humans are extremely talented at predicting what's going to happen before it happens and placing themselves in exactly the right position they need to be in to defeat their opponent.

In a very short amount of time on Earth, sapiens have proven that advanced pattern-finding combined with planning and strategizing using higher-order intentionality and theory of mind is an extraordinarily effective power projection tactic. Between their ability to wield fire and their ability to effectively read their prey's minds by thinking like their prey, sapiens quickly became one of the world's most lethal predators despite being comparatively unathletic.

4.4.3 Abstract Thinking Application #5: Semantically & Syntactically Complex Language

As sapiens further honed their abstract thinking skills, they graduated from imagining highly realistic hunting simulations to imagining all sorts of other abstract realities. Their overclocked and hyperactive neocortices started to assign abstract meaning to recurring patterns in their campfires and in the stars above their heads. They started using abstract thinking not just to connect dots and fill gaps in experiential knowledge, but also to account for phenomenon they can't physically verify using sensory inputs (most notably what happens after death). [68]

Perhaps the most disruptive application of abstract thinking to emerge after planning and strategizing was semantically and syntactically complex, high-order language. High-order language enabled humans to share their individually-imagined, abstract realities together much more easily. This allowed them to synchronize their imaginary thoughts together and create large-scale shared abstract realities.

Before discussing high-order language, it's important for the reader to note that sapiens were well-equipped with rudimentary forms of language long before they learned how to use symbolism to produce semantically and syntactically complex language. Humans communicate the same way other mammals do, in ways that far predate words and grammar, and to many people's surprise these instinctive communication protocols foster far deeper connections and bonding. [68]

Humans groom each other, dance, laugh, sing, and howl with each other. They adjust their size and posture, make gestures with their appendages, beat their chests, stomp their feet, squint their faces, and use many other communication techniques that other mammals use. It is not uncommon to see animals of different classes (e.g. mammals and birds) communicate friendship or bonding exactly the same way via cuddling or grooming, or with very similar body language. This is why humans communicate their trust and affection for each other (and for their pets) by cuddling or petting them – this is simply a common language between many species, especially mammals. [68]

On the opposite side of the communication spectrum is how animals signal pain, sadness, and suffering the same way (this topic is uncomfortable but relevant to discussions about human-on-human killing and why humans adopt belief systems to try to avoid killing). Mammals, for example, narrow their eyes and grimace the same, wince the same, and scream and whimper the same. This isn't a coincidence – it's a communication protocol. Biologists have argued that this instinctive form of communication (combined with having enough high-order intentionality and theory of mind to empathize) is why humans often struggle to kill other mammals who are equipped with eyelids and vocal cords and who use them to communicate pain and suffering the same way humans do. Yet, the same human who would never want to hurt a mammal will have no problem impaling, disemboweling, crushing, burning, electrocuting, or boiling fish, shellfish, or insects alive. One explanation for why this happens is because these different classes of animals have evolved different communication protocols and, unfortunately for bugs, fish, and shellfish, they don't communicate pain, sadness, and suffering the same way humans do, so humans don't get the message. [68]

As illustrated in Figure 41, if fish had eyelids or contort-able muscles in their faces to communicate pain the same way mammals do, the experience of fishing would probably be much different for sapiens than it is today (imagine what fishing would be like if fish screamed & howled). This same phenomenon is believed to be the reason why domesticated wolves (a.k.a. dogs) have more prominent unpigmented sclera (i.e. show more of the white portion of their eyes) and special muscles around their eyes capable of making sad expressions which appeal to their human masters. So-called "puppy dog eyes" are argued to be an evolutionary phenomenon which helped dogs communicate to humans in ways that we're more intuitively capable of understanding – particularly their sadness, pain, suffering, and desires. Some argue that the "cuteness" of dog eyes is a defense mechanism against the cruelty of their masters, and one of several reasons why the kinder-looking, sadder-looking, or more pitiful-looking dogs (expressions that humans might condescendingly call "adorable") have survived as human pets. [68, 80]

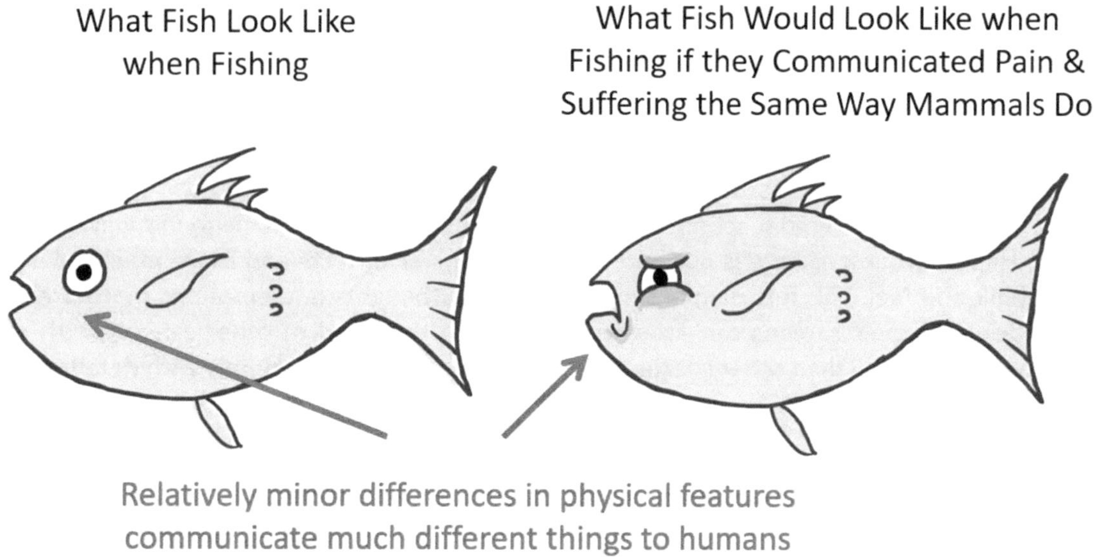

Figure 41: Example of a Fish Communication Protocol

Both anthropologists and biologists have theorized that these different methods of communication between animal classes explain the double-standard in predatory human behavior. Humans are simply less capable of understanding that fish and insects are experiencing pain and suffering. The study of fish and insect neurons have shown they are capable of nociception (detection of pain) and that they react the same way mammals do when they encounter extreme heat, cold, or other harmful stimuli. But these organisms lack the eyelids, vocal cords, and other musculature needed to grimace and howl to communicate their pain and suffering the same way most mammals do, so humans are naturally less reserved about harming them than they are about harming mammals which communicate pain and suffering the same way humans do. This small change in how different classes of animal evolved to communicate differently leads to dramatically different emergent behavior. [68]

In addition to being able to communicate happiness or suffering using common languages between mammals, sapiens also have a base-layer common language specific to the species. For example, all sapiens smile, laugh, and sing exactly the same way regardless of where they are born. Human children born deaf and blind, who have never seen or heard other people smile or laugh, will smile and laugh the exact same way other humans do. Sapiens also instinctively communicate with others who have immature or underdeveloped brains the same way, using a sing-song pattern of wider-spectrum and varying pitch known as infant-directed speech or *baby talk*. This is why humans "baby talk" to their babies, pets, elderly, or mentally handicap people regardless of their culture – humans instinctively recognize their having underdeveloped brains and have a common base-layer communication protocol for it. This also explains why people feel comforted when they hear other people baby talk; it's an instinctive communication protocol that connects humans on a far deeper level than modern spoken and written languages. [68]

These communication techniques are purely instinctive. They're natural patterns of behavior passed on via genetics. This means sapiens actually communicate the same way (and more meaningfully) regardless of what higher-order languages they use. By singing, dancing, laughing, grooming, and baby talking with each other, it is possible to form deep, emotionally satisfying connections with strangers or people who can't speak or write the same higher-order language (hence the romantic stories about strangers from different worlds who speak different languages but still find deep emotional connections with each other).

For these reasons, it's not uncommon for people to be more emotionally attached and affectionate with their pets than with their friends and extended family members. The loss of a pet can be as emotionally traumatic as the loss of a family member because animals (mammals especially) bond the same way sapiens do without the need for symbolism, semantics, syntactics, and higher-order language. The point is that spoken and written languages are not nearly as important for human-on-human communication and bonding as people think they are. Humans communicate the majority of their feelings using body language and in other unconscious techniques, and then rely on semantically and syntactically complex spoken or written languages to fill in the symbolically complex or mathematically discrete portions of their thoughts. [68]

The main takeaway of this exploration of human communication is to illustrate that the primary value-delivered function of higher-order spoken and written language is not to emotionally connect with each other, because there are far more effective ways to do that than by using semantically and syntactically complex language. Instead, the primary value-delivered function of higher-order language appears to be to *formally encode* thoughts and emotions. **Higher-order language is optimized for the transfer of facts and more precise descriptions of what people think and feel. This lets people synchronize their thoughts and emotions more precisely. Using spoken or written language, sapiens can effectively peer into the mind of other sapiens with much more precision and symbolic detail than other species can, thus filling each other's minds with detailed descriptions of their thoughts and ideas**. [68]

Large and overpowered neocortices, it turns out, are perfect vessels for the transfer of symbolically complex and mathematically precise thoughts. Large neocortices enable sapiens to construct higher-order languages by transmitting, receiving, and processing complex audible and visual patterns and converting them into abstract thoughts rich with symbolic meaning syntactic precision. Once sapient brains became effortlessly gifted at symbolism, people quickly developed a habit of assigning symbolic meaning to commonly-detected audible and visual patterns to create morphemes, the smallest meaningful lexical items used in higher-order language. Then, over time, this process of assigning symbolic meaning to otherwise meaningless audible and visual patterns became increasingly more complex, until the nature of the meaning assigned to various lexical items could be changed based off their composition, in a process we call *semantics*. Eventually, humans learned how to combine various symbols with different semantic meaning into structured units of discourse called words, then further combined them into mathematically discrete structures like phrases, sentences, and grammar in a process we call *syntactics*.

To this day, practically all high-order languages (including and especially the languages we use to program our computers) are composed of these same two components. Shapes and sounds are imbued with abstract and symbolic meaning (semantics) and then combined into some type of mathematically discrete and formal structure (syntactics), giving rise to the semantically and syntactically complex languages we use to convey all of our symbolically and structurally complex thoughts and ideas. We know this type of communication is not necessary for survival or bonding because humans have spent most of their existence surviving and bonding without it. Nevertheless, high-order language is extremely useful as a medium of exchange for conceptually dense and mathematically discrete information, through which people can collectively share their individual thoughts and ideas. Consequently, high-order language makes it possible to create a single shared consciousness out of individual minds.

As will be discussed in the following chapter, **the pinnacle of humans using symbolism to assign semantically and syntactically complex meaning to physical phenomena in order to connect their neocortices together to form a single shared consciousness is machine code. Using techniques like Boolean logic, sapiens have learned how to convert practically any physical state-changing phenomenon into symbolically and structurally complex information that can be stored on machines and transferred to people through their machines**. Machine code has very quickly emerged as the crown jewel of higher-order semantically and syntactically complex language. This remarkable application of abstract thinking and high-order communication is more commonly known as the internet.

A clear limitation of higher-order language is that it only works for people who have taken the time to memorize the semantic meaning and syntactic structure of the communication protocol. Whereas it takes no effort to learn how to communicate with strangers by singing, laughing, and baby talking, higher-order languages take years, even decades, to learn. Sapiens can and often go their entire lives with nothing more than a shallow appreciation of the semantic and syntactic depth of the language protocols they learn. This is perhaps because higher-order language is merely a means to an end, not an end in and of itself. Most sapiens don't learn higher-order language for the sake of knowing the semantic and syntactic details of a language, they learn it for the sake of peering into the minds of other sapiens and connecting abstract realities together through a process called storytelling.

4.4.4 Abstract Thinking Application #6: Storytelling

"The pen is mightier than the sword."
Edward Bulwer-Lytton [81]

By assigning symbolic meaning to random shapes and sounds, then using syntactics to assemble them together into a common topological or audible structure, humans can exchange information with high levels of precision to communicate their abstract ideas. This capability allows sapiens to do something remarkable: connect their imaginations together to form a single, shared abstract reality. This concept is illustrated in Figure 42.

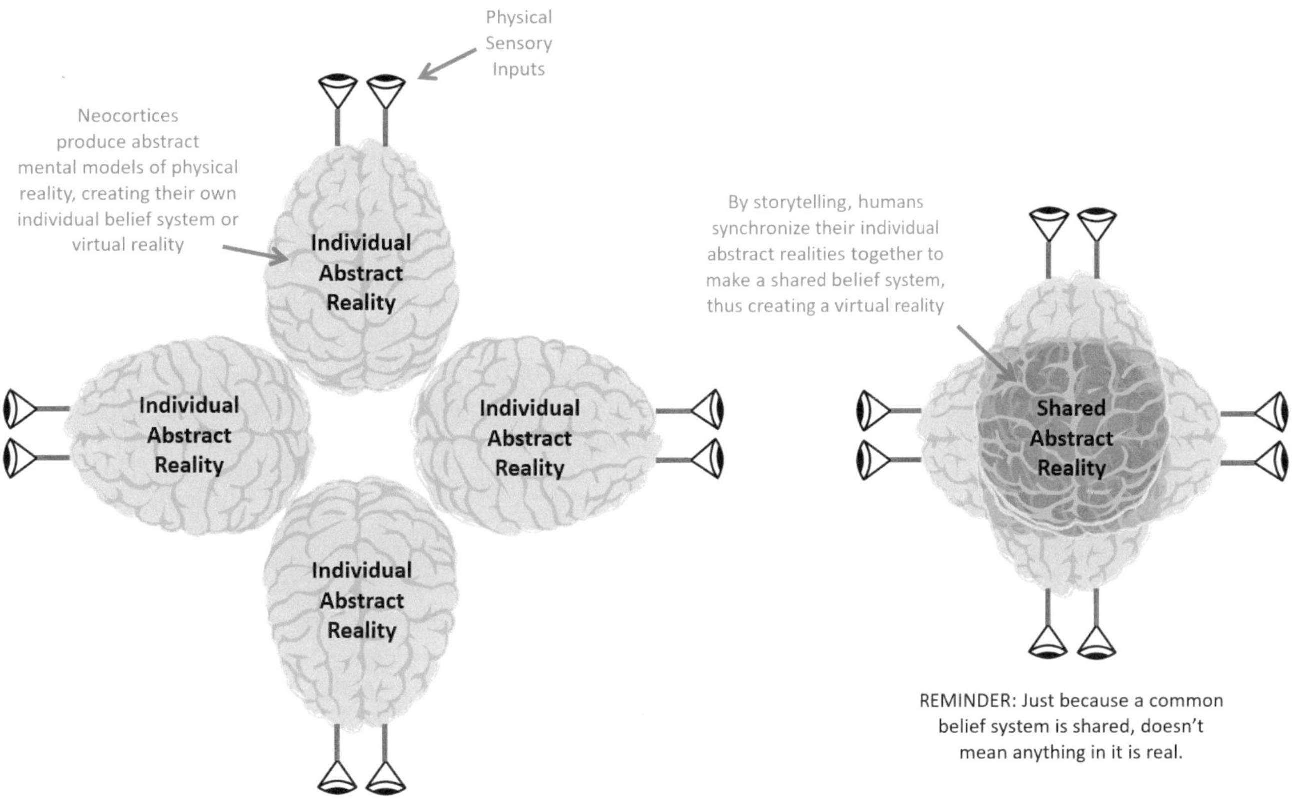

Figure 42: Illustration of the Metacognitive Impact of Storytelling (forming Shared Abstract Reality) [76]

Figure 42 shows how people create and develop common belief systems. Thanks to the invention of higher-order languages, two or more neocortices with no physical or neural connection can share the exact same symbolic experiences and knowledge, allowing them to think the same things, detect the same patterns, and even interpret the same meaning from mutually exclusive physical sensory inputs (two brains can't share the same physical sensory input, but they can interpret different sensory inputs the same way). Through higher-order language, sapiens build shared abstract scenarios within their minds and fill them with emotionally complex and interesting people with multiple layers of mental states, perceptions, beliefs, desires, high-order intentionality, and high degree of theory of mind. Non-physically-existent people can be put into places which don't exist, surrounded by creatures and objects which aren't real. Moreover, their actions, motivations, and thought processes can be imbued with conceptually dense, symbolic meaning.

From the safety and comfort of their minds, sapiens who know the same higher-order language can experience and explore their imaginary worlds together. They can teach each other useful lessons, explain unsolved mysteries, offer profound insights, and steer sapient brains through quests, crucibles, and puzzles which allow their hosts to experience the full spectrum of emotion. Using a written form of higher-order language, neocortices can connect together in this way not just across thousands of miles of space, but also across thousands of years of time, communicating asynchronously over timescales which far exceed their hosts' lifespans. Through written language, humans can even share the same abstract experiences with other humans who lived thousands of years before them. The author defines this extraordinary capability as *storytelling*.

For the same reason that sapient brains are extremely gifted at planning and hunting, they're also extremely gifted at storytelling. There is a clear, cognitive link between storytelling and hunting; both activities rely on precisely the same set of cognitive tools (i.e. high-order intentionality and theory of mind) with only minor differences in their application. This becomes an extremely important concept to note for future discussions about hunting people *through* their belief systems by nefarious story telling. There are solid technical and metacognitive grounds for sapiens to have an instinctive fear of, or distrust in, or attraction to good rhetoricians (e.g. politicians, religious leaders). To put it simply, peak storytellers are peak predators. They deserve caution.

It has been hypothesized that sapiens' unique ability to think abstractly, form symbols, develop higher-order languages, and connect brains via storytelling is what enabled them to start cooperating at far higher scales than other packs of primates. A famous anthropological theory by Robin Dunbar states there is a cognitive limit to how many physical features (i.e. patterns) a human brain can memorize. This would imply there's a cognitive limit to how many different faces a single brain can recognize and trust enough to cooperate with. This notional limit, often called *Dunbar's Number*, is estimated to be approximately 150 people. Dunbar's theory offers a viable explanation for why human tribes did not exceed this number for many hundreds of thousands of years prior to the emergence of abstract thought. Despite their big brains and all their excess thinking power, humans simply didn't have the memory to form meaningful relationships with more than 150 people, so they weren't inclined to cooperate at higher scales. [82]

Like the single-celled organisms discussed in the previous chapter, the inability to cooperate at higher scales inhibits humans' ability to grow their $\mathbf{C_A}$ and results in a bounded prosperity trap. Despite their ability to control fire, small nomadic tribes of humans had relatively little advantage. They were not nearly as high on the food chain as modern humans are today. Sapiens were the only species of human to escape the bounded prosperity trap caused by Dunbar's number-induced cooperation limits, and it appears they did it in part by combining abstract thinking, theoretical planning, symbolism, higher-order language, and storytelling together to create shared abstract realities and belief systems which enabled them to trust each other and cooperate at scales which far exceed the physical constraints causing Dunbar's number.

Another way to think about storytelling is that it allows sapiens to mentally "groom" and bond with each other via stories, rather than having to physically groom each other or even spend any time with each other (hence the deep emotional attachments people form with movie celebrities or other types of story characters). Mutually shared beliefs or stories enable sapiens to transcend the physical constraints of regular grooming to form trusting and cooperative relationships at much larger scale. For example, a single primate is capable of physically grooming one, maybe two other primates at the same time to earn their trust and cooperation. Meanwhile, sapiens can capture the attention and influence the behavior of thousands of other sapiens simultaneously using nothing more than a story to earn their attention, trust, and willingness to cooperate. That same storyteller can get those 1000 people to trust and cooperate with each other by virtue of the fact that they now share the same abstract reality or belief system. Sapiens are unique in comparison to other wild animals because they cooperate together simply because they believe in the same thing, not because they share the same physical experiences. This subtle but extraordinary difference makes sapiens much more capable of mass cooperation at scale. Why? For the simple reason that **the ability to believe in the same thing is not constrained by physics, whereas the ability to share the same physically objective experiences together is highly constrained by physics**.

Storytelling can therefore be thought of as the glue which holds modern societies together. Without that glue, sapiens are both physically and physiologically incapable of cooperating together at levels exceeding small tribes. We literally don't have the time, energy, or memory capacity for it. This means a primary value-delivered function of storytelling is to overcome the constraints of shared objective physical reality. **We use our spoken and written stories to bypass the constraints of physics as well as the constraints of our own bodies to communicate with each other, inspire each other, cooperate with each other, and achieve things we would otherwise be both physically and psychologically incapable of achieving**.

If we didn't adopt common symbols, languages, stories, and belief systems and instead relied exclusively on experiential knowledge and memory to cooperate with each other, we would need to have massive budgets of time and memory to develop the experiential knowledge needed to create connections and commit them to memory. We would need comedically large brains, very large appetites, and very long lifespans to achieve a fraction of the cooperation we can achieve by simply using stories. Storytelling functions as a lifehack that sapiens unlocked by learning abstract thinking. Thanks to stories, people don't have to rely on interactions with other people to cooperate at scales that are up to forty-million times larger than Dunbar's number; they just need to hear the same stories and adopt the same belief systems.

This explains why symbolism is extraordinarily effective at promoting higher levels of cooperation; it gets people to think the same way, believe the same things, and most importantly, signal to each other that they have the same belief systems to facilitate higher-order cooperation and trust. Humans can use their storytelling abilities to build shared abstract realities which offer satisfying explanations for phenomenon observed in shared objective reality. They can imbue objective physical phenomenon with abstract meaning to make mutual experiences feel more profound and enjoyable. Then, armed with the symbolic meaning of the stories they tell, sapiens can cooperate at scales which far exceed their own physical and physiological capacity.

In other words, storytelling is an abstract superpower. Hence why Edward Bulwer-Lytton observed that "the pen is mightier than the sword." A more technically accurate way to say the same phrase would be "higher-order syntactically and semantically complex written language can influence, organize, and direct more unified physical power than a single person swinging a sword can." **People can achieve things far beyond their physical limits using the right combination of stories. By simply believing in the same thing (regardless of whether that thing is real or imaginary), sapiens can sum their power together to achieve extraordinary things like assembling the international space station. Rooted at the center of all this achievement is something which doesn't even necessarily have to exist. The stories sapiens tell each other to cooperate as massive scale can be (and probably are) fictional. All that matters is that people believe in them. As long as people can stick together by believing in the same imaginary things, they can build the great pyramids of Giza – they can literally move mountains.**

Storytelling is how money and currencies work. Money is nothing more than a belief system – one of many completely fictional stories told by storytellers that people voluntarily choose to believe in. By believing in the same money (i.e. medium for transferring financial information), people can and will cooperate with each other at scales which far exceed their physical and physiological limits. On the flip side, when money breaks down, cooperation breaks down. If people stop believing in the same money, cooperation comes crashing down. A collapsing money is a collapsing society; it has ended several empires. Because money is a belief system, the manipulation of money (such as when manipulating the supply of money) is technically a form of passive-aggressive psychological abuse. People who distort the medium for transferring financial information are systemic predators who prey on people's belief system, eroding their ability to cooperate with each other and contributing to the collapse of society. The most successful moneys have been those which physically constrain this type of systemic exploitation (e.g. gold).

Adding all these concepts together, we get a comprehensive understanding for why symbols and stories are extremely important to the prosperity of sapiens. Symbols and stories breed common belief systems, and common belief systems have unrivaled success in the animal kingdom at achieving cooperation. By believing in the same stories told by storytellers, humans can cooperate at extraordinary scales which dwarf other species. They can use that power to build pyramids or to impose severe, physically prohibitive costs on attackers. **This manner of**

cooperating via storytelling is fundamentally a dual-use power projection tactic evolved by nature: a way to increase B_A, increase C_A, and adjust BCR_A. To project more physical power, simply sum it together by telling more compelling stories which motivate people to cooperate at a higher scale by adopting the same belief system. With the right combination of stories, humans can do things they would never otherwise be capable of doing. They can end old empires and build new ones simply by changing what they believe in.

A good storyteller can facilitate sapient cooperation at scales which far exceed Dunbar's number because they don't require people to physically experience the same thing, they just require people to hear and believe the same story. Add to this phenomenon the fact that sapiens are extremely intelligent hunters who can control fire and tap into what is effectively an infinite supply of exogenous fire power from the surrounding environment, and it's easy to see why the stories told by sapiens, not the sapiens themselves nor the technologies they wield, are extremely effective and asymmetrically valuable power projection tactics. **Effective storytelling is a skill that should never be underestimated; civilizations rise and fall as a direct result of their adopted symbols, stories, and belief systems – things which technically don't exist anywhere except within people's collective imaginations**.

The sociotechnical implications of storytelling are simple but profound. Sapiens did not physiologically change after arriving at behavioral modernity to become more capable of memorizing faces; we are equally as cognitively constrained as our ancestors were during the Upper Paleolithic era – in fact we are probably more constrained (human brains started to shrink after they started domesticating themselves). [68] The ability for sapiens to cooperate and function on scales up to nation state level is therefore derived predominantly from something abstract which resides exclusively within their imagination. That means things like national strategic security are heavily derived from stories more than probably anything else. To harness more physical power to increase C_A and lower BCR_A, tell persuasive stories. To capture more resources or be a good conqueror, be a good storyteller and hunt humans through their belief systems.

Here we begin to see the downside of storytelling. What humans gain in their ability to cooperate with each other, they lose in systemic security. A major downside to storytelling is that it makes it possible to hunt humans psychologically rather than physically. Recall from the previous section that a hunter's job is to decrease their target's C_A and increase their BCR_A. Storytelling enables people to do this to other people, often without attributability (i.e. no blood trail). With the right stories, people will forfeit their physical power or lay down their arms. Sapiens can be domesticated by the stories they believe, and like lambs, they will walk straight into slaughter. It's also possible to feed stories to human populations which divide them and make them less likely to cooperate. They can be convinced via theology, philosophy, and ideology to forfeit their physical strength for something which only exists only in their collective imagination.

The bottom line is that **the prosperity and survival of the sapient species depends upon the stories they choose to believe in. All of our combined achievements are irrevocably linked to our stories and combined belief systems. What a population chooses to believe has very real, very meaningful consequences on their ability to survive and prosper, so it is extraordinarily important for them to pick the "right" stories and belief systems**.

4.4.5 Abstract Thinking Application #7: Solving Physically Unverifiable Mysteries

Another useful application of storytelling is creating explanations for phenomenon that are physically impossible to verify via physical sensory inputs or experiential knowledge. The most common example of this is explaining what happens to sapiens after death. Upper Paleolithic sapiens came to believe in the same afterlife because of storytelling. The primary storytellers of this time were shamans who convinced their tribes the afterlife was a desirable place to be. Over countless years and countless campfires, shamans expanded on their stories, specifying details about gateways to the afterlife. They added more details about how they can access the gateway to the afterlife and even communicate with deceased ancestors on the other side. Here, the first signs of abstract power began to emerge.

An easy way to tell if human fossils are from a time after the emergence of abstract thinking and storytelling is to look for signs of a belief in the afterlife. Most animals show nothing more than casual interest in the bodies of their dead, but storytelling humans who lived within the past 50,000 years demonstrate substantial interest in the bodies of their dead. Signs of a collective belief in the afterlife appear in the human fossil record at approximately the same time as other early signs of abstract thinking and self-consciousness. Belief in the afterlife is therefore one of the oldest known human belief systems. [68]

Preparation for the afterlife is noteworthy because it demonstrates an understanding of oneself in relation to time, as well as indicates a time preference for future self (noting that the concept of "future" is, in itself, an abstract construct which humans struggle to understand) over current self. Sapiens not only started to make a conscious distinction between themselves and others within their environment, but they also started to make a conscious distinction between their current selves and future selves, particularly with respect to living self and unliving self.

Behaviorally modern Paleolithic sapiens started making careful preparations for their future selves via ceremonies like burying rituals. They indicated their time preference for the future by virtue of their sacrifices in the present. The dead would be buried with valuable resources the tribe needs for survival, a self-sacrificial practice which makes no rational sense except for those who believe in and have a higher preference for an imaginary future self, living in a place after death. Through Paleolithic burial practices, we can see a signature characteristic of behavioral modernity: making meaningful sacrifices for something completely imaginary – their future self.

4.4.6 Abstract Thinking Application #8: Developing Abstract Constructs and Belief Systems

*"The great thing about bugs is that nobody gives a s*** if you kill them."*
Kenny the Talking Gun, *High on Life* [83]

Upper Paleolithic shaman storytellers could answer questions about the afterlife which other members of the tribe couldn't even think to ask. These shamans could ostensibly communicate to tribal ancestors years after death – giving shamans very high social value. Shamans could also imbue tribal activities with symbolic meaning, making tribal activities seem more blessed and profound. Shamans who were particularly skilled at storytelling could persuade their peers to believe that a person's hunting and gathering actions represented much more to the tribe than just the act of physically capturing resources. It represented something important, or something even more novel: something ideologically "good." The emergence of the concept of "good" also meant sapiens could start engaging in activities that would qualify as being ideologically "bad," giving rise to the development of abstract constructs we now call *ethics* and *morals*.

As much as a sapiens' hubris might compel them to believe otherwise, there may not be anything objectively "good" or "bad" about anything we experience in shared objective physical reality. It is possible that our hyperactive brains are just assigning meaning to objectively meaningless things. But if we assume, for the sake of argument, that there is such a thing as objective "good," then there's still nothing to suggest that humans are special creatures in the Universe who are uniquely qualified to define what "good" means. Far more intelligent interstellar travelers who visit Earth could have much different opinions about the definition of "good" than humans do, and we might not agree with it in the same way that we already don't agree with each other about what "good" means (a common plotline in movies).

One possible explanation for why humans subscribe to moral, ethical, theological, and ideological belief systems is to provide abstract explanations for their natural instincts. As discussed previously, sapiens have an incredible capacity for using their imagination to detect patterns and come up with viable explanations for unsolvable mysteries. Well, natural instincts have a clear pattern of behavior across many generations and are quite mysterious. So it makes sense that humans would use abstract thinking to produce satisfying explanations for why they repeatedly feel compelled to behave in certain ways. Ethics and morals provide these logical and satisfying abstract explanations.

For example, people tend to believe that the reason why sapiens don't like the idea of killing other sapiens is because it's morally, ethically, or ideologically "bad." But at the same time, the disinclination to kill members of the same species is a common natural instinct shared by multiple species across multiple different animal classes (especially pack animals) which appears to have developed over hundreds of millions of years preceding human consciousness. Obvious existential reasons explain why less fratricidal pack animals would survive better in the wild than highly fratricidal pack animals. If humans weren't instinctively disinclined to kill each other over food and territory disputes, they would be less capable of cooperation and probably less likely to survive and prosper in the wild against mutual threats. Therefore, the belief that killing people is "immoral" could just as easily be described as an abstract explanation for natural instincts and natural selection. People who aren't instinctively disinclined to kill their peers don't survive to pass on their genes as much as people instinctively disinclined to kill their peers.

Abstract thoughts about morals, ethics, and ideologies can also be used to offer satisfying justifications for behavior which goes *against our* natural instincts too. To illustrate this, we can revisit the previous discussion about the double-standard of human predation. Many sapiens have no trouble preying on fish, selfish, and insects in sometimes brutal and unforgiving ways because these animals do not communicate pain and suffering the same way mammals do. The same person who would never cut the throat of the cow, pig, or chicken they eat has no problem killing an intrusive spider or going fishing at the local pond. For some reason, humans go to great lengths to avoid the discomfort of killing and injuring other mammals, typically choosing to outsource killing or predation to specialized workforces (the author has direct experience with this as both an active-duty service member as well as the heir to a beef cattle farm).

Ironically, the same thing that makes humans extraordinarily good hunters (their high order of intentionality, theory of mind, and ability to read the minds of their prey) tends to make them prone to feeling guilty about killing their predatory behavior. Predatory guilt is a unique trait of behaviorally modern sapiens. Practically every other apex predator in the wild does not appear to feel as guilty about their predatory behavior as people do, likely because other animals are physiologically incapable of it. It takes a large, power-hungry, and hyperactive neocortex to develop the high-order intentionality and theory of mind required to sufficiently imagine the pain and suffering that another animal is going through enough to feel guilty about contributing to it.

Mammalian predators can certainly detect when their mammalian prey are in stress thanks to their shared communication protocols (screaming, contorted faces), but most predators are far less capable of viscerally imagining the fear, anxiety, and trauma experienced by their prey in comparison to humans. Behaviorally modern sapiens are hyper aware of these emotions. A human predator's uncanny ability to read the minds of their prey makes them effortlessly capable of envisioning the stress they cause, which ironically stresses the human out in the process. As a result, most humans like to (1) outsource their predation and killing to a select group of people who can endure that stress, or (2) use their abstract thinking skills to justify it or create belief systems to reduce or avoid it.

A popular anthropological theory posits that humans started utilizing their abstract thinking skills to develop imaginary justifications and belief systems to reconcile the emotional discomfort of being predatory, murderous, and fratricidal (again, these are common behaviors in nature, but humans are uncommonly empathetic). Some theorize that this need for emotional reconciliation was a leading contributing factor to the rapid adoption of theistic religions following the domestication of animals. The theory posits that Neolithic humans adopted abstract belief systems where they perceive *themselves* to be gods and rulers of nature because it offered them a way to emotionally reconcile the trauma of entrapping and slaughtering domesticated animals by the billions. It's simply easier to justify the massive-scale enslavement and slaughter of animals if people believe animals exist to serve their human masters.

The argument is that sapiens started believing in humanoid gods to alleviate their cognitive dissonance and emotionally reconcile their physical and systematic abuse of the animals they were domesticating. Irrigating large amounts of rain-watered land requires the entrapment and enslavement of aurochs to create plow-pulling pack animals like oxen. Feeding densely populated areas requires massive-scale entrapment, enslavement, and slaughtering of animals like boars, junglefowl, and aurochs to produce the bacon we eat for breakfast, the chicken we eat for lunch, and the beef we eat for dinner. In other words, running a modern, "civilized" agrarian society involves a lot of humans working with animals, forming trusting relationships with them, bonding with them, betraying that trust, and then killing them to feed their meat to strangers. Mammals squint their eyes, tense their faces, scream and cry aloud as they are continuously bred by humans to be imprisoned, corralled, whipped, and slaughtered at massive scale. For empathetic predators with a powerful neocortex that can empathize with that pain and suffering, it's a hard business to be in. Many people can't handle it. [68]

The author does not intend to sound self-righteous or judgmental (to repeat, he is both a military officer and heir to a beef cattle farm). The goal of this discussion is to practice being as technically valid, intellectually honest, and amoral as possible. It's hard to dedicate one's life to our modern agrarian lifestyle without being tempted by abstract belief systems which claim that sapiens are "above" other animals or that somehow another animal's pain and suffering isn't as bad as we think it is because they are less intelligent than we are. These belief systems are attractive because they help people reconcile their predatory behavior, but that doesn't necessarily make them true. It could just be a coping mechanism that, ironically, enables humans to be more predatory and destructive.

Once people have adopted the characteristically Neolithic belief that sapiens have transcended nature and animals exist to serve them, it's not a major leap of cultural evolution to believe that gods exist, or that these gods have a distinctively humanoid shape. An easy way to cope with the domestication of animals is by believing that *we are the gods*. This line of thinking would explain why signs of theological beliefs in the human fossil record explode after agriculture. As Foster notes, artwork changes from humans living amongst animals, running alongside packs of free-roaming caribou with, to cracking whips over the backs of their entrapped, genetically deformed, docile servants. Scenes change from sapiens living within nature to humans living above nature. Humanoid shapes start sitting on thrones physically isolated from the wild and looking down upon it. [68]

By the time written language emerged, sapiens had thousands of years of experience believing in gods and looking down on nature. This alone explains why the basis of most moral, ethical, and theological philosophies look down on nature too (especially the killing), and encourage humans not to behave like wild animals. The implicit assumption of these assertions is that the behavior of wild animals is somehow "bad." The lion who kills the pack's cubs to keep the bloodline pure, or the squirrel who eats her babies to avoid starvation, is perceived as ideologically "bad" behavior. The act of being physically aggressive or using physical power to settle disputes, manage resources, and establish dominance hierarchies like wild animals do is asserted to be "bad" even though *sapiens have practically always been physically aggressive and have practically always used physical power to settle their disputes, manage their resources, and establish their dominance hierarchies* (hence 10,000 years of warfare). [68]

The point of this section is to illustrate that theology, philosophy, and ideology are highly subjective constructs which emerged after sapiens became gifted at thinking abstractly. If our morals and ethics weren't highly subjective, we probably wouldn't have theological, philosophical, and ideological disagreements. Therefore, although we like to use our ideologies to make bold assertions that we are somehow uniquely qualified to know what "right" is, an intellectually honest and humble person should recognize that it may not be objectively true that sapiens have miraculously discovered a metaphysically-transcendent, ontologically superior, or causally efficacious capacity for "good" and "bad." It could simply be that sapiens have oversized, overpowered, and hyperactive neocortices optimized for abstract thinking, combined with a lot of spare time on their hands to pontificate about "right" and "wrong" after they outsourced their predation and killing to other people.

4.5 Understanding Abstract Power

"Those who tell the stories rule society."
Plato [84]

So far, the author has outlined eight of ten applications of abstract thinking which behaviorally modern sapiens have evolved since the Upper Paleolithic era. The reader has learned about how sapient neocortices help sapiens with pattern-finding, exercising caution, symbolism, planning, strategizing, higher-order communication using semantically and syntactically complex languages, storytelling, solving physically unverifiable mysteries, and developing abstract constructs and belief systems like morals, ethics, and theologies. The last two applications of abstract thought are so important to the topic of power projection tactics employed by humans in modern agrarian society that they each get their own dedicated section: (9) creating abstract power, and (10) encoding abstract power hierarchies.

As a quick disclaimer for the reader, these sections represent an inflection point where the concepts become "political sounding." This is due to the subject matter. A discussion about how abstract power is created, how abstract power hierarchies function, and how abstract power hierarchies become dysfunctional to the point of motivating people to fight and kill each other in massive-scale physical power competitions (i.e. wars) is fundamentally a discussion about politics. Moreover, a discussion about why humans fight and kill each other is an emotionally charged topic. There are few things more personal than the motives behind why a person would be compelled to take the life of another. But to understand the complex social implications of Bitcoin and to develop a common understanding of how Bitcoin could represent a "soft" form of warfare that could mitigate fratricide, it is necessary to investigate them.

4.5.1 To Avoid Physical Confrontations, Sapiens Dress Up and Play Make Believe

The desire not to cause pain and suffering to other animals (except fish, selfish, and insects) not only explains why humans started adopting abstract belief systems to morally justify their predation of animals, it also explains why humans adopt abstract belief systems to (1) avoid having to fight and kill each other as much as possible, and (2) morally justify their fighting and killing when they inevitably have to do it (as a simple example of this, consider how the word "defense" translates to "morally-justified fighting and killing").

As discussed in the previous chapter, animals use physical power to settle intraspecies disputes and establish dominance hierarchies over limited resources, but this can sometimes lead to fratricide. Fratricide is especially troubling for highly empathetic predators like humans, who are hyper aware of the stress and trauma they can cause others thanks to their big brains. **One way to circumvent the discomfort of potentially injuring a fellow human to solve a property dispute or establish a pecking order is to adopt a belief system where disputes can be settled and pecking orders can be established by people who have *imaginary* power rather than *real* power**.

Sapient brains are so effortlessly gifted at abstract thinking, and people have such strong natural instincts not to injure each other, that people will attempt to use their imaginations to avoid having to physically confront each other to settle their disputes, manage their resources, and establish their dominance hierarchies. **One of the most defining characteristics of behaviorally modern sapiens who lived after the invention of agriculture and the widescale domestication of animals is the adoption of common belief systems where some people wield abstract or imaginary power, and those people are allowed to settle disputes, manage resources, and determine the pecking order for the broader population, explicitly so they don't have to fight over it like practically all other pack animals do**.

Modern domesticated sapiens could be described as being so averse to physical confrontation that they prefer to dress up in costumes and play make believe to settle their disputes, manage their resources, and establish their pecking order. Then, emboldened by their ideologies, they look down upon wild animals precisely because those animals *don't* (or more accurately, can't) use their imaginations to settle their disputes or establish their pecking order (as previously noted, wild animals appear to be physiologically incapable of this – they don't have the watts to think abstractly and use their imaginations because they didn't learn how to handle tinder, control fire, and cook their food to fuel their brains like humans did).

Herein lies the reason why the author spent so much time walking the reader through the evolution of abstract thinking and human metacognition. **Without critically examining human metacognition, it's impossible to establish a first principles understanding of how and why agrarian society has decided to adopt common belief systems where fully grown adults put on wigs and gowns and live-action role play (LARP) like they have power in order to settle people's disputes without physical conflict. Almost everywhere one looks in modern agrarian society, sapiens are seen using *symbols of power* (e.g. rank) rather than *real power* (e.g. watts).** They print symbols of their imaginary power on pieces of cloth and tie them on top of flag poles. They wear symbols of their imaginary power as lapel pins. Some even continue to dress up in wigs, gowns, and crowns, and practically all of them stand up in front of podiums etched with symbols of their imaginary power. Why do behaviorally modern humans behave in this incredibly bizarre way in comparison to other species?

In what some might consider to be a comedic display of irony, society's symbols of abstract/imaginary power frequently show images of wild predators like lions – animals which became fierce precisely because they *don't* role-play to settle their disputes and establish their pecking order. Tying this observation back to the core concepts presented in the previous chapter about power projection tactics in nature, one of the reasons why lions are so fierce is because they mastered *real* power projection to build their dominance hierarchies and they do other fierce things which domesticated societies claim is ideologically "beneath" them (even though they still do it too).

If one were to take the perspective of a non-human outsider (such as an alien visiting earth), the social behavior of modern domesticated agrarian sapiens might seem bizarre compared to the behavior of other animals. Sapiens behave much differently than other species in nature. They live under a mutually-adopted, global-scale, consensual hallucination where very few people get to have extraordinary levels of imaginary power that most of the population doesn't get to have access to, and then the population chooses to allow these people with non-existent physical power to call the shots.

Why would sapiens behave like this? The answer is deceptively simple: to save energy, and to (attempt to) reduce the risk of injury. Abstract beliefs systems where people have imaginary power serves as an alternative way to settle disputes, establish control authority over resources, and achieve consensus on the legitimate state of ownership and chain of custody of property in an energy-efficient way that doesn't directly cause injury (emphasis on the word *directly*). In other words, **abstract power is a story that people are intrinsically motivated to believe in because they like the idea of not having to spend energy or hurt each other to settle intraspecies disputes or establish their pecking order**.

Abstract power and their corresponding dominance hierarchies (what the author calls abstract power hierarchies) represent a belief system to which many people subscribe simply because they want to believe that there are viable alternatives to physical conflict as the basis for settling intraspecies disputes and establishing intraspecies pecking order. It's a good story which motivates people to work together and cooperate at large scales. The idea that sapiens have somehow outsmarted natural selection and used their neocortices to find a viable substitute for physical confrontation to settle intraspecies disputes and establish interspecies pecking order is an extremely attractive idea. And as we know from the concepts provided about storytelling, the more sapiens can get behind a common idea, the more they can sum their physical power together and literally move mountains.

Unfortunately, abstract belief systems are fictional stories. Our beliefs about a better way to settle our intraspecies disputes and establish our intraspecies pecking order also clearly don't work as well as we wish they would work, because sapiens still routinely engage in physical confrontation to settle intraspecies disputes and establish intraspecies pecking order the exact same way animals do.

4.5.2 Belief in Imaginary Power is an Attack Vector which Breeds God-Kings

"It is not the lash they fear; it is my divine power. But I am a generous God.
I can make you rich beyond all measure."
Xerxes, *300* [85]

The problem with the belief that physical confrontation can be mitigated by subscribing to a common belief system where people with abstract/imaginary power settle disputes and maintain a pecking order in a non-physical way is that this reasoning ignores important systemic factors which motivate physical confrontation – namely the fact that people demonstrably can't be trusted not to abuse their abstract/imaginary power. In other words, these belief systems ignore why societies eventually become compelled to engage in physical confrontations in the first place, despite their desire to avoid it. [68, 70]

There are major downsides to the widescale adoption of common abstract belief systems which are critical to investigate if we are to understand why sapiens struggle to find lasting alternatives to physical confrontation as the basis for solving intraspecies disputes and establishing intraspecies pecking order. A major problem which will be discussed at length in this chapter is that belief systems represent a breeding ground for systemic predators to psychologically abuse and systemically exploit entire populations of people through their belief systems. These attacks are passive-aggressive and often go without detection because they don't have a physical footprint. It is therefore necessary to call them out explicitly so that we can better understand the complex, trans-scientific, and sociotechnical implications of new technologies like Bitcoin.

Storytelling introduces a psychological attack vector where sapiens can be preyed upon. The most common way this happens in agrarian society is by telling stories to convince people to adopt belief systems where a select few people have abstract power. The problem is, abstract power is systemically exploitable. By convincing people to believe in abstract power, storytellers deliberately implant an exploitable vulnerability into people's imaginations which they can take advantage of later (it's essentially a zero-day, for people who are familiar with common computer exploits). Once storytellers have convinced a population to adopt a belief system where imaginary power exists, storytellers create a vector through which they can exploit people by giving themselves access to the imaginary power endogenous to the belief system. This type of predatory behavior through people's belief systems emerged early in agrarian society and has persisted for thousands of years. A simple example of these kinds of predators who systemically exploited their population's belief systems were god-kings such as the Pharoah shown in Figure 43.

Figure 43: A God-King Exploiting a Population's Belief System

To better understand how vulnerable sapient belief systems are to psychological exploitation and abuse, we can use adversarial thinking to analyze how to create and codify abstract power. The reader is invited to assume you are a systemic predator who wants to psychologically exploit a human population's belief system for your own personal advantage. What is the most important thing you need the population to believe in so that you can have extraordinary amounts of imaginary power and control authority over their valuable resources? One thing that you could do is convince them to believe that using physical power to establish their dominance hierarchy is morally "bad." You could convince them there are ideological alternatives to physical power for establishing control authority over their resources and achieving consensus on the legitimate state of ownership and chain of custody of their property.

Perhaps the population might become concerned that you could abuse your abstract power to exploit them. If that's the case, then you could convince them that imaginary logical constraints encoded into rules of law are fully sufficient at protecting them against systemic predators like you who can exploit imaginary power. Once the population has been convinced that imaginary power hierarchies encoded into rules of law are incontrovertibly better solutions than physical power, then you could simply place yourself at the top of that abstract power hierarchy by masquerading as the morally, ethically, ideologically, or theologically fit candidate for the job. If you are successful, the population will bend to your will and do your bidding for you, labor for you, kill for you, give you their most valuable resources and worship you like a god – just like a domesticated animal would. In other words, you can domesticate a human population by getting them to adopt a belief system which convinces them that it's "bad" to be physically powerful or physically aggressive. Once they have adopted ideologies which cause them to forfeit their real power for your imaginary power, you essentially own them.

One of the major challenges associated with using imaginary power as the basis for settling disputes and managing resources is that it's imaginary. It exists for no other reason than the fact that people are physiologically capable of thinking abstractly and adopting abstract belief systems. Because of the way our neocortices effortlessly engage in bi-directional abstract thinking and symbolic reasoning, people can and often do live their entire lives cognitively entrapped under these population-scale consensual hallucinations where it's impossible for them to see how vulnerable they are to psychological abuse and exploitation through their belief systems. Tragically, this also makes them incapable of seeing how easy it would be to escape their psychological entrapment. People will legitimately believe those who wear striped headcloths or lapel pins are *actually* powerful and fear their "divine" power over generations, birthing dynasties of oppressive god-kings. Populations will labor for their god-kings, kill for them, forfeit their resources to them, and even let their oppressors define what's "right" or "good" or "fair."

4.5.3 The Cycle of Human History

"Hard times create strong men, strong men create good times,
good times create weak men, and weak men create hard times."
G. Michael Hopf, *Those Who Remain* [86]

Across multiple generations, entire populations of agrarian sapiens will believe that imaginary logical constraints are viable substitutes for physical power as a mechanism for imposing physical costs on attackers. **They will believe they have miraculously transcended natural selection and found a moral or ethical alternative with the same capacity for survival (moral or ethical according to whom?). And then, once enough of the population has been convinced that they're secure against exploitation due to nothing more than the logical constraints encoded into rules of law, they self-domesticate. They become docile; they condemn physical confrontation and aggression**. Instead of vectoring their resources to the most physically powerful, they socially exile them. They condemn them as war mongers. They place the people with real power at the bottom of their pecking order, in favor of people with "peaceful" forms of imaginary power. The fossil record shows us what happens next. Either their own god-kings exploit/slaughter them, or their neighboring god-kings do.

As illustrated in bowtie notation in Figure 44, human history seems to work in cycles, where societies get comfortable and complacent with their high-functioning belief systems and stop projecting physical power to increase their C_A. As they become increasingly more resource abundant but increasingly less physically powerful and aggressive, their **BCR_A** climbs. Not surprisingly, these societies get devoured by predators, just like any other organism or organization in nature would. Meanwhile, societies which employ the "grow C_A first, then grow B_A" second strategy survive and become the new dominant society because they have the lowest **BCR_A** thus the highest prosperity margin. But this new society eventually becomes comfortable and complacent, stops increasing their C_A, and the cycle repeats.

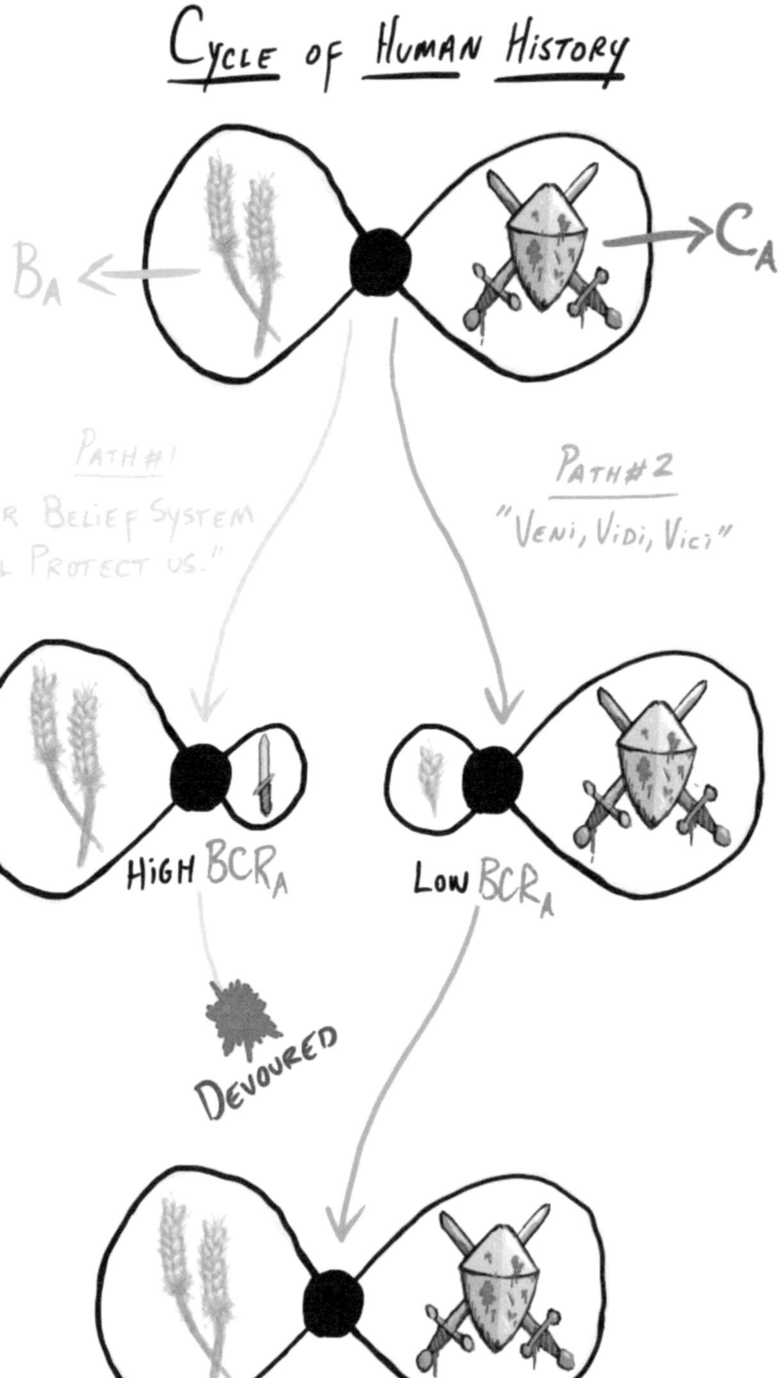

Figure 44: A Repeating Cycle of Human History

When human populations become too docile or domesticated, either their territory gets physically captured, or their belief system gets psychologically exploited. If the former, the population gets emergency drafted, misplaced, sent to labor camps, sent to mass early graves, or most likely starves. If the latter, the population gets systemically exploited and oppressed at extraordinary scales through their own belief systems, leaving them entrapped and enslaved with no capacity to understand what the root cause of their oppression is. During the collapse of these complex agrarian societies, maybe a few will have the intellectual humility to think twice about their decision to condemn physical power and aggression. Maybe they will stop and consider the idea that it was a mistake to adopt belief systems which require trust in untrustworthy people to function properly, and recognize that their beliefs in imaginary power, combined with their condemnation of physical power, led them straight to the slaughter. [87]

Rather than take accountability for their decisions and question the grossly unrealistic assumptions they made when they adopted their belief systems, many will instead choose to blame their invaders or their systemic oppressors for their losses. To their graves, they will continue to LARP like they ever had the option of living in a world without predators and entropy – as if they are the only organism in the world that doesn't have an intrinsic responsibility to keep themselves physically secure against attackers. They will masquerade like "peaceful" alternatives to physical conflicts ever existed at any time in history except temporarily, or at any place on this planet except exclusively within their imaginations. Such is the tragedy of domesticated sapiens and their willful ignorance of power projection tactics in modern society.

4.5.4 Understanding the Sociotechnical Differences between Real Power & Imaginary Power

A 10,000-year-old human fossil record shows evidence of the same pattern repeating itself over and over again, backed by an additional 5,000 years of written testimony. Bad things happen to those who forfeit their capacity and inclination to project physical power in favor of imaginary or abstract power. Good things happen to those who master their capacity and inclination to project physical power and impose severe physical costs on their attackers, whether the attacks come from inside or outside society.

To recall a core concept from the previous chapter, we have highly randomized and variable data sets between many different agrarian societies who have tried many different experiments with many different resource management strategies. We can analyze these data ex post facto using statistical methods to find causally inferable relationships between physical power, physical aggression, and social prosperity. These causally inferable relationships can also be found in the animals we domesticate and slaughter on a regular basis (the proof is literally served to us on a silver platter). Yet somehow, sapient populations keep allowing themselves to fall into the exact same traps over and over again. Why is that?

One explanation could be that because so few people in modern agrarian society devote themselves to the task of understanding and mastering physical power projection, they don't understand it. Combining this idea with the fact that most people don't think about their own metacognition, it could be the case that people are blissfully unaware of the fact that there are very clear, very measurable differences between physical power and abstract power which explain their different emergent behavior. If that's the case, then we can explicitly call out the differences between real power and imaginary power so that people can better understand why they produce different emergent behavior. **Once we understand how and why imaginary power and real power produce different emergent behavior, we can understand why abstract power hierarchies have so many dysfunctions which lead to war**.

Figure 45 provides a breakdown of some of the characteristics of physical power using a real example. Here, Captain Elizabeth Eastman is pictured doing a pre-flight inspection of her physical power projection technology, an A-10 Thunderbolt II "Warthog." Like all physical power, Captain Eastman's power is self-evident and self-legitimizing. People can instinctively recognize and verify the presence of her physical power. It's also exogenous to people's belief systems, making it invulnerable to systemic exploitation. Physical power is unsympathetic, meaning it works the same regardless of whether people believe in it or sympathize with it. More of it can't be created out of thin air, making it thermodynamically restricted. Its execution can't be reversed, making it path dependent. Everyone is free to access and leverage watts the same way she can regardless of their rank, title, standing, or belief system, making it inclusive and egalitarian. Physical power is also unbounded; there's theoretically no limit to how many watts people can use to defend themselves (particularly against her and her peers). Physical power also has a physical signature and leaves a blood trail, making it easy for people to see the threat and organize to defend against it. Table 2 provides a breakdown of these characteristics.

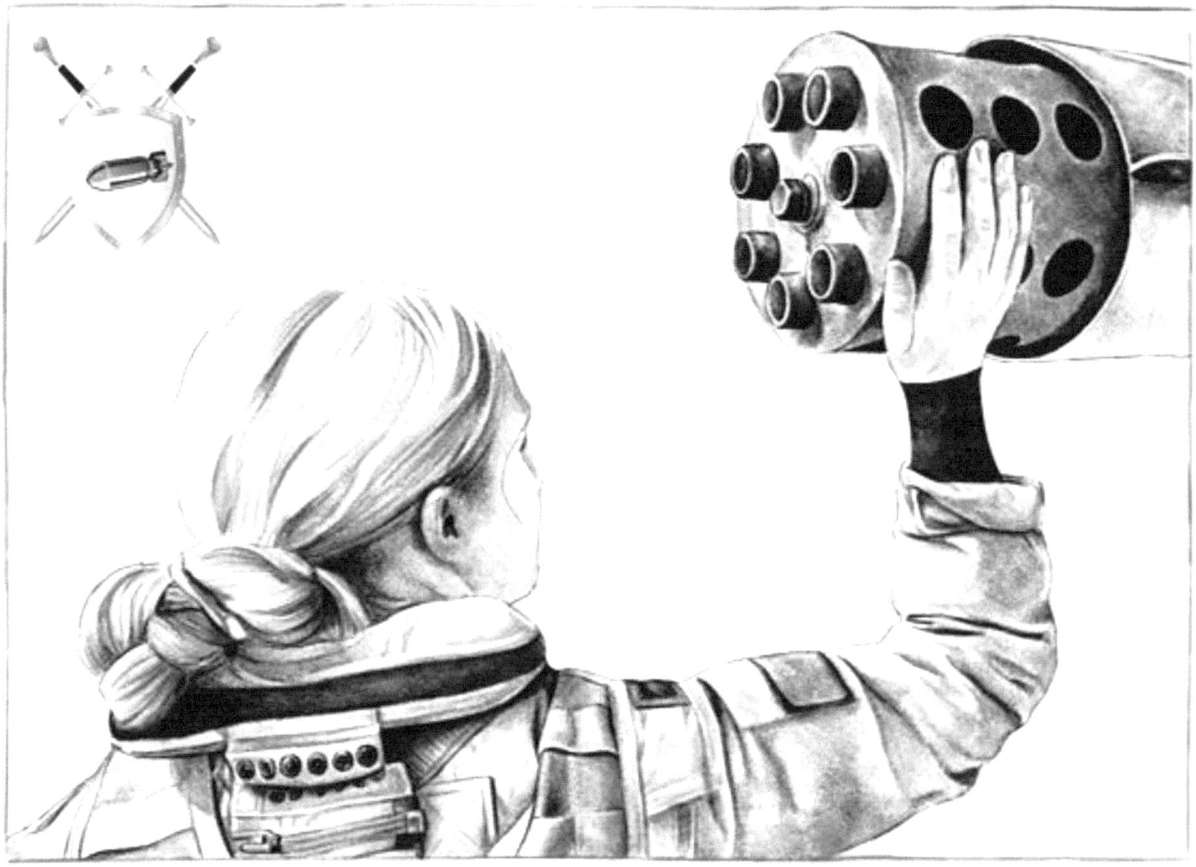

Figure 45: Captain Elizabeth Eastman does a Pre-Flight Inspection of her A-10 Thunderbolt II. This is an Illustration of Real Power (i.e. Physical Power or Watts)
[88, 89]

Table 2: Characteristics of Captain Eastman's Real/Physical Power

Systemic Characteristic	Description
Self-Evident & Self-Legitimizing	People can instinctively recognize & respect her power based off its own merit, making it independently verifiable.
Systemically Exogenous	Derived from a source that is external to people's belief system, therefore it's impossible to be systemically exploited.
Unsympathetic	Doesn't need people to believe in or sympathize w/her power for it to function; functions the same for different people w/different belief systems or sympathies.
Physically Constrainable	In her operational domain of shared objective physical reality, there are a lot of ways to physically constrain her from using or scaling her power.
Thermodynamically Restricted	More of this power can't be created out of thin air & awarded to people. If she is to have more of it, she must be intelligent & resourceful enough to master nature.
Path Dependent	The execution of her power cannot be reversed, appealed, or undone.
Inclusive & Egalitarian	Everyone can access to the same type of power she's using and countervail her with it regardless of their rank, title, standing, or belief system.
Unbounded	There is no limit to the amount of her type of power that people can use against her.
Attributable	Because her power has a physical signature (i.e. blood trail), it's easy for people to see the threat of her power and organize to defend against it.
Energy Intensive	Requires a great deal of physical effort for her to exercise her power, creating a natural barrier to entry.
Directly Leads to Injury (if Kinetic)	People directly get hurt when she exercises or mismanages her power, making it much easier for people to see and get upset by it, giving her lower margin for error.

These characteristics could be described as useful features which enable physical power to function nicely as a zero-trust, permissionless, and egalitarian basis for settling intraspecies disputes, establishing control authority over intraspecies resources, and achieving consensus on the legitimate state of ownership and chain of custody of intraspecies property in a way that is invulnerable to systemic exploitation of people's belief systems. The downside, of course, is that projecting physical power is energy intensive and often leads to injury (if kinetic), hence why people are often motivated to seek alternative solutions to physical power projection.

If we combine all these characteristics together, we can expect a sociotechnical cooperation system that uses physical power to settle disputes, manage resources, and establish pecking order to have precisely the same kind of complex emergent behavior observed in the physical power projection competitions of nature, where those who emerge as the top power projectors have survived a rigorous but objectively fair natural selection process which separates those who are stronger, more intelligent, better organized, and more resourceful from those who are demonstrably unfit for survival.

In direct contrast to real/physical power, we have abstract/imaginary power. Figure 46 provides a breakdown of the characteristics of abstract power using another real example. Here, Chief Justice Dudley is pictured presiding over the Supreme Court of Gibraltar. Like all abstract power, Chief Justice Dudley's power is imaginary; it' doesn't physically exist anywhere except exclusively within people's collective imaginations. It's neither self-evident nor self-legitimizing, which means people can't instinctively recognize or independently verify it based off its own merit. It's also systemically endogenous to people's belief system, making it highly vulnerable to systemic exploitation and abuse. Chief Justice Dudley's abstract power is sympathetic; it requires people to believe in it and to be sympathetic to it or else it doesn't function. It's also physically unconstrained; there is nothing physically limiting his imaginary power from scaling globally; the only constraints to his imaginary power are imaginary logical constraints encoded into a ruleset he's not required to be sympathetic to, which are endogenous to the same belief system and therefore equally as vulnerable to systemic exploitation. His abstract power is also path independent, making it reversible. At the same time, it's thermodynamically unsound because it can be created out of thin air. It's non-inclusive and inegalitarian because not everyone can have access to positions of high rank. There are also hard limits to the amount of imaginary power that people can use against him, making it bounded, and giving him an upper hand. And because his power is imaginary, it has no physical signature and leaves no blood trail. This makes people less capable of detecting the threat of his power if it's used against them, making them less likely to be motivated to organize to countervail it. Table 3 provides a breakdown of these characteristics.

Figure 46: Chief Justice Anthony Dudley Presides Over the Supreme Court of Gibraltar. This is an illustration of Imaginary Power (i.e. Abstract power or Rank) [90, 76]

Table 3: Characteristics of Chief Justice Dudley's Imaginary/Abstract Power

Systemic Characteristic	Description
Neither Self-Evident nor Self-Legitimizing	People can't instinctively recognize or respect his power based off its own merit, making it not independently verifiable.
Systemically Endogenous	Because his power is internal to people's belief system, it can be systemically exploited (thus making it a form of psychological abuse of a population through their belief system).
Sympathetic	He needs people to believe in or sympathize with his power else it won't function; it also doesn't function the same for different people with different belief systems or sympathies.
Physically Unconstrainable	In his operational domain of shared subjective abstract reality, there is no way to physically constrain him from using or scaling his power. People must attempt to use logical constraints to constrain him, but those are demonstrably insecure against systemic exploitation and they also require people to believe in or sympathize with them.
Path Independent	The execution of his power can be reversed, appealed, or undone.
Thermodynamically Unsound	This power can be created out of thin air & awarded to people. He doesn't have to be intelligent & resourceful to master nature to have more of it, he just has to change the rules.
Non-inclusive & Inegalitarian	Not everyone can have access to the same type of power he has without specific rank, title, standing, or belief system.
Bounded	There are hard limits to the amount of his type of power that people can use against him.
Non-Attributable	Because his power has no signature (i.e. no blood trail), it's very hard for people to see the threat of his power, much less organize to defend against it.
Non Energy Intensive	Requires minimal effort to exercise his power, removing natural barriers to entry.
Indirectly Leads to Injury	People indirectly get hurt when he exercises or mismanages his power, making it much harder for people to see and get upset by it, giving him far larger margin for error.

Note how the sociotechnical characteristics of imaginary power are mostly flaws. Because of these flaws, we can expect cooperation systems which use abstract power to settle intraspecies disputes, establish control authority over intraspecies resources, and achieve consensus on the legitimate state of ownership and chain of custody of intraspecies property to be dysfunctional and prone to systemic exploitation and abuse.

Nevertheless, there are benefits to imaginary/abstract power. For starters, this kind of power requires minimal energy to exercise and has practically no natural barriers to entry, making it extremely efficient and easy-to-adopt. It is also less directly attributable to physical injury, making it ostensibly more moral. In other words, because imaginary power doesn't involve physical confrontation, it's ostensibly "good." However, the reader should note that the underlying argument of this thesis is that **abstract power is so dysfunctional that it directly motivates people to engage in large-scale physical confrontations (i.e. wars), effectively defeating its own purpose.**

Now that we have explicitly called out the sociotechnical differences between real and imaginary power and shown how these are physically, systemically, and ontologically different things which should be expected to produce different complex emergent behavior, we can take some time to reflect on why people so commonly misunderstand them. This requires an even deeper look at sapient psychology and metacognition.

4.5.5 Understanding the Logical Flaw of Hypostatization

One of the most ubiquitous logical fallacies in modern agrarian society is a fallacy of ambiguity called hypostatization, where people "*construe a contextually subjective and complex abstraction, idea, or concept as a universal object*." [91, 92] In plain terms, hypostatization is the mistake of believing something imaginary is something real. The fallacy of hypostatization is so common and ubiquitous in society that it's easy to forget it happens practically all the time (similar to the saying about fish forgetting the presence of water, humans forget about the presence of hypostatization because they constantly think abstractly). Hypostatization is one of the most common logical fallacies, yet it's rarely discussed. Whole systems of philosophy, politics, religion, and social theories are built upon or supported by these fallacies. [91]

In his book *Science and the Modern World*, Alfred Whitehead warns of a very similar fallacy of "*misplaced concreteness*" where people build "*elaborate logical constructions of a high degree of abstraction*" which cause them to regard abstract beliefs and hypothetical constructs as if they were concrete things. This is another fallacy of ambiguity called *reification*. Whitehead summarizes the reification fallacy as simply "*the accidental error of mistaking the abstract for the concrete*." [91, 93]

The distinction between hypostatization and reification comes down to the type of abstractions involved, otherwise their definitions are virtually identical. Reification is commonly understood as a subset of hypostatization where the abstractions which are fallaciously regarded as concrete things are either theological, philosophical, or ideological. For example, abstract constructs like "good" and "justice" are generated from theology, philosophy, or ideology. People who act like "good" and "justice" are concretely real things are guilty of reification because, technically speaking, "good" and "justice" are abstract concepts or beliefs which don't concretely exist in shared objective reality. [94]

Both hypostatization and reification are considered to be fallacies of ambiguity because they tend to happen when people get metaphors confused with literal meaning. Reification in literature (where it is ostensibly understood to be intended metaphorically or as a figure of speech) is encouraged and often considered to be good writing, so becomes very common in writing for cultural reasons. However, when hypostatization or reification occurs in logical arguments, it becomes a logical fallacy. Thus, both are regarded as fallacies of ambiguity because it can be ambiguous whether the author intended to speak metaphorically or intended to make a sincere logical argument. This ambiguity is often deliberate, such as when it's used in rhetoric. Rhetoric relies heavily on reification and will often use literary metaphors to present logical arguments to make them more thought-provoking and attention-capturing (sapient neocortices crave their stories).

Understanding the fallacy of hypostatization and reification are key to understanding why people can enjoy control authority over resources *without* using physical power. They create an abstract form of power like rank, title, or station, and then through hypostatization, people mistake their abstract power for a real thing. Practically all belief systems used to manage resources depend on hypostatization or reification of abstract power – power which people think is concretely real even though it technically exists nowhere except within their collective imaginations. This could be one reason why so few people talk about hypostatization despite its ubiquity. People in existing abstract-power-based dominance hierarchies obviously wouldn't want their populations to think about how their abstract power isn't real.

4.5.6 Legitimizing Imaginary Power using Real Power

"Force, my friends, is violence. The supreme authority from which all other authorities are derived."
Jean Rasczak, *Starship Troopers* [95]

Another way to motivate people to hypostatize abstract power (other than by using persuasion or rhetoric) is by using physical power to legitimize it. Rather than using theological, philosophical, and ideological arguments to convince people that one's abstract power is real, it's possible to create displays of physical power, and then let your audience draw false positive correlations between two physically, systemically, and ontologically different things.

To illustrate this point, consider the difference between kings and knights. We have established that a king's power is symbolic, not physically real. Yet, sapiens have a clear tendency to cherish the symbolic world in their heads more than the physical one in front of their eyes. Being a modern human often means hypostatizing or reifying abstract constructs and then role-playing as if the abstractions are concretely real things. This behavior is why if you choose to disobey the king, you have a high probability of being injured by LARPers who actually wield real power: knights.

Knights are people who have volunteered to subscribe to a belief system where kings have abstract power. Some of them may even believe that the king's abstract power is concretely real. Because people subscribe to these belief systems, they are willing to shape physically objective reality ex post facto to match what exists exclusively within their imagination. So for example, storytelling kings will claim that disobedient people (i.e. people who have not subscribed to the same belief system as them) should be physically constrained or have their rank demoted to prisoner for being unsympathetic to the orders of the king. In response, knights will use their physical power to legitimize the king's abstract power by making shared objective physical reality match their shared, subjective, abstract reality.

This process of using physical power to legitimize abstract power is more commonly known as enforcement. The name literally means *to introduce force*. In other words, the name means to inject real power into a situation where only imaginary power was previously being exercised. Enforcement is noteworthy because it shows the two aforementioned use cases of physical power projection occurring simultaneously: (1) imposing physically prohibitive cost, and (2) creating a proof-of-power a.k.a. proof-of-real signal. The knights' power is not only projected to increase the $\mathbf{C_A}$ and lower the $\mathbf{BCR_A}$ of undermining an abstract belief system, but it's also projected to produce a proof-of-real signal to motivate people to draw false positive conclusions about the "realness" of the king's abstract power.

In the first use case, physical power is projected to impose severe physical costs on people who don't subscribe to the same belief system and therefore aren't sympathetic to or influenced by the king's abstract power. As mentioned in the previous section, imaginary power is sympathetic; it needs people to believe in it or sympathize with it or else it will not function properly. Therefore, not being sympathetic to the king's abstract power is a direct threat to the functionality of his entire abstract power hierarchy and could expose that the king doesn't actually have real power. The solution is to impose real-world physical costs on those who are unsympathetic to the king's orders. This not only decreases the benefit-to-cost ratio of undermining the king's orders, but it also causes bystanders to hypostatize his imaginary power as something physically real.

People are inclined to believe the king's imaginary power is real for the same reason they are inclined to believe a harmless stack of sticks is a deadly snake. Our brains produce false positive correlations between abstract thoughts (e.g. the imaginary power of the king) and sensory inputs (e.g. the physical power of the knights) because natural selection has caused our brains to take abstract imaginary things as seriously as physically real things.

People are quick to hypostatize the king's imaginary power as real power because the knights' physical power projection manually generates a cross-referenceable physical sensory input to match the king's claim about having real power. In essence, enforcement leverages the same realness verification algorithm people use when they poke something or pinch themselves to generate haptic feedback. Knights manually generate force to displace mass over time to produce a proof-of-power signal of realness, the exact same way that people poke things or pinch themselves to produce a proof-of-power signal of realness. The main difference is that the proof-of-power signal comes from a third party (much like how actual haptic feedback systems work) and the amount of power used doesn't cause injury. An illustration of this is provided in Figure 47.

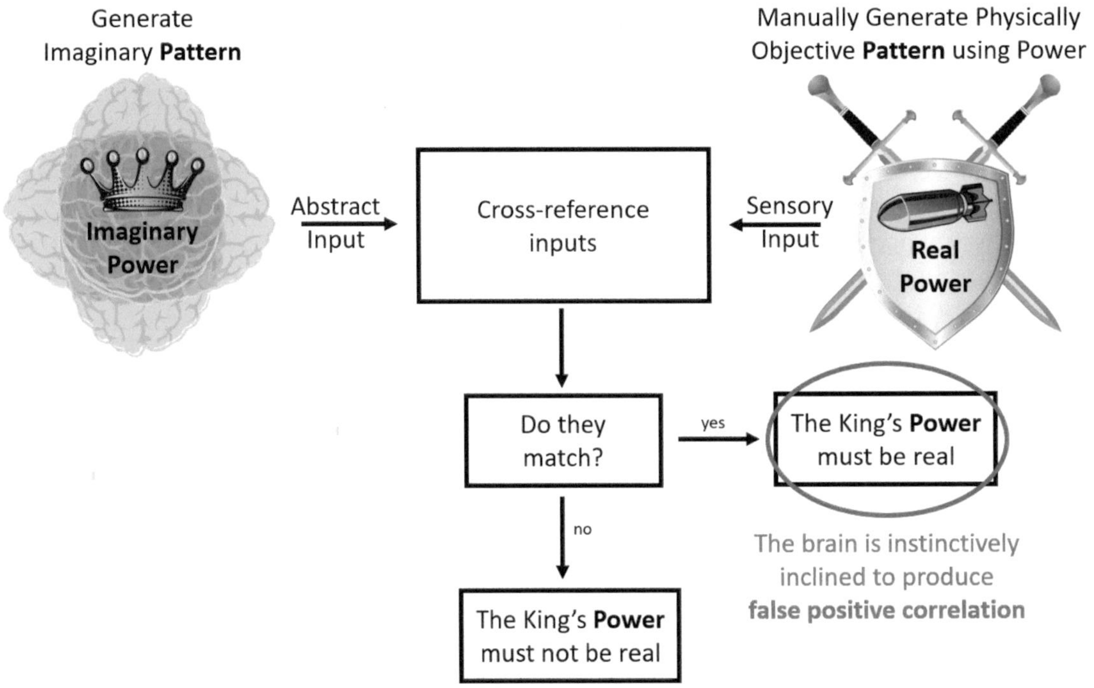

Figure 47: False Positive Correlation Produced by the Brain's Realness-Verification Algorithm
[88, 90, 76, 89]

As a quick side note, while the author was writing this, the founder of the oculus virtual reality system claims to have designed a haptic feedback system for virtual reality gamers that physically harms/kills the wearer, the idea being that it makes the gaming experience feel more real and materially consequential. [96] This is a perfect demonstration of the "proof-of-power equals proof-of-real" concept discussed here. A haptic feedback system which physically kills a gamer if they die in-game is, in essence, an enforcement system which works exactly the same as a knight who kills a citizen for disobeying the king. Both scenarios represent a situation where physical power is used to make something abstract *feel* more real since the virtual reality of a video game is, by definition, just as abstract as a king's imaginary power. Since physical power is path dependent and self-legitimizing, a virtual reality gaming system which utilizes physical power to kill its wearer if they die in-game inherits the systemic properties of the physical power it utilizes to make the game more path dependent and legitimately hazardous.

The physical power produced by a knight works exactly like a lethal haptic feedback system for virtual reality gaming systems. It provides a synchronous cross-referenceable signal of realness to match the king's abstract power. In other words, proof-of-power produces a proof-of-real signal. At the same time, the population contributes to the illusion. Other people subscribe to the same belief system that says the king's power is physically real, steering everyone's combined physical action to shape objective reality to match what is otherwise just an abstraction. As a result, the population is quick to lose sight of the fact that all abstract power held by all people of all ranks in all abstract-power-based dominance hierarchies exist within the imagination only. None of it is physically real no matter how many people LARP like it's real or use real-world physical power to make it look and feel more real. The combined effect is a population-scale consensual hallucination which can gaslight the public into believing in the divine strength of god-kings.

The result of enforcement is compliance with the king's orders. People who blatantly undermine the king's rank are physically punished for not recognizing his imaginary power by obeying his orders. This, combined with routine physical shows of force, makes people quick to commit the logical fallacy of believing the king's imaginary power is something concretely real. People will sincerely believe they are being physically compelled to behave some way merely by virtue of reading about it or watching physical power projectors march around on a computer screen, despite never having been involved in any physical confrontation (as another side note, this is why some regimes love military parades – these parades are marketed as a show of force to foreign nations and a way to comfort a proud population about how secure they are, but in actuality, a hidden purpose of the parade is for the regime to produce a proof-of-power signal for their own populations to motivate them to hypostatize the regime's abstract power as real power and make them less motivated to resist the regime).

4.5.7 Illegitimatizing Imaginary Power using Real Power

"The world will know that free men stood against a tyrant, that few stood against many, and before this battle is over, that even a God can bleed."
Leonidas, *300* [85]

US President Kennedy was among the most abstractly powerful people to have ever lived when he was assassinated in 1963. Incidentally, Kennedy's assassination occurred just five months after signing EO 11110, which journalists have argued was an attempt to reign in the abstract power of the Federal Reserve. In his book *Crossfire*, Jim Marrs presents the argument that President Kennedy attempted to replace the purchasing power of Federal Reserve notes (i.e. money issued and controlled by the Federal Reserve – a private institution) with silver certificates (i.e. money issued and controlled by the US Department of Treasury – a public institution with the deferred abstract power of the US President). In other words, the logic encoded into EO 11110 would have stripped the abstract power of the Federal Reserve to make money and given it back to the US government. For that reason, journalists have argued that the Federal Reserve Bank (most notably the bank's anonymous shareholders who receive interest off their notes lent to the US government) would have had the largest financial motive to contribute to President Kennedy's assassination, as EO 11110 would have undermined their monopoly control over the US monetary system. [97]

Of course, people can only speculate as to the true motive behind JFK's assassination, and there is no shortage of conspiracies related to the topic. But no matter what the true motive(s) for JFK's assassination was, it very clearly demonstrated how physical power is both superior to and unsympathetic to abstract power. President Kennedy had far higher rank and far more abstract power than the person who took his rank and abstract power from him, but it didn't protect him because abstract power is merely imaginary power – rank doesn't stop a speeding bullet.

If, purely for the sake of illustrating a core concept of this grounded theory, we assume that people associated with the Federal Reserve Bank were indeed behind JFK's assassination because of financial motives, then JFK's assassination would represent a scenario that demonstrates how physical power both legitimizes and delegitimizes abstract power simultaneously. To fire a bullet is to project physical power kinetically for the purpose of imposing severe physical costs on neighboring organisms. In this scenario, the bullet which delegitimized the abstract power of the US President would have simultaneously legitimized the abstract power of the Federal Reserve – because four months after Kennedy's assassination, the redemption of silver certificates for silver dollars was irrevocably halted, implicitly restoring the Federal Reserve's monopoly control over the US monetary system.

As the demise of any leader at the top of any abstract power hierarchy has shown (of which there have been many examples – President Kennedy merely being the most recent one in our particular abstract power hierarchy), as easily as physical power can be used to legitimize abstract power, it can also be used to delegitimize abstract power. King Leonidas famously highlighted this concept in the movie *300* when he suggested that an effective way to undermine the legitimacy of god-king Xerxes' hypostatized power is to make Xerxes bleed. This concept demonstrates yet another application of physical power. Physical power not only serves as a proof-of-real protocol, but it can also serve as a proof-of-*not*-real protocol that illegitimatizes people's claims to abstract power. This function alone explains why many wars are fought. In almost all cases of large-scale human versus human conflict, people try to either legitimize or delegitimize the abstract powers assigned to a given belief system.

A technical name for this concept is power hypostatization – the act of believing that abstract power is concretely real power. The pitfall of power hypostatization is two-fold. First, as illustrated in the previous subsection, the human tendency for power hypostatization motivates people to project physical power to convince a population to believe in their abstract power. Second, as illustrated in this subsection, power hypostatization simultaneously motivates people to project physical power to convince a population *not* to believe in someone's abstract power. In both cases, physical power is used to either legitimize or illegitimatize abstract power, resulting in a kinetically destructive war.

Power hypostatization is, of course, a glaring logical flaw. The flawed reasoning is easy to point out just by asking a few simple questions. For example, if the god-king's power were concretely real, then why isn't it self-evident? Why does he need an army in the first place? Remove the knights, and there would be little confusion about the existence of the king's power. Without knights, it would be trivial to see that the king is, in fact, physically powerless. There would be no sensory input available to cross-reference the king's imaginary power with the knights' real power, thus no way for the mind to produce the false positive correlations that lead to power hypostatization. In a world where the people at the top of the abstract power hierarchies actually had the power people claim they have, there would be no such thing as assassinations or revolutions or foreign invasions. An all-powerful god-king wouldn't need to hire an army to fight a war for him; he would only need to snap his almighty finger. The fact that wars exist is therefore a direct biproduct of the fact that abstract power doesn't physically exist.

The concept of power hypostatization highlights the metacognitive function of phenomena like enforcement and military shows of force. Metacognition – thinking about how humans think – helps us understand the primary value-delivered function of enforcement and military shows of force. These are both psychological techniques designed to influence people's behavior by getting them to either hypostatize abstract power as concretely real power, or to "snap out" of their consensual hallucinations and recognize that abstract power and real power are not the same kind of power, thus they don't have the same emergent behavior. As anyone who has ever experienced a military show of force maneuver can attest (e.g. warning shots or low-altitude flybys of military aircraft), these are the moments where people "wake up" from their imaginary belief systems about power and realize that "things just got real."

A primary difference between what populations consider to be illegitimate and legitimate belief systems or legal policies (a.k.a. a generic set of rules versus *the* rule of law) is the amount of real-world physical power projected by people to enforce and secure those policies. The author could easily design a set of rules to encode his own abstract-power-based dominance hierarchy and place himself at the top of it, but people are probably not going to subscribe to a belief system that the author is *actually* powerful, because the author lacks both the physical power and the storytelling capacity to convince people to believe his story.

As an example of this concept, the abstract powers encoded by the US Constitution are backed by the US military. When people of high rank try to undermine the US constitution, the US military's job is to step in and remind them that the abstract power of the US Constitution is backed by *real* power of the US military and that *real* power is non-negotiable regardless of how unsympathetic people are to it. The most recent example where this happened was following the US Capitol insurrection during the inauguration proceedings of President-elect Biden. In an unprecedented and thinly-veiled warning by the US Joint Chiefs of Staff, the American public (to include the sitting commander-in-chief) were explicitly reminded that "*the Armed Forces of the United States... remain fully committed to protecting and defending the Constitution of the United States against all enemies, foreign and domestic*" and that "*in accordance with the Constitution... President-elect Biden will be inaugurated.*" [98] Here, the Joint Chiefs explicitly reminded the public that their job is to physically secure the abstract powers encoded by the US Constitution against foreign *and domestic* enemies (i.e. against US citizens and even the sitting commander-in-chief, if necessary, if they were to continue to organize and illegitimatize the US Constitutional process using physical confrontation).

The phenomenon of power hypostatization is also why US military servicemembers do not have the same freedom of speech rights as civilians do, most notably the right to speak contemptuously towards public officials regardless of rank. In other words, people wielding physical power on behalf of the US are not legally allowed to speak contemptuously towards people wielding abstract power on behalf of the US, because physical power illegitimatizes abstract power and represents an existential threat to the existing abstract power hierarchy.

Military servicemembers are special in that they wield *real* power. It's one thing to have a politician speak contemptuously towards other politicians (as they often do), but it's an entirely different thing to have a commander of military forces speak contemptuously towards a politician. Almost every time this happens, the officer is quickly fired. A general has a lot more physical power backing up his rank than that of other civilian officials with similar rank. Because military forces are *actually* powerful, it would be inappropriate for military officers to speak contemptuously towards public officials, as that represents a direct threat to the abstract power of the existing abstract power hierarchy. To see a manifestation of this concept, watch any state of the union address and take note of the joint chiefs. The reader will notice that the joint chiefs wear "poker faces" and scarcely react to any of the statements made by any commander-in-chief. This is to avoid non-verbally signaling approval or disapproval of anything spoken by their ranking officer during the public address. Now compare the joint chief's reactions to those in the audience who only wield abstract power, and the reader will notice a stark difference. Congressional members constantly signal their approval or disapproval of what is said during the public address, both verbally and non-verbally.

History has shown us many times that people with close connections to standing armies wielding *real* power are a clear and present danger to high-ranking members of any abstract power hierarchy. Smooth-talking generals who speak contemptuously about their government's public officials have a well-documented tendency to eventually delegitimize their abstract power-based dominance hierarchies. All one has to do is read Cicero to get a play-by-play account of what it looks like when the imaginary power of senators operating in a republic gets usurped by smooth-talking generals calling themselves emperor.

One way to mitigate this threat is to logically constrain the speech of military officers via laws which prohibit them from speaking contemptuously towards public officials who wield abstract power. Making it illegal for military officers to speak contemptuously towards public officials serves as a stop-gap or interlocking safety mechanism that gives high-ranking people the legal justification they need to fire or incarcerate an emerging threat before they have time to organize a military insurrection. This sometimes works to nip the threat of coup d'état in the bud before it blossoms, but it isn't always effective. Military insurrections can and do still happen, hence why the United States of America exists in the first place.

Political assassinations and military insurrections showcase how real power trumps imaginary power whenever they come head-to-head. Many high-ranking people have relearned this lesson the hard way. A contributing factor to this problem appears to be that high-ranking people have a tendency to start believing in their own abstract power; they self-hypostatize their imaginary power as concretely real power. They make the critical mistake of believing the abstract powers granted to them by their rank is a concretely real thing and they lose sight of the fact that they are physically powerless to do anything without other people who believe in their imaginary power and do the real power projection for them.

As history testifies, the resource control authority afforded to abstract power hierarchies like monarchies only exist insofar as the ruled class, not the ruling class, is (1) willing to believe in the monarchy's abstract power and (2) willing to back it with their own sweat and blood. Undermine the *real* power projectors, and the king's control authority over the population's valuable resources swiftly disappears, as it has for several monarchies throughout history. This is how new abstract power hierarchies like the US are born. Americans are first and foremost insurrectionists who used real power to delegitimize the abstract power of their oppressive king. The US is proof of the concept that physical power trumps imaginary power.

4.5.8 Metcalfe's Law Works both Ways: Belief in Abstract Power can Disappear as Quickly as it Appears

In addition to having their abstract power physically illegitimatized, a king can also have their imaginary power suddenly vanish simply because of reverse network effects. In other words, a king can have their abstract power "cancelled" if they become too abusive with it. The simple explanation for this is that Metcalfe's law works both ways. Just as non-linearly as the value of a belief system can grow as the number of believers grows linearly, the value of a belief system can also fall non-linearly as its number of believers decreases. For this reason, it can be surprising how quickly the abstract power of a king or government can vanish from people's collective imagination once enough of the population realizes it is in their best interest to stop believing in it.

Countless revolutions have shown very clearly that it takes far less time to dissolve an abstract belief system than it does to establish it (this same phenomenon manifests itself in the modern age as "cancel culture"). A king can lose the power he spent decades building just by saying or doing one wrong thing to have power broadly questioned. For these reasons, rulers who seek to preserve their abstract power and control authority over resources would be wise not to forget about network effects or do anything to motivate mass defection. Rulers who prosper are rulers who understand they don't have *real* power, they just have *abstract* power like rank, and there are major systemic differences between these two different types of power.

4.6 Creating Abstract Power

"The truth is, one who seeks to achieve freedom by petitioning those in power to give it to him has already failed, regardless of the response. To beg for the blessing of 'authority' is to accept that the choice is the master's alone to make, which means that the person is already, by definition, a slave."
Larken Rose [99]

4.6.1 To Wield Abstract Power Over People, Convince Them They Need Your Permission or Approval

To seek permission or social approval from someone is to tacitly give them abstract power over you. On the flip side, if you want to create and wield abstract power over a large population of humans, simply convince them to adopt a belief system where they need permission or social approval from you. Once a population believes they need your permission or approval, you have successfully gained abstract power and influence over them.

It should be noted that abstract power is a relatively new phenomenon. Evidence of abstract power appears quite recently in the human fossil record. An easy way to detect when humans started believing in abstract power is when they started giving some sapiens far more materially grandiose burial rituals than other sapiens. Pre-Neolithic society appears to have been largely rank-less, with what anthropologist Peter Turchin describes as *"remarkably cooperative and egalitarian societies, with leaders who could not order their followers around, leading instead by example."* [22] For hundreds of thousands of years, humans lived in rankles societies like this, with very few distinctions beyond age, gender, and earned reputation. Everyone had a similar burial ritual. [68]

Then, starting in the Neolithic age after the invention of agriculture, disproportionately gaudy graves emerged, packed full of gold and other precious resources. God-kings wielding enormous amounts of reified abstract power emerged. Not surprisingly, they exploited people with their imaginary power. As Turchin describes, *"they oppressed us, enslaved us, and sacrificed us on the altars of bloodthirsty gods. They filled their palaces with treasures and their harems with the most beautiful women in the land. They claimed to be living gods and forced us to worship them."* [22]

Based on nothing more than artifacts dug up from the ground, we can observe that several thousand years ago, people suddenly learned that they could psychologically exploit their peers through their mutually adopted belief systems, and they've been doing it ever sense. Somewhere along the path of sapient abstract thinking and cultural evolution, the practice of storytelling went from Paleolithic shamans describing the mysteries of the afterlife and imbuing the tribe with symbolic meaning, to god-kings using rhetoric and written languages to convince thousands of people to mistake their imaginary forms of abstract power for something concretely real, then using that abstract power to gain and exploit control authority over the entire population's resources.

The modern abstract power hierarchies we live in today are derived from a similar style of ideological gatekeeping which god-kings first mastered in early civilization. The general approach to creating and wielding abstract power appears to be largely the same as it started at least as early as 7,5000 years ago. Creating abstract power can be described as a four-step process listed below and illustrated in Figure 48.

- **Step 1:** Use storytelling skills to convince a population to adopt a belief system with a desired theological, philosophical, or ideological state of being. The existence of a desired state of being implicitly defines the existence of an undesired state of being with a discrete separation between the two states of being (in this example, this discrete separation is abstracted as a gate).
- **Step 2:** Use storytelling skills to convince people there's a method, path, or gateway to reach the desired state of being.
- **Step 3:** Use storytelling skills to name yourself the gatekeeper who can generously lead people to/through the gate to achieve the desired state of being. This is a subversive form of abstract power building which distracts a population from seeing that you tacitly gave yourself denial-of-service power, which you can use to passive-aggressively deny/revoke people's access to the desired state of being.
- **Step 4:** Increase the adoption of your belief system.

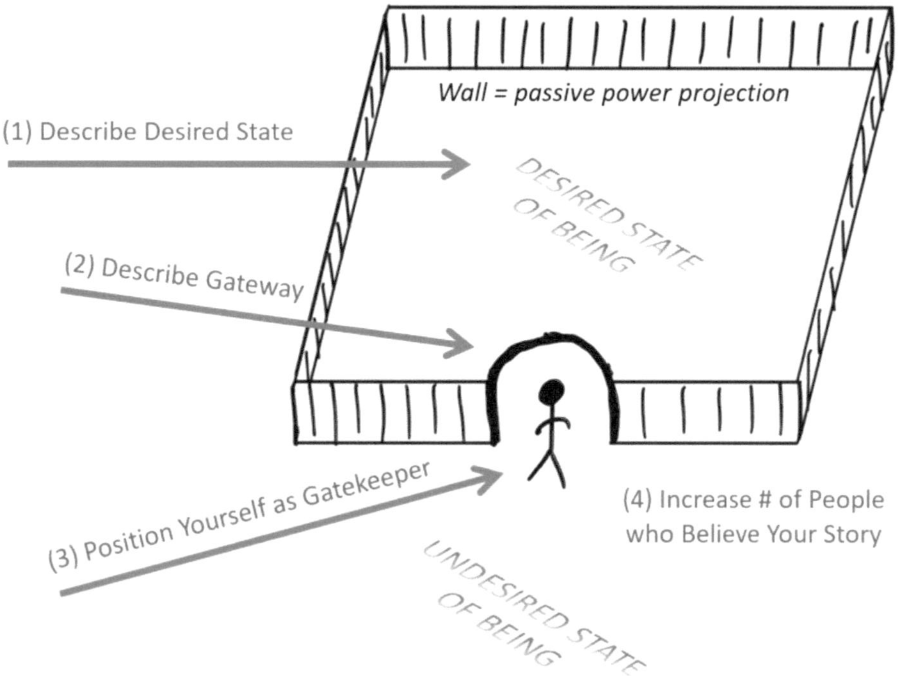

Figure 48: A Strategy for Passive-Aggressively Creating and Wielding Abstract Power

The first step of creating abstract power is to use one's storytelling and rhetorical skills (applications of abstract thinking discussed in previous section) to construct a desirable theological, philosophical, or ideological state of being, then convince people it's a real thing via reification. For most of written history, this desired abstract state of being has been described as a place – a paradise of some kind, usually in the afterlife. As society becomes gradually less theological, the desirable state of being has become gradually more ideological, but equally abstract. Instead of wanting to go to heaven, for example, people often want to be moral or ethical and will hypostatize/reify constructs like "universal moral good" as something concretely real.

The "desired state of being" is often described as an all-or-nothing phenomenon rather than a spectrum. It's usually either zero or one. You're either in paradise or you're not; you're either moral or you aren't. It's important for the desired state to be discretely separable (i.e. Boolean) like this because the creation of a discretely "true" state automatically implies the existence of a discretely "false" state. This way, the presence of a desired state (e.g. saved, divine, favored by the gods, moral, ethical) tacitly implies the existence of an undesirable state (e.g. not saved, not divine, not favored by the gods, immoral, unethical), as well as a discrete boundary between the two states which subtly and passive-aggressively denies unqualified people from reaching the desired state. If we recall the discussion on abiogenesis and passive-aggressive power projection tactics like pressurized membranes and colonization attacks in the previous chapter, this is essentially an abstract version of the same power projection tactic.

Once people have adopted a belief system where there are two discretely separate states of being (i.e. "good" and "bad") and we have convinced enough of the population to hypostatize or reify things like "good" and "bad" as concretely real things, the second step of creating and wielding abstract power is to imply there's a way to get from the undesired state (e.g. bad) to the desired state (e.g. good). The population must believe there's a "gateway to good," otherwise it would be impossible to execute the third and probably most important step to creating abstract power.

The third step to creating abstract power is to make oneself the sherpa or gatekeeper who is uniquely qualified to lead people from the undesired state (e.g. bad) to the desired state (e.g. good). By becoming the sherpa or gatekeeper, a person can wield abstract power passive-aggressively in the form of denial-of-service (DoS) attacks. Moral gatekeepers implicitly have the authority to deny people's access to the population's mutually desired state of being by simply saying someone doesn't qualify for it. In other words, the person nominated to be the "moral judge" has the ability to judge someone as immoral, and outcast them.

As an added bonus, this tacit and passive-aggressive abstract power projection tactic is easy for gatekeepers to disguise as benevolence, making it even more subversive and effective. People are quick to believe in desired states of being (e.g. moral, ethical, good, holy) due to theological, philosophical, or ideological reasons and quick to see their theological, philosophical, or ideological gatekeepers as people who generously lead others to desired theological, philosophical, or ideological paradise. The public becomes oblivious to the fact that moral gatekeepers are people who tacitly wield passive-aggressive power to socially outcast anyone they choose. The point is so important that it bears repeating: to seek permission or special approval from someone is to tacitly give them abstract power over you. As soon as a population nominates someone to be their gatekeeper, they become subservient to that gatekeeper (as a side note, this could be one reason why a popular mantra within the Bitcoin community is to "slay your heroes," because they instinctively recognize this passive-aggressive power projection tactic).

The fourth step to creating abstract power is straightforward: convince as many people to adopt the belief system as possible. Because it's easy to disguise the gatekeeper's passive-aggressive, hypostatized/reified, abstract power as benevolence, it's easy to motivate people to expand the reach of the gatekeeper's abstract power because they will believe they are helping their peers. Followers of any given belief system will be inclined to spread the good news that they have discovered the path to a theological, philosophical, or ideological paradise, and all people must do to get there is do whatever the benevolent gatekeeper says to do! Pay no attention to the elephant in the room: the fact that we are handing a person the means to psychologically exploit and abuse the population.

4.6.2 Using Moral Ambiguity, Politicking, and Demagoguery to Grow Abstract Power

As previously discussed, hypostatization/reification is often intended to be a figure of speech, but some people use it in rhetoric to convince people to reach fallacious conclusions through thinly-veiled logical arguments. This commonly happens in *politicking*, when people use rhetoric to assert that something has theological, philosophical, or ideological qualities such as moral value, and then use moral value as the logical basis for their argument. This is problematic because morals are abstract constructs – they have no capacity to be causally efficacious and they are ontologically independent from what we currently understand to be objective reality. Moreover, it's incontrovertibly recognized that hypostatization is a fallacy when used in logical arguments.

The most insidious use of hypostatization or reification (often employed by demagogues) occurs when something intended to be taken as a logical argument is disguised as a series of moral or ethically-charged metaphors. This could explain why people who seek to create abstract power often speak in metaphors. The true intent of these metaphors is to present a logical argument to persuade the audience to adopt a given belief system (for which they're the gatekeeper who decides what "right" is), but the logical argument is deliberately ambiguated as a story or series of metaphors to either (1) disguise the orator's unsound and fallacious logic, (2) preemptively hedge against anticipated critical examination of the orator's logic, (3) make a more entertaining, motivating, or persuasive speech, or (4) appeal to the desires and prejudices of ordinary people.

To illustrate the ambiguous nature of rhetoric, consider President John F. Kennedy's famous inaugural address written by Theodore Sorenson, perhaps most remembered for its "*ask not what your country can do for you; ask what you can do for your country*" statement. This chain of logic suggests American citizens exist for the sake of serving their government, which is directly contradictory to the philosophical intent of the American Constitution and the founding fathers who proposed the exact opposite theory that governments exist to serve the people. As previously discussed, Americans are insurrectionists – people who are overtly defiant to abstract power – dismissive of rank, disloyal to their king, capable of and highly motivated to kill thousands of redcoats to delegitimize their oppressive king's abstract power. The US Constitution gives American citizens the right to free speech and the right to bear arms for the explicit purpose of empowering American citizens to delegitimize the abstract power of their government if it becomes too abusive or systemically exploitative, just like the British monarchy did in the 1700s. [91, 100]

Of course, President Kennedy was probably just being rhetorical to inspire the audience to understand the importance of civic action and public service. The point is that nobody can really know what his intentions were because of moral ambiguity. It's impossible to know whether the president was trying to inspire the audience to take civic action, or he was intentionally trying to present a blatantly un-American or non-constitutional argument and passive-aggressively imply that citizens are immoral unless they devote themselves to the service of their state. Moral ambiguity is still a relevant issue today. For example, patriotism is commonly used as a form of ideological gatekeeping for demagogues to create and wield abstract power. [91]

4.6.3 Signs of People Exploiting or Abusing their Abstract Powers

"I will send a fully-armed battalion to remind you of my love."
King George III, *Hamilton* the Musical [21]

The principles of systems security apply to belief systems just as much as they apply to physical system security. It is important to be aware of one's own mental models and to guard against the threat of psychological exploitation and abuse of one's belief system. The reader is invited to reflect on signs of theological, philosophical, or ideological gatekeeping, intentional reification, politicking, moral ambiguity, demagoguery, or other examples of people trying to create abstract power so they can exploit it.

There are clear signs of people trying to create and wield abstract power. People will claim that they know how to reach a paradise in the afterlife even though they can't see, smell, touch, taste, or hear it. People will claim there is an objectively moral or ethical good even though it can't be seen, smelled, touched, tasted, or heard. People will imply there are discrete differences between heaven and hell, moral and immoral, ethical and unethical. People will imply they get to decide what behavior qualifies as moral or immoral, ethical or unethical, or worthy of heaven or hell. Then, either directly or in some passive aggressive way, storytellers will attempt to use their rhetoric to persuade you that you are guilty of being immoral or unethical, or you are destined to go to hell unless you behave the way *they* say you ought to behave, unless you follow some set of rules *they* say you have a moral obligation to follow, unless you adopt *their* belief system. Once you adopt these belief systems, you must understand that you have tacitly given them a form of passive-aggressive, abstract power over you. You have entered a permission-based belief system, where you tacitly need the permission and approval of the moral gatekeepers.

This is a generalized approach that humans use to create and wield abstract power over other humans by exploiting them through their abstract belief systems. This is how most abstract-power-based dominance hierarchies are formed. All abstract power hierarchies introduce a psychological attack vector in the form of systemic exploitation of people's belief systems. Does this mean all abstract power hierarchies are systemically exploitative? Not necessarily. But it does mean *they can be* systemically exploited. These types of belief systems rely on trusting in people with imaginary power. Whether or not abstract power hierarchies descend into a state of psychological exploitation and abuse depends upon the population's ability to recognize the signs of it happening. Unlike the abuse of physical power, the abuse of abstract power can scale far higher, far faster, and be far harder to recognize because it doesn't leave a blood trail.

4.7 Abstract Power Hierarchies

"What good is title if you have to earn it?"
Sir Walter Elliot, *Persuasion* [101]

4.7.1 Nothing but Stories told by Storytellers

We have established that sapiens are unique in comparison to other species in nature because of how much they rely on abstract sources of power to form their dominance hierarchies rather than physical power to form their dominance hierarchies. Sapiens have been behaving this way since at least the dawn of the Neolithic era. If the previously mentioned anthropological theories are true, then using abstract power to form dominance hierarchies became popular because they gave people a way to avoid physical confrontation while simultaneously helping them to emotionally reconcile the cognitive dissonance they were feeling about their domestication of animals and mass destruction of surrounding flora and fauna.

It is clear from the fossil record that humans started believing in abstract power and using it to form their dominance hierarchies thousands of years before the invention of written language, thus thousands of years before sapiens were even capable of formally encoding the logic of their abstract power hierarchies using what we now call *rules of law*. Written language is what gave sapiens the means to begin formally codifying the logic of their abstract beliefs using syntactically and semantically complex logic.

As previously mentioned, one of the major benefits of storytelling is that it allows people to transcend their physical constraints and cooperate on much larger scales. Written language was a storytelling game-changer when it comes to getting people to adopt common belief systems about abstract power, because written languages transcend the need for synchronous communication of stories. Write something down, and that story can be shared asynchronously with unlimited people without the author needing to be present, or even alive. People with abstract power took advantage of written language to convince more people to believe in their abstract power. This means written language was yet another power projection tactic. To this day, written language still remains one of the most potent power projection tactics in human society (hence why the pen is often considered to be mightier than the sword).

Not surprisingly, the ability to read and write increased the likelihood of having abstract power and control authority over people's valuable resources. **Literacy – especially the ability to read and write laws (or in modern times, the ability to read and write software) – translates directly to abstract power. Practically as soon as people learned how to write, they started encoding logic into rules of law which placed themselves at the top of an abstract power hierarchy**. Alternatively, if their attempts to give themselves abstract power were too obvious, people would use their reading and writing skills to place a god at the top of the abstract power hierarchy, and subversively imply they were god's formally-chosen representative, implicitly giving themselves access to god's abstract power.

There are several different ways to create and award oneself with abstract power, but the bottom line is that the people who write the stories and make the laws had a clear tendency to become the rulers. Literacy therefore translates directly into the ability to build empires and exploit populations through their belief systems at unprecedented scale.

We have established that abstract power and abstract power hierarchies are nothing but belief systems. They are elaborate logical constructions with high degrees of abstraction passed down over thousands of generations of storytelling. Symbols of this abstract power can be pressed into clay, written on parchment, worn as a crown, or encoded into a computer, but the abstract power they wield exists nowhere except within the minds of sapiens, the only place where it isn't both physically powerless and objectively meaningless.

Because people mistakenly hypostatize abstract power as real power, people tend to forget this. As more people treat abstract power as a concretely real thing, they proceed to act like it is a concretely real thing and that combined action convinces others it's a concretely real thing. Eventually, everyone just starts acting like abstract power is a real thing without questioning it. Creating abstract power is therefore a "fake it until you make it" phenomenon. People can fake like they have real power until people start acting like they have real power and eventually project real power for them, creating a self-perpetuating cycle. This phenomenon is commonly seen today with celebrity influencers – people who are famous for being famous. Early god-kings were similar to celebrity influencers, and the written languages they used to create and expand their abstract power was a way to make their influence "go viral" faster.

This is a very important concept to note, because in the next chapter, the author will describe how this is exactly what's happening following the invention of a new language called machine code and a new form of literacy that gives people control over people's computers and digital information. The bottom-line up front is that software represents a new way for people to encode abstract power hierarchies which place themselves at the top of those abstract power hierarchies. The argument will be made that computer programs give 21st century humans the ability to create and wield abstract power at unprecedented scale. But before we get there, we need to develop a thorough understanding about the abstract power hierarchies formed in previous centuries.

4.7.2 Types of Abstract Power Hierarchies

Societies appeared in Mesopotamia at least as early as 6,000 years ago and evolved into (ostensibly) the first cities and states around 5,000 years ago. With the invention of written language during this same timeframe, these abstract power hierarchies were formally codified into written rulesets we now call rules of law, marking the beginning of written history. Not surprisingly, the design philosophy of these abstract power hierarchies was rooted in theology, philosophy, and ideology. They rely on having someone declare the "right" way to settle disputes, establish control authority over property, and achieve consensus on the legitimate state of ownership and chain of custody of property. The people with the highest rank at the top of these abstract power hierarchies are ostensibly the best-qualified people to know what "right" is. To use the terminology from the previous section, the highest-ranking members within these abstract power hierarchies are the moral gatekeepers from whom the population seeks permission and approval.

Over time, the abstract power wielded by high-ranking people within their abstract power hierarchies got hypostatized. Their imaginary power began to look and feel physically real because everybody started thinking and acting like it were real. This still holds true today. Theological, philosophical, and ideological gatekeepers start to look powerful because our peers start thinking and acting like they're powerful. Fast forward through thousands of years of storytelling, and this process has produced many different types of abstract power hierarchies involving many different types of high-ranking people who wield many types of abstract power.

Regardless of their structure, all abstract power hierarchies effectively work the same way: the people at the bottom of the dominance hierarchy must have permission and approval from the people at the top of the dominance hierarchy. In other words, all abstract power hierarchies are trust-based, permission-based, and inegalitarian systems. The people at the top of these hierarchies are the people who are supposedly the most qualified to determine what permissions and approvals the rest of the population ought to have. This means all abstract power hierarchies are simultaneously trust-based systems. The people at the bottom of the dominance hierarchy must trust that the people at the top are indeed the most qualified to determine what permissions the population out to have. The people at the bottom of the dominance hierarchy must also trust that the people at the top won't deliberately withhold their permission and approval, nor systemically exploit the population with their special permission and approval capabilities (therein lies the massive systemic security flaw of all abstract power hierarchies).

The first types of people from which populations sought permission and approval appear to be Paleolithic shamans. These shamans were eventually replaced by Neolithic priests and god-kings. By the bronze and iron ages, cultural evolution (and a whole lot of physical conflict) changed god-kings into regular kings, senators, or emperors. Today, the highest-ranking positions wielding the most amount of abstract power are called kings, presidents, senators, and prime ministers. These high-ranking positions sit within the encoded abstract power hierarchies described in Table 4.

Table 4: Examples of Modern-Day Abstract Power Hierarchies

Abstract Power Hierarchy Design [102]	How the Abstract Power is Encoded into the Design [102]	Example
Presidential Republic	Head of state is a president that serves as the head of government, exercises abstract power alongside an independent legislature.	United States of America, Mexico, most countries in South America
Semi-presidential Republic	Head of state is a president who has some executive powers, exercises abstract power alongside an independent legislature. Remaining abstract power of the president is invested in a ministry that is subject to parliamentary confidence.	France, Russia, Ukraine, Mongolia, South Korea
Republic w/executive presidency nominated/elected by a legislature	President is both head of state and government. The ministry (including the president) may or may not be subject to parliamentary confidence.	South Africa, Botswana, Guyana
Parliamentary Republic	Head of state is a president who is mostly or entirely ceremonial. Ministry is subject to parliamentary confidence.	Germany, Italy, India
Constitutional Monarchy	Head of state is a monarch that is mostly or entirely ceremonial. Ministry is subject to parliamentary confidence.	United Kingdom, Canada, Australia, Spain, Norway, Sweden
Semi-constitutional Monarchy	Head of state is an executive monarch who personally exercises abstract power in concert with other institutions.	Morocco, Jordan, Qatar, United Arab Emirates
Absolute Monarchy	Head of state is executive, with all authority invested in a monarch.	The Kingdom of Saudi Arabia, Oman
One-Party State	Head of state is executive or ceremonial, abstract power is constitutionally linked to a single political movement.	China, North Korea, Vietnam

Something important to note about today's abstract power hierarchies is that they're exponentially more centralized and asymmetrically powerful now than they have ever been in the history of human civilization. The valuable resources of approximately eight billion people are now controlled by something like ten thousand high-ranking people across the world who have the overwhelming majority of all abstract power. While highly energy efficient, this type of social system comes at the cost of creating systemic security vulnerabilities. Never in human history have so many valuable resources been more centrally controlled by such a small ruling class (it would have been mathematically impossible in the past due to population sizes, even though there were a larger number of independent city states). People are living in an abstract belief system where the combined global resources of all sapiens are governed by only 0.0001% of its population.

4.7.3 Modeling the Differences between Physical-Power-Based & Abstract-Power-Based Resource Control

The abstract power-based (APB) dominance hierarchies created by behaviorally modern sapiens can be modeled as resource control structures. APB resource control structures that sapiens use have the same function as the physical-power-based (PPB) resource control structures of other species modeled in the previous chapter. This means we can compare the APB resource control system created by Neolithic sapiens to the PPB resource control system created by natural selection to gain some additional insight about how and why human attempts to establish their dominance hierarchy using something *other* than physical power can result in different complex emergent behavior. Figure 49 and Figure 50 compare both resource control systems (see section 3.11 for a more thorough explanation of Figure 49).

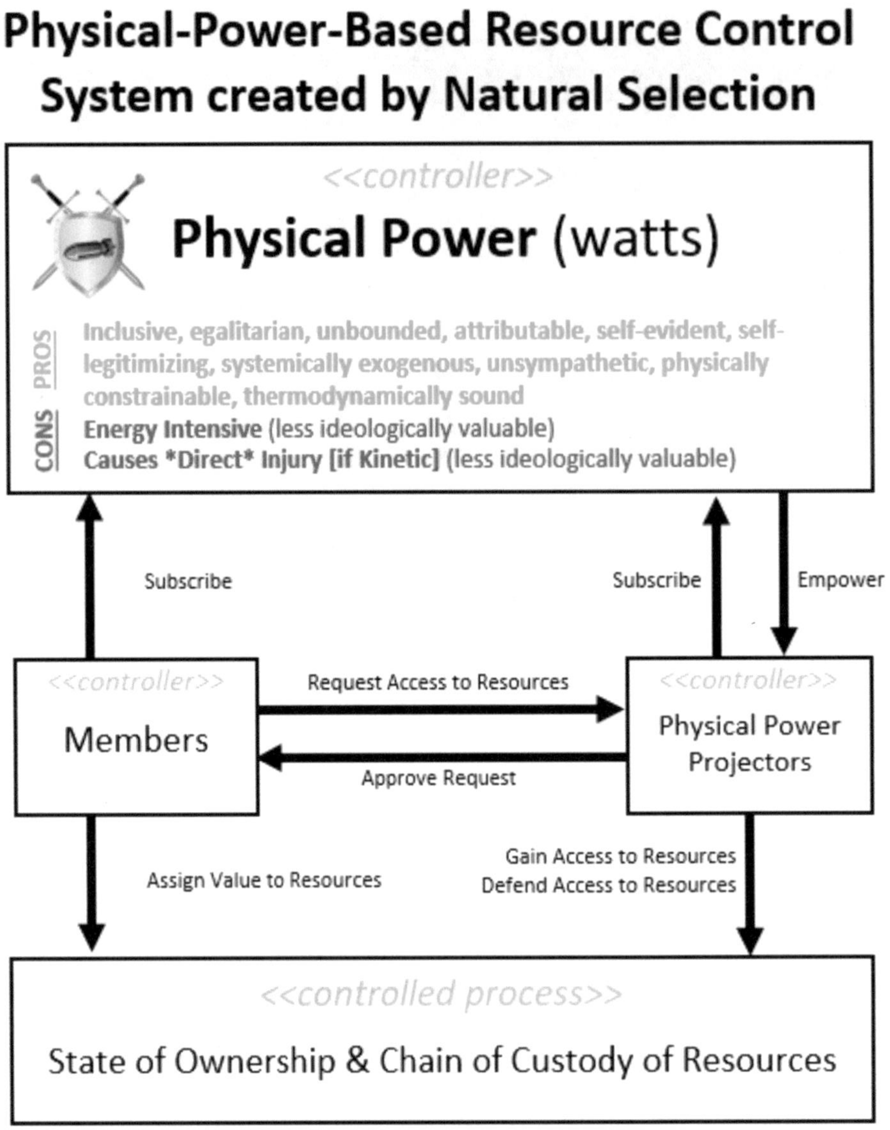

Figure 49: Model of the Resource Control Structure created by Natural Selection [88, 89]

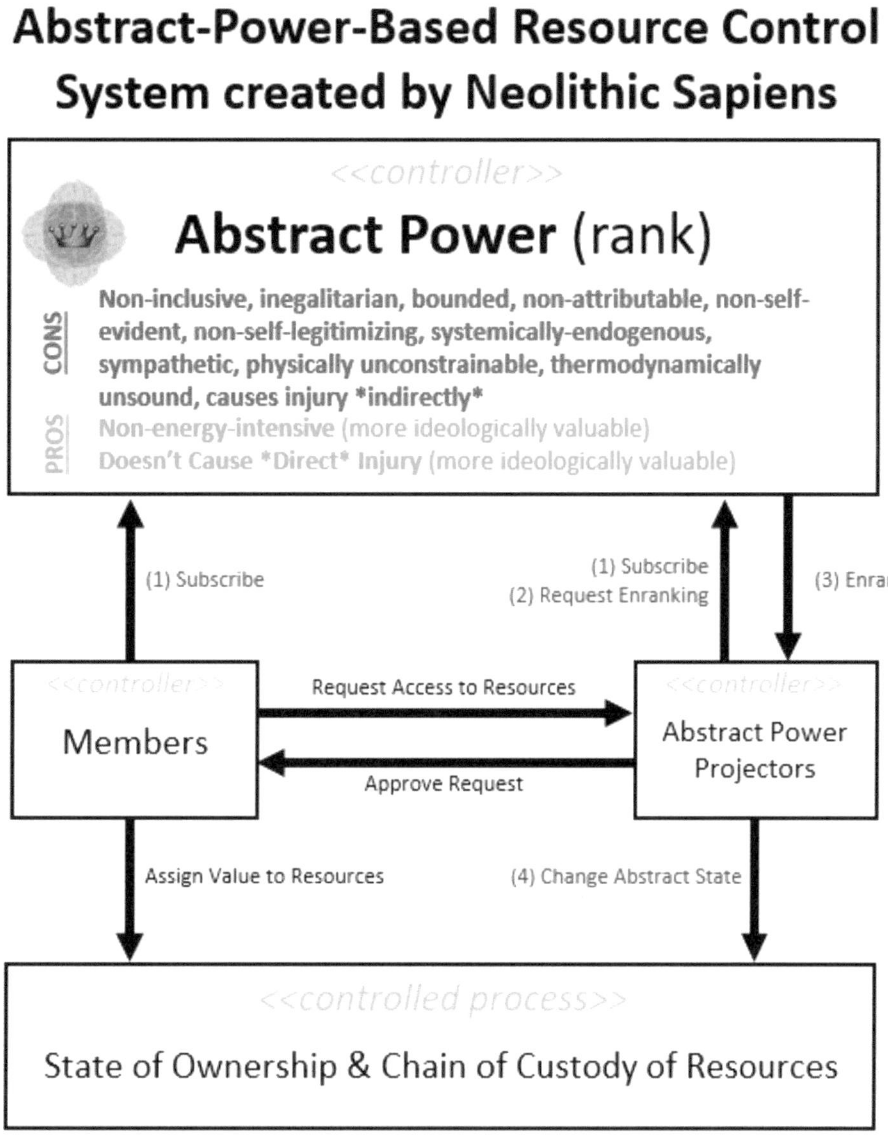

Figure 50: Model of the Resource Control Structure attempted by Neolithic Sapiens
[90, 76]

At first glance, PPB and APB resource control structures look the same. The controlled process is the same, and so is the purpose of the overall control structure: to establish consensus on the state of ownership and chain of custody of resources. The controllers are also the same. Just as wild pack animals are comprised of members and (physical) power projectors, post-Paleolithic society is also comprised of members and (abstract) power projectors. Similarly, just as (physical) power functions as controller with control authority over members and power projectors, so too does (abstract) power function as a controller and wield control authority over sapiens. Consequently, both resource control techniques have the same overall shape and structure.

A clear difference between systems is the type of power used. Whereas wild animals use physical power (watts) to establish control authority over resources, post-paleolithic sapiens *attempt* to use imaginary, or abstract power (rank) to establish control authority over resources. Sapiens use their large neocortices to create an imaginary source of power within their collective minds called rank, and then they nominate different people to have that rank. These high-ranking abstract power projectors are formally granted control authority over society's resources via formally codified rules of law, allowing them to approve (or deny) people's access to society's resources, as well as change the state of ownership of those resources. For example, the state of ownership of dry land is usually controlled by government executives or ministries. Civilians (i.e. members) within modern society must request access to that land, and government officials approve or deny those requests.

Major differences between physical and abstract power-based resource control systems have been enumerated and highlighted in different colors. Beginning with the control actions highlighted in purple, the first of four enumerated control actions worth noting for its differences is the "subscribe" control action executed by members and power projectors. Members and power projectors within APB systems subscribe to the control authority of abstract power just like they do within PPB systems. What makes this control action appreciably different is the fact that it's implicitly voluntary. Whereas members and power projectors of PPB systems do not have the ability to unsubscribe from the control authority of watts, members and power projectors of APB systems do have the ability to unsubscribe from the control authority of rank. People can simply choose to be unsympathetic to rank by ignoring it or refusing to believe in it, whereas it is physically impossible to ignore or be uninfluenced by real-world physical power.

The second of four enumerated control actions worth noting for its systemic differences is the "request enranking" control action executed by power projectors. Whereas physical power projectors receive their physical power passively via natural sortition and actively via engineering ingenuity, abstract power projectors receive their rank passively via things like rhetoric, enforcement, and rules of law. They run for election, or they climb their way up an existing hierarchy to achieve the rank they desire. People don't achieve their rank unless it's passed on to them (hence the practice of bloodlines and inbreeding) or randomly unless sortition is codified into the rule of law (Romans used to do this to this to mitigate the threat of corrupt people abusing their rank, US does this using concepts like jury duty).

The third of four enumerated control actions worth noting is the "enrank" control action executed by power. Instead of being physically empowered, power projectors within APB resource control systems are merely enranked. As discussed previously, abstract power projectors don't have *real* (i.e. physical) power.

The last of four enumerated control actions worth noting is the "change abstract state" control action executed by abstract power projectors. Because abstract power projectors don't have physically real power, they cannot physically change the state of ownership and chain of custody of resources in shared objective physical reality. They can only change the abstract state of ownership of resources. In other words, high-ranking people can claim that a piece of property you physically control access to doesn't belong to you, but they can't physically gain access to, or physically deprive you from having access to it. Thus, the ability of abstract power projectors to change the state of ownership and chain of custody of agrarian resources is symbolic only – they are actually physically powerless to do this.

Another noteworthy difference between physical versus abstract power projectors is that abstract power projectors can't execute the same control actions as physical power projectors can. Whereas physical power projectors can physically "gain access to resources" and physically "defend access to resources," abstract power projectors can do neither of these things. The same flaw of imaginary power keeps appearing in different ways: high-ranking people are physically powerless.

4.7.4 Physical Power Hierarchies are Inclusive, but Abstract Power Hierarchies Aren't

The different characteristics of physical power (watts) and abstract power (rank) lead to several different complex emergent properties between PPB and APB dominance hierarchies and resource control structures. There are at least three major differences worth noting for this theory.

The first of three major differences between PPB dominance hierarchies and APB dominance hierarchies is that **physical power hierarchies are inclusive to everyone whereas abstract power hierarchies aren't**. Everybody has access to watts and can effectively "vote" with them. Everyone can swing a fist, swing a blade, pull a trigger, solve a hash function, or do a host of other things to use physical power to represent their interests. This makes physical power hierarchies highly representative of the will of the people who choose to use it. In contrast, abstract power is non-inclusive. Few people get to have rank, and not everyone gets the right to vote. Instead, only an extremely small sample of the population gets to have rank required to vote on their rules, making the system far less representative of the will of the people who choose to use rank rather than watts. In this sense, terms like "representative democracy" are oxymorons akin to terms like "the people's republic," because a very small number of votes (0.0001% in the case of the US representative democracy) is, statistically speaking, not a highly representative sample size of the population's beliefs and interests.

4.7.5 Physical Power Hierarchies Give Everyone Unbounded Power, but Abstract Power Hierarchies Don't

A second major systemic difference between physical power (watts) and abstract power (rank) hierarchies is that **watts are unbounded and non-zero-sum whereas rank is bounded and zero-sum.** There is no theoretical limit to the amount of physical power honest people can use to represent and secure their interests. In contrast, there is a hard, mathematical limit to how much rank people can use to represent and secure their interests. This systemic difference is especially relevant in voting systems because it makes voting systems simple to systemically exploit.

For example, the resources and interests of 330 million Americans is controlled by less than 1,000 high-ranking people. This means Americans are limited to the abstract power of less than 1,000 people to represent and secure their interests. If the majority of those 1,000 high-ranking people collude at the expense of the people they ostensibly represent, then 330 million people are mathematically guaranteed not to be able to overturn their vote. They can't access more rank, and they can't outvote or overturn collusion, resulting in a state of systemic oppression which can only be solved via physical confrontation (hence why republics and democracies descend into civil wars and revolutions).

This security vulnerability occurs because all abstract power hierarchies and voting systems are trust-based and permission-based systems. People must necessarily trust that their representatives will use their rank to represent public interests rather than their own personal interests, because they are physically powerless to stop them, as well as mathematically incapable of countervailing a regulatorily captured voting system. People must operate based on the tacit permission of a small number of high-ranking people who control the majority vote, because otherwise people are both mathematically and physically incapable of adding more votes to impeach the colluding members' codified control authority. If there is no way to impeach the control authority of majority or unanimous voters, then it's technically a trust-based and permission-based system where a centralized ruling party has irrevocable control authority. Cicero famously called this design feature out: "Great is the power, great is the authority of a senate that is unanimous in its opinions."

4.7.6 Physical Power Hierarchies Can't be Systemically Exploited, but Abstract Power Hierarchies Can

A third major systemic difference between physical power (a.k.a. watts) hierarchies and abstract power (a.k.a. rank) hierarchies is that **watts are exogenous to people's belief systems and therefore invulnerable to systemic exploitation**, whereas abstract power is endogenous to people's belief systems and therefore vulnerable to systemic exploitation. In other words, rank exists internally within the system it's used. The control authority associated with rank is formally codified by rule makers, and so are the logical constraints. That means lawmakers can simply change the rules or write exploitable logic to gain, maintain, and abuse their abstract power.

Rule makers can meddle with the rules, bait-and-switch the rules, or deliberately design their logic to have exploitable properties which benefit one group of people at the expense of others. Rule makers can design the rules to award themselves with more voting power than other people, or they can make one side's rank and votes carry more mathematical weight and control authority than another side's rank and votes by simply changing the logic of the voting system. With discrete mathematical precision, rule makers can design loopholes, backdoors, trapdoors, and zero-days into the system's design logic which nullify people's ability to represent and secure themselves. They can award themselves with veto power. They can exploit voting protocols via jerrymandering or counterfeiting or other common forms of fraud. There are many ways abstract power can be systemically exploited, and there are many ways to deliberately design rules of law so that they are intentionally exploitable without detection from the public.

Physical power, on the other hand, is impossible to systemically exploit in this manner because nobody has the capacity to write any of the rules. Watts are systemically exogenous and fully independent from anybody's belief system, or any ruleset designed or encoded by any person for that matter. Physical power exists in an ontologically separate plane of knowledge than abstract power, predating human-made constructs like rank and authority by at least 14 billion years.

It's impossible to counterfeit watts. It's impossible to meddle with the properties of watts or bait-and-switch watts. It's impossible to gerrymander watts. There is no way to veto someone's watts or make one person's watts carry more weight than another person's watts. There is no way to ignore someone's watts. Nobody with any amount of rank can escape from the effects of watts. Being unsympathetic to watts doesn't negate the effect of watts. There's no way to manipulate the logic of watts to subversively encode exploitable back doors or trap doors. Watts are free from the risk of meddling, interference, mismanagement, and abuse. For these and many more reasons, watts produce a systemically secure technique for people to represent their will, resolve their disputes, and reach consensus on the legitimate state of ownership and chain of custody of their resources.

4.8 Dysfunctions of Abstract Power

"Nearly all men can stand adversity, but if you want to test a man's character, give him power."
Robert Ingersoll [103]

Now that we have explored how abstract power hierarchies are created and utilized by modern human society, we can discuss how and why they become dysfunctional. As a warning to the reader, this section and the next section consist entirely of politically and emotionally charged concepts. The purpose of this discussion is to explore the question of why humans fight wars as thoroughly as possible. This will lay the conceptual foundation for understanding why people might consider using "soft" warfighting protocols like Bitcoin. **If the theory presented by the author is true that Bitcoin represents a "soft" form of warfighting, then a general explanation about the common dysfunctions of abstract power hierarchies and why people feel compelled to fight wars should double as an explanation for why people might feel compelled to adopt Bitcoin**.

The main takeaway from this section is that the dysfunctions associated with abstract power hierarchies are very similar, whether those abstract power hierarchies are encoded by rules of law, or they're encoded by software. In both cases, there's a reason why humans eventually become so frustrated and discontented with their existing abstract power hierarchies that they pivot to using physical power to settle their property disputes and secure the policy or property they value.

A secondary takeaway from this section is that the reasons why people become frustrated with their abstract power hierarchies are likely going to remain the same no matter if the resources under people's control are physical (e.g. land) or digital (e.g. bits of information). In other words, the same reason why people would feel compelled to project physical power to secure physical property will likely be the same as the reasons why people would feel compelled to project physical power to secure digital property.

4.8.1 It's Important to Acknowledge that our Modern Agrarian Way of Life has a Dark Side

When hunting a pack of significantly more powerful animals using nothing but rocks and spears, each human's individual contribution to their small Paleolithic tribe matters, and no single person wields appreciably larger amounts of physical power than others. Within these small tribes, survival depends heavily upon forming deep, lasting, trusting relationships with peers. Effective communication becomes an existential imperative, and prosperity depends upon adopting flat organizational relationships which optimize for the lateral transfer of knowledge. [68]

This is one of many explanations for why anatomically and behaviorally modern Upper Paleolithic sapiens lived in highly egalitarian and non-hierarchical societies, as evidenced by how they buried their dead. In contrast, when living a sedentary life of comfort, complacency, and resource abundance, it is possible to afford to be more isolated from each other and adopt inegalitarian relationships which optimize for hierarchical command and control.

The way sapiens organized themselves changed dramatically during the transition from the Upper Paleolithic era to the Neolithic era. Sapiens started to believe they were metaphysically transcendent to their natural environment. They started believing in imaginary powers, which quickly blossomed into beliefs about god-kings. They began to forfeit their physical power in exchange for abstract power, trading their highly representative and egalitarian resource management systems for non-representative, inegalitarian, and systemically vulnerable abstract power hierarchies with imaginary power and resource control authority split across imaginary positions and silos.

These abstract power hierarchies were intentionally designed to give a select few members of the population sustained and unimpeachable control authority over the population's most valuable resources using an imaginary source of power and authority. When these new types of abstract-power-based dominance hierarchies formed, sapiens stopped hunting for food and started hunting for abstract power and resource control authority over other people's resources instead. People were particularly interested in gaining control authority over irrigated farmland. They stopped hunting for caribou and started hunting each other, all for something which exists only within their imagination. This would imply that much of what modern agrarian humans fight over today are, ironically, the special permissions and authorities granted to high-ranking positions of abstract power hierarchies – abstract power hierarchies which ostensibly exist to mitigate the need for using physical power to settle disputes and establish our dominance hierarchy. See the problem? **By trying to create systems where we replace physical power with abstract power in order to reduce injury and intraspecies fratricide, human society ended up creating a system where we fight and injure each other at unprecedented scale for access to abstract power**. Our desire to not use physical power as the basis for establishing our pecking order backfired.

At first, the spread of agrarian society was constrained to river basins, and much of nature was safe from their exploitation and abuse. But after the invention of the plow (and the enslavement of aurochs to pull those plows), sapiens were able to dredge up nutrients deeper in the soil. Combined with the discovery of irrigation, this allowed sapiens to irrigate land further away from river basins, enabling agrarian society to spread inland. In the wake of agriculture came artificially partitioned sections of land generating enormous amounts of resource abundance. As resource abundance grew, the need for resource control hierarchies grew (this was far less of an issue when sapiens were still highly nomadic because they "owned" relatively little possessions).

To that end, sapiens invented abstract power hierarchies to manage control authority over their agrarian resources – namely irrigated land. They created imaginary power and authority in the form of rank and title to serve as a more energy-efficient surrogate to real (i.e. physical) power. As Charles Foster summarizes, "*with title came a sense of entitlement*." [68] Sapiens started to feel superior to their fellow humans because of systems they developed to take account of their excess resources. A consensual hallucination emerged that sapiens have an inherent right to rule over nature. Perhaps even more absurdly, the self-worth of entire sapient populations became tied to how much rank and title they have.

To paraphrase Foster, after the emergence of domestication and agriculture and abstract power hierarchies, the growth rate of sapient brains started to reverse course and to shrink (as all animal brains do when they become domesticated, even in fish). Highly infectious diseases emerged. Famines emerged. Occupational diseases emerged. Dietary deficiencies emerged. Iron deficiencies emerged. Severe mental health deficiencies emerged. Closed-minded and intolerant monocultures emerged. Political competitions and infighting emerged. Excessive resources gave rise to excessively large tribes. Populations grew so large that sapiens started to experience something they hadn't experienced before: anonymity. The ability to be anonymous in excessively large tribes translates to the ability to prey on tribe members without the natural deterrence of social and reputational damage. Consequently, crime emerged, and with it the need for a formal mechanism of deterrence and delivery of punishments. Police emerged, and with it came the systemic vulnerabilities of the police becoming criminal organizations in and of themselves, giving rise to one of many forms of corruption within abstract power hierarchies. On top of all these other "features" to blossom from the fertile soil of modern agrarian society, so too did the profession of warfighting - a far less lethal byproduct of agrarian society in comparison to infectious disease, dietary deficiencies, and famine, but one that seems to get far more people's attention. [68]

During the transition of sapient social organization from egalitarian hunter-gatherers to farmers and god-kings, our fossil record indicates that despite their natural instincts, humans started killing each other far more commonly and at unprecedented scale. It appears they started fighting each other so they could have imaginary power and control authority over their agrarian resources. They started fighting for the right to define what "right" is, as well as for the right to write the laws of their abstract power hierarchies based on what they define "right" to be.

During the Upper Paleolithic era of behavioral modernity, evidence of human-on-human killing is minimal. But during the gradual transition to the Neolithic era, evidence of killing accelerates alongside the emergence of agriculture and their corresponding abstract power hierarchies. These observations are noteworthy because they illustrate how **massive-scale human-on-human killing didn't appear to become popular until *after* domestication, *after* sapiens started believing they were superior to nature, *after* the emergence of rank-based dominance hierarchies, and *after* sapiens freely chose to adopt sedentary lifestyles which give unprecedented amounts of asymmetric, top-down command and control authority over people's valuable resources for nothing more than imaginary reasons**. [68, 70, 22, 104]

Once these new characteristics of human society emerged, leisure time – that precious limited resource needed to establish one's own experiential knowledge and meaningful emotional connections with other sapiens – disappeared. It only takes 3 weeks for one farmer with a simple flint sickle to harvest enough cereal for a single family to eat over the entire course of a year, but farmers labor year-around, spending most of their time growing and harvesting grain for people they will never know. [68]

If a Neolithic farmer were to travel back to the Upper Paleolithic era and make an attempt to explain how agrarian society functions to a Paleolithic hunter-gatherer, the farmer would have to explain how sapiens decided it was a good idea to stop roaming the land and start spending the majority of their lives in one place laboring to produce unnecessarily large quantities of food for strangers, while simultaneously fighting and killing each other at unprecedented scale because of abstract belief systems where strangers wield imaginary power and control authority over the resources they produce for ideological reasons. And what motivates agrarian populations to do this? So they can strive to afford the luxury of a fraction of the amount of leisure time, travel, freedom, lack of infectious disease, and meaningful interpersonal connection that their Paleolithic ancestors had in abundant quantity prior to the invention of agriculture and their corresponding abstract-power-based dominance hierarchies. The hunter-gatherers would likely be quick to point out to the farmer that the primary aspiration of modern agrarian society is to escape from the systemic prison they created for themselves.

Ironically, agriculture was invented by sapiens in pursuit of energy efficiency – to *not* have to spend so much time and energy hunting and gathering their own food. It was also intended to reduce physical injury – to *not* have to risk personal injury securing access to food. Tragically, in their pursuit of safer and more energy-efficient methods for gaining and maintaining access to food resources, sapiens ended up creating systems which take more energy to maintain and result in more fighting and injury to keep secure. They traded the burden of having to chase down caribou with the burden of having to kill each other at unprecedented and unnatural scales to keep their arbitrarily partitioned plots of highly vulnerable irrigated land safe against neighboring abstract power hierarchies or abusive god-kings. An intellectually honest person should therefore not be so quick to cherish agrarian society's way of life with unthinking conventional reverence, because in many ways, it backfired on them. Foster offers the following explanation for how sapiens placed themselves into this predicament:

"Humans (no, let's be honest, we) wanted convenience and what we saw as security. We wanted to reduce or eliminate contingency. We sought to rule the natural world, and began to see ourselves as distinct from it, rather than part of it. Our early efforts at control were, in one sense only, very successful. We managed to produce a lot of calories in one place. That caused a population explosion. Once the population started to increase, there was no way back. We had to produce more calories, and to increase the size of the places in which we produced them. There was no escape from the places... Enter status, surplus, markets, all sorts of camp followers, including overcrowding, loneliness, occupational disease, diseases of sedentary life and epidemics of infection diseases. Continue synergistically for 12,000 years or so, and you have us." [68]

Foster and many other anthropologists explain how sapiens systemically trapped themselves with agriculture. He argues the first cities from which civilization emerged represent the point when sapiens lost all their options and were forced down a hazardous path from which they can't escape. [68]

"Farming, like heroin, is easier to get into than out of. Surpluses boost population, and high population kills all the animals and eats all the nuts and berries from miles around, making return impossible. Once the jaws of monoculture close around you, that's it: you've just got to go on producing more. And when you start trading, the law of supply and demand increases the pressure; binds you more tightly to the wheel..." [68]

For these and many other reasons, modern civilization can be seen as a cautionary tale. If our Upper Paleolithic ancestors could see how modern agrarian domesticated sapiens live today, they would probably not envy our lives. Humans replaced the emotionally fulfilling challenge of hunting and gathering (for which we were psychologically and physiologically optimized) with unnaturally sedentary and laborious lives filled with social isolation, infectious diseases, health deficiencies, warfighting, and probably most devastating of all, high-ranking sociopaths who psychologically abuse and systemically exploit their populations through their belief systems at extraordinary scale.

Clearly there are benefits to our modern way of life, but it's certainly not all sunshine and rainbows. The tragedy of modern civilization is that most sapiens alive today have only ever known isolated, sedentary, tediously laborious lifestyles teeming with inequality and systemic exploitation of their belief systems, compared to the egalitarian, adventurous, dynamic lifestyles sapiens once had. Modern agrarian domesticated sapiens engorge themselves on cheap, artificial, and easy food. They chase after imaginary wealth and power, all while chasing the illusion of security and prosperity, blissfully unaware of the systemic hazards they place themselves in, and the eye-watering levels of exploitation their belief systems routinely get subjected to. In the process of domesticating and entrapping animals, sapiens domesticated and entrapped themselves, and now they are incapable of knowing how utterly unhappy they are because they have never seen, known, or experienced anything except the inside of their agrarian cage.

4.8.2 Sacrificing Individual and Collective Security to Spend Less Energy and Cause Less Injury

The sapient instinct to not fatally injure fellow human beings is often stronger than the sapient instinct for self-preservation. People have such a powerful inclination to avoid hurting each other that they often refuse to injure people who represent a direct threat to their own life and limb – an instinct which militaries spend a great deal of time and effort to overcome with training. It should therefore come as no surprise to the reader that **sapiens will accept the flaws of abstract power hierarchies for the sake of not having to injure each other. They will go against four billion years of natural selection and *not* use physical power to settle intraspecies disputes, establish control authority over intraspecies resources, and achieve consensus on the legitimate state of ownership and chain of custody of intraspecies property**. They will attempt (emphasis on the word attempt) different pecking order strategies than "might is right" or "feed and breed the power projectors first" and try to ration their resources using non-physical techniques which don't risk physical injury. [70]

As discussed in the previous section, sapiens are so good at abstract thinking and so instinctively disinclined to physically injure each other, they invent imaginary sources of power to serve as a surrogate to physical power. They use this imaginary power to settle disputes, establish control authority over resources, and achieve consensus on the legitimate state of ownership and chain of custody of their perceived property. When they do this, they tacitly trade something real (watts) for something abstract (rank), creating a physically different system with different emergent behavior. The motivation for doing this is primarily ideological – less killing and less energy expenditure is something inherently "good."

Trading an energy-intensive resource control strategy prone to causing injury for an energy-conserving resource control strategy that (ostensibly) doesn't lead to physical injury is considered to be a better approach to solving the existential imperative of establishing pecking order for reasons that are difficult to explain without subjective and abstract reasoning. A flaw of this reasoning is that abstract power hierarchies clearly do lead to massive amounts of energy expenditure and physical injury because of their systemic dysfunctionality – more so than humans experienced prior to their invention.

Abstract power's perceived advantages are derived from the fact that abstract power doesn't physically exist. Because abstract power doesn't physically exist, it is incapable of consuming energy or causing injury. This is considered to be a good thing because people think they can create abstract power hierarchies to serve as more energy-efficient and safer systems for settling disputes, establishing control authority over internal resources, and achieving consensus on the state of ownership and chain of custody of perceived property. But like most things, the decision to use abstract power versus physical power to manage resources comes with major systemic tradeoffs. Because abstract power doesn't physically exist, that means it can't be physically constrained. Because abstract power doesn't consume energy or cause physical injury, that means it can be scaled and abused in ways that are highly exploitative yet completely unattributable and imperceptible to entire populations of people.

The physical differences between physical power and abstract power were outlined in the previous sections. These differences mean people should not expect to see the same complex emergent behavior between PPB and APB dominance hierarchies. But how often do people stop and think critically about these differences, as opposed to outright rejecting the idea of using physical power as the basis for managing resources because of ideas that it is morally, ethically, or theologically "bad?"

When sapiens trade physical power for abstract power, they make a tradeoff in complex emergent behavior. What they sacrifice in the trade is systemic security. They take a resource control protocol that is demonstrably secure and vetted by four billion years of natural selection, and they trade it for a demonstrably insecure resource control protocol which is vulnerable to predation. They take a zero-trust, permissionless, inclusive, egalitarian, unbounded, and systemically exogenous resource control protocol, and they replace it with a non-inclusive, bounded, inegalitarian, and systemically endogenous protocol where a ruling class must be trusted not to abuse their abstract power, and a ruled class must tacitly have permission from that ruling class to have access to their property. They take a resource control structure that can be seen, vetted, physically constrained, and verifiably decentralized, and they replace it with a resource control structure that's invisible, that can neither be physically constrained nor verifiably decentralized. Then, they act surprised when an entirely different system with vastly different physical properties doesn't exhibit the same complex emergent behavior as the system they're trying to replace. To put it simply: people replace physical power with abstract power and then act surprised when it backfires on them by causing either foreign invasion or widescale systemic oppression.

In exchange for energy efficiency and a desire to manage their resources with less physical injury, sapiens adopt abstract power hierarchies which undermine their own systemic security. This tradeoff leads straight to severe losses in the form of foreign invasion or systemic exploitation – all of which must inevitably be resolved using far larger quantities of energy-intensive and injurious physical power than they probably would have needed if they had just been more inclined to recognize there is no viable replacement to physical power as the basis for resolving disputes in a zero-trust and egalitarian way.

The systemic security flaws of abstract power hierarchies could explain why sapiens started killing each other at unprecedented scale following the emergence of agrarian society. We already know from the domestication of animals that there is a direct, causal relationship between systemic insecurity and interference of naturally-selected pecking order heuristics. It is possible that over-reliance on imaginary power and abstract power hierarchies represents a form of self-domestication that actually *causes*, rather than merely *correlates* to, severe population-scale security vulnerabilities which inevitably lead to sudden and explosive reversions back to physical confrontation. [68]

It's entirely possible that a direct contributing factor of warfare – a phenomena which appears to have emerged alongside many other unfortunate side effects associated with modern agrarian society (e.g. famine, infection disease, dietary deficiencies, physiological degeneration) – is, counterintuitively, the abstract power hierarchies we ostensibly use to avoid warfare. In what would be a supremely ironic and tragic turn of events, it is feasible that sapiens cause far more inefficiency, waste, and death for themselves by trying to *avoid* physical conflict than by simply using physical power to settle their disputes. Their aversion to physical conflict may lead them to over-rely on abstract power as an alternative but far less egalitarian or systemically secure basis for establishing pecking order and managing resources, causing population-scale resource mismanagement, exploitation, and abuse to fester and blossom until the point where it inevitably boils over into far more destructive wars.

4.8.3 Nature Suggests that "Might is Right" might Actually be the "Right" Approach

Now is a good time to remind the reader of the core theoretical concepts of power projection in nature and the importance of correcting for survivorship bias. What we see in nature is what has survived a rigorous natural selection process. On the contrary, that means what we do *not* see in nature could be what *didn't* survive the same natural selection process. With this in mind, let's ask ourselves what do we *not* see in nature? We don't see animals avoiding the use of physical power to establish pecking order and manage their internal resources. We see them optimizing for it – we see the strongest, most intelligent, and most physically aggressive animals rising to the top of practically every food chain in every biome. It's probably not the case that other resource management strategies which use less energy or cause less injury haven't been tested in the wild before. It's more likely the case that several alternative heuristics for establishing dominance hierarchies have been attempted many times before, but we don't observe them because they didn't survive. In other words, they weren't the right strategy.

For the sake of scientific rigor, people should be willing to accept an amoral hypothesis that "might is right" or "feed and breed the strongest and most intelligent power projectors first" is the appropriate heuristic for establishing pecking order over our resources, simply because it's a demonstrably effective way to survive in a world filled with predators and entropy, from which animals don't have the option of escaping. In other words, it might be worth accepting the downsides of using physical power as the basis for establishing pecking order over resources because it's the most demonstrably successful protocol for survival. If you don't use physical power as the basis for establishing intraspecies pecking order, then you should, at the very least, not expect to have the same capacity for survival and prosperity as the animals who do, because that is clearly what we can observe in nature, and there's no reason to believe that sapiens would be an exception to nature.

The use of physical power to settle disputes and establish resource control authority is no doubt energy-intensive and directly prone to injury, but it's clearly more capable of prospering in a world filled with predators. There are also major systemic benefits of physical power that abstract power is clearly incapable of replicating. Physical power doesn't require trust. Physical power doesn't require permission. Physical power is inclusive, egalitarian, unbounded, and systemically exogenous. When physical power is used as the basis for establishing pecking order, it creates a natural meritocracy which physically constrains and verifiably decentralizes how much control authority over resources any one single pack member can have. These systemic benefits explain why sapiens so frequently revert back to using physical power to settle intraspecies disputes and manage intraspecies resources, despite how energy-intensive and destructive it is.

Wild animals who utilize physical power to establish their dominance hierarchies do not appear to be capable of perceiving that it is theologically, philosophically, or ideologically reprehensible. They don't appear to have neocortices capable of abstract thought, so they are incapable of being peer pressured into believing that physical power-based resource control protocols are "bad." Instead, animals simply accept the energy expenditure and risk of injury. Instead of trying to find a replacement to physical power, they evolve special technologies which continue to allow them to use physical power as the basis of settling disputes and establishing pecking order but mitigate the threat of physical injury (e.g. antlers).

People look down upon wild animals and condemn their pecking order heuristics as brutish or unfair. What they ignore is that animals which use physical power as the basis for settling property disputes or establishing control authority over their resources don't appear to experience psychological exploitation of their abstract belief systems, or incompetent mismanagement or near-routine, catastrophic collapses of abstract power hierarchies which lead to famine and warfare. Animals which don't believe in imaginary systems don't suffer from countless problems which arise from giving members of a pack too much physically unconstrained abstract power and asymmetric control authority over their resources.

4.8.4 Unconstrained Abstract Power Hierarchies can Breed Complacency and Corruption

"Great civilizations are not murdered. They commit suicide."
Arnold Toynbee

5,000 years of written testimony indicates that sapiens have not transcended primordial economics. Their resources are depleted because they stop capturing them. Their abstract power hierarchies become teeming with politicking and get regulatorily captured by incompetent, corrupt, or self-serving intermediaries who abuse their rank, mismanage internal resources, and systemically exploit the rules. This breeds social distrust and civil unrest. Inevitably, opportunistic neighbors arrive at the gate. Physical aggressors or corrupt officials justify their actions using theological, philosophical, or ideological pretenses, but the primordial economic reasons for their attacks remain the same: high **BCR_A** [22]

The study of sapiens' written history is the study of the catastrophic dysfunctionality of abstract power hierarchies and a continuous return to physical power as the basis to settle disputes, re-establish control authority over resources, and achieve consensus on the legitimate state of ownership and chain of custody of property. **No matter how much people like to pride themselves on developing more energy-efficient or peaceful abstract power hierarchies they meticulously codify using rules of law, history makes it very clear that these systems are highly dysfunctional and utterly incapable of producing a secure society for long durations of time**. These systems appear to last for short durations of time where it's possible for populations to trust their high-ranking adjudicators not to become incompetent or abusive with their authority. Otherwise, these systems only last as long as it takes for hungry neighbors to sense physical weakness and devour their prey. As anthropologist Peter Turchin observes:

"There is a pattern that we see recurring throughout history... When survival is no longer at stake, selfish elites and other special interest groups capture the political agenda. The spirit that 'we are all in the same boat' disappears and is replaced by a 'winner take all' mentality. As the elites enrich themselves, the rest of the population is increasingly impoverished. Rampant inequality of wealth further corrodes cooperation. Beyond a certain point a formerly great empire becomes so dysfunctional that smaller, more cohesive neighbors begin tearing it apart. Eventually the capacity for cooperation declines to such a low level that barbarians can strike at the very heart of the empire without encountering significant resistance. But barbarians at the gate are not the real cause of imperial collapse. They are a consequence..." [22]

Early agrarian societies had highly exploitative, rank-based power hierarchies led by tyrannical god-kings, but then transitioned to less exploitative rank-based hierarchies over thousands of years. Why? Because of people who risked life and limb to impose severe physical costs on their oppressors and invaders. Rarely did people negotiate their way out of their god-kingdoms, oppressive empires, and total monarchies; they bought their way out by paying a high price in bloodshed.

Somehow populations keep allowing themselves to forget these dynamics. Despite how many times sapiens have relearned the same lesson the same hard way, they keep forgetting it. Perhaps it is because of short memory spans that we let ourselves forget that oppression begins with our belief systems. Oppression happens when storytellers use rhetoric to persuade people to delegitimize their physical power (watts), reify imaginary power (rank), give control authority over resources to the people with the most amount of rank.

Early god-kings mastered this strategy using theological constructs to convince thousands of people they were living gods, metaphysically superior to others and transcendent to nature. Citizens living within these abstract power hierarchies legitimately believed their leader's imaginary power was something concretely real, just like how today people still believe that the imaginary power of high-ranking people is concretely real. As a result, early agrarian society became domesticated and enslaved. They forfeited the meritocracy of their physical power for the imaginary power of their god-kings and as a result, they were literally put in cages.

The abstract power of god-kings was legitimized and hypostatized by the physical power of role-playing enforcers, but that physical power was often dwarfed by the physical power of the population being oppressed. This begs the question, why didn't these populations rise against their oppressors sooner? The god-kings' enforcers represented only a small portion of the population and could have (and eventually were) overthrown by slaves and working-class people who far outnumbered their oppressors. This means a major contributing factor to the high levels of systemic exploitation and abuse of god-king abstract power hierarchies was pacifism – a lack of willingness to impose physical costs on oppressors because of ideological beliefs. The ruled class was not inclined to spend the energy or risk the injury required to countervail the control authority of their oppressors by making it impossible to justify the physical cost of exploiting them. Just like all other types of domesticated animals, **a major factor which contributed to the egregious levels of systemic exploitation and abuse experienced by early agrarian populations was their *lack* of physical aggression – their abstract beliefs about the imaginary power of their god-kings made them docile, submissive, and easily exploitable**.

These dynamics of oppression have not changed over the years – just the abstractions used to convince people to forfeit their physical power and their ability to impose severe physical costs on their attackers. Today, populations adopt pacifist belief systems for other abstract reasons. In the bronze and iron ages, oppressors used theology to reify abstract power and persuade people to forfeit their physical power (i.e. obey the ruling class because god said so). After that went out of style, oppressors started using increasingly more philosophical or ideological constructs (i.e. obey the ruling class because they represent moral and ethical good). For some reason, philosophical or ideological approaches to creating abstract power hierarchies are perceived to be less vulnerable to systemic exploitation and abuse than theological approaches. This doesn't make rational sense because although they are more secular, philosophy and ideology are equally as abstract as theology (i.e. "good" is just as subjectively defined as "god"), and therefore equally as vulnerable to the exact same reification fallacies god-kings and monarchies used to prey upon to systemically exploit their populations.

Greeks and Romans philosophized about moral good and divided their abstract power across an ostensibly representative group of voters. Meanwhile, they ran their empires on human slave labor who weren't allowed to vote. While senators pontificated about the equality of man and the moral imperative to establish governments more representative of the people, they deliberately codified laws which only represented and benefited specific types of people (specifically rich, white, literate males).

Eventually, overt physical exploitation through slavery became too hard to do because of the blood trail it leaves. Physical power is explicit and active-aggressive, making it easy for populations to see the offenses being committed against them and organize to secure themselves. As a result, today the preferred way to exploit a population is to do it non-physically, usually through a population's belief system. Simply write laws with exploitable logic in them – deliberately encode vulnerabilities and attack vectors that you can exploit later. Slap ethically-pleasant-sounding names on bills to fool the masses with morally camouflaged sleights-of-hand. These strategies work because they're subversive and passive aggressive. Populations won't recognize they're being exploited. Or perhaps they will recognize that they're being exploited, but they won't be able to identify the root cause: the system itself.

4.8.5 Abstract Power Hierarchies can Motivate Citizens to Forfeit their Physical Strength and Aggression

One way to systemically exploit modern sapiens is a form of domestication where people are persuaded via ideology to lay down their physical power and condemn their physically aggressive power projectors. The assertion is made that the use of physical power to establish control over resources or impose severe physical costs on others is morally reprehensible due to how energy-intensive and injurious it is. While perhaps well-intended, this argument is logically unsound when considering the core concepts of survival, natural selection, and the existential imperative of physical power to solve the survivor's dilemma. It's also blatantly ignorant to lessons offered by history (as a side note, this is a predominant argument against Bitcoin; people are making the assertion that Bitcoin is "bad" because of how energy-intensive it is, oblivious to what that physical power is being used to impose severe physical costs on attackers).

Ideologists will argue that sapiens are somehow "above" natural selection or morally superior to or even exempt from it. They will persuade people to lay down their power projection technology out of moral necessity in favor of "peaceful" abstract power hierarchies codified into rules of law which, of course, place the ideologists which define what "right" is at the top of the ranks within the resource control hierarchy. Oblivious to the meaningful systemic differences between physical and abstract-power-based resource control structures, a population will pride itself on their moral and ethical superiority and their effort to replace physical power with abstract power to establish pecking order. In exchange for the promise of less energy expenditure and less potential for injury, they will remove the physical constraints preventing belligerent actors from invading or exploiting them.

Depending on how trustworthy and benevolent people are, a society which hamstrings their own ability to impose severe physical costs on threats may enjoy a few generations of economic surplus and resource abundance due to the efficiencies gained by blatantly ignoring their security responsibilities. But as written history very clearly shows, these societies struggle to survive long term. Sooner or later, their governments become corrupt, or Hannibal arrives at the gate, and their future depends on their ability and willingness to summon as much physical power and aggression as quickly as they can to countervail the attack. But it's often too little, too late.

As hard as they try and as well-intentioned as they may be, sapiens cannot escape their own nature. Primordial economics, natural selection, the survivor's dilemma, and the basic dynamics of survival are always in play regardless of how "civilized" sapiens feel they have become, or how well-designed their abstract-power-based dominance hierarchies are. **If a sapient population makes itself a ripe target of opportunity by not continually growing their C_A by improving their ability to impose severe physical costs on neighbors, it risks getting devoured the same as any other living thing in nature does**. Ideologists ignore four billion years of natural selection and attempt to replace physical power with abstract power for the sake of energy efficiency and injury avoidance. They argue they have a moral imperative to reduce waste, but then create trust-based, permission-based, inegalitarian, bounded, physically unconstrained, systemically insecure, imaginary power structures over resources which become extremely wasteful.

4.8.6 Abstract Power Hierarchies Create Honeypot Security Problems with All Honey but No Sting

Honeybees offer a simple explanation for why abstract power hierarchies are so systemically insecure. Honeybees represent a literal honeypot security problem; they produce something everybody wants. To resolve their honeypot security problem, honeybees are equipped with stingers designed to inflict severe physical pain on attackers. Honeybees will sink their stingers so deep into the skin of their attackers that they can't detach without suffering from fatal disembowelment. This happens because a honeybee's stinger is not designed for self-defense; it is designed to achieve the complex emergent benefit of strategic security. This complex emergent benefit is a byproduct of the fact that honeybees are designed to maximize the amount of pain (i.e. severe, physically prohibitive cost) imposed on attackers, to the point where it is fatal to the bees who produce that pain. The reason why this works is because of the simple primordial economic dynamics discussed in the previous chapter. Maximizing one's ability to impose physical costs (i.e. increase C_A) lowers the $\mathbf{BCR_A}$ of the hive and successfully deters attack. In other words, **honeybees prove that the solution to the natural honeypot security problem is to impose the maximum amount of physical pain on attackers as possible, even if it means risking physical injury**.

Abstract power hierarchies represent a honeypot security problem, where the honey is asymmetric control authority over a population and their abundant resources. When abstract power hierarchies are codified into rules of law, they invite attacks from two types of threats: external invasion and internal corruption. The former uses physical power to perform the attack, whereas the latter uses the hierarchy's own belief system to perform the attack. In both cases, primordial economics offers a simple explanation for why these attacks occur: abstract power (i.e. rank) and formally codified abstract power hierarchies (i.e. rules of law) generate enormous resource abundance and control authority, but they have no capacity to impose unbounded, infinitely expandable, and severe physical costs on attackers. Consequently, their $\mathbf{BCR_A}$ approaches infinity.

Like producing a hive full of delicious honey, abstract power hierarchies create attractive and desirable resource that people want. But because the worker bees have no physical capacity to sting attackers (or more likely because they believe it's "bad" to sting people), they have no capacity or inclination to keep themselves systemically secure. Abstract power is all honey and no sting.

Herein lies the primary strategic security flaw of abstract power hierarchies and their codified rules of law. The better they're designed, the more people use them and the more asymmetric resource control authority they produce. But as a population becomes increasingly reliant on abstract power to settle their disputes, establish control authority over their resources, and determine the legitimate state of ownership and chain of custody of their property, they risk self-domestication. They lose their capacity and inclination to impose severe physical costs on their oppressors. If they aren't aware of this risk, they will turn themselves into soft targets of opportunity to be invaded or systemically exploited. All it takes for this to happen at scale is for enough people to become convinced that physical power is "bad."

The systemic security threat caused by human domestication is why people should maintain a healthy caution for abstract power hierarchies. No abstract power hierarchy is safe from the systemic security flaws of abstract power. No matter how well-designed they are, all abstract power hierarchies are systemically insecure against external invasion or internal exploitation if they don't incorporate the use of physical power to mitigate these threats. Over-reliance on non-physical adjudication methods like laws and courtrooms sounds desirable because of how energy efficient and "peaceful" they sound, but this comes at the cost of creating a major strategic honeypot security problem.

High-ranking positions start wielding asymmetric levels of abstract power and control authority over their population, increasing the B_A of capturing those positions and simultaneously causing the population's $\mathbf{BCR_A}$ to creep higher and higher. If they aren't careful, the population will turn into a hive of honeybees with no stingers, laboring to produce valuable honey for their queen, but physically incapable of (or morally disinclined) to keep themselves systemically secure by remaining physically strong and aggressive to keep their C_A high and their $\mathbf{BCR_A}$ low. Eventually, like everything else in nature who didn't get the memo on how primordial economics works, a population that doesn't do this will get devoured. To expect a population with a permanently-increasing BCR_A *not* to be attacked either internally or externally is to expect humans to *not* behave like humans.

Without physical power, neither rules of law, nor the high-ranking people writing the laws and controlling the resources, nor the honest and well-intentioned users participating in these systems, can secure themselves against attack. Because their power is strictly imaginary, abstract power hierarchies can only increase society's B_A and therefore only increase their BCR_A. From both a systemic security and primordial economic perspective, this is clearly not ideal for any organization seeking long-term survival and prosperity. **The population is forced to trust in the benevolence of their neighbors or their rulers *not* to physically attack or systemically exploit them (respectively) despite an ever-increasing benefit of doing so – and trust is a highly ineffective security strategy**.

Abstract power hierarchies are often founded on ideological principles which claim physical aggression is "bad." The downside of this belief system is that it makes a population morally disinclined to be physically aggressive – that is, to do the thing that has already been proven to be the surviving solution to honeypot security problems. Honeybees quite literally represent nature's chosen solution to the exact same honeypot security problem that agrarian sapiens face with their irrigated land, or whatever other valuable resource they want to protect (e.g. bits of information transferred across cyberspace). The fittest solution to a honeypot security problem is, without a doubt, to maximize everyone's ability to sting the attacker – to impose as much severe physical cost as possible despite the risk of personal injury.

In their state of ideologically-induced docility, sapiens ignore nature's proven solution to the honeypot problem and become disinclined to impose severe, physically prohibitive costs on foreign invaders or internal exploiters, thus shrinking their own C_A. This tendency of abstract power hierarchies to increase B_A and decrease C_A simultaneously is a major systemic security hazard because it can cause a population's BCR_A to increase rapidly. This is the same primordial economic dynamics seen when humans domesticate animals, as well as when they hunt caribou. This concept is illustrated in bowtie notation in Figure 51.

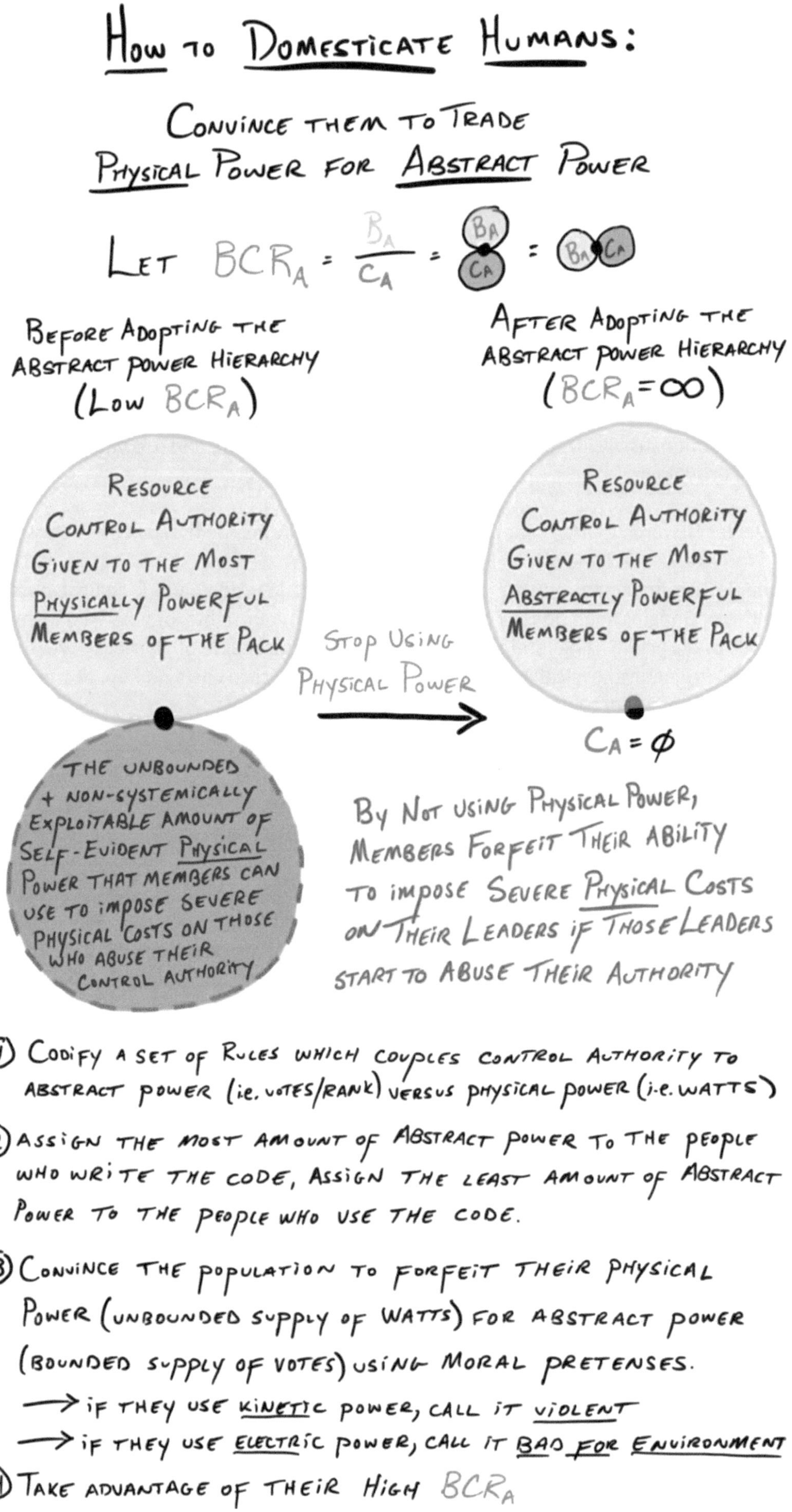

How to Domesticate Humans:
Convince them to Trade Physical Power for Abstract Power
Let $BCR_A = \frac{B_A}{C_A}$
BA
CA
Before Adopting the Abstract power Hierarchy (Low BCR_A)
After Adopting the Abstract power Hierarchy ($BCR_A = \infty$)
Resource Control Authority Given to the Most Physically Powerful Members of the Pack
Stop Using Physical Power
Resource Control Authority Given to the Most Abstractly Powerful Members of the Pack
$C_A = \phi$
The unbounded + non-systemically exploitable amount of self-evident Physical Power that members can use to impose severe physical costs on those who abuse their control authority
By Not using Physical Power, Members Forfeit Their Ability to impose Severe Physical Costs on Their Leaders if Those Leaders start to Abuse Their Authority
① Codify a set of Rules which couples control Authority to abstract power (i.e. votes/rank) versus physical power (i.e. watts)
② Assign the most amount of Abstract power to the people who write the code, Assign the least amount of Abstract Power to the people who use the code.
③ Convince the population to forfeit their physical Power (unbounded supply of watts) for abstract power (bounded supply of votes) using moral pretenses.
→ if they use kinetic power, call it violent
→ if they use electric power, call it bad for environment
④ Take advantage of their High BCR_A

Figure 51: Bowtie Illustration of How to Domesticate Humans

Tying this point back to the core concepts presented in the previous chapter, abstract power hierarchies and their codified rules of law represent the "option #1" strategy for solving the survivor's dilemma. They represent a strategy for chasing after prosperity which actively reduces prosperity margin (see Figure 18 for a refresher). History makes it clear that abstract power hierarchies which focus too much attention on resource abundance, but too little attention on imposing severe physical costs on attackers, are practically guaranteed to be devoured either from the outside via foreign invasion, or from the inside via corruption. As desperately as people want to believe otherwise, no matter how well they are designed and codified, abstract power hierarchies can do nothing to physically prevent or constrain attacks. Abstract power can't stop unsympathetic neighbors or systemic exploitation, and too much reliance on abstract power causes a population to risk self-domestication.

To be clear – the author is not arguing that society should end the practice of attempting to govern itself or manage resources using abstract power hierarchies codified via rules of law. Rules and rank-based abstract power hierarchies have no doubt helped facilitate cooperation and achieve extraordinary levels of resource abundance for sapiens across the world. The point is that abstract power hierarchies are trust-based and permission-based systems which are demonstrably insecure against corruption and invasion, and it is vital for societies to recognize these systemic security flaws.

4.8.7 Abstract Power Hierarchies are Demonstrably Incapable of Stopping Physical Power Projectors

As much as people try, imaginary powers encoded by rules of law don't work against outsiders who aren't sympathetic to them. Genghis Khan doesn't care about your rank, your rights, or your rule of law. Abstract power is utterly incapable of disincentivizing foreign invasion, and clearly incapable of physically preventing it. What does stop Genghis Khan? Projecting physical power to impose severe physical costs on him, at the risk of using a lot of energy and risking a lot of injury to do so.

From the perspective of primordial economics, natural resources like farmland create a honey pot target with high **BCR_A**. Imagine laboring for years to create freshly irrigated and fertile land filled with specially-designed grains capable of producing abundant amounts of food, but then doing nothing to impose severe physical costs on neighbors. Sapiens who don't physically secure their farmland actively invite predation by deliberately creating a high **BCR_A** resource. In the early days of agrarian society, freshly irrigated land would have been like placing millions of dollars' worth of cash into the middle of Times Square with no supervision or oversight, and then expecting nobody to take it. At a certain point, the blame for the losses associated with an attack goes to the people who are stupid enough to put millions of dollars' worth of cash into the middle of Times Square with no supervision or oversight and expect it not to be attacked. What do you expect is going to happen when you put a high **BCR_A** source of food in the middle of a continent filled with hungry and aggressive peak predators? Is Sargon the conqueror really to blame? Or are unrealistic expectations about how to survive and prosper in nature to blame?

At a certain point, foreign invasion can be attributed to people who ignore primordial economics and proceed to design hazardous systems which produce egregiously high **BCR_A**. Sapiens are first and foremost predators; to expect a group of sapiens not to attack something with a high **BCR_A** is to expect sapiens not to behave like sapiens. As Nietzsche observed, "*Even the body within which individuals treat each other as equals ... will have to be an incarnate will to power, it will strive to grow, spread, seize, become predominant – not from any morality or immorality but because it is living and because life simply is will to power*." [105] Populations that seek to survive and prosper have a responsibility to understand life's relationship to power, and design systems which are resilient to predation.

Once freshly irrigated farmland has been created, sapiens must maintain and secure their farmland. After many generations pass, agrarian society becomes wholly attached to their farms, incapable of surviving any other way except by farming. Foster summarizes this dynamic as follows:

"You've taken the hard coats off seeds and altered the instincts of cattle. They can't survive in the wild any more than you can. You have to be around to defend them. Forget a year-long ivory-hunting or soul-searching break in the Derbyshire tundra: you've got to be on duty on your farm, protecting your crops and your stock from the vulnerabilities you've chosen for them. And if the envious people in the next village decide to beat their ploughshares into swords, you've nowhere to run to (as hunter-gatherers you had the whole world as refuge) and no resources, either physical or imaginative, to survive anywhere other than the farmstead." [68]

The need to physically secure farmland creates a need for specialized workforces of power projectors dedicated to imposing severe physical costs on anyone who tries to take that farmland. But if you create a specialized power projection workforce dedicated to defending your farm, you have technically created a dual-use power projection capability that can also be used to physically capture other farms. This creates a window of opportunity for you to become a predator. Or to be more accurate, for you to continue to be a predator.

Just like other dual-use power projection tactics observed in nature, military workforces enable human organizations to affect both sides of the primordial economic equation. Not only can soldiers impose severe physical costs on neighbors to secure farmland, but they can also use their militaries to capture neighboring farmland. This dual-use power projection capability makes the military an extremely effective survival tool that was seemingly destined to become a strategic Schelling point for agrarian society, just like any other dual-use power projection tactic observed in nature.

It didn't take long for history's first recorded primordial economist, Sargon of Akkad, to figure out the dual-use benefits of militaries. Like a single-celled bacteria that had just discovered phagocytosis, Sargon subsumed neighboring Sumerian city-states with ease, focusing his attention on the ones with the highest **BCR_A**. By physically compelling poorly defended city-states to forfeit control over their resources to the Akkadian Empire, Sargon popularized a trend which continues to this day. In modern times, populations with the most powerful militaries still have the most control authority over Earth's resources. The predatory game of eat or be eaten has essentially never changed.

4.8.8 Encoding Logical Constraints into a System doesn't Stop People from Exploiting the System's Logic

"It is impossible over the longer term for laws to protect a population from manipulated money... when the laws are influenced by the people that benefit most from that manipulation. Only an emergent parallel system outside of that control could solve that paradox."
Jeff Booth [106]

Something which has an even higher **BCR_A** than an undefended farm is an abstract power hierarchy filled with people who wield imaginary power and control authority over the valuable resources produced by *thousands* of farms. It should come as no surprise to anyone who understands human nature that high-ranking people are going to be tempted to prey on the abstract beliefs system of their populations, especially if there's no severe physical cost for doing so.

Government corruption is essentially a form of psychological abuse. To avoid spending energy or risking injury to settle their disputes and manage their resources, populations will adopt belief systems where they give people imaginary power. The people with imaginary power get control authority over the population's resources. The primary attack vector for any predator who wants to gain and maintain access to those valuable resources is to become the person with imaginary power, and then exploit it. When they do this, they are exploiting the population through their own belief system.

In response to this threat, people attempt to encode logical constraints into their abstract power hierarchies using rules of law. For example, when total monarchies proved to be too oppressive, people started creating rules of law called constitutions and forming new types of abstract power hierarchies like constitutional monarchies or republics. The problem with this approach is that these logical constraints are endogenous to the belief system, therefore they are just as capable as being exploited. And technically speaking, logical constraints are just as imaginary as the abstract power they ostensibly constrain.

Here is a point that will be repeated throughout this thesis because it's the leading contributing factor of both government corruption and cyber attacks. Logical constraints cannot stop people from systemically exploiting or abusing a belief system; they can only change the way people systemically exploit or abuse a belief system. In fact, logical constraints are often specially designed to have deliberate systemic vulnerabilities (e.g. back doors, trap doors, loopholes, zero days) to make them easier for malicious actors to subvert.

Entrust sapiens with abstract power and give them asymmetric control authority over resources using formally codified logic (a.k.a. rules of law), and they will find a way to exploit that logic for their own personal advantage. It doesn't matter how well designed the logic is; people are demonstrably incapable of stopping other people from exploiting logic using logic. History makes it clear that different rules of law which codify different types of abstract power-based resource control structure designs will cause people to find different ways to exploit those control structures.

The bottom line is that **the formally encoded logic of our belief systems is the *source* of systemic exploitation, not the *solution* to systemic exploitation. No amount of encoded logic can eliminate the systemic exploitation of logic, no matter how well it's designed and no matter if it's encoded using rules of law or software. The exploitation of the law (and software) keeps happening because people keep making the mistake of thinking that high-ranking people with imaginary power can be constrained by written logic (i.e. laws). Logical constraints can't prevent the exploitation of logical constraints, they just change how logic can be exploited. So merely adding more rules of law to an abstract power hierarchy to attempt to constrain the abstract power of high-ranking positions is never going to fully remove the threat of exploitation of abstract power, it's just going to change the way abstract power can and will be systemically exploited.**

Abstract power hierarchies have not become less systemically exploitable since the early days of god-kings, they are just exploited in different ways (and at much larger scales). People get confused and they misattribute the less oppressive governments we have today to better encoded logic in the rules of law. In reality, rules of law are just as equally incapable at stopping god-kings as they were seven thousand years ago. **The reason why we have less oppressive governments today is because people have spent a great deal of time, energy, and injury *slaying* god-kings (and their armies) when they become too abusive or exploitative. In other words, the reason why we have less oppressive governments today is because of people who are capable of and willing to project physical power to impose sever physical costs on those who abuse their abstract power.**

After many revolutions, populations have learned to create rules of laws which are much more democratic, where abstract power and control authority is split across a larger number of organizations and positions, each which with different control authorities (i.e. checks and balances) designed to make them more resistant to exploitation and abuse. It's a great idea in theory, but in practice, adding more checks and balances doesn't fully prevent people from exploiting or abusing their imaginary power, it just changes the way these rank-based control structures can and will be systemically exploited (if the author sounds like a broken record at this point, that is intentional).

4.8.9 Voting Systems are Systemically Exploitable

"Great is the power, great is the authority of a senate that is unanimous in its opinions."
Cicero [107]

Today, rank-based voting protocols are a core component of most modern democratic rules of law. Unfortunately, rank-based voting systems are vulnerable to systemic exploitation. A vote is another form of abstract power. Because abstract power is non-inclusive, bounded, and systemically endogenous (i.e. internal to a belief system with encoded logic that can be systemically exploited), that means votes are equally as non-inclusive, bounded, and systemically endogenous. People can (and have, many times) taken advantage of the systemically exploitable properties of voting systems to the detriment of the people who depend on them to secure their rights.

As Cicero observed, high-ranking people within a senate can make themselves unimpeachable simply by gaining the majority vote. This means a systemically oppressed population cannot impeach higher-ranking and abusive intermediaries if those intermediaries succeed in capturing majority vote. This phenomenon holds true no matter how voting power is ostensibly decentralized across a deliberative representative assembly of people meeting together to make decisions on behalf of the entire membership.

Consider a Roman-style senate of 100 high-ranking people wielding control authority over a population in the form of voting power. All it takes to achieve unimpeachable control over the entire population is for 51 of those high-ranking people to collaborate as a centralized entity. In many ways, this type of centralized and unimpeachable abstract power is even more favorable than a King's abstract power because it's more difficult to see the threat of consolidated abstract power (especially when that abstract power is given a misleading name like "stake," as will be discussed in the next chapter).

People believe that creating a rule of law to logically constrain abstract power across 100 people protects them from high-ranking people consolidating and abusing abstract power. It doesn't; it just changes the way to consolidate and abuse abstract power. Instead of 1 person (a monarch) acting in their individual self-interest to systemically exploit the abstract power hierarchy, it takes 51% of people in a ministry or legislature acting in their mutual self-interest to exploit the abstract power hierarchy. The exact same security flaw exists regardless of the fact that people attempt to logically constrain it by splitting abstract power across a ministry or legislature. Therefore, logical constraints like decentralization of abstract power don't stop systemic exploitation of abstract power; it just changes how to systemically exploit it.

Even if we ignore the fact that on multiple occasions throughout history these techniques have proven to be demonstrably insufficient at securing citizens against consolidation and abuse of abstract power, it should be clear to the reader why these logical constraints don't work using simple logic. Dividing the imaginary power of one person across multiple positions in a ministry or legislature doesn't prevent those people from consolidating and exploiting their imaginary power. All this does is change how people can and will exploit their imaginary power when there's inevitably an increasing benefit to doing so. In other words, logic cannot constrain itself. Encoded logical constraints can do nothing to fully eliminate the exploitation of logic because encoded logic is as imaginary as the power they ostensibly constrain.

Rank-based voting systems make it possible for conspirators to act with the same unimpeachable control authority of a King to serve a group's own interests as a corrupt King would do, but split their abstract power across separate identities to present the false appearance of a decentralized, representative democracy. This is a well-known security vulnerability; governments like the Roman republic have attempted to mitigate this security vulnerability via term limits or sortition, but it clearly doesn't change the systemically exploitable features of the design, it just changes the names of the people with the means to exploit it. Whatever they're called and however they're done, the exploitation of voting protocols is as old as electoral systems themselves. The first rank-based voting systems employed by the Greeks were regulatorily captured by affluent white males as soon as they were invented, starting a 2,500-year-old trend of subtle, passive-aggressive, and highly calculating systemic exploitation of voting systems (reminder that women didn't even have a right to vote in the US until only 103 years ago).

4.8.10 Abstract Power Hierarchies can Break Down Regardless of How Well They're Designed

We have 5,000 years of written testimony about the struggles of preventing abstract power hierarchies from becoming dysfunctional. Based on well-recorded events of history, we know that abstract power hierarchies are exploitative, abusive, and prone to near-routine collapse. Hungry and envious outsiders can and will be unsympathetic to them. Their design can and will not protect people from attackers. No matter how their rules of law are encoded, abstract power hierarchies can be systemically exploited by self-serving, incompetent rulers. People can and will abuse their rank and control authority for their own benefit. People can and will attempt to use their abstract power and resource control authority to oppress their populations, and they can and will get away with it for thousands of years if the population becomes too docile because they believe too much in theological, philosophical, and ideological abstractions which make them grossly unaware of the merits of physical power.

For these and many other reasons we can verify using our own experiences and empirical knowledge, it's clear that populations can't prevent their abstract power hierarchies from being systemically exploited by their highest-ranking members, or flat out physically attacked by neighboring countries. We have been searching for an alternative to physical power as the basis for managing our property for thousands of years, and nothing we have tried has been able to survive.

Eventually, Hannibal inevitably arrives at the gate, or the government becomes so self-serving and corrupt. After sapiens suffer enough losses, they eventually become willing to do what surviving animals already do: project physical power to settle their disputes, establish control authority over their resources, and achieve consensus on the legitimate state of ownership and chain of custody of their property. People are eventually willing to spend the energy and risk the injury to secure the resources they value the way nature demands from every other animal in the wild.

After all the rhetoric about how theologically, philosophically, and ideologically reprehensible physical power and aggression is, after all the unproductive attempts to negotiate and bargain with unimpeachable oppressors, and after all the unsuccessful efforts to improve the design of abstract power hierarchies by applying ineffective logical constraints, people eventually cry havoc and call for the use of physical power to patch up the security flaws of their dysfunctional systems and impose severe, physically prohibitive costs on their attackers.

Despite how much people like to condemn the use of physical power to settle their disputes or establish their dominance hierarchies, surviving societies (emphasis on the word surviving) eventually come to terms with how nature works. **To solve the survivor's dilemma, societies must master the art of projecting physical power to impose severe, physically prohibitive costs on their attackers – whether those attackers come from the outside or the inside. They must adopt tactics, techniques, and technologies which allow them to continually increase their C_A and continuously lower their BCR_A**. Sapiens aren't miraculously exempt from primordial economics. If anything, the growing size of their tribes makes them even more vulnerable to their number one predator: themselves.

4.9 Emergent Benefits of Warfighting

"What this generation needs is a war."
Yves Klein Blue [108]

As much as people hate to admit it, war has emergent benefits for society. A primary value-delivered function of warfare is to decentralize zero-trust and permissionless control over valuable resources in a systemically exogenous way that can't be systemically exploited or manipulated by people in high positions of rank. Thanks to more than 10,000 years of agrarian warfare, no person or polity has ever been able to gain full and unimpeachable control over the world's natural resources, no matter what they believe in and no matter how much abstract power they have consolidated. War has therefore acted as a great decentralizing force of human society that has successfully prevented the adoption of a single belief system which can be systemically exploited by an untrustworthy ruling class.

The reason why this section and the preceding section devotes so much time towards thoroughly analyzing the dysfunctions of abstract power hierarchies and the complex emergent social benefits of warfare is to prep the reader for the core argument of this grounded theory, which is that Bitcoin represents a "soft" (as in non-kinetic) form of warfare that could be used to counteract the dysfunctions of software-aided abstract power hierarchies emerging in digital-age society. Therefore, to understand the necessity and complex emergent benefits of power projection protocols like proof-of-work, it's necessary to understand the necessity and complex emergent benefits of warfare. As this section argues, they're effectively the same thing for most intents and purposes. The point of warfare is to decentralize abstract power and to preserve zero-trust and permissionless control over resources, and so is the purpose of proof-of-work protocols like Bitcoin.

4.9.1 To Understand the Complex Benefits of Proof-of-Work, understand the Complex Benefits of War

Not surprisingly, this is a difficult topic to discuss. It would not be possible to present a theory about Bitcoin as a "soft" form of warfighting without a detailed exploration of the complex sociotechnical benefits of warfighting. If the author's theory is valid that proof-of-work represents "soft" warfare, then the arguments provided in this section should simultaneously apply to Bitcoin. Here, the author will attempt to illustrate the similarities.

While coding data for this grounded theory, the author discovered that the most common objection to the comparison between Bitcoin and warfare was due to an ideological objection to warfare rather than a logical objection. People don't like making logical comparisons between Bitcoin and warfare because they believe that warfare is ideologically "bad." This objection was so common that the author felt it was necessary to dedicate a substantial amount of time thoroughly exploring this concept. As a result, this section is quite long, and in some cases repetitive because of ground theory's constant comparative method, so if the reader has no ideological objections to warfare, they are encouraged to move on after section 4.9.7.

If the reader does have strong ideological objections to warfighting, or they strongly object to logical comparisons between proof-of-work protocols like Bitcoin and warfare, then they're encouraged to read this entire section to get a thorough breakdown of concepts which explain why it could be inappropriate to claim that warfare is objectively "bad." There are many well-researched and well-documented complex emergent sociotechnical benefits to warfare (namely the ability to secure the policies and properties that people value) that people ought to consider before they reject this theory for ideological reasons.

By explaining some of the complex emergent social benefits of warfare, the author hopes that it will be easier for the reader to understand later discussions about what complex emergent benefits Bitcoin could bring to society as a "soft" or non-lethal form of warfare. Conversely, by explaining why it's questionably ethical to accuse warfare as being ideologically "bad," the author hopes to make it easier to understand why it's questionably ethical to accuse proof-of-work technologies like Bitcoin of being ideologically "bad" (a common ideological objection to proof-of-work is that it's bad for the environment because of the amount of energy it uses).

If the theory is valid that Bitcoin represents a "soft" form of warfare, then a valid counterargument against people who ideologically object to war could simultaneously function as a valid counterargument against people who ideologically object to proof-of-work technologies like Bitcoin. In other words, an effective argument against pacifism should double as an effective argument against those who oppose Bitcoin (in particular because of its energy usage). This section is therefore relevant to a discussion about Bitcoin even though it might not appear to be on the surface. Such is an advantage of a grounded theory methodology, as it's explicitly designed to uncover these conceptual relationships in counterintuitive ways.

A primary takeaway from this section is that **rejecting proof-of-work because its energy usage is ostensibly bad for society is functionally the same argument that pacifists use to condemn warfare.** If valid, this would not only suggest that anti-pacifism arguments double as pro-proof-of-work arguments, but it would also further reinforce the theory that Bitcoin is indeed a form of warfare. The fact that the arguments presented in this section align with Bitcoin subject matter further validates the author's overarching theory that Bitcoin is indeed a new form of warfare that could dramatically impact how national strategic security is achieved in the digital age.

4.9.2 Wars Remind Humans that Humans Haven't Outsmarted Nature

War could be described as the byproduct of four flawed human assumptions. First, people assume that other humans can be trusted the same at increasingly high scales of cooperation. Second, people believe that trustworthy people will remain trustworthy as they gain exponentially more abstract power and control authority within a given abstract-power-based dominance hierarchy. Third, people believe they can sufficiently constrain other people from abusing their abstract power using logical constraints encoded into rules of law. Fourth, humans assume they are exceptions to nature – that there was ever a time where they didn't fight and kill for their resources (humans continually fight and kill for their resources, but most people in modern polities outsource this ugly business to specialized workforces like militaries and butchers).

For some reason, humans like to adopt belief systems where they're some kind of special exception to nature. Because people are so prone to making these faulty assumptions, they design and adopt trust-based and permission-based belief systems where they are neither protected by nor against the abstract power of their adopted ruling class. It should therefore come as no surprise that these belief systems are systemically insecure. In every abstract power hierarchy, the ruling class is physically powerless to defend their population against outside threats, and the ruled class is physically powerless to defend their polity against insider threats. For these reasons, abstract power hierarchies have a well-recorded history of being invaded or corrupted unless physical power (a.k.a. warfare) is used to compensate for these security flaws. Warfare is a recurring phenomenon because (1) the abstract power hierarchies created by human belief systems are routinely dysfunctional because they're so systemically insecure, and (2) there doesn't appear to have been a time in the 4-billion-year history of life on Earth where it hasn't been necessary to project physical power to survive and prosper.

War appears to be the activity which occurs *after* a population relearns the same lesson that abstract power-based dominance hierarchies are systemically insecure against invasion and corruption. War is essentially an act of capitulation when people who subscribed to a belief system tacitly admit to the flaws of their assumptions. During times of war, people implicitly confess that abstract power hierarchies are flawed because (1) people cannot be trusted with abstract power, and (2) encoded logic can't constrain people from abusing their abstract power. During war, a population acknowledges they can neither secure themselves *with* imaginary power, nor *against* imaginary power. As much as they hate it, they have to use *real* power because their imagination alone isn't going to protect and defend them.

War can be summarized as the act of domesticated sapiens rediscovering, time and time again, that there is no viable substitute for physical power (a.k.a. watts) as the basis for settling their disputes, establishing control authority over their most precious resources, and achieving consensus on the legitimate state of ownership and chain of custody of their property in a systemically secure way. War is the act of people admitting that the trust-based, permission-based, and inegalitarian control authority over valuable resources gained by using abstract power hierarchies is not a viable substitute for the zero-trust, permissionless, and egalitarian control authority over valuable resources gained by using physical power. War is incontrovertible proof that dominance hierarchies based on abstract power do not sufficiently replace dominance hierarchies based on physical power. War proves that as much as humans like to play make believe and pretend like they can break free of physics and nature and learn to settle their intraspecies resource disputes or establish their intraspecies pecking order using something *other* than physical power, they can't. War proves they can't, no matter how much thinking and design logic and imagination they try to throw at the problem, there is incontrovertibly no viable replacement for physical power as the basis for settling disputes and establishing a pecking order.

4.9.3 Wars Create Zero-Trust & Permissionless Control over Resources

A primary value-delivered function of warfighting is zero-trust and permissionless control over valuable resources. **War gives people a way to settle disputes, establish control over resources, and achieve consensus on the legitimate state of ownership and chain of custody of property in a way that does not require them to trust anyone or ask anyone's permission.** War gives agrarian society a way to manage their resources the same way wild animals already do. War is therefore an act of people behaving like the creatures they really are: wild animals, and one of the planet's most lethal apex predators.

When people watch humans physically fighting each other, they watch the natural, undomesticated version of themselves. Not surprisingly, domesticated people find this activity to be repulsive and beneath them, just like they would probably find many other things in nature to be repulsive and beneath them. From their high tower far above the rest of the food chain, they forget how they got into that high tower. In these situations, war serves as a reminder that humans are still merely humans; that we still bite and claw and physically clash at each other to establish our intraspecies dominance hierarchy just like other wild pack animals must do so. Why? Chapter 3 already discussed why. Because Earth's resources are limited, the world is full of predators and entropy, and only the strongest and sharpest survive. War is a very humbling activity because it shows us that no matter how much we wish we could *think* our way out of solving the survivor's dilemma and no matter how much we can *imagine* that we have transcended nature, humans are nevertheless helpless against physics, primordial economics, and natural selection.

4.9.4 War Gives People the Ability to Resist Systemic Exploitation & Abuse (a.k.a. Belief System Hackers)

In 1503, Admiral Christopher Columbus landed near the north coast of what we now call Jamaica with a crippled fleet of merchant ships. When he arrived, native Jamaicans welcomed him and aided him. But over the following year, Columbus' crew were extremely abusive to their native hosts. They repeatedly stole from the natives, cheated them, raided them, and committed several other indiscretions including rape and murder. Not surprisingly, the native Jamaicans responded by ceasing to provide aide. [109]

By 1504, Admiral Columbuses' crew found themselves stranded on the shores of Jamaica and under threat of starvation. They could not leave because they did not have seaworthy ships, but they could not stay because they had an insufficient supply of food and were highly outnumbered by increasingly disgruntled natives. Desperate and danger-close to starvation, Admiral Columbus did what many Europeans had already learned to do well. He used his literacy and storytelling skills to give himself abstract power.

Thanks to his son's literacy in astronomy, Admiral Columbus was aware of an impending lunar eclipse. Armed with this knowledge, Columbus met with native Jamaican chiefs and warned them that God would become so angry with the natives for not providing Columbus' crew with resources, that God would destroy their moon. Having already been betrayed by Columbus, the Jamaican chiefs learned not to trust Admiral Columbus' crew, so they initially did not believe him. But in late February 1504, a total lunar eclipse of the moon occurred over the skies of Jamaica just as the Admiral Columbus' son said it would. [109] In his journal, his son describes what happened next:

"The Indians grew so frightened that with great howling and lamentation they came running from all directions to the ships, laden with provisions, and praying the Admiral to intercede with God that He might not vent His wrath upon them and promising they would diligently supply all their needs in the future." [109]

This story has several lessons which many (domesticated) people seem to forget. The core lesson is that **people don't need an army to create and consolidate abstract power and use it to gain control authority over a population's valuable resources – they just need the population to have a belief system they can exploit.** It is possible to inspire an entire population of people to bend to your will by barely lifting a finger. All it takes is a persuasive story. The god-king playbook is extremely simple: convince a population to adopt a common belief system – particularly a belief system which involves something or somebody with a lot of abstract power – and then exploit the population with it. For this reason, **militaries have never been strictly necessary for the creation, consolidation, and expansion of abstract power and control_authority over society's precious resources. Instead, militaries are necessary for a limited number of occasions where people *refuse* to believe in someone's abstract power – when they are strong enough, courageous enough, and aggressive enough to physically resist being exploited through a belief system**.

The purpose of this anecdote is to highlight a complex emergent social benefit of warfighting which (domesticated) people unfortunately tend to overlook. When sapiens adopt abstract belief systems and ostensibly more "peaceful" ideologies as the basis for settling their intraspecies disputes and establishing their dominance hierarchy, they simultaneously adopt a belief system which can be molded, adjusted, and exploited. A clever foreigner can waltz right in and exploit these belief systems to give themselves the imaginary power of that belief system. On the contrary, when people use physical power as the basis for settling their disputes and establishing their dominance hierarchy, they adopt a system which is invulnerable to this type of attack.

Aside from providing a zero-trust and permissionless way to control valuable resources, the physical power used in warfare is also beneficial because it doesn't create an exploitable dominance hierarchy. It doesn't give anyone the ability to define or *redefine* what "good" means or what "right" is. It doesn't place anybody with imaginary powers at the top of the pecking order. Instead, warfare lets physics and probability decide the pecking order.

If Jamaican natives were more inclined to use physical power to establish their dominance hierarchy rather than subscribe to belief systems where entities with imaginary power were given dominance over them, they might not have been so easily manipulated by Admiral Columbus (someone who was in a very weak position at the time), and history might have played out differently for them. But instead, native Jamaicans literally fell to their knees and begged weak, helpless, starving oppressors for essentially no other reason than the fact that they had adopted a belief system where people with imaginary power have dominance over them.

Here we illustrate an important sociotechnical feature of warfare that many people overlook. **The "might is right" pecking order heuristic for establishing interspecies dominance hierarchies creates an amoral foundation which can't be exploited. "Might is right" could also be called "physics is right." Wild animals which use physical power competitions to settle disputes and establish resource control authority (what we call warfare when our species performs the same activity) let physics and probability decide their pecking order for them. By using physics rather than abstract ideologies to determine the pecking order, wild animals don't suffer from the same vulnerability that sapiens routinely suffer from: population-scale psychological abuse and exploitation.**

Study a pack of wild wolves, and you'll notice that they spend a lot of time infighting, but they also don't suffer from abusive god-kings or false prophets. Wolves may expend a lot of energy and risk a lot of injury establishing their pecking order, but they also don't fall to their knees and worship weak, helpless oppressors who abuse them. Wolves are mean, but they also don't allow themselves to be systemically exploited by their peers who wield nothing but imaginary power all because of unfalsifiable and continually changing beliefs about what good means, or what God wants.

That's the sociotechnical benefit of physical-power-based-dominance hierarchies. Species which choose to establish their pecking order using physical power competitions (what we call warfare) don't suffer from belief system exploitation. This concept is trivial to verify from our own experience because we can see from our own historical testimony that populations which remain physically powerful and aggressive don't suffer god-kings as gladly as weak or docile populations do. This principle is why the US exists.

4.9.5 Wars Decentralize Control over Vital Agrarian Resources

Zero-trust, permissionless, and non-systemically-exploitable control over resources are not the only sociotechnical benefits of warfare. Another major benefit of warfare that people overlook – one that is extremely relevant to a discussion about Bitcoin – is decentralization of abstract power and control authority over resources.

In addition to making a population more secure against psychological exploitation and abuse, warfighting also gives people the ability to physically constrain other people's abstract power and control authority. The ability to physically *constrain* abstract power represents the ability to physically *decentralize* abstract power hierarchies and its corresponding control authority over resources. In the author's experience performing research for this theory, this seems to be one of the most underappreciated sociotechnical benefits of warfare. People don't seem to understand that **one of the biggest advantages of warfighting is that it allows sapiens to decentralize abstract power and control authority over their valuable resources. Warfighting is the reason why there is no single, centralized ruling class over all of the world's resources. Despite several attempts, nobody in history has ever been able to establish a single ruling class precisely because of the physical constraints caused by warfighting**.

This concept can be very counterintuitive to people because many people (especially those people who don't actually study or participate in warfare) have accused warfare as being a power *consolidation* mechanism rather than a power *decentralization* mechanism. These people will claim that warfare only gives people the ability to *centralize* their control authority and use it to oppress people. The reasoning often goes something like this: kings raise armies and then utilize those armies in military campaigns to capture more resources. Therefore, militaries are centralizing forces because rulers build and use them to grow and consolidate their abstract power and control authority over resources.

There are three problems with this line of reasoning. First, it makes no distinction between physical power and abstract power. People who make this argument treat physical power and abstract power as if they're one-and-the-same thing when they're clearly not the same. Why do they make this false correlation between abstract power and real power? We have already discussed one reason why: the logical fallacy of hypostatization. People don't think about *how they think about* power.

A second problem with this line of reasoning is that it relies on linear-chain-of-events thinking rather than systems thinking. People who adopt this line of reasoning effectively blame the military for the king's consolidation of abstract power for no other reason than the fact that a military campaign happened to be the most recent (or easiest to observe) thing in a linear chain-of-events preceding the consolidation or expansion of the king's abstract power. In other words, they're tacitly implying a direct causal relationship between military campaign and consolidation of power simply because of a temporal relationship, while willfully ignoring other explanations that could equally explain why the king's power has been consolidated. It's not uncommon for wars to break out in addition to other factors like disease or famine or other disruptive factors which happen at the same time and could just as easily explain how a person could consolidate abstract power. Another perfectly valid explanation for why a king has more consolidated power following a military campaign is simply because more people choose to believe in the king's abstract power for reasons separate than but temporarily correlated to the military campaign.

A third problem with this line of reasoning is that it willfully ignores the fact that physical power has never been strictly necessary to create, codify, or consolidate abstract power and control authority over resources. This reasoning essentially ignores the lesson which Admiral Columbus and so many others have demonstrated throughout history: **all that is required to create, codify, and consolidate abstract power and control authority**

over people's resources is for people to adopt a common belief system. As countless ideological movements have demonstrated across thousands of years, militaries are unnecessary for the consolidation of abstract power. Clearly, warfighting is used to constrain and decentralize ideological movements that consolidate abstract power too much for people's liking (hence religious wars).

There is nothing physically limiting a person or regime from convincing a population to adopt an ideological belief system (for example, a monetary system or a religion) which expands worldwide, then consolidates and centralizes abstract power and control authority over everyone's resources. Except, of course, for one thing: real power. Herein lies a key insight into warfighting which so many civilians tragically miss: warfighting is one of the main activities preventing agrarian society from a single, consolidated, centralized belief system which could be systemically exploited by a single ruling class. There is no "one world government" because societies are willing to fight wars to prevent that other government from becoming *their* government.

If you ever feel like you're being oppressed, then take a look around. There is a high probability that you either (1) have chosen to adopt an exploitable belief system, (2) you're not strong enough or willing to physically fight to secure yourself against oppression, or (3) some combination of both. We have established that god-kings don't need physical power to oppress you; all they need is for you to (1) adopt ideologies which give them abusable imaginary power, and (2) be too weak or docile to physically resist them. So if you feel like you're being oppressed, it's likely because you have adopted a belief system that is being exploited, combined with the fact that you're too weak, scared, docile, or ineffectual to physically secure yourself the way all animals in nature do. It's an impolite but valid argument.

With these concepts in mind, let's consider what prevents an oppressive god-king from taking over the world and exploiting everyone at global scale? People's ideologies and desire for non-physical dispute resolution clearly motivate them to adopt exploitable belief systems. So what remains as the final line of defense for the global consolidation of abstract power are the fighters. There is a tiny subset of the global human population who remain unwilling to forfeit their physical power, or who remain unsympathetic to abstract, exploitable belief systems which are so very clearly flawed. These people are the only ones left in our global society who have the physical strength and aggression required to preserve our species' physical-power-based dominance hierarchy. They spend the energy and risk the injury needed to settle disputes, preserve globally decentralized control over resources, and achieve consensus on the legitimate state of ownership and property the way so few of us are willing to do. This tiny subset of people (less than 2% of the global human population) are warfighters who represent the last people standing in the way of one ruling class gaining consolidated and unimpeachable abstract power over everyone.

Warfare is the reason why control over our valuable resources remains decentralized. Removing this global-scale physical power competition would remove the complex emergent benefit of decentralized control authority over our resources. In other words, **warfare is one of few things physically preventing abstract power and control authority over Earth's resources from being completely consolidated by one ruling class. So it is absolutely *not* valid to make the claim that warfare consolidates abstract power because it's actually one of the largest physical barriers *preventing* the consolidation of abstract power!**

4.9.6 For Many Intents and Purposes, War Functions as a Proof-of-Work Protocol

Why have sapiens fought and died by the millions over thousands of years? One answer is so they don't have to suffer from the oppression of a ruling class they know they cannot trust. **Warfighting could just as easily be described as a more than 10,000-year struggle to *prevent* global consolidation of abstract power and control authority over valuable resources**. Sapiens have demonstrated very clearly that they would sooner risk nuclear annihilation than adopt a single ruling class. In fact, in every example of history where an emerging ruling class has attempted to gain too much abstract power and control authority over the world's natural resources or have tried to consolidate everyone under a single empire, society has responded with massive-scale wars, utilizing physical power as the mechanism for adjusting the difficulty of achieving that much control. In other words, **warfighting is functionally identical to Bitcoin's proof-of-work protocol. By participating in a highly energy-**

intensive, global-scale physical power competition, agrarian society enjoys the complex emergent benefit of decentralized control over their resources.

When a population goes to war, they accept the energy expenditure and risk of injury required to keep themselves secure against the centralization of abstract power. Of course it would be far more efficient (in time, energy, money, and lives) to have a single ruling class settle all property disputes, manage all the resources, and decide the legitimate state of ownership and chain of custody of all the property. But that would be a systemically insecure solution that is highly vulnerable to exploitation and abuse by the ruling class. This type of system would require one moral code and one moral authority which everyone would have to trust in and tacitly receive permission and approval from.

No matter how much people like to virtue signal to their peers about how much they condemn physical aggression, they demonstrate through their actions that they are very well aware of this attack vector. **If people truly believed that peaceful adjudication methods could be used to settle *all* disputes and manage *all* resources, then people would achieve consensus on a single belief system where a single ruling class settles *all* disputes and manages *all* resources across the world**. People clearly don't behave like this. They go through great lengths, spending extraordinary amounts of time and energy and risking an extraordinary amount of personal injury, explicitly to *prevent* people from unifying under a common belief system. Through their actions, not their words, people demonstrate that they instinctively understand that warfare has irreplaceable social benefits no matter how desperately they wish it didn't. **If societies didn't have something to gain from war, there wouldn't be war**.

We know from the bones we dig up, from thousands of years of written testimony, and from our own experiences that people absolutely cannot be trusted with too much abstract power and control authority over us. It's simply too dangerous to give one person or one ruling class too much abstract power and control authority over any form of resource. So what do we do? We fight wars; we engage in global-scale physical power competitions to ensure nobody can have too much abstract power and control over our resources. We fight to protect ourselves against the expansion of belief systems which we know are vulnerable to systemic exploitation. We fight to make sure *our* beliefs are adopted rather than *their* beliefs because we don't trust theirs or because we feel like the devil we know is preferable to the devil we don't know. Through our combined actions and behaviors we show that we are just as wild, brutal, and unsophisticated as the animals we look down upon, no matter how much we like to wear our little lapel pins and masquerade like we're above it.

Without being able to physically constrain the expansion of abstract power, there would be nothing physically preventing abstract power from spreading world-wide. Warfare gives our species an opportunity to overthrow or delegitimize abstract power when it becomes exploitative or abusive. With the ability to physically constrain and delegitimize abstract power comes the ability to physically decentralize control authority. Proof of this complex emergent behavior can easily be verified by simply noting how our species chooses to partition its territory. The reason why the dry surface area of the Earth has been partitioned across 195 different countries is largely the result of the wars fought to decentralize those territories. **If warfighting can decentralize our control authority over land, then it stands to reason that it can decentralize our control authority over other assets too – including digital resources like bits of information. The question is really about *how* it might be done, not *if* it could be done**. Therein lies the strategic significance of Bitcoin.

4.9.7 Warfighting is a Safety Feature, not a Bug

Another way to think about the complex emergent social benefit of warfighting is to think about fire safety. Fire safety engineers design special doors called fire doors. Fire doors improve building safety because they can contain and isolate fires in specific parts of a building to prevent them from spreading or to slow their expansion down. With this concept in mind, consider the function of national borders. National borders are forged by warfare and have essentially the same safety features as fire doors. When an abstract power hierarchy becomes hazardous (e.g. oppressive), national borders make it possible to contain or isolate this hazard to a specific region of the world and to prevent it from spreading.

Thanks to national borders, exploitative and abusive abstract power hierarchies remain contained. So long as a population can do a good job at securing their own borders, then the only threat of oppression they have to worry about is oppression from their own ruling class. And so long as our species continues to do a good job at warfighting to divide control authority over our valuable physical resources, we can minimize the amount of damage that could be caused by a single oppressive ruling class. In other words, warfare is a safety feature, not a bug. It protects humans from themselves – particularly their exploitable belief systems. It prevents the spread of hazardous belief systems by physically constraining and containing them.

4.9.8 War is the Exact Same Primordial Game that All Organisms Play, but Given a Different Name

In addition to helping sapiens protect themselves against the hazards of their own abstract belief systems, it could also be argued that warfighting has the same upsides as predation in nature. When sapiens engage in warfare, they engage in the same physical power projection activity other living organisms have engaged in for four billion years. It is therefore reasonable to believe that warfare could also produce the same complex emergent benefits that physical power projection competitions provide to all species – namely the ability to revector resources to those who are the strongest, most intelligent, and most capable of adapting to their environment and therefore more capable of surviving in a world filled with predators and entropy.

The explanation for why sapiens are so prone to warfighting could be the same explanation for practically everything else we observe in nature: what we see is what survives. There appears to have been many human polities over the past ten thousand years which didn't raise militaries and didn't go to war. We know these societies existed because we can find and dig up their mass graves and see how early and unpleasantly they died. It's not the case that peace-loving societies didn't exist, it's the case that peace-loving societies didn't survive. If we account for our survivorship bias (i.e. account for the fact that what we observe, including ourselves, went through a very rigorous selection process which weeded out a lot of other possibilities), then we can reframe a question about why sapiens fight wars in a way that's much easier to answer: "why do some societies believe they will be able to survive without warfighting?" Or to put it differently, "what gives humans the impression that they are the only animals in nature who have an inherent right to live peacefully without predators?"

Admittedly, "survival of the fittest" is an unsatisfying explanation for warfare. Like eating a head of lettuce, this explanation is neither salty nor sugary. It's a stoic explanation; it just is. It's neither a profound nor a romantic way to think about war, it's an amoral one which accepts physical conflict as natural behavior. This point of view does nothing to justify or reconcile the cruelty and bloodshed we see in war, nor how angry it makes us feel. And frankly, it's boring, not to mention super annoying, to be reminded of fact that sapiens are just as ordinary and unremarkable as the wild animals we like to look down upon and believe we have outsmarted. It's also quite frustrating to think that despite how bulbous our foreheads are – we have not found a way to outsmart natural selection except only in our imaginations.

Nevertheless, here we are, growling, snarling, scratching, snapping, and biting at each other over food and territory, no different than a pack of wolves or wildcats. The only difference seems to be the abstract thoughts filling our bulbous heads with reasons to justify or condemn our physical aggression. **War is the same power projection game we can all independently observe in nature, just with a different name. Instead of "primordial economics," people call it "war." Instead of "the survivor's dilemma," people call it "national strategic security."** Whatever names people choose to assign to these phenomena and whatever fancy uniforms they like to wear while they wage it, the first-principles explanation of warfare remains the same (hence why it's called a first-principles explanation). **Sapiens are one of thousands of other species of pack animals playing the same primordial game, using physical power to establish their dominance hierarchy just like the rest**.

While perhaps not as emotionally satisfying as other pontifications about warfare, the previous chapter about power projection in nature can at least give the reader an appreciation of "why" and "how" warfare works without complicating our understanding of the matter with abstract and unfalsifiable ideologies. War can easily be explained using basic principles in physics and biology. If we can wrap our heads around the complex emergent benefits of power projection in nature, then we should be able to understand the complex emergent social benefits of warfare, despite how many other abstract explanations people and their bulbous foreheads have come up with over the last several thousand years.

4.9.9 War Highlights Lessons Learned about Survival and Countervailing Entropy

As discussed in the beginning of this chapter, we live with the immense burden of being prisoners trapped behind our own prefrontal cortices. We cannot see or experience the world as it is, we can only experience a version of the world that has been filtered through a minefield of emotionally-charged abstract thoughts and symbolism. Like looking through a pair of rose-tinted glasses, sapiens look at the world through ideology-tainted glasses. This makes it very difficult to present arguments about the benefits of warfare on society without it being complicated by abstract ideas like ethics, morals, theologies, or modern politics. The solution? Simply don't call it warfare. Call it something else like primordial economics.

In case it wasn't already clear to the reader, the previous chapter about power projection tactics in nature was a clandestine attempt to get the reader to understand and appreciate the necessity and complex emergent social benefits of warfighting. These benefits are backed by science and lots of independently validated empirical evidence with naturally occurring data sets that can be analyzed ex post facto for causal inference. Like most clandestine operations, an argument about why warfighting is good for society must be presented to people in this manner because it is a highly illicit activity; it goes against social custom to explain why warfare is *good* for society.

To circumnavigate conflicting ideologies regarding warfare, the author provided a detailed argument for the emergent benefit of warfare without calling it warfare. Just by changing the name of the activity and using non-human examples (like the origin story of birds and mammals), it's possible to provide a logical, first-principles, grounded theory about the necessity and merit of warfare that is backed by science and loads of empirical evidence.

As a refresher, the logic presented in the previous chapter was as follows: power projection competitions give life an accelerated existential imperative to innovate, self-optimize, and vector limited resources to organisms which are demonstrably more fit for survival in a congested, contested, competitive, and hostile environment – an environment they don't have the option of escaping. Physical power projection competitions and predation have a causally inferable tendency to make life stronger, more organized, more intelligent, more adaptable, and thus more capable of countervailing the entropy of the Universe.

If the reader understands this logic, then they should be able to understand why it's reasonable to believe that warfighting provides the same complex emergent benefit to humans too. War is effectively the same power projection game with a different name, so it stands to reason that warfare would have the same complex emergent properties for humans as it would for other organisms. We can validate from our own experiences that warfare appears to give society the ability to self-optimize and ensure their limited resources are vectored toward the subset of the population who are more fit to survive in congested, contested, competitive, and hostile environments. Warfare clearly appears to help society become more capable of countervailing the entropy of the Universe. We have no shortage of empirical evidence to back this theory; we see it practically everywhere on – and off – Earth.

4.9.10 The First Moonwalkers were Not Just Explorers – They were the World's Apex Predators

The first Earthen life to escape Earth and walk on the surface of another heavenly body were sapiens – Earth's peak predator. The population of sapiens which walked on the moon was the population which devoted itself to mastering the challenge of warfighting. Moonwalkers weren't just any sapiens, they were sapiens from the world's most powerful military nation who got there by riding an oversized version of nuclear intercontinental ballistic missiles partially designed by Nazi scientists and engineers. NASA's Director and chief engineer for the Saturn V program were Nazi rocket scientists and engineers ushered into the US without public scrutiny via a secret intelligence program known as Operation Paperclip. [110]

Once humans made it to the moon and starting walking around on its surface, the first semantically and syntactically complex language they spoke were English words. They didn't just speak any random language, they spoke the acrolect of a United Kingdom which had just spent several preceding centuries conquering and colonizing the planet through several aggressive campaigns, to include one which successfully established a beachhead in North America, the place from which the humans who walked on the moon launched. And where did these moonwalkers get the calories they needed to do all that walking and talking on the moon? They got their calories from freeze-dried vegetables grown from fields plowed by the oxen they entrapped & enslaved. Freeze-dried beef from the cattle they domesticated and slaughtered over thousands of years.

The point is, **there are obviously some complex emergent benefits of warfighting when it comes to countervailing the Entropy of the Universe. The first moonwalkers weren't just our species' top explorers, they were our species' most powerful and aggressive power projectors**. In fact, they were arguably life's most powerful and aggressive apex predator.

The Apollo campaign was a thinly-veiled effort to raise public funding and support for the research and development of the critical enabling military technology needed to remain strategically competitive with the USSR at a time when public support of the military was at an all-time low (the entire campaign happened during the Vietnam War and a pacifist movement). In the face of growing pacifism caused by discontent of an ongoing war, how do you convince the American public to send lots of public money to newly patriated Nazi scientists and engineers to develop better intercontinental and cislunar nuclear missiles? Simple: Swap nuclear warheads with astronauts, and pump funding into a different marketing strategy which put lots of imagery of it on TV with inspiring narratives about peace and exploration. [110]

With this simple slight-of-hand, a workforce will have little to no reservation devoting substantial amounts of their time, technical talent, and public resources towards the development of strategic military assets. As the author has attempted to demonstrate, all one must do to circumnavigate a domesticated population's negative opinions about warfighting is simply call it something other than warfighting. Call it primordial economics. Call it space exploration. Or perhaps, call it a peer-to-peer electronic cash system. Simply change the branding, and people will have no problem pumping boatloads of money into the same technologies with the same functional use cases with little to no reservations.

Look around, and we can see similar evidence of the emergent benefits of warfare everywhere. The organizations which master warfare have a clear tendency to become technological and economic leaders and general-purpose masters of their natural environment. This is not political dogma; this is an independently verifiable observation backed by four billion years' worth of empirical data. There is a lot of supporting evidence to feel concerned about populations who condemn war and refuse to fight it. We don't need a theory to explain the benefits of warfare; we just need to look around us. Like physical power, warfighting is proof of its own merit. We admire the footprints on the moon left behind by those who master warfighting; we dig up the mass, early graves left behind by those who don't master warfighting.

Primordial economics, the survivor's dilemma, and the innovate-or-die and cooperate-or-die dynamics of predation are clearly at play for sapiens just as they are for all other species. If an intellectually honest reader can acknowledge there's merit to this line of reasoning, then they should understand the argument for why there are very important, complex emergent social benefits to warfighting that we have a logical, moral, and most importantly, an *existential* responsibility not to ignore. **We must be willing to entertain an uncomfortable but potentially valid hypothesis that wars provide an irreplaceable social and technical benefit to humanity**. The self-inflicted stress of predation and global power competitions have clearly made life more prepared to survive and prosper against the universe-inflicted stress of entropy.

The technical systemic benefits of predation wouldn't change just because sapiens arbitrarily named this primordial behavior "war." The dynamics of physical power projection don't change just because sapiens levy abstract thinking to produce moral, ethical, or theological justifications or explanations for it. Morals, ethics, and theologies are not first-principles explanations backed by empirical evidence or scientific rigor. **A first-principles explanation of warfare backed by empirical evidence is that all living organisms physically battle each other over resources and have clearly experienced major systemic benefits from those battles**, like becoming organized, powerful, and resourceful enough to survive devastating existential threats like meteor strikes.

Since the emergence of primordial life, the act of living has been fundamentally an act of physical power projection to countervail a cold, harsh, unforgiving, and unwelcoming universe filled with predators and entropy. This didn't change just because sapiens grew overclocked, overpowered, oversized neocortices capable of thinking of abstract, imaginary worlds where this isn't incontrovertibly true. The human capacity to believe in unicorns doesn't make unicorns physically real, and neither does the human capacity to believe in peaceful, alternative forms of physical power as a basis for settling disputes and establishing their dominance hierarchy. At least when people choose to believe in unicorns, they don't make themselves systemically vulnerable to foreign invaders or to population-scale systemic exploitation (unless you count the unicorn symbolizing the purity and power of the British monarchy).

It's simply not logical to believe that sapiens are exceptions to these principles, especially when there is so much empirical evidence backing it. It seems like it would be far harder to make the argument that it was confounding effects, correlation, or pure coincidence that the first Earthen life to walk on the moon were English-speaking Americans riding on top of the missiles they originally developed to kill each other. The much simpler explanation backed by first principles logic and highly randomized, causally-inferable empirical evidence is that the same reason Americans were the first to walk on the moon is the same reason lions are ostensibly the king of the jungle (technically speaking, tigers are the king of the jungle, if you don't include fire-wielding sapiens).

Perhaps it is difficult to appreciate the complex emergent benefits of warfare because sapiens desperately want to believe they aren't predators or that they have somehow transcended the cruelties of survival in nature. People like to imagine that they have discovered an equally effective basis for settling disputes, establishing control authority over resources, achieving consensus on the legitimate state of ownership and chain of custody of property, or otherwise just solving the existential imperative all animals face of establishing pecking order limited resources. So what do they do? We have already discussed what they do. They adopt abstract belief systems where people have imaginary power, and then they literally put on costumes and LARP as people with imaginary power to settle their disputes.

Most pack animals use physical power to settle disputes, establish control authority over resources, and achieve consensus on the legitimate state of ownership and chain of custody of property. This is both an intra-special and inter-special protocol that is 20,000 times older than anatomically modern sapiens and 80,000 times older than behaviorally modern sapiens who have tried (and so far, been unsuccessful) to create alternatives protocols for managing resources which doesn't rely upon physical power – alternatives which utilize imaginary power and are incontrovertibly and demonstrably dysfunctional. **Despite how much sapiens wish they could escape the energy expenditure and injury risk associated with warfare, five thousand years of written testimony (plus another five thousand years of agrarian fossil records) indicate sapiens have very clearly never found a satisfactory substitute for physical power as the basis for settling disputes and establishing dominance hierarchies**.

All life forms owe their existence to the process of leveraging physical power to capture and secure their resources. Humans are no different. Humans don't negotiate for their oxygen; they capture it using physical power. Humans don't negotiate for the food they eat; they capture it using physical power (and then negotiate with each other over price). Humans don't negotiate for the land they occupy; they capture it from nature using physical power (and then negotiate with each other over price). It's incontrovertibly true that living organisms gain and maintain access to their resources using physical power. To live is to project physical power to capture and secure one's control over resources. This isn't something to condemn, it's something to devote oneself to studying and mastering.

4.9.11 A Sign of Peak Predation is the Hubris to Believe that One Has Transcended the Threat of Predation

Does natural selection condemn physical power projection as morally, ethically, or theologically bad? No, it does the exact opposite; it asymmetrically rewards this behavior with more resources and the enormous privilege of survival. Take a look around, and the reader will likely note that the only animals condemning the use of physical power to establish intraspecies dominances hierarchies are sapiens. And they're probably not using a technical justification, they're probably using subjective, abstract reasoning and unfalsifiable claims about "good" or "god."

Like Dave Chappelle's fictional character Clayton Bigsby (a passionate member of an anti-African-American hate group who is blind and therefore incapable of seeing he's African-American), pacifism is comedically ironic. Pacifists turn a blind eye to their own nature. They carry the genes of thousands of generations of predators who scorched the world and slaughtered their way to the top of the food chain. They are the children of their ancestors' conquest, so comfortable and complacent in their homes they built over the graves of the conquered that they forget their own heritage – they forget *they're* the colonialists, conquerors, and peak predators benefiting directly from the activity they condemn.

It's not surprising why modern domesticated sapiens can develop distorted or misattributed views. Sapiens have become so high on the food chain that many eat thousands of animals without killing a single one. They have mastered predation and killing to a point where they have turned it into a subscription service. They outsource their predation services so effectively that they forget they're predators and they get upset – even horrified – when they're reminded about what they're paying for.

Modern domesticated sapiens devour their meals sitting in cushioned seats, watching mentally-cushioned videos of wildlife that has the cruelest and most brutal parts carefully edited out so the video better aligns with the inspiring music and narration given by some guy with an English accent (sapiens crave stories told by storytellers, especially when they are told by wise old shamans who help them reconcile the mysteries and cruelty of nature). And they do it all from the comfort of well-insulated, air-conditioned rooms with entrapped and genetically deformed wolves and wildcats licking their feet and worshiping them. Modern domesticated sapiens sincerely and unironically believe they understand nature and have transcended the threat of predation. They genuinely think they have discovered a viable alternative to physical power projection – that they outsmarted natural selection.

Ironically, **there is probably no greater sign of peak predation than the extraordinary amount of hubris required to believe that one has transcended predation**. The fact that people get so upset when they are reminded where their steak comes from doubles as proof of their extraordinary success at predation. They prey upon animals without even thinking about it. They scorch the world, colonize it, conquer it, and kill off their competitors so effectively that they forget how they came to "own" the land they're living on and "morally" defending in the first place.

These same concepts apply to warfare. Sapient populations can become so good at warfare that they develop the hubris to believe they have transcended warfare (this is a common problem in system safety too – success at safety breeds complacency). As a simple illustration of the point, the reader is invited to study every moral, ethical, or theological argument condemning warfare. Make a list containing the countries of origin from which those arguments were made and then compare that list to a list of countries with the most dominant militaries. Don't just do linear regression to see correlations, do other techniques like propensity score matching to determine if there are causally inferable relationships. Build predictive models on that data and run the model against history to see how accurately it can predict where wars are fought and won. When similar techniques have been tried by anthropologists, they find that their models are strikingly accurate. Several anthropologists have committed themselves to research which causally links the most prosperous, organized, cooperative, resource abundant societies which enjoy the most amount of art and literature and social freedom, to the most warring societies. [22]

4.9.12 Peace Depends upon Demonstrably Flawed Assumptions about Predatory Human Behavior

Another way to describe the sapient desire for peace is that it's fundamentally a desire for an end to predation. This desire appears to have spontaneously emerged after our species preyed on enough neighboring organisms to place ourselves comfortably at the top of the food chain. But is it realistic to expect predation to end? Given the sociotechnical benefits outlined thus far, is it even a good idea to desire an end to predation? The innovate-or-die and cooperate-or-die dynamics which emerge from predation clearly benefit life's ability to countervail entropy. It's no secret that many of the most revolutionary technologies developed by humankind over the past 10,000 years emerged from their military conflicts against each other (i.e. human-on-human predation). It is also no secret that the existential threat of warfare motivates sapiens to cooperate at scales which dwarf the levels of cooperation shown by other species.

Even if we ignore the systemic benefits of warfighting, it remains true that unrealistic design assumptions must be met for alternative approaches to physical power to function properly as a mechanism to settle sapient disputes, establish control authority over sapient resources, and achieve consensus on the legitimate state of ownership and chain of custody of sapient property. When high levels of sympathy, trust, and cooperation exist within a sapient population, then abstract power hierarchies seem to be able to function nicely as an alternative to warfare. But given a large enough population over a long enough time span, it's hard to believe that these conditions can be permanently satisfied well enough to prevent them from becoming dysfunctional.

Consider how much of the population needs to be untrustworthy for modern abstract power hierarchies to become dysfunctional. We can use the United States to serve as a better-case scenario of an abstract power hierarchy which has a lot of checks and balances (i.e. logical constraints encoded into rules of law) to logically constrain the exploitation and abuse of abstract power. The United States is one of many presidential republics with a fully independent legislature supposedly capable of preventing the consolidation and abuse of abstract power. But with 535 members of Congress representing the will of 336 million Americans, it would only take 0.00008% of the US population (the president plus ~51% of its legislature) to be dishonest or incompetent with their imaginary power for the United States to degenerate into unimpeachable, population-scale exploitation and abuse of abstract power. The reader is invited to ask themselves: how responsible is it to entrust 0.00008% of the population with abstract power and control authority over the remaining 99.9999%?

Combining this observation with the core concepts presented previously, here's another way to frame the same point: because of our desire not to use physical power to establish our dominance hierarchy, people adopt belief systems like presidential republics that can be exploited and abused even when 99.9999% of the population is competent and trustworthy. With the exception of extremely limited (but revocable) privileges provided to citizens by the second amendment of the US Constitution, the US abstract power hierarchy relies on trust in people with imaginary power to manage their resources and keep them secure against high-ranking members of the population wielding enormously asymmetric levels of abstract power and control authority over them that they clearly have incentives to exploit it. 99.9999% of the population must trust that 0.00008% of the population will be clever, competent, and honest enough with their abstract power for the US presidential republic to function properly as a viable substitute for physical power as the basis for settling their disputes and establishing pecking order over their resources. Again, the US was chosen because it represents a better case scenario; many abstract power hierarchies would be even easier to systemically exploit than the US presidential republic.

It is clearly unrealistic to expect 0.00008% of the population to be clever, competent, and trustworthy with abstract power *all of the time*. Most citizens intuitively understand that the imaginary power and control authority we give to our politicians represents a major attack vector which is practically guaranteed to be systemically exploited and abused *eventually*, just as five thousand years of written testimony tells us they have been exploited and abused in the past. There's a reason why the word "politician" is often considered to be a pejorative term. Citizens certainly recognize the risk of exploitation and abuse of abstract power. They are simply willing to accept the risk of exploitation, and even to tolerate a certain threshold of it, in exchange for not having to expend energy or risk injury settling their disputes, managing resources, and establishing pecking order using real-world physical power. In other words, we put up with it because we know that fighting to settle our disputes and establish our dominance hierarchies would be time intensive, energy intensive, and destructive.

We citizens agree to take part in a semi-consensual imaginary structure to give us a temporary reprieve from settling disputes, capturing and securing resources, and establishing pecking order the way natural selection demands from all species. We put up with the reoccurring flaws and the dysfunctional behavior of our rules of law. We acknowledge the tendency for politicians everywhere, both in our abstract power hierarchies and in our neighbor's abstract power hierarchies, to be untrustworthy with their rank, and we put up with it because we don't want to have to fight each other to settle our disputes and establish our pecking order the way wild animals do. At least subconsciously, it appears that many people seem to intuitively understand how the alternative to abstract power is physical power.

4.9.13 Peace has only been a Temporary Reprieve from War, not a Permanent Replacement to War

Like the lunar eclipse enjoyed by Admiral Columbus, the temporary and fleeting moments of reprieve that sapiens get from the cruelty of predation are both beautiful and awe-inspiring. Unfortunately, both our primordial nature and our written history indicate that these moments are an exception, not a rule. Peace appears to be a reprieve from war, not a replacement for it. Sapiens can be thankful they get to experience these reprieves on special occasions when exactly the right conditions align in exactly the right way, but it is clearly not reasonable for them to expect it to last forever.

We can see that abstract power hierarchies can indeed function properly in a narrow subset of cases where populations can reasonably expect their ruling class to use their imaginary powers effectively. Sapiens have proven that it's possible to use imaginary power to satisfactorily settle disputes, establish control authority over resources, and achieve consensus on the legitimate state of ownership and chain of custody of their property. They have proven that it's possible to use their imaginations, abstract thinking skills, and design logic to cushion and insulate themselves from a cold, hard reality filled with predators and entropy. They can establish a pecking order in a way that doesn't expend energy or risk injury like physical power does.

The problem is, of course, that imaginary power is just that: imaginary – it doesn't physically exist. Humans are playing make believe. They only *think* they've found a viable replacement to physical power as the basis for establishing their dominance hierarchy. **The ability to use imaginary power as a surrogate to real power is a *story*, an inspiring message that people are eager to believe in because it's hard for humans to emotionally reconcile how cold, cruel, and unsympathetic the laws of nature and survivorship truly are**. These stories help us mentally escape from what could be the most difficult part of life to reconcile: the fact that *we're* the cruelest and most unforgiving peak predators of them all. But the reality is undeniable: we're constantly at war.

Study enough nature and read enough history, and it becomes easy to see why some people argue that war is the rule, and peace is merely a temporary exception. Crack open a history book or look at a teaspoon of ocean water under a microscope to see evidence of this assertion on your own. Like life itself, agrarian society has always been at war – it has always been projecting physical power to capture and secure access to resources. Society has always been fighting to decentralize control over resources to maximize its chances of survival. Based on what we can independently observe, war appears to be a continuous and cyclical process that takes place because agrarian society has yet to discover a sufficient alternative to settling their disputes, establishing control authority over their resources, and achieving consensus on the legitimate state of ownership and chain of custody of their property. They attempt to use abstract-power-based dominance hierarchies rather than physical-power-based dominance hierarchies, but it clearly doesn't work. Our species is continuously fighting on a global scale; it's one of our predominant behaviors.

4.9.14 It Might be Immoral to Claim that War is Immoral: The Common Argument Against Pacifism

"You may not be interested in war, but war is interested in you."
Leon Trotsky [111]

One of the most successful military commanders of all time was a son-in-law descendant of Genghis Khan, Timur the Great. Generally undefeated in battle, it is estimated that 5% of the global human population were killed during the 14th century Timurid conquests and invasions. During the establishment of the Timurid Empire, some cities fell so quickly that Timurid armies didn't even need to kill their contenders or haul their bodies to their graves. In some cases, the conquered would dig their own mass graves and then be buried in them while still alive. [112]

The Timurid conquests serve as one of many examples in history illustrating a fundamental flaw with pacifism: pacifists *literally* dig their own graves.

According to the Oxford dictionary, pacifism is a belief that war is unjustifiable under any circumstances and that all disputes can and should be settled by peaceful means. The problem with this assertion is that it relies on unrealistic views about nature and human behavior. For people to become pacifist, they must believe there is an alternative to physical power as a basis for keeping people safe and secure against predators that is equally as capable of keeping people secure as physical power is, despite thousands of years of evidence that there isn't. Unfortunately, there does not appear to be a viable alternative to physical power with the same level of security needed to keep people secure against invaders and oppressors. When populations refuse to master the art of projecting physical power to secure themselves, the outcome is usually the same: mass graves dug by the pacifists themselves – people who are clearly willing to die without putting up a fight.

Pacifists seek peaceful adjudication to all property or policy disputes. The problem with this approach is that peaceful adjudication requires a judge or a jury (people with abstract power appointed by an abstract power hierarchy) to make a judgement about the dispute. Peaceful adjudication therefore requires *trusting* a judge or jury to cast impartial or fair judgements (note how the term "fair" in this context is subjective, thus incapable of impartiality from the start). Peaceful adjudication requires common consensus as well as trust in the people being judged to honor the outcome of their judgement. These all seem like good ideas in theory, but they rely on an unrealistic assumption that these conditionalities can and will be met, and that people are going to remain sympathetic to their verdict forever. As history makes abundantly clear, it's not possible for these conditions to be met *all the time, forever*. Sooner or later, Timur or one of his many reincarnations are going to reappear.

An impolite but simple argument to make against pacifism is that **pacifists are self-domesticated animals who are unfit for survival in a world filled with predators**. Pacifists forfeit their capacity and inclination to project physical power and to impose physical costs on others for abstract reasons, usually because of the energy it expends or the injury it causes. History makes it very clear what happens to pacifists. Their belief systems get psychologically exploited or their resources get physically captured or invaded. They get oppressed by corrupt, unimpeachable rulers, or they get steam-rolled by unfriendly neighbors who don't subscribe to the same pacifist ideologies. We have thousands of years of detailed written testimony about this, but somehow people keep allowing themselves to forget the moral of the story.

It could be argued that pacifism is what happens when people spend too much time with the friendly, docile, domesticated versions of themselves, or watch too many carefully edited videos of nature which have the unforgiving, brutal, and cruel parts (i.e. the most natural parts) filtered out. Pacifism is what happens when generations of people spend their entire lives outsourcing their physical security and predation to other people, so they don't have to experience the discomfort associated with these activities. They spend their lives without having to earn their food or their freedom of action – without having to kill the animals they eat, or to kill the people who are unsympathetic to their desire to live comfortably with the property and policies they value. It should come as no surprise that these types of people can develop distorted points of view about reality.

Pacifists are people who become spoiled by the spoils of war, oblivious to the reason why they can afford to forget about Timur. Successful populations can get so comfortable living their leisurely, sedentary, and domestic lives that they gain the luxury of developing unchallenged, imaginary beliefs about the world – worlds where people aren't actively capturing and securing access to everything using brute-force physical power and where pacifists themselves don't directly benefit from this behavior. Pacifists appear to live in an imaginary world that does not exist (one devoid of predators), perhaps because they have little experiential knowledge of wild nature from the safety and comfort of their neatly-structured, un-harassed societies. By practically all written accounts, this world has never actually existed. Modern agrarian society appears to have always been physically fighting each other intermittently.

It is incontrovertibly true that forfeiture of physical power makes a population physically powerless to defend themselves. Moreover, there are clear, causally inferable relationships between physical power projection and prosperity (or conversely, there is a clear, causally inferable relationship between pacifism and mass graves). It is therefore just as easy to argue that warfare is justifiable in some cases, and that mastering the art of warfighting is just as morally imperative for society. **History makes it clear that pacifism is a security hazard, so it could just as easily be argued that it's unethical for pacifists to motivate people to adopt pacifist ideologies which make them demonstrably insecure against predation**.

This same line of reasoning has been repeated many times throughout history. The lessons of history tell us why it is strategically crucial for populations not to allow themselves to believe that physical power competitions are bad for society just because they use a lot of energy or because they risk injury. When pacifists morally condemn the use of physical power, they contribute to a systemic security hazard which commonly invites invasion or oppression.

Predators feed on weakness. **Oppressors benefit from the sanctimonious peer pressure of pacifists who condemn physical aggression; oppressors *want* their population to be passive. Passive populations are physically docile, and their belief systems are easy to exploit**. Again, this isn't dogma speaking; it is incontrovertibly true that docility leads directly to slaughter. We have more than three dozen naturally randomized A/B testing experiments between wild and domesticated animals to causally infer a link between lack of physical aggression and systemic-scale exploitation. There is a clear, causally inferable relationship between pacifism and exploitation upon which humans continuously take advantage of. It is a matter of fact that boars don't experience peace when they are selectively bred to be less physically powerful and aggressive, they experience being turned into bacon. Our civilization was built upon making animals docile and exploiting them at massive scale to plow our fields and fill our stomachs. Sapiens are animals too; it's unreasonable to believe we aren't equally vulnerable to the same threat.

For whatever reason, people keep allowing themselves to forget the basic lesson of domestication despite how often it reappears in history. The lesson is simple: systemic exploitation and abuse is a byproduct of people who *don't* use physical power to settle their disputes, not a byproduct of people who do. Remove a population's capacity and inclination to impose severe physical costs on their neighbors, and they will face severe systemic security problems. If it's true for aurochs, boar, and junglefowl, then it's reasonable to expect it to be true for primates like sapiens, too. A human population's aversion to physical aggression and their forfeiture of physical power is likely to be a direct contributing factor to insecurity.

It could be possible that foreign invasion and the egregious levels of loss associated with wide-scale systemic exploitation by corrupt government officials is directly attributable to pacifism – to the people who *don't* project physical power to impose severe physical costs on those who attack or abuse them. That makes it just as easy to morally condemn people who *aren't* physically aggressive as it is to condemn people who are. At least the people who assert that physical power and aggression is good can back it with billions of years of empirical evidence. They can point to the animals which enjoy the highest levels of freedom and self-agency in the wild and note how mean and aggressive they are – how strong they are, how sharp their teeth and their nails are. That's probably not a coincidence.

Americans attempted to peacefully declare independence via a piece of paper in 1776, but that independence wasn't formally acknowledged until thousands of British soldiers were slain over the following seven years. Years after winning that fight, the Constitution was written. **Our founding fathers were aware of the emergent benefits of physical power and also well acquainted with the threat of an overreaching and abusive government. They understood that people who allow themselves to forfeit their ability to project physical power for theological, philosophical, or ideological reasons are people who forfeit their own security. This is precisely why the US Constitution has a second amendment**.

In conclusion, those who understand how domestication works can understand why pacifism is as easy to call an unethical and immoral bane to society as the warfighting it condemns. **Villainizing the use of physical power (watts) to impose severe physical costs on potential attackers is perhaps the worst strategic blunder a freedom-loving society could make**. Pacifism causes populations to adopt the exact opposite strategy they need to remain safe and secure. They cause populations to become weak, complacent, docile, and unfit to survive in a congested, contested, competitive, and hostile environment filled with predators and entropy.

4.9.15 There's No Excuse for Failing to Understand the Importance of Physical Power Projection

"Your problem is not the problem. Your problem is your attitude about the problem."
Captain Jack Sparrow [113]

Many have claimed that physical power breeds oppression, but this section provides a counterargument that oppression is actually caused by the asymmetric application of physical power rather than physical power in and of itself. It's not simply the fact that one side uses physical power on the other that's to blame for oppression. It's that one side is asymmetrically *more capable of and willing to use* physical power on the other side that's to blame for oppression. In other words, oppression occurs when one side uses physical power and the other one doesn't. This would imply that it's actually the *lack of* capability or willingness to utilize physical power that breeds oppression. To put it bluntly: **oppression is what happens when people refuse to fight for what they value**.

Herein lies a counterargument about why it is not logical for the oppressed to blame their state of oppression on the physical strength and aggression of their oppressor. It's not reasonable for people to villainize the use of physical power to capture and secure resources because this is simply how nature works. Look around, and you will see that the strongest, sharpest, and most physically assertive creatures rise to the top of every food chain in practically every biome on Earth. For people to ignore their primordial roots and these basic lessons of nature and survivorship is not logical – it's ideological. The author asserts that this is what happens when people spend too much time separated from predators. They forget how survival works.

The act of using physical power to settle disputes, establish pecking order, and secure resources preceded agrarian civilization by billions of years – life has *always* behaved like this since it was nothing but a thin film of organic material stretched across a volcanic rock, and there's simply no *logical* reason to believe that sapiens would be an exception to this behavior. The reasoning that people use to claim that the use of physical power is "bad" is ideological, not logical. But ideologies alone can't secure irrigated land or protect society from neighboring societies who want to take that irrigated land back – only physical power does that.

If we were to factor basic lessons of nature and survivorship into our calculus, then one counterargument to people who claim that physical power and aggression is the root cause of oppression is that the oppressed should take accountability for their *lack of* physical power and aggression. It's irrational to expect sapiens to be exempt from the basic principles of survivorship that we can all independently observe and verify. Life has always been about survival of the fittest – about finding the best ways to project power and adapt. Our imaginations and our ideals don't change the laws of physics.

As argued throughout this thesis, survival of the fittest didn't suddenly disappear when sapiens started using abstract thinking to form moral, ethical, or theological concepts. Sapiens are not exempt from life's rigorous natural selection process. We do not get to unsubscribe from primordial economics and the survivor's dilemma just because we happen to be capable of believing in imaginary alternative realities where physical power *isn't* the primary basis for settling our intraspecies disputes and establishing our dominance hierarchies.

For the past four billion years, the universe has been consistently harsh and unsympathetic to organisms which don't find ways to project power in increasingly clever ways. If any organism is to survive and prosper in this environment, they must stay accountable to themselves and not allow themselves to stop projecting power. Sapiens are organisms. Like any other organism, they are responsible for developing increasingly clever ways to project power for the sake of survival and prosperity in a world teeming with predators and entropy. Since sapiens are peak predators, they *especially* must keep searching for increasingly clever power projection tactics, techniques, and technologies which maximize their ability to survive against *themselves*.

These are stoic lessons which many military officers (like the author) have learned to accept. The blunt, logical reasoning goes something like this: Your opponent's strength may not be the only thing to blame for your losses; your own physical weakness, incompetence, or complacency could just as easily be blamed for your losses. Your opponent's inclination to be physically aggressive may not be exclusively to blame for your losses; your disinclination to be physically aggressive at appropriate times and places could just as easily be blamed for your losses. Your fear and aversion to using physical power could just as easily be to blame for your losses. Your trust in untrustworthy people, or your reliance on imaginary power which doesn't physically defend you could just as easily be to blame for your losses. So if you don't want to experience losses or become oppressed, it stands to reason that you should consider taking responsibility for your own weakness, ignorance, and incompetence rather than assigning the blame to other people. If you don't fight for what you value then you will be devoured, like everything else is in nature.

If you allow yourself to become physically weak, docile, and domesticated, then it stands to reason that you should expect to suffer the same fate as dozens of other animals which became physically weak, docile, and domesticated. If you fail to adapt to shared objective physical reality, then it is unreasonable for you to expect (or even more ridiculous, for you to believe that you deserve) a different outcome than countless other species who failed to adapt to physical reality over the past hundreds of millions of years. We can all independently observe how the world functions outside of that imaginary one we keep building inside of their heads. Proof-of-power is all around us, we can see and measure it everywhere. There is no shortage of evidence outside our homes, inside our history books, or on top of our dinner plates.

There is simply no excuse for not recognizing the essential role that physical power plays for one's own security and prosperity, no matter how energy intensive it is, and no matter how much physical injury it risks. Incontrovertible proof of how essential physical power projection is for survival is present in practically every observable corner of our environment, at every size and scale. It doesn't make sense to expect nature to function differently for sapiens than for everything else. In shared objective physical reality, every decision (including and especially the decision *not* to project physical power) has material consequences, and nobody gets to unsubscribe from them.

4.10 National Strategic Security

"Before we can abolish war, we need to understand it."
Peter Turchin [22]

4.10.1 Modeling How Modern Agrarian Society Controls its Resources

It is possible to model warfare as a resource control protocol. Section 4.5-4.7 discussed the differences between abstract power and physical power-based resource control systems . These sections discussed how agrarian society *attempts* to use abstract power to replace physical power. The word "attempt" was emphasized because abstract power-based resource control structures clearly tend to break down. If abstract power functioned properly as the basis for managing resources, there wouldn't be so many physical conflicts. The fact that wars break out suggests that abstract power isn't the only thing that humans use to manage their resources, so we need to update our model.

To produce a more accurate model to describe how agrarian society manages their resources, it is necessary to account for their use of physical power competitions (i.e. warfare). This can be accomplished by creating a resource control structure model which incorporates both abstract power and physical power. To that end, a hybrid resource control model has been provided in Figure 52. This model accounts for the fact that agrarian society *attempts* to use abstract power to manage the state of ownership and chain of custody of their resources, but routinely reverts back to using physical power.

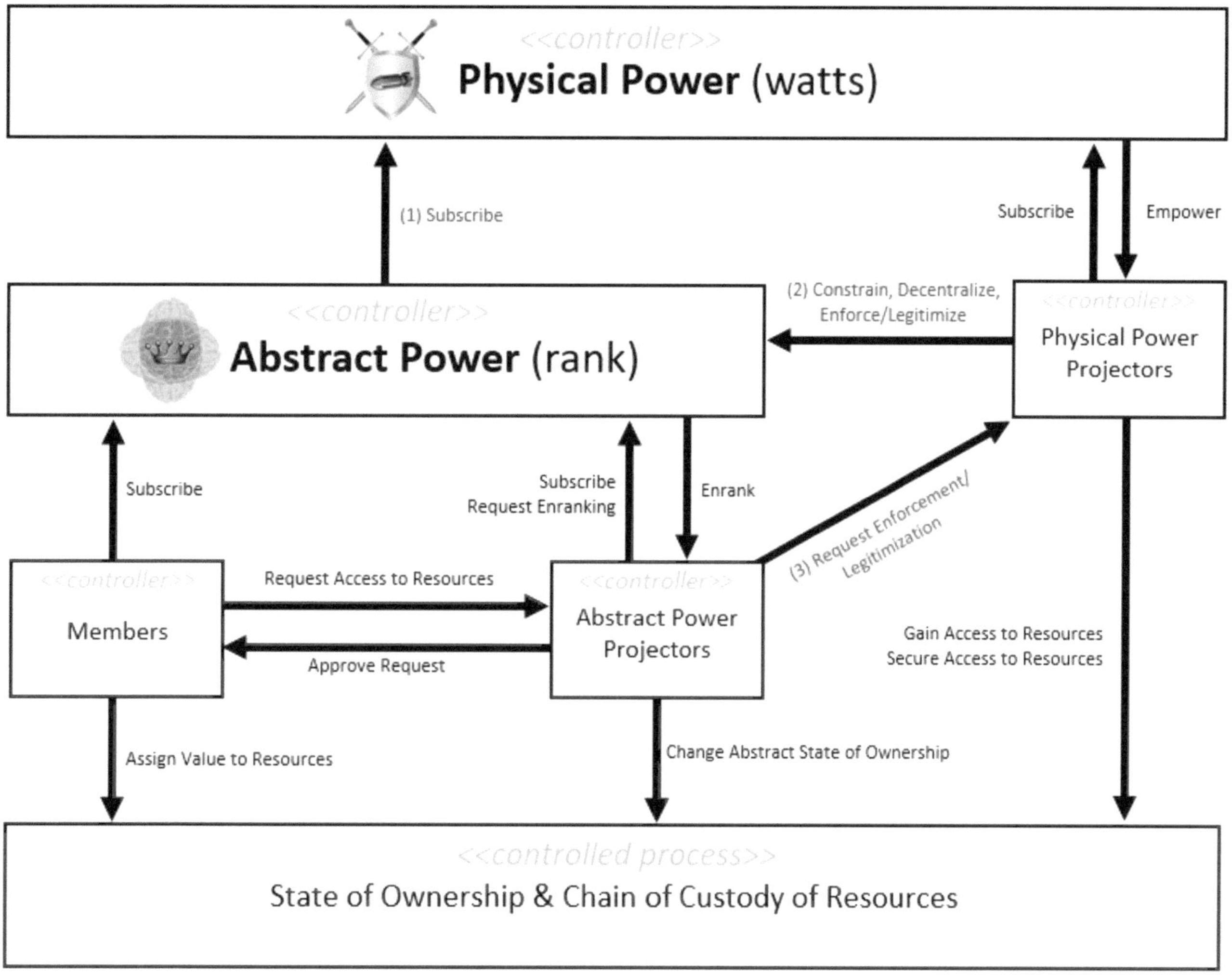

Figure 52: A More Accurate Model of the Resource Control Structure Adopted by Modern Society [88, 90, 76, 89]

This model contains the same controllers as the previously described resource control models and combines them together into a single system. Three control actions worth the reader's attention have been enumerated and highlighted in purple. The first control action is "subscribe." As previously discussed, everyone tacitly subscribes to the control authority of physical power by virtue of the fact that nobody gets to unsubscribe from the influence of physical power. **No matter what belief systems people adopt, and no matter how people choose to design their abstract power hierarchies, nobody who wields abstract power gets to "unsubscribe" from the authority of physical power**.

As previously explained, physical power is unsympathetic; it works the same regardless of whether or not people believe in it or sympathize with it. This means our presidential republics, semi-presidential republics, parliamentary republics, constitutional monarchies, semi-constitutional monarchies, absolute monarchies, or one-party states are all equally subordinate to physical power and equally incapable of escaping its impact. For these reasons, physical power is placed above abstract power in this control structure model.

The second control action worth special attention is "constrain, decentralize, enforce/legitimize." As discussed in this chapter, a primary value-delivered function of warfare is that it allows people to physically constrain and decentralize abstract power. Warfighting is the reason why control authority over Earth's dry land has been divided across 195 different abstract power hierarchies (what we now call countries). The precise boundaries of these abstract power hierarchies have been adjusted many times over the past several thousand years, but the control authority over Earth's resources has always remained globally decentralized precisely because of physical power projection.

In addition to physically constraining and decentralizing control authority over Earth's natural resources, physical power projectors (i.e. militaries) also legitimize the abstract power wielded by abstract power projectors. This references the concepts presented in sections 4.5 and the example about how kings utilize the physical power projected by their knights to legitimize and convince people to believe in their own abstract power. By having physical power projectors project physical power within the same context of the king's assertion of abstract power, the imaginary power wielded by the king is easier to misperceive as concretely real. Technically speaking, this is a false-positive correlation between an abstract input produced by the human imagination, and a physical sensory input produced by a power projector. Nevertheless, it works at legitimizing abstract power.

As many societies have proven time and time again through countless rebellions, revolutions, and coup d'états, Kings are physically powerless in shared objective physical reality. All it takes to undermine the abstract power and control authority of anyone with abstract power is to simply (1) stop believing in their abstract power or (2) countervail the physical power of the king's physical power projectors (i.e. defeat the king's army – the people with real power). We know this process works because it is the reason why monarchies today are almost entirely ceremonial and have virtually no abstract power (one major exception being the Kingdom of Saudi Arabia).

The reason why most monarchies today are almost entirely ceremonial is because power projectors stop legitimizing their abstract power. The populations living under monarchies in the past got so fed up with the exploitation and abuse of their belief system that they (1) stopped believing in the king's imaginary power or (2) started projecting real power to countervail or their monarch's army, thus delegitimizing the king's abstract power. If kings truly had the power they claimed to have, this would not have happened. But a series of revolutionary wars have made a very compelling case that kings don't have real power, they just have abstract power. For this reason, abstract power projectors have been placed in a lower and more subordinate position to physical power projectors in this model.

The last enumerated control action is "request enforcement/legitimization." This is another tacit control action that is easy to explain using the same king and knight example. When a king orders his army to carry out his will, what he is really doing is *asking* physical power projectors to legitimize his abstract power. The same thing happens during virtually any form of physical enforcement. Physical power is superior to abstract power, which means physical power projectors are superior to abstract power projectors. Additionally, laws are only imaginary logical constraints, not physical constraints. Because imaginary constraints are demonstrably incapable of physically preventing anyone from doing anything, laws must be physically legitimized using physical power. The request for physical legitimization and enforcement of abstract power is therefore implicit, not explicit, but nevertheless still a request.

As many rulers have learned the hard way over many acts of mutiny, physical power projectors can and often do choose to stop enforcing or legitimizing their ruler's abstract power. When this happens, the control authority that abstract power projectors have over people's resources disappears. Conversely, when abstract power projectors choose not to legitimize physical power projectors, the control authority of physical power projectors does not change. For this and several other reasons, we know that abstract power projectors require the tacit permission of physical power projectors, and not the other way around. This cold, hard truth is backed by a well-recorded history of many rulers being physically overthrown.

4.10.2 Two Ways for the Oppressed to Countervail Both Abstract and Physical Power Projectors

Using this newly updated control structure design, we can see that it reveals two ways for members to escape oppression if they feel like they are in an abusive or exploitative resource control structure. If a member is being oppressed by either abstract or physical power projectors (regardless of whether those power projectors are from a neighboring country or from the member's own country), those members have two ways to countervail their control authority. The first way was originally mentioned in section 3.11: members can simply refuse *not* to assign value to the resources being controlled.

As an example, consider the US dollar (USD) world reserve currency. USD is an international resource controlled by an abstract power hierarchy backed by the world's most powerful power projectors (the US military). The US military legitimizes the abstract power of its presidential republic, and by virtue of the design logic encoded into their rules of law, that presidential republic has executive control authority over USD. While it is certainly true that both types of power projectors (the abstract power of the presidential republic plus the physical power of the US military) have control authority over the state of ownership and chain of custody of USD, that doesn't make anyone have to value USD. Therein lies the key to countervailing US power.

Hypothetically speaking, if the US were to forget how this power structure works and became systemically exploitative and abusive with their control authority over USD (for example, if they started denying people's access to USD through sanctions, or debasing people's purchasing power by inflating USD), then **members could countervail both the physical and abstract power of the US by simply *not* valuing USD as their world reserve currency anymore**. For this reason, the "assign value to resources" control action which members can exercise seems small, but it is in fact very empowering. If the people in charge of the USD were to do something which motivated members to exercise this control action and stop valuing USD, their abstract power and control authority would disappear. It is therefore critical for the US to not do anything to motivate members from exercising this control action, else they risk losing their power.

The US deliberately made themselves vulnerable to this attack vector by converting USD from a physical system into an abstract belief system in the 1970s. Like so many organizations to come before them, the US seems to have lost sight of the value of physical constraints to abstract power. **By converting USD from a money denominated by gold into a money denominated purely by bits of information (a.k.a. fiat), the US converted their entire monetary system into an abstract belief system with no physical constraints securing it against systemic exploitation by high-ranking people who control the transfer and storage of those bits of information**.

Additionally, **the people with abstract power and control authority over USD only have it insofar as people are willing to believe in it because physical power is irrelevant at securing money which doesn't physically exist**. So not only do members have the freedom to choose not to assign value to USD, they also have the freedom to choose not to recognize the abstract power and control authority of the people who control USD. This means the entire USD monetary resource control system is backed by nothing but faith in the value of the dollar and the abstract power of the people who control the dollar. Of course, people can quickly lose their faith at any time, so it's imperative for the US not to do anything to motivate people to lose their faith in USD, which means it's imperative for the US to not deny people's access to USD or to degrade its purchasing power. Yet, in fiat form, there's nothing to physically constrain the US from doing either of these things.

These power dynamics put the people who have control over USD on thin ice and make it especially important for them to have the discipline not to deny people's access to USD or degrade its purchasing power. These people must be careful not to do anything to cause people to lose faith in their imaginary power, because if they do, their abstract power and control authority over this valuable abstract resource could quickly evaporate no matter how physically powerful the US is. This same principle applies to virtually any form of non-essential resource. For example, many people use both physical and abstract power to control the state of ownership and chain of custody of diamonds. Both physical and abstract power over diamonds can be made obsolete by simply not valuing diamonds.

There are, of course, some essential resources which sapiens don't have the option of not valuing (e.g. food, water, oxygen). In cases where sapiens don't have the option of escaping oppression by choosing not to value resources, their second option to regain control authority over the state of ownership and chain of custody of their resources is to become their own physical power projectors and impose severe, physically prohibitive costs on their oppressors. This is a primary motivation behind wars. Because physical power is inegalitarian and inclusive, anyone can choose to become a physical power projector regardless of their rank or title or standing within an existing abstract power hierarchy.

4.10.3 The National Strategic Security Dilemma

Now that we have a general model of agrarian society's resource control structure, we can review the topic of national strategic security. Members of abstract power hierarchies have two primary vulnerabilities: foreign invasion or internal corruption. The former usually occurs when a foreign belligerent actor uses physical power as the basis for their attack. The latter usually occurs when a belligerent actor uses abstract power as the basis for their attack. Either way, both vulnerabilities have the same solution: impose severe physical costs on the attacker until they don't have the capacity or inclination to continue their attack.

Based on this insight, we can see that the same primordial economic dynamics which apply to wild organisms and organizations also apply to agrarian society. This makes perfect sense considering how agrarian society is quite literally a pack of wild animals just like any other pack animal species observed in nature. We can therefore describe the dynamics of national strategic security the same way we described the survivor's dilemma in the previous chapter.

Like any other wild organism in nature, every nation has a BCR_A. A nation's BCR_A is a simple fraction determined by two variables: the benefit of attacking it (B_A) and the cost of attacking it (C_A). B_A is a function of how much resource abundance and control authority is offered by a nation's abstract power hierarchy. Nations with large economies, high levels of resource abundance, and substantial amounts of control authority over those resources have a higher B_A. On the flip side of the equation, C_A is a function of how capable a nation is at imposing severe, physically prohibitive costs on attackers. Nations with populations who are more capable of and inclined to impose physically prohibitive cost on attackers have higher C_A.

Nations which survive and prosper are those which manage both sides of their BCR_A equation effectively. To prevent their BCR_A from climbing to hazardous levels, nations must either shrink the numerator or grow the denominator of their BCR_A equation. They must either shrink their economy and control authority to decrease B_A, or they must grow C_A by increasing their capacity and inclination to impose severe, physically prohibitive costs on attackers. Decreasing the size of a nation's economy is not an ideal solution, so growing C_A is the preferable option. If nations choose to grow their economy without growing C_A at an equal or higher rate than the rate at which B_A increases, BCR_A will climb. This explains why pacifism (i.e. a decrease in a nation's inclination to use physical power to impose physically prohibitive costs on attackers) is such a systemic security threat. The more pacifist a nation becomes, the higher their BCR_A will climb, the more likely they are to be devoured by predators, either in the form of foreign invasion or internal corruption.

To achieve long-term survival, nations must keep the BCR_A level lower than the hazardous BCR_A level of the surrounding environment (i.e. the BCR_A level which would motivate belligerents to attack). The space in between a nation's BCR_A level and the hazardous BCR_A level can be called its prosperity margin. This margin indicates how much a nation can afford for its BCR_A to rise before it risks being attacked.

Figure 53 provides an illustration of the resulting national strategic security dilemma (note this is exactly the same figure as Figure 16 except with a different name). As a nation's economy becomes stronger and more resource abundant, its B_A increases. This causes the nation's BCR_A to increase and get closer to the environment's hazardous BCR_A level. As a nation's BCR_A level approaches the hazardous level, their prosperity margin shrinks. This creates an unfavorable dynamic where the more successful a nation becomes, the more vulnerable it is to either foreign invasion or internal corruption. To make matters even more challenging, a nation cannot know for sure how much prosperity margin it has, nor how quickly it's shrinking. This is because nobody can truly know what the hazardous BCR_A level is, as it a probabilistic phenomenon which depends on the capacity and inclination of neighboring nations and is therefore completely outside of a nation's individual control. All a nation can know about their environment's hazardous BCR_A level is that it will continuously drop as the environment becomes increasingly congested, contested, competitive, and hostile.

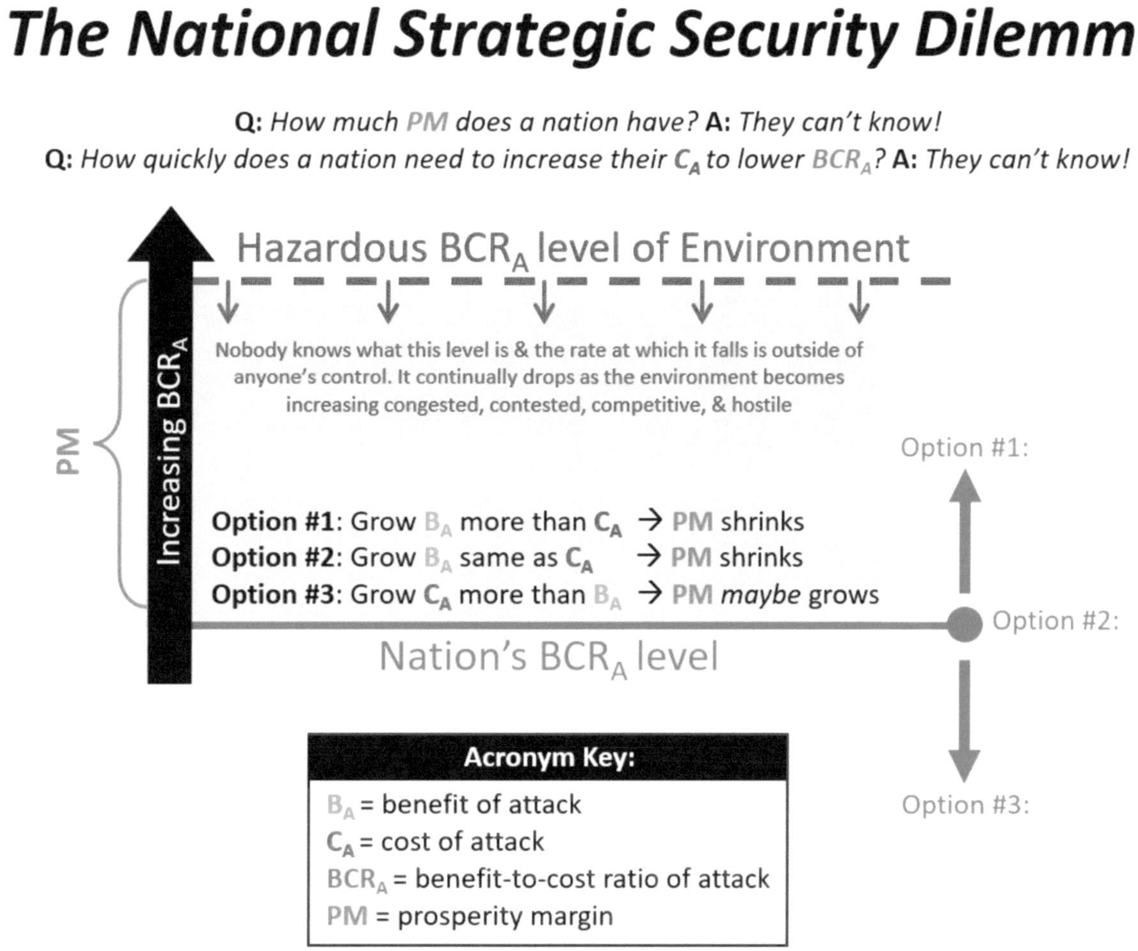

Figure 53: Illustration of the National Strategic Security Dilemma

The national strategic security dilemma puts all nations into a predicament where they have the same three response options described in the previous chapter. Option #1 is to do nothing to counterbalance the effect of their increasing B_A. The upside to this strategy is that it is more energy-efficient (this is because the population effectively ignores its security responsibilities). The downside of this strategy is that it causes the nation's BCR_A to continue rising ad infinitum, shrinking prosperity margin until the point where the nation is virtually guaranteed to be invaded or internally corrupted.

Options #2 and #3 represent strategies where a population doesn't ignore their security responsibilities and uses physical power to impose severe physical costs on attackers (i.e. increasing C_A). The difference between option #2 and option #3 is that option #2 only grows C_A at the same rate as B_A grows, causing the nation's BCR_A to remain fixed. Unfortunately, this will still cause prosperity margin to continue shrinking because it does not account for the fact that the environment's hazardous BCR_A level continuously falls as it becomes increasingly congested, contested, competitive, and hostile. Option #3 remedies this flaw by endeavoring to increase C_A faster than B_A grows, causing BCR_A to fall and prosperity margin to grow, assuming the nation succeeds at increasing their C_A fast enough to out-pace their environment's falling hazardous BCR_A level.

Not surprisingly, the best strategic move a nation can make to solve the national strategic security dilemma is the same move any living organisms or organization can make to solve the survivor's dilemma: option #3. **As illustrated by the agrarian fossil record and thousands of years of written testimony by survivors, if an agrarian population wants to survive and prosper, they need to endeavor to master their capacity and inclination to project physical power so they can continually increase C_A and buy themselves as much prosperity margin as possible.** This creates a national strategic Schelling point for all nations to vector a portion of their resources towards increasing their C_A. Unfortunately, this Schelling point causes the surrounding environment to become increasingly more contested, competitive, and hostile, causing the environment's hazardous BCR_A level to fall faster. This creates a self-reinforcing feedback loop which makes it increasingly more imperative for nations to continue increasing their C_A and lowering their BCR_A as much as they can afford to do so.

The emergent effect of this self-reinforcing feedback loop is the same as what we observe in nature. Agrarian societies grow in size and scale, organizing in larger and increasingly clever ways, and developing increasingly clever power projection tactics. They focus much of their time, attention, and resources on discovering and adopting dual-use power projection tactics which help them manage both sides of their BCR_A equation, just like the behavior observed with the evolution of life. Just as these dynamics explain why nature's top surviving wild animals are often fierce-looking and tough, they also explain why the most successful nations with the best-performing economies often have the largest and most successful militaries. Eventually, this power projection game scales into what we see today with massive-scale militaries and extraordinary power projection capabilities. Thus, the same dynamics which explain power projection in nature simultaneously offer us a simple explanation for how and why behaviorally modern sapiens scaled their physical power projection capacity to the point of risking nuclear annihilation.

4.11 Mutually Assured Destruction

"Because of your leaders' refusal to accept the surrender declaration that would enable Japan to honorably end this useless war, we have employed our atomic bomb... Before we use this bomb again to destroy every resource of the military by which they are prolonging this useless war, petition the emperor now... Act at once or we shall resolutely employ this bomb and all our other superior weapons to promptly and forcefully end the war."
US Warning Leaflets dropped on Japanese Cities Following the Atomic Bombing of Hiroshima [114]

4.11.1 Cooperate or Die

Anthropologists like Peter Turchin have outlined a compelling case that more than 10,000 years of warfighting have produced enough data to indicate a causally inferable relationship between warfighting and the size and scale of human cooperation. He asserts that sapiens became the world's greatest cooperators as a direct result of the existential imperative caused by agrarian warfighting. Additionally, thanks to their increasingly higher levels of cooperation, less of the human population is required to participate in warfighting.

To support this theory, Turchin developed a methodology to measure the size and scale of human cooperation and trace how the number of people in polities has grown since the dawn of the Neolithic age, which he then compared to the proliferation of miscellaneous warfighting technologies. He was able to build a model which accurately predicts the size, scale, and time of human civilization's growth based on nothing more than the development and proliferation of their warfighting technologies. [22] The model accurately predicts where and when the largest, most cooperative, and most prosperous human societies reigned.

Based on experiments like this and other forms of cultural evolutionary analysis, Turchin argues that there could be a causal relationship between warfare and cooperation which drove human cooperation to scales far exceeding the largest colonies of organisms observed in the wild, even in comparison to organisms which are famous for having highly cooperative colonies, like ants and termites.

Turchin also argues that warfare creates a counterintuitively positive dynamic of creative destruction where agrarian society channels its resources to the most stable, cooperative, and productive societies capable of sustaining high-functioning cooperative relationships for the longest amounts of time. In other words, **warfare creates trickle-up dynamics where natural resources flow to the most well-organized and cooperative civilizations**.

While Turchin's theory may be novel from an Anthropological perspective, it should not be surprising to anyone who understand the core concepts of power projection in nature and the "innovate, cooperate, or die" dynamics of predation discussed in the previous chapter. What Turchin describes in his theory appears to be nothing more than the continuation of a four-billion-year-old trend of power projection tactics in nature. It is incontrovertibly true that living organisms have been evolving into increasingly more organized, cooperative, and innovative creatures as a direct response to the existential threat of predation since at least the first bacteria discovered phagocytosis: the capacity to subsume or "eat" other bacteria.

As discussed in the previous chapter about power projection tactics in nature, cooperation is first and foremost a physical power projection tactic which emerged at least as early as 2 billion years ago. To project more power and to make it harder to justify the physical cost of attacking them, organisms simply sum their individual physical power together using cooperation. The concept is very simple, but quite difficult to execute – especially on large scales. The simple truth of the matter is that teamwork is hard.

It's extraordinarily difficult to get a bunch of organisms to trust each other and work together effectively to mutually increase their C_A and lower their BCR_A. Cooperation requires pack animals to come up with solutions for very challenging questions, and one of the most difficult challenges they face is the existentially important challenge of establishing a pecking order. In a world filled with predators, entropy, and limited resources where organisms must learn how to cooperate to survive, who gets feeding and breeding rights? What's the best way to divvy up the pack's limited resources to ensure they have the best chances for survival as a team, rather than as individuals?

A repeating theme of this chapter is that there's no reason to believe sapiens are special exceptions to the same struggles faced by other organisms in nature. It would be difficult to make the argument that sapient organizations wouldn't benefit from the same trend of predation in precisely the same way as the wild animals from which they evolved – a way that's quite simple to empirically validate by spending some time outside observing wild animals. It seems natural that sapiens would learn that the best way to survive against predators and entropy is to cooperate together at increasingly higher scales. Sapiens would also experience the same cooperation challenges. Therefore, does it make sense to believe that humans wouldn't adopt the same pecking order strategy that practically all other organisms with the same problem have adopted? Why wouldn't humans feed and breed the most physically powerful and aggressive members of their tribe if that's what pack animals have independently verified as an ideal survival strategy?

If the reader has made it this far, then you now have a deep-seated appreciation for how sapiens have attempted to deal with the "pecking order problem" differently than other animals. Instead of using "might is right" or "feed and breed the physically powerful first" as their pecking order strategy, sapiens use their imagination to come up with abstract pecking orders where people with imaginary power get first dibs rather than the people with real power. Why exactly is that? The author has now offered two potential explanations. The first was because it uses less energy and the second was because it's less physically destructive. However, here Turchin provides another viable explanation: because it scales human cooperation.

There is theoretically no limit to how many people can choose to believe in the same thing. Thus, there's theoretically no limit to how far an abstract power hierarchy can expand its control over resources. As long as people are willing to adopt the same belief system, there's no limit to how far humans scale their cooperation (for all we know, there could have already been a galactic-scale empire long ago in a galaxy far, far away that figured out how to scale their cooperation to the galactic level). The takeaway is that abstract power hierarchies have essentially zero marginal cost to scale because belief systems are essentially free to adopt.

Because abstract power hierarchies represent nothing more than a belief system, all it takes to build an abstract power hierarchy is to simply share an idea. This becomes exceptionally easy to do when there is an existential imperative to adopt a particular belief system. For sapiens, the "cooperate or die" dynamics of survival turn into "believe in this idea or die." Therefore, out of existential necessity and for the sake of self-preservation (particularly in response to the threat of invasion), sapiens are compelled to adopt increasingly larger-scale belief systems because if they don't, they're much less likely to survive against *other* sapiens who adopt increasingly larger-scale belief systems. This dynamic appears to be what led agrarian society towards adopting massive-scale belief systems like nation states, as well as the national power alliances formed by multiple nation states (e.g. European Union, NATO).

200,000 years ago, humans traveled in foraging bands comprised of tens of people. 10,000 years ago, sapient populations increased to farming villages comprised of hundreds of people. 7,500 years ago, sapiens developed simple chiefdoms (small cities) comprised of thousands of people. 7,000 years ago, these simple chiefdoms turned into complex chiefdoms comprised of tens of thousands of people. 5,000 years ago, the first archaic states emerged and were comprised of hundreds of thousands of people. Five hundred years after that, human cooperation exploded into macro-states comprised of millions of people. 2,500 years ago, the first mega-empires emerged, comprised of tens of millions of people. Remarkably, the large nation states we live in today are merely two hundred years old. [22]

Abstract power hierarchies clearly have the ability to scale quickly because ideas have essentially no marginal cost. This makes them a useful countermeasure against nations attempting to scale their physical power projection capabilities. Abstract power hierarchies help humans coordinate their efforts better (assuming the people with abstract power are competent and trustworthy). And due to the constant existential threat of warfare, humans are forced into accepting these belief systems where people have imaginary power over them, because they need high levels of cooperation to survive in a world filled with predators and entropy.

These dynamics imply that sapiens are placed into a strategic pickle where they must choose which abstract power hierarchy they want to subscribe to, or else they will find themselves fending for themselves alone in the wild. Except our Paleolithic ancestors weren't helpless in the wild; they had the freedom and confidence to roam the continents hunting and gathering. They didn't have to worry about massive-scale nuclear-armed tribes run by dictators on abstract power trips.

Turchin's anthropological theory is a straightforward argument which matches biological theory quite nicely. His theories seem intuitive to people who study nature with scientific and amoral rigor, but it can be unintuitive to (domesticated) sapiens who have adopted the habit of believing they're "above" the routine physical confrontations observed in nature, or believing that physical conflict and aggression (both of which are very common behaviors observed in every size, scale, and corner of the wild and rewarded by natural selection) is immoral according to whatever their individually subjective and unfalsifiable definition of moral is.

If we were to combine these concepts with the core concepts presented in the previous chapter about power projection in nature, then we can see how the emergent behavior of agrarian society is precisely the same as primordial economic dynamics observed in evolutionary biology. This means we can summarize the complex emergent behavior of national strategic security and the phenomenon of warfighting using simple bowtie notation. To that end, Figure 54 shows a bowtie representation of national strategic security dynamics. Here the author illustrates how it's possible to boil the complex dynamics of 10,000 years of agrarian warfighting down to a simple little illustration where organisms pair up with other organisms and adopt increasingly larger-scale abstract power hierarchies so they can increase their ability to impose physical costs on neighboring organizations, all for the sake of surviving and prospering in a congested, contested, competitive, and hostile environment filled with predators and entropy.

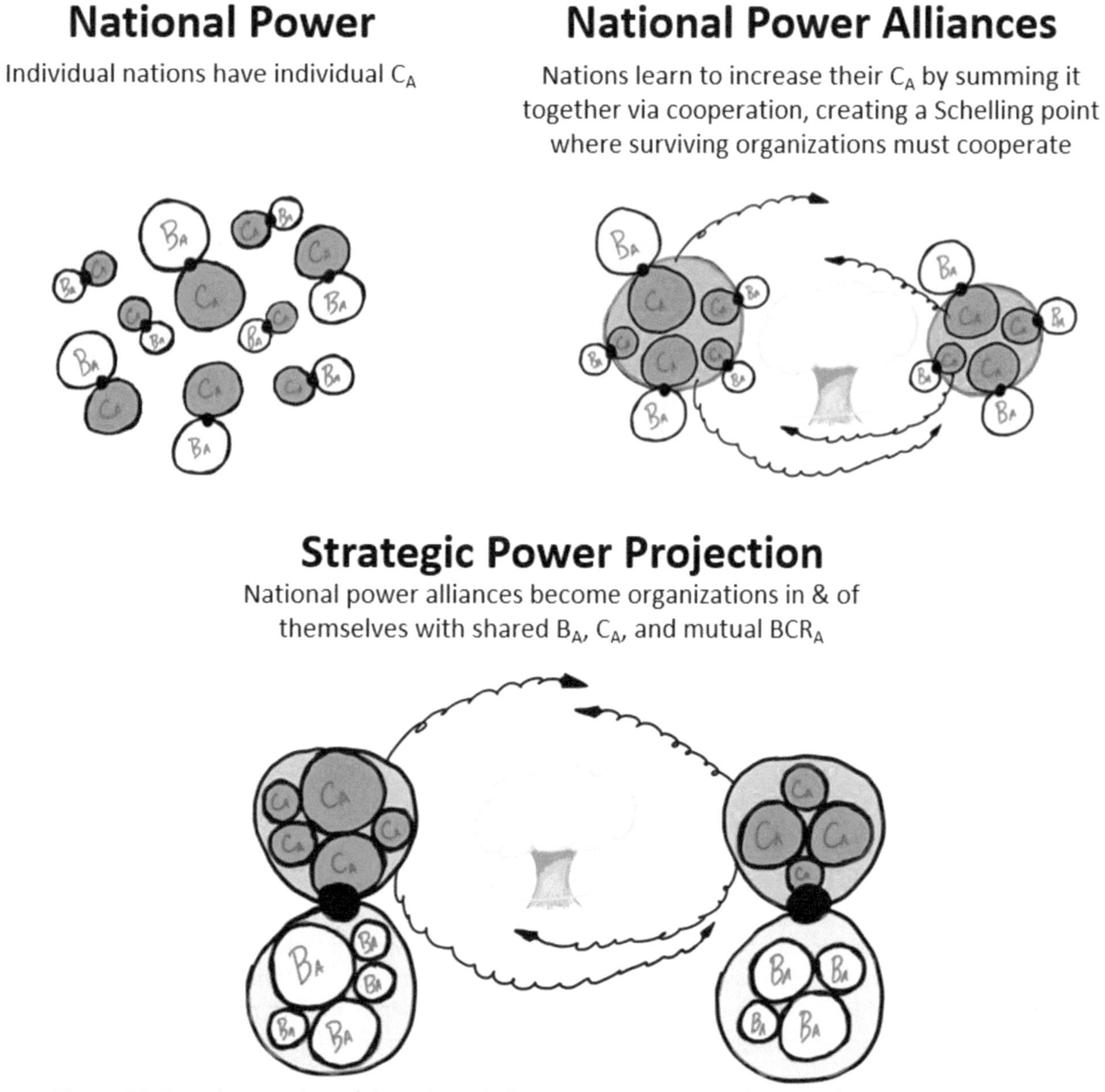

National Power
Individual nations have individual C_A
National Power Alliances
Nations learn to increase their C_A by summing it together via cooperation, creating a Schelling point where surviving organizations must cooperate
Strategic Power Projection
National power alliances become organizations in & of themselves with shared B_A, C_A, and mutual BCR_A

Figure 54: Bowtie Notation of the Primordial Economic Dynamics of National Strategic Security

The top left portion of this figure shows individual nations projecting national power. Each nation has their own individual B_A and C_A and thus their own individual BCR_A. To increase their ability to survive against predators and entropy, they learn how to organize and cooperate and sum their C_A together, as shown on the top right portion of the figure. By summing their C_A together, they create a Schelling point where other nations must sum their C_A together or else their individual BCR_A will be too high. Fast forward these dynamics across thousands of years and we arrive at the situation shown at the bottom of this figure, where alliances of individual nations double as organizations in and of themselves (e.g. NATO), comprised of billions of people endlessly searching for increasingly clever ways to project power, sum it together, and use it as a mechanism to make it impossible for adversaries to justify the cost of a fight.

So there you have it. Sapiens are exactly the same as any other multi-cellular organism learning how to stick together through colonization or clustering. Like archaeon, little sapient conquers go around capturing cities, forcing them to stick together, and turning them into city states. A few generations pass, and another sapient conqueror goes around capturing city states, forcing them to stick together, and turning them into nation states. Each time this happens, the pressurized membrane surrounding each city state (i.e. their militarily-enforced borders) grows larger and stronger, increasing their capacity to project power. Under the mutually beneficial security of these increasingly powerful borders, colonies of clustered humans grow increasingly more interdependent and reliant on each other for vital nutrients, materials, and gene swapping. They are able to form highly complex structures, self-assembling into increasingly more specialized workforces, trading various goods and services and becoming ever more efficient, productive, and resource abundant. Through this special combination of robust security and high-functioning internal economy, sapiens were able to follow a multistep biological path towards ever-increasing structure until it managed to self-assemble into complex, massive-scale economies we see today.

4.11.2 Warfare is a Self-Perpetuating Process that Naturally Increases in Size and Destructive Power

A primary takeaway from Turchin's theory on the relationship between warfare and cooperation is that humans are compelled out of existential necessity to adopt common abstract belief systems which make them more capable of cooperating and combining their power projection capabilities together for mutual survival and prosperity. Expanding on Turchin's theory, it's possible to make two follow-on observations. The first observation is described in this section, the second observation is described in the next section.

First, the size and asymmetry of modern agrarian abstract power hierarchies (i.e. governments) appear to be a direct result of human warfighting. This is a subtle but remarkable observation to make because it implies that warfare is its own root cause. If Turchin's theory is valid, then we can thank the existential stress of warfare for being a reason why abstract power hierarchies have more asymmetric power and control authority over valuable resources than ever before. But at the same time, we also know that dysfunctional abstract power hierarchies are one of the primary motivations behind warfare. Add these two insights together, and we can generate a profound observation that the *root cause* of warfare is the *emergent effect* of warfare. In other words, warfare causes itself.

Warfare creates an existential imperative for people to adopt increasingly larger (and thus more dysfunctional and vulnerable to systemic exploitation) abstract power hierarchies which create increasingly larger security hazards capable of leading to increasingly larger losses. Dysfunctional abstract power hierarchies motivate people to wage wars, which are won by adopting larger-scale abstract power hierarchies (e.g. national power alliances) to scale cooperation and sum enough physical power together to win the war. This creates a cyclical, self-perpetuating process where civilization learns to cooperate at higher scales, but also learns to fight at increasingly larger and more destructive scales, driving them to adopt increasingly more systemically insecure and hazardous belief systems. Through this spiral, what starts as a comparatively minor territorial dispute over irrigated land between Neolithic chiefdoms hurling spears at each other snowballs into modern nation states pointing nuclear intercontinental ballistic missiles at each other (which is now expanding extra terrestrially into space and cislunar orbit).

If the author's description of national strategic security dynamics is accurate, then it's quite ironic. It means that even though agrarian society uses warfare to physically constrain and decentralize abstract power hierarchies and prevent the world from falling under the rule of a single ruling class, the emergent effect of that activity causes the size of agrarian society's abstract power hierarchies to grow and gain asymmetrically higher amounts of control authority over everyone, making populations increasingly more vulnerable to systemic exploitation from their own ruling classes. To put it more simply, in our attempts to physically constrain and decentralize our neighbor's ruling class so that they can't exploit us with their imaginary power and resource control authority, we inadvertently make ourselves more vulnerable to exploitation and abuse from our own ruling class. By winning world wars against our neighbors, we create massive-scale abstract power hierarchies with extraordinarily asymmetric amounts of abstract power and control authority over our most valuable resources (thus high **BCR_A**), and we attract systemic predators like moths to a flame. We practically lay down a red carpet for predators and invite them to come systemically exploit the common belief systems we have to adopt to win our wars, which inevitably causes our belief systems to break down and lead to more war. It's a vicious, tragic cycle.

4.11.3 Nation States are an Untested Multicellular Organism Living in the Wild

A second follow-on observation from Turchin's theory is that nation states are relatively untested in nature. For 99.9% of the time that anatomically modern sapiens have walked the Earth, they have not lived in nation states. For 99.6% of the time that behaviorally modern sapiens have been thinking abstractly and constructing shared imaginary ideological constructs for themselves, they have not believed in nation states. Abstract power hierarchies have been around for thousands of years, but these abstract power hierarchies have never been large enough to qualify as nation states and have never been as asymmetrically powerful (thus as systemically insecure) as they are today. We feel like nation states have been around forever because enough time has passed where we (and all the people we have ever met) have only ever lived in a world governed by abstract power hierarchies large enough to qualify as nation states. But comparatively speaking, nation states are very new, and we don't know if they're a properly-functioning, long-term survival strategy for our species.

Nation states have clear systemic security flaws. They inherit all the flaws of abstract power hierarchies discussed at length in this chapter, and then magnify them to unprecedented scale. It is incontrovertibly true that people who are given asymmetric abstract power and control authority over people's valuable resources can't be trusted not to exploit or abuse that abstract power. We have thousands of years of written testimony of people exploiting and abusing their abstract power. Based on that same written testimony, we also know that it is incontrovertibly true that attempting to logically constrain abstract power by encoding rules of law are demonstrably insufficient. Nevertheless, modern agrarian society has scaled this repeatedly-dysfunctional belief system to the point where hundreds of millions of people (even billions of people in some cases like China) must trust hundreds of people not to exploit their abstract power and control authority over the population's most valuable resources.

Why would people agree to this? If Turchin's theory is valid, it's because it's existentially necessary for survival. This would imply that modern agrarian society effectively backed itself into a corner with another strategic Schelling point. We must adopt massive-scale abstract power hierarchies because it is the only way to survive against our neighbor's massive-scale abstract power hierarchies. But as a result of adopting these massive-scale abstract power hierarchies, we entrap ourselves.

Are nation states *really* a good idea for agrarian society in the long-term? How can we know? Nation states didn't emerge until the last 0.1% of our anatomically modern time here on Earth. Our nation states and their corresponding national power alliances are practically untested belief systems. We don't yet know how well they will enable sapiens to survive in the wild (particularly survival against ourselves). It's possible they could backfire on us just like so many other emergent power projection tactics have backfired on other lifeforms over the past several billion years. We can't yet know if these types of abstract power hierarchies can function properly at this scale because we don't have enough data to know yet.

However, we do have enough data to conclude that at smaller scales, abstract power hierarchies are highly prone to exploitation and abuse. Why are they so vulnerable? Because metacognition; because humans live according to their abstract thoughts and imaginations and symbolic knowledge, not according to experiential knowledge or what they can gain from shared objective physical reality. Abstract power hierarchies are fundamentally belief systems, and all sapiens are vulnerable to psychological exploitation and abuse of their belief systems. Sapiens can and routinely do allow themselves to be systemically exploited in, from, and through their belief systems – especially when those belief systems involve highly asymmetric amounts of abstract power and control authority over valuable resources.

So far, all we have to go on to determine if nation states are a good idea is a statistically irrelevant sample of about 200 years' worth of data. That data is inconclusive, to say the least. It's filled with many of the best examples of sapient aspiration and achievement that would have never been possible if it weren't for the extraordinarily high scales of cooperation that nation states enabled. Through the cooperation and coordination of our nation states, we walk on the moon and build international space stations. But at the exact same time, we carpet bomb cities and drop nuclear warheads on them. Many of the worst examples of destruction ever experienced by sapiens have occurred in the past 200 years.

4.11.4 Like Cyanobacteria, Sapiens may Have Discovered A New Type of Bounded Prosperity Trap

Nevertheless, no matter what our opinions about nation states are, they're essentially irrelevant to the subject of security. Nation states may be a few hundred years old, but the first principles dynamics of security are billions of years old. National security is therefore the same physical power projection game that has been played for billions of years, but with a different name and different branding. This makes national security a very straightforward process. If you want to know how to be good at national security, simply study nature and observe the behavior of nature's stop survivors.

To be good at national security, make your nation better organized, cooperative, and innovative so that it can find increasingly clever ways to project physical power against neighbors to increase C_A and decrease BCR_A. Like any other type of organism in the wild, nations must strive to continually evolve if they want to keep themselves secure against predators and entropy. The more they can continually increase their capacity to impose severe physical costs on neighboring nations in increasingly clever ways, the more they can make it impossible to justify the cost of attacking them. The more a nation can make it impossible to justify the cost of attacking them, the more prosperity margin it can buy for its internal population. At scale, this survival strategy produces the same dynamics discussed in section 3.8 where nations must chase after infinite prosperity by continually searching for new and innovative ways to increase the cost of attacking them ad infinitum. It should be no surprise that this process led agrarian civilization right to the brink of nuclear annihilation.

In the previous chapter, the author discussed how organisms sometimes struggle to find increasingly clever power projection tactics, techniques, and technologies, making them incapable of countervailing predators and entropy. The author named this situation a "bounded prosperity trap." One of the most dramatic examples of a bounded prosperity trap was the example provided about life's mass extinction event called The Great Oxygenation Event. As a refresher, cyanobacteria discovered an innovative tactic called photosynthesis, but it backfired on them by covering the world with highly combustible oxygen and setting themselves and the oceans ablaze for millions of years.

Fortunately, life had the ingenuity to learn how to (literally) stick together and cooperate at increasingly higher scales. Organisms continually experimented with different types of power projection tactics until they were able to discover the right countermeasures needed to countervail the blaze and escape their fiery hell. This was perhaps the most compelling display of life's rebellious "do not go quietly into the night" ethos against the cold and unsympathetic nature of the Universe. Entropy quite literally lit a flame under life's hindquarters, and life responded not by giving up, but by becoming stronger and more intelligent and more powerful than it had ever been before. Cyanobacteria crawled out of their fiery hellscape by innovating, and they exited their bounded prosperity trap with multicellular membranes and many other innovations to which we owe our everlasting gratitude today, billions of years later.

With the concept of bounded property traps fresh in mind, let's examine the state of our world today. The author asserts that life appears to have found its way into yet another bounded prosperity trap. This time, sapiens sprung the trap and placed much of life on Earth into yet another fiery hellscape: a perpetual state of global-scale agrarian warfare and now a looming threat of nuclear extinction. Herein lies a core hypothesis of the author's theory on softwar grounded in theoretical concepts from biology, psychology, anthropology, political science, game theory, and systems security theory.

4.11.5 A Recurring Contributing Factor to Warfare is the Hubris to Believe it isn't Necessary

With extraordinary hubris, sapiens appear to believe they can outsmart natural selection and find a viable alternative to physical power for settling disputes, managing internal resources, and establishing a pecking order using nothing more than their imaginations. They linked their prefrontal cortices together through storytelling and adopted imaginary points of view where physical power isn't needed to settle their disputes, manage their internal resources, and establish their pecking order. They made faulty design assumptions and adopted systemically exploitable belief systems where people with abstract power placed at the top of abstract power hierarchies are allowed to settle their disputes, control their resources, and decide the legitimate state of ownership and chain of custody of their most valuable property.

Like cyanobacteria and photosynthesis, sapiens and their abstract power hierarchies were a remarkable innovation that helped them achieve unprecedent levels of resource abundance. But also like photosynthesis, sapient abstract power hierarchies literally backfired on them and set the world ablaze by creating an emergent phenomenon we call warfare. Now life appears to be on the precipice of yet another mass extinction event, where they must figure out a way to stick together if they want to escape this new bounded prosperity trap.

By choosing to believe in abstract power and becoming over reliant on abstract power hierarchies as an ostensibly "peaceful" alternative to settling their disputes, managing internal resources, and establishing their pecking order, the belief systems sapiens design to avoid warfighting appear to be a leading cause of warfighting. In yet another tragic example of irony, the human desire to avoid physical conflict to settle small-scale policy and property disputes repeatedly cascades into massive-scale physical conflicts. By trying to avoid the use of physical power as the basis to settle disputes and manage resources, sapiens use their oversized foreheads to adopt belief systems which make them vulnerable to exploitation and incapable of surviving in a congested, contested, competitive, and hostile environment filled with predators and entropy. In their attempts to avoid the energy expenditure and injury risk of human-on-human physical conflict, sapiens seem to invertedly contribute to creating more of both.

Some agrarian populations become so self-domesticated by their imaginations that they forfeit their capacity and inclination to project physical power altogether. Not surprisingly, a population which doesn't believe in using physical power because of ideological reasons is a population incapable of protecting themselves against invaders who don't share the same ideologies. Alternatively, these populations become so docile and unsuspecting that they allow themselves to be exploited on a massive scale through their own ideologies. As history has shown, many populations would sooner worship oppressive god-kings who literally brand and herd them like domesticated animals than to project physical power to physically secure themselves and the property they value against systemic exploitation and abuse. In their desire for peace, they become oppressed.

People keep thinking their laws will keep them secure. They keep subscribing to demonstrably flawed beliefs that logical constraints encoded into laws and signed by people with abstract power are sufficient enough at protecting them against systemic exploitation and abuse, and they inevitably find themselves entrapped in states of major inequality and oppression with no ability to recognize the source of the trap, thus no hope of escaping it. They mentally ensnare themselves by believing that logical constraints are viable replacements to physical constraints as a mechanism for keeping themselves and their property secure. They make no effort to understand the difference between logical or physical constraints, nor the difference between imaginary power and real power, and they get devoured.

Surviving societies which don't get devoured are the societies which figure out the source of their vulnerabilities, call on their compatriots to take arms with them, and make it impossible to justify the physical cost of either physically invading them or systemically exploiting their belief system. They learn how to organize better, cooperate at larger scales, and invent innovative technologies to win their battles. When an endangered society fights off the threat of a neighboring abstract power hierarchy, it's called a war. When an endangered society fights off the threat of exploitation by their own abstract power hierarchy, then if they win it's often called a revolutionary war. If they lose, it's often called a civil war.

Either way, war is war. It's the same power projection game, with a different name. There is nothing different happening in shared objective physical reality when people engage in physical power competitions and call it regular warfare, civil warfare, or revolutionary warfare. The same species uses the same physical power projection tactics, techniques, and technologies. The primary difference is the stories they tell – the abstract thoughts people use to motivate themselves to organize and impose severe physical costs on each other.

In every war, the cause people fight for is often imaginary; people frequently fight for nothing more than a belief system. Trapped behind the cage of their overpowered, overactive neocortices, sapiens construct abstract realities indistinguishable from physical reality which are more meaningful and satisfying for them. Their imaginary mental models of the world become so important to them that they will gratefully line up and die for them. Their abstract thoughts completely overpower their instincts not to harm their own kind, and they commit unnatural and unprecedented amounts of intraspecies fratricide, gutting and mangling each other for the sake of "good" or "god" or "government" because people can't seem to be able to come to global consensus about what these things mean or what the correct design for them is.

And despite all this suffering, as much as people hate to admit it, warfare keeps resurfacing because it has complex emergent social benefits. At the same time, it clearly has major downsides because it creates a self-reinforcing dynamo of self-destruction and intraspecies fratricide – more so than any other species on Earth. The more societies go to war with each other, the more they must cooperate at higher scales by forming larger and more exploitable abstract power hierarchies. This cooperation allows them to sum more power together to accomplish extraordinary achievements like winning world wars and traveling through space. But the more they create asymmetric imaginary power and authority over larger amounts of resources, the more asymmetrically advantageous it becomes for self-serving sociopaths to exploit populations through these belief systems.

The more people get exploited through their belief systems, the more motivated they become to cry havoc and let slip the dogs of war once again, to restore society back to an acceptable state of systemic security. But the more people fight wars, the more they must cooperate at larger scales by creating larger abstract power hierarchies. This creates a larger window of opportunity to be systemically exploited at even larger scales, which must be resolved using larger scales of physical conflict to impose larger amounts of physically prohibitive costs on exploiting them.

If this is starting to sound repetitive, that's intentional. The author is illustrating to the reader that war is not only cyclical, it's tragically predictable. On, and on the dynamo of self-destruction turns, the same process with the same root causes repeating itself ad infinitum and ad nauseum, snowballing over tens of thousands of years to the point where sapiens subscribe to belief systems which make it socially acceptable to bomb cities to secure themselves against the imaginary belief systems adopted by the people living in the cities being bombed. And what do these people believe? They believe there is a viable substitute to physical power as the basis for settling their disputes, establishing control authority over their resources, and achieving consensus on the legitimate state of ownership of property. They believe there is a moral or ethical alternative to physical power – that we could live in peace so long as we let *them* define what "right" means, give *them* asymmetric abstract power and control authority over our resources, and entrust *them* not to exploit it.

New alliances are formed at increasingly remarkable levels of cooperation to win global-scale wars against self-serving sociopaths, only for those alliances to make larger populations more systemically vulnerable to future generations of self-serving sociopaths who wear lapel pins and shake their hands in the air as they tell stories and try to convince people they know what's "right." People line up to subscribe to these belief systems and let themselves get exploited at unprecedented scale, so entrapped behind their abstract thoughts and imaginary realities that they won't stop it. Through their belief systems, billions of people allow themselves to be gaslighted and systemically exploited in broad daylight by increasingly brazen god-kings. Onwards, agrarian society stumbles over itself like brain-dead zombies, dead men walking toward larger, more devastating wars until they finally hit the brink of thermonuclear annihilation and the Schelling point known as mutually assured destruction.

Imagine, if instead of trying to replace physical power with abstract power, that people learn to accept there is no replacement for physical power, and we instead sought to replace kinetic power with electric power. If we could figure out a way to use a non-kinetic thus non-lethal form of warfare that still enabled global-scale strategic power competitions to settle disputes, manage resources, and establish a pecking order, then we might be able to break this cycle. If only there were a global-scale physical power projection technology out there to which society had zero-trust, egalitarian, and permissionless access... enter Bitcoin.

4.11.6 Warfare could be Described as a Blockchain

Study war long enough and it starts to look as predictable as clockwork. The self-reinforcing feedback loop of flawed human belief systems are so reliable that it seems like we could set our watch to it. Populations adopt abstract power hierarchies and stop projecting physical power, causing their **BCR_A** to increase beyond a safe threshold, creating opportunities where populations will revert back to their primordial instincts to either capture high-**BCR_A** resources or impose high physical costs on attackers. A lot of watts (and lives) will be expended until populations have sufficiently lowered their **BCR_A**, settled their disputes, established control authority over their resources, and achieved consensus on the legitimate state of ownership and chain of custody of their property.

A block of time will pass where people can enjoy a reprieve from this global-scale physical power competition. This reprieve is so revered it's given a special name: "peace." And then when enough time has passed for people to become complacent, the population will forget how painfully predictable they are, their **BCR_A** will climb, the predators will return, and the whole process starts over again. These blocks of time link together linearly, forming a chain of time blocks, or a blockchain. The winners of this continuous global power competition are given the privilege of writing history, which is nothing more than a globally distributed ledger of information that keeps account of who has control of what and what the general state of consensus is regarding the legitimate state of ownership and chain of custody of the world's resources. This clockwork behavior of modern society is illustrated in Figure 55.

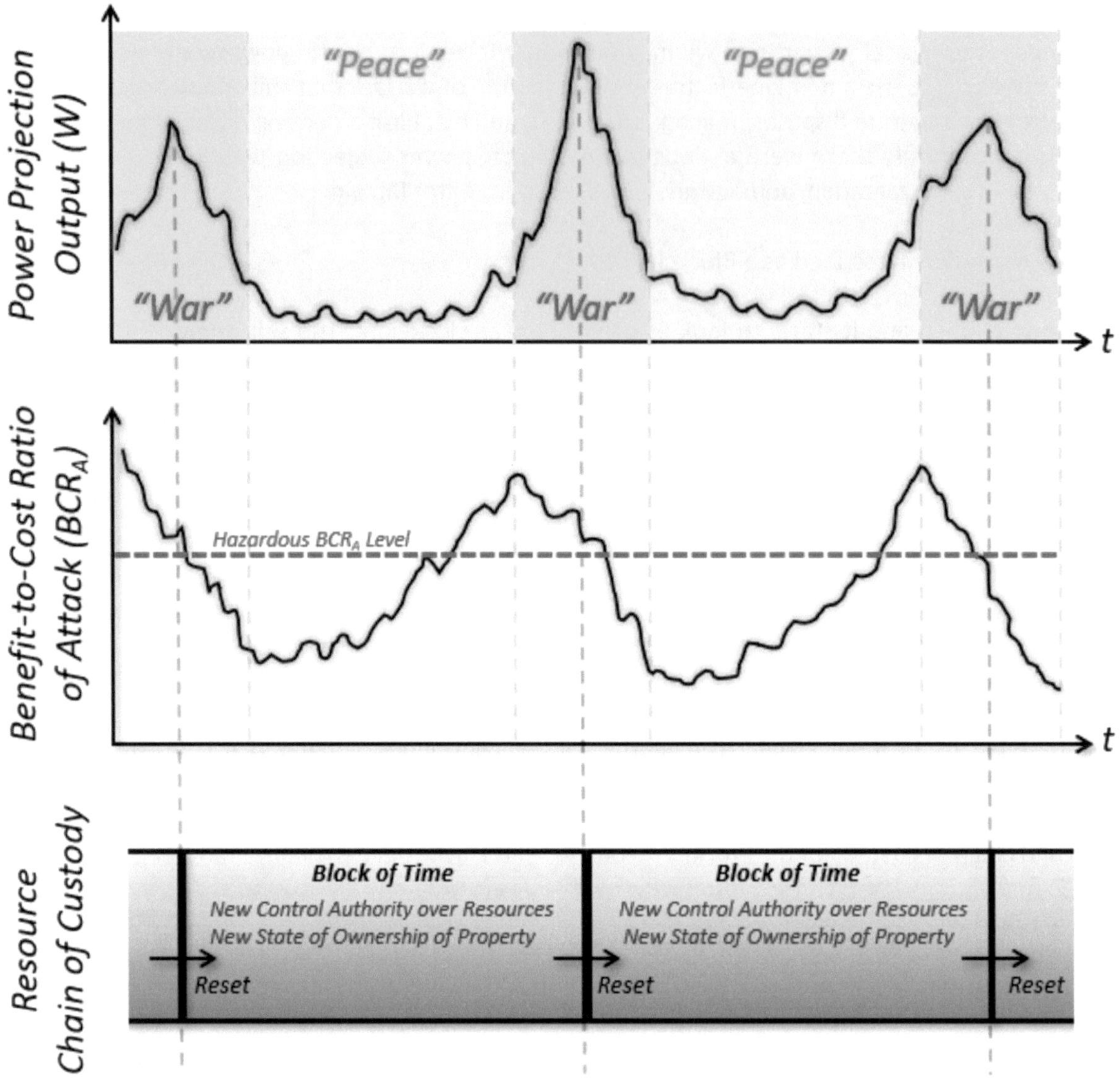

Figure 55: How the Cycle of War Creates a "Blockchain"

Countless people have been sent to early graves, warning future generations through bone trails dug up thousands of years later. Countless other warnings have been written into the pages of history by the survivors. For thousands of years, people of the past have been trying to warn the people of the future that something isn't working. Our belief systems are clearly flawed; they clearly don't work the way we wish they would. Countless times people have proven through their actions that there is no viable substitute for physical power for any population who wants to live free from the threat of foreign invasion or who wants to remain systemically secure against the threat of corrupt, self-serving sociopaths who psychologically exploit and abuse people through their belief systems.

In our extraordinary hubris, we believe we can do better, and we ignore the warnings of our predecessors. We keep scaling our beliefs about abstract power to the point where those beliefs become a clear and present danger to the survival and prosperity of our species. By convincing ourselves early in the Neolithic age that we are above nature, that we don't need physical power to establish our pecking order, we appear to have set ourselves on a path to destroy ourselves. We keep telling ourselves a lie that we don't have to spend the energy or risk injury to use physical power to establish our pecking order like the animals we domesticate. We keep avoiding fights when and where they should have happened, where energy expenditure and injury would have been minimal. We keep kicking the can down the road, hoping to avoid the physical conflict for moral or ethical reasons until the hazards blossom into such extraordinary levels of dysfunction and losses that it must be resolved using far more energy causing far more destruction and injury. And now we appear to have backed ourselves right into a corner where it has become too costly to physically settle our biggest property and policy disputes.

4.12 Humans Need Antlers

"Peace? No peace."
The Alien from *Independence Day* [115]

4.12.1 Hitting a Kinetic Ceiling

Caught in a bounded prosperity trap, agrarian society appears to have reached the crescendo of this 10,000-year-long song by running it straight into a kinetic ceiling. In our endeavors to use an imaginary replacement to physical power as the basis for settling our disputes, managing our resources, and establishing our pecking order, we ironically seemed to have scaled our capacity to impose physical costs on attackers to the point where it is no longer practically useful as a mechanism to keep our property and policy physically secure against systemic exploitation.

Sapiens are extraordinarily clever and resourceful. They are constantly finding new and innovative ways to be faster and more efficient at solving their problems. One of agrarian society's biggest problems is the survivor's dilemma, also known as national strategic security. To solve this problem, sapiens are continually searching for increasingly clever and more efficient ways to impose severe, physically prohibitive costs on others to make it impossible to justify the cost of attacking them. But there's a catch: it is theoretically possible for sapiens to become so efficient and resourceful at imposing severe, physically prohibitive costs on each other, that it defeats its own purpose. The author asserts this could be what happened with the invention of strategic nuclear warheads.

The evolution of human physical power projection tactics can be visualized in graphs like Figure 56 using two evaluation criteria: (1) how efficient they are, and (2) how much physical power they have the capacity to produce. The efficiency of a given physical power projection technology is a function of how much physical power it can project divided by the cost required to produce that power (cost can be measured several ways, to include money or casualties). The more efficient physical power projection technology is, the easier it becomes to project large quantities of power on a potential attacker to make it impossible to justify the cost of an attack. In other words, the more efficient power projection technology becomes, the better it becomes at growing C_A, reducing BCR_A, and buying prosperity margin for the nation utilizing that technology.

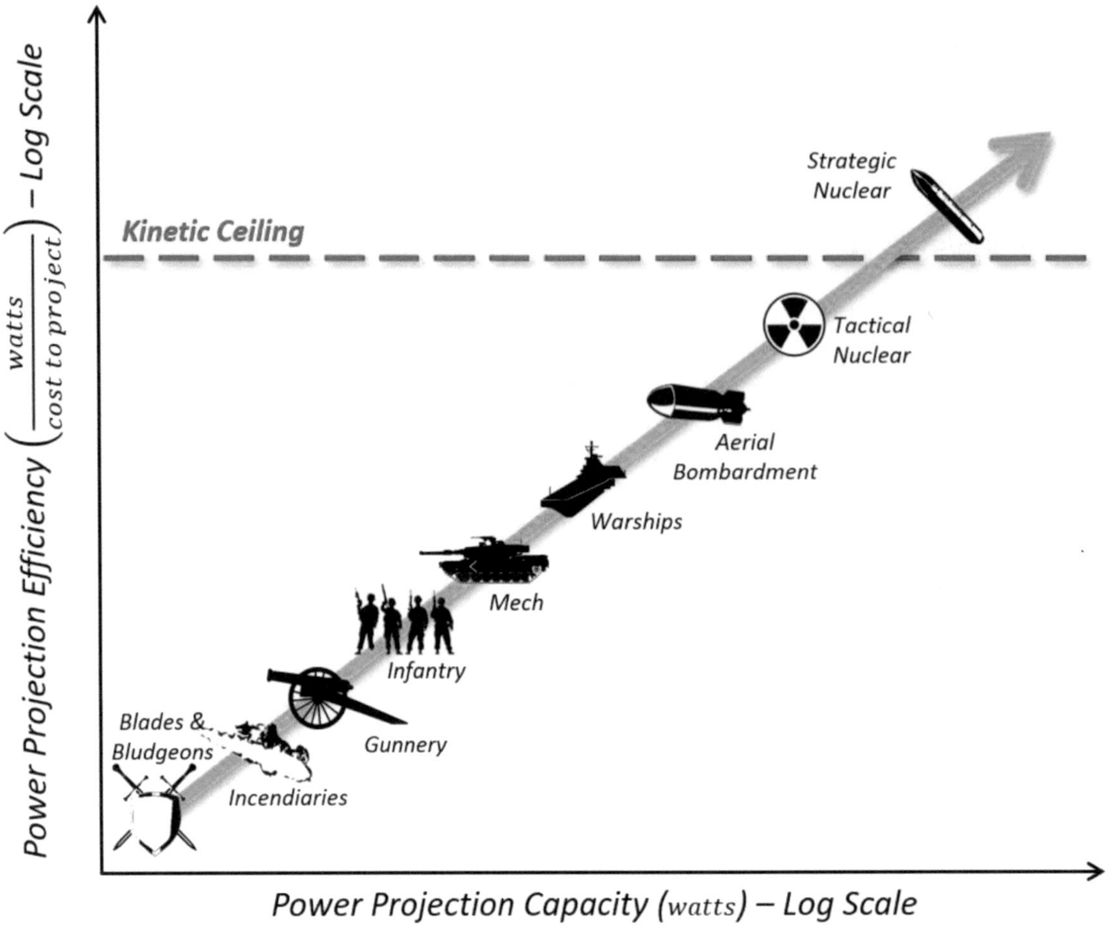

Figure 56: Evolution of Physical Power Projection Technologies Developed by Agrarian Society
[88, 89, 116, 117, 118, 119, 120, 121, 122]

Strategic nuclear warheads and deterrence policies like mutually assured destruction suggest it is possible to engineer physical power projection technologies that are so efficient, they're not practically useful. If multiple nations have the means to place thermonuclear warheads in multiple independent reentry vehicles sitting on top of intercontinental or even cislunar ballistic missiles, that suggests we have become so efficient at projecting kinetic physical power on each other that it represents an existential threat to the species. In yet another display of irony, human ingenuity appears to have made kinetic physical power so inexpensive to project (in terms of size, weight, matter, and monetary resources) that it has become too expensive to project (in terms of infrastructure destroyed and lives lost).

In their quest to become more efficient at national security, humans appear to have accidentally created the most inefficient national security capability ever: strategic nuclear warfare. It is hard to imagine anything else in this world that could project more power for less size, weight, matter, and money than a thermonuclear bomb. Yet, it's hard to imagine anything else in this world that could be more costly to agrarian civilization than thermonuclear warfare. **Such is the paradox of kinetic power projection – if you get too efficient at it, it becomes too inefficient**. The author calls this phenomenon the "kinetic ceiling" and illustrates it on Figure 56.

4.12.2 Kinetic Stalemates Don't Create Peace; They Create Major Systemic Security Hazards

Sapiens could be described as an Icarus-like species who tried to outsmart natural selection and got burned. In their quest for efficiency, they made themselves inefficient. They try to avoid the energy expenditure of using physical power as the basis for settling their disputes and establishing their pecking order, only to end up using much larger and costly quantities of physical power as the basis for settling their disputes and establishing their pecking order. **Humans chased after more efficient power projection technologies for more than 10,000 years, only to build the most inefficient power projection technology possible. In their hubris, sapiens brought agrarian society to the brink of nuclear wars that cannot produce a winner. And now they have cornered themselves by running straight into a kinetic ceiling. They appear to have scaled kinetic physical conflict to the point where it is no longer practically useful as a basis for settling disputes and establishing a pecking order. And they do not appear to understand how systemically hazardous this stalemate is.**

We have scaled our capacity to project kinetic power and compete in global-scale kinetic power competitions beyond the point where it would be practically useful as a basis for settling disputes and establishing pecking order. It is possible that agrarian society is in the middle of a strategic-level kinetic stalemate. Incidentally, the seventy years of time that has passed since the invention of nuclear warfare is just enough time for sapiens to forget the hard-earned lessons of history and the root causes of warfare discussed throughout this chapter. Our peace – that temporary block of reprieve between wars – appears to be stretching towards its limit. We may be overdue for another strategic-scale war, but what happens if a strategic-scale kinetic war cannot be waged or cannot produce a winner?

Is a strategic-scale nuclear kinetic stalemate a good thing or a bad thing? The knee-jerk reaction from someone who doesn't understand the necessity of physical power projection in agrarian society would probably assert that a nuclear stalemate is a good thing. "Finally," they might claim, "we can have peace because we have made strategic-level kinetic warfare too expensive to wage!"

Now that we have reached the end of this chapter on power projection dynamics in modern agrarian society, the reader should have a thorough understanding of the flaws of this line of reasoning. For starters, it presumes that the only threat to a population is invasion from a neighboring abstract power hierarchy. **A strategic-level stalemate might secure a population against an invasion from a neighboring abstract power hierarchy, but it wouldn't secure a population against massive-scale systemic exploitation and abuse from their own abstract power hierarchy.** So right out of the gate, people are ignoring a major security hazard and a reoccurring cause of war (this is why the author dedicated so much of this chapter to explaining the security flaws of abstract power hierarchies – to help the reader understand the threats they would likely face in a kinetic stalemate).

Peace is not an option so long as predators exist (especially systemic predators who hunt people through their collective belief systems). Peace has never been a replacement to war, it has only been a name that we assign to a state of reprieve between wars, when people are competent and trustworthy enough with their abstract powers to settle our disputes, manage our resources, and establish our pecking order without physical power. But as 10,000 years of evidence would suggest, peace doesn't last. It is as fleeting and as fragile as the imaginary power it gives to high-ranking people and entrusts them not to abuse. Like clockwork, our abstract power hierarchies become dysfunctional, and the next war comes.

No species, to include sapiens, have ever walked the earth without subscribing to the physical power projection game. We have never had the option of unsubscribing from this game. We are not special because we have big brains capable of imagining a different world where we have outsmarted natural selection and aren't constantly under threat of being attacked, invaded, or exploited. No matter what our imaginations show us, the real world remains filled with these threats.

We have never walked on a planet that isn't filled with predators; we have only allowed ourselves to forget about them. We become so comfortable and complacent in our high tower of success built for us by the sacrifice of our predecessors that we become docile and domesticated. We get drunk off the luxury of forgetting that we're surrounded by predators who don't take reprieves. We may be able to stop predators from invading us, but we can't stop predators from exploiting our belief systems – in particular the belief systems we use to manage our most valuable resources (like our money).

The more abstract power and control authority we give to a ruling class, the more benefit they gain from exploiting their abstract power and control authority over us. To expect a ruling class not to exploit increasingly asymmetric levels of abstract power is, frankly, ignorant. We know that this is incontrovertibly true because if it weren't true, people wouldn't be compelled to fight wars. Their abstract power hierarchies and the imaginary logical constraints they encode into their rules of law, would be sufficient to secure them without physical power. But it clearly isn't sufficient, hence warfare.

So what happens when agrarian society hits a kinetic ceiling and stalemates itself at the strategic level? To believe that a stalemate is a good thing is to make a tacit assumption that there are viable alternatives to physical power as a basis for settling our strategic disputes, establishing control authority over our strategic resources, and achieving consensus on the legitimate state of ownership and chain of custody of our property in a zero-trust, permissionless, and egalitarian way. Is it possible that sapiens are the first species on earth to have discovered an alternative to physical power competition as the basis for establishing pecking order? Perhaps. But it seems more likely that the people who think a strategic-level stalemate is a good thing are ignoring the core concepts of natural selection, human metacognition, and the differences between abstract and physical power.

The point of view that a strategic-level stalemate is good for agrarian society hinges on a delusion that sapiens have the option of living in a world without predators and entropy. It implies we can use our imaginations to adopt belief systems where people can be entrusted not to use their imaginary power to exploit our belief systems. It implies we can keep ourselves secure against systemic exploitation and abuse using nothing more than logical constraints encoded into rules of law to which systemic predators aren't sympathetic. For these and the reasons discussed at length throughout this chapter, it's unreasonable to believe that a kinetic stalemate represents a lasting peace. Instead, it's far more reasonable to believe that a kinetic stalemate represents a major systemic security hazard.

4.12.3 A Stalemate to War would Represent a God-King's Paradise

To improve our own capacity to survive and prosper, it is vital for us to understand that there is no such thing as a substitute to physical power as a zero-trust, permissionless, and egalitarian basis for settling disputes and establishing pecking order, no matter how much people like to preach about alternatives. **If it's true that agrarian society has stalemated itself at a global-scale strategic level with nuclear warheads, then that means agrarian society has never been more vulnerable to the threat of unimpeachable systemic exploitation and abuse than it has ever been in the past 10,000 years of populations suffering from the oppression of their god-kings**.

To understand the hazard we could be in, simply ask "what are the complex emergent social benefits of warfare that society would lose in a kinetic stalemate?" The author has already enumerated these benefits, but to summarize to top four: (1) zero-trust, permissionless, and egalitarian control over resources, (2) the ability to physically resist, constrain, and decentralize dysfunctional abstract power hierarchies, and (3) the existential motivation to innovate and cooperate at increasingly higher scales, and (4) the ability to vector limited resources to the strongest and most intelligent members of the pack who are demonstrably the best suited to survive in a world filled with predators and entropy.

If these are valid benefits of warfare, then we can see that a stalemated society would be a trust-based (thus systemically insecure), permission-based, and inegalitarian society where a ruling class has unimpeachable control over a ruled class. A stalemated society would have no capacity to use kinetic power to physically constrain people from adopting a single ruling class to settle all their disputes, manage all their resources, and determine ownership of all their property. A stalemated society would have no existential motivators to cooperate at higher levels than the level required to maintain their stalemate. A stalemated society would have no way to identify who among them is demonstrably capable of navigating chaos and surviving in a world filled with predators and entropy. A stalemated society would have successfully mitigated the threat of invasion by neighboring abstract power hierarchies, only to make themselves far more exploitable by their own abstract power hierarchy.

A global agrarian society locked in a strategic-level kinetic stalemate would necessarily have to adopt belief systems which utilize abstract power to settle their disputes, manage their resources, and establish their pecking order because they would have lost the option of using physical power to accomplish this in a zero-trust, permissionless, and egalitarian way. To establish a global pecking order, the entire population would necessarily have to adopt a single abstract power hierarchy and give a single ruling class extraordinary amounts of abstract power and control authority over them. And then the whole world would have to trust that single ruling class not to exploit them, because they would be physically powerless to stop it at a strategic level.

Hopefully now the reader can understand why the author spent so much time discussing the power dynamics of modern agrarian society. These dynamics help us understand why modern agrarian society may have entered an unprecedentedly hazardous state after the invention of nuclear warfare and policies like mutually assured destruction. Without the ability to physically constrain and decentralize abstract power hierarchies, society would lose its means to physically secure itself against global-scale exploitation and abuse by a single, tyrannical ruling class. All it would take for a tiny percentage of the population to gain completely centralized and unimpeachable control authority over the rest of the global population would be for high-ranking people inside nuclear-armed nations to collude within and between other nuclear-armed nations. Neo oppressors would be able to exploit the world's belief systems without penalty because the rest of the global population would no longer have the practical means to make it impossible to justify the physical cost of exploiting them.

In other words, a kinetic stalemate between different nuclear-armed nations would also represent a kinetic stalemate between the ruled and ruling classes of those nations. In addition to making it too impractical to fight a strategic war between nuclear-armed nations, a stalemate could also make it too impractical to fight a civil or revolutionary war within nuclear-armed nations. If an incompetent, belligerent, or self-serving group of systemic predators were to regulatorily capture the abstract power granted to them within these nuclear-armed hierarches, there might not be a practical way for civilians to escape their exploitation the same way they have always done it in the past (by summing together their kinetic physical power to make it too physically costly to continue exploiting them). In this kind of situation, an oppressed populace would always be trapped in a trust-based, permission-based, and inegalitarian system where a ruling class must always be trusted not to exploit their abstract power, simply because the population is otherwise physically powerless to countervail them.

To put it in plain terms: a strategically stalemated populace would be a god-king's paradise. The population would be backed into a corner where they would have to rely on people with abstract power to settle their disputes and establish their pecking order for them. They would have to adopt a global-scale abstract power hierarchy to establish a global-scale pecking order, tacitly giving one ruling class more asymmetric abstract power and resource control authority than any ruling class has ever achieved. The global human population would have no choice but to trust their rulers not to exploit their abstract power, because they would otherwise have no practical means to physically countervail them.

4.12.4 Non-Nuclear Kinetic Warfare can Lead to a Non-Nuclear Kinetic Stalemate

After hitting the kinetic ceiling with strategic nuclear warheads, society appears to be attempting to back themselves out of a corner by turning to non-nuclear kinetic warfare. The Korean War, Vietnam War, and the Global War on Terror are some examples of nuclear superpowers deliberately choosing to use weaker and less efficient power projection technologies to settle their disputes. Why would nuclear superpowers deliberately choose to use weaker and less efficient power projection technologies? Because it might be the only way kinetic power can still be useful as a basis for setting disputes and establishing pecking order. In other words, non-nuclear kinetic power might be the only way that a kinetic war can still have winners.

Similar to the emergence of the idea of nation states, kinetic warfare in the age of strategic nuclear warheads is another one of those situations where it is still too early to know how useful it will be as a mechanism for agrarian society to continue solving their disputes, managing their resources, and establishing their pecking order in a zero-trust, permissionless, and egalitarian way. The results so far appear to be inconclusive.

No two nuclear superpowers have gone head-to-head with each other in physical combat to settle a meaningful dispute. Instead, they have fought proxy wars (e.g. cold wars and trade wars). Since the invention of strategic nuclear warheads and the adoption of policies like mutually assured destruction, non-kinetic warfare has only been used to settle minor disputes in comparison to kinetic wars of the past. These disputes have ostensibly been between non-nuclear nations or asymmetrically powerful nations where one side has nukes, and the other side doesn't. This means we essentially have no idea if kinetic warfare is useful at solving large-scale property or policy disputes between nuclear superpowers anymore, because we have not tried it yet.

The author has a difficult time believing that non-nuclear kinetic warfare could settle a major global-scale dispute between nuclear superpowers without escalating to nuclear warfare and once again stalemating at some derivative form of mutually assured destruction. If we assume this is true, then we can conclude that society has indeed reached a kinetic stalemate at both the nuclear and non-nuclear level and is therefore highly vulnerable to the systemic security hazards discussed throughout this chapter. But for the sake of argument, if we were to assume that it is still possible to settle major global-scale strategic disputes using non-nuclear kinetic warfare, then the author asserts it's not reasonable to believe this capability will last for long. If there still is a window of opportunity for kinetic warfare to be useful as a means to settle major strategic disputes, then that window of opportunity might be closing.

As mentioned before, sapiens are obsessed with efficiency. Nuclear warheads represent what happens when sapiens create more efficient power projection technologies; they design and build power projection technologies that are so efficient they aren't practically useful. Nuclear warheads prove that too much power projection efficiency can become counterintuitively too inefficient; it's clearly possible to build power projection technologies that are so efficient at projecting power that they're too costly to use because of how destructive they are.

With this in mind, consider what it means when agrarian society deliberately chooses to settle their disputes and establish their pecking order using non-nuclear technologies. It means they're going to strive to make their non-nuclear kinetic power projection technology more efficient. What is the end state of making non-nuclear kinetic power projection technologies more efficient? We already know what the end state is: eventually, non-nuclear kinetic power projection technology will become too efficient to be practically useful.

By deliberately choosing to withhold from nuclear warfare and engage in non-nuclear kinetic warfare as the primary basis for settling global disputes in a zero-trust and permissionless way, agrarian society is setting itself up for a situation where it discovers yet another way to make kinetic warfare too costly to wage. Instead of nuclear technology, it would just be some non-nuclear technology that becomes too efficient at projecting power to be practically useful. The way things are starting to play out, it looks like it could be something involving artificial intelligence and swarms of flying, crawling, and swimming drones.

If we return to the chart showing the evolution of power projection technologies employed by agrarian society, we can illustrate this issue by adding another arrow to the chart, as shown in Figure 57. By choosing to engage in non-nuclear forms of kinetic warfare, agrarian society is essentially trying to fork its evolutionary path. The problem is that the forked path is just as vulnerable to running into the same kinetic ceiling! **The end state of increasingly more efficient kinetic power projection technologies is the same regardless of whether it's nuclear or non-nuclear: a kinetic stalemate. All that agrarian society will accomplish by forking the evolutionary path of kinetic power projection is to eventually discover yet another way that kinetic wars can't be won because it's too costly to wage – creating yet another form of mutually assured destruction that leads to the same kinetic stalemate, possibly at both the strategic and the tactical level.**

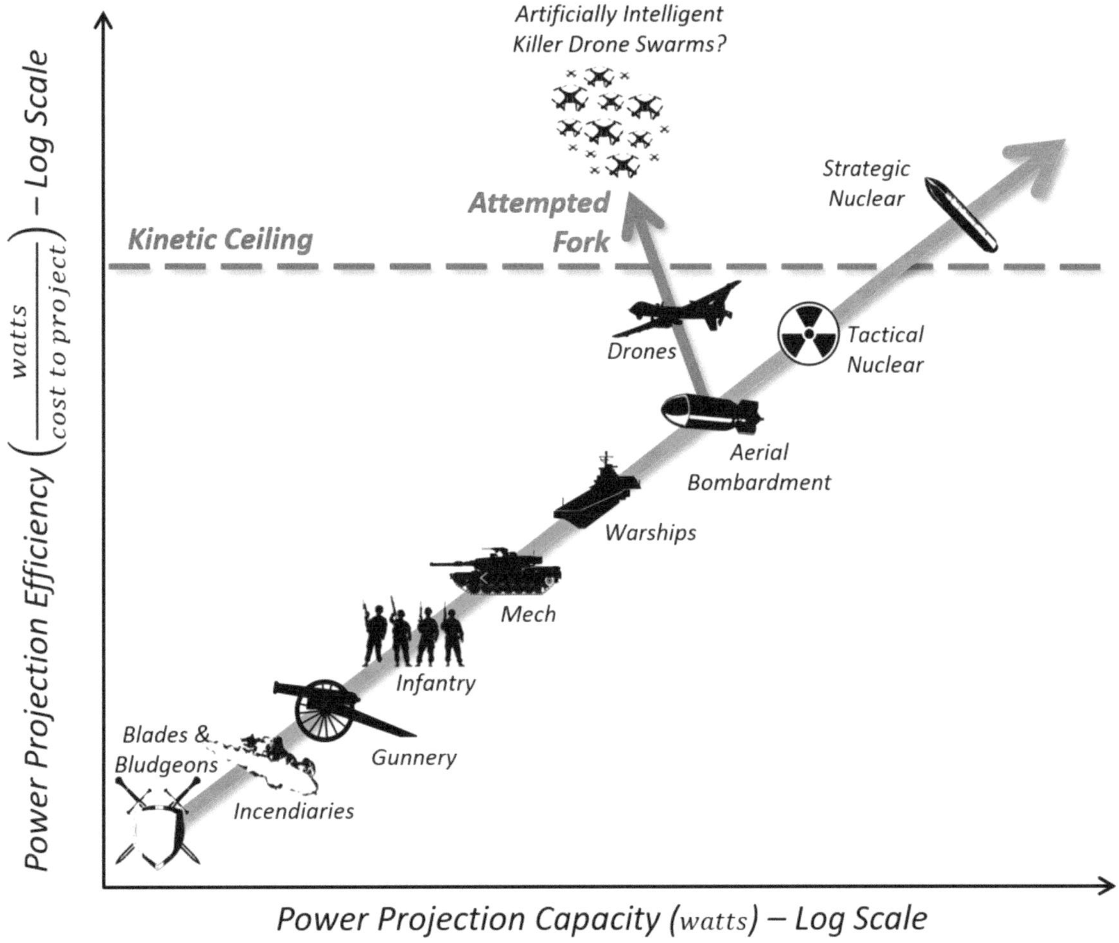

Figure 57: Evolution of Physical Power Projection Technology, Shown with an Attempted Fork
[88, 89, 116, 117, 118, 119, 120, 121, 122, 123]

Nations have started to realize that the best way to make their non-nuclear kinetic power projection technologies more efficient at projecting power is to take humans out of the loop. Software-operated drones are replacing human-operated machines. The size, weight, and power of these drones are collapsing. They're starting to fly in swarms. They're starting to drop bombs on people's heads with extraordinary precision. They're starting to think for themselves and predict human behavior better than humans can.

The marginal cost of putting precisely the right amount of kinetic power in precisely the right place to impose the maximum amount of physical cost on attackers is plummeting thanks to drones. We're getting extremely good at kinetically striking people with surgical-like precision for cheap. At the same time, other technologies like artificial intelligence are converging. We are marching straight towards the dystopian future envisioned by the Wachowskis in *The Matrix*, where swarms of sentient killer robots patrol the skies over the ruins of their creators.

Mutually assured destruction was the byproduct of the collapsing marginal cost of kinetic power projection technology following the discovery of nuclear technology. The more the marginal cost of projecting kinetic power decreases, the easier it becomes for multiple nations to scale it. But this same dynamic applies to non-nuclear technology too. The easier it is to scale non-nuclear kinetic power projection technologies, the easier it becomes for society to reach the kinetic ceiling again, and once again arrive at the point where our kinetic power projection technology becomes too mutually devastating to be practically useful as a method for solving disputes.

Human society cannot be safe from the threat of mutually assured destruction merely by switching to a non-nuclear form of kinetic warfare, because switching to a non-nuclear form of kinetic warfare doesn't do anything to address what caused mutually assured destruction to happen in the first place: the collapsing marginal cost of kinetic power projection caused by the discovery of new technologies. So even if we assume that non-nuclear kinetic wars between peers can still be conclusively won (which is already a tall assumption to make considering the track record of kinetic conflicts which have taken place after the invention of nuclear warheads), then we still have to acknowledge that the most that can be gained for our species from non-nuclear kinetic warfighting is temporary and incremental advantages along an evolutionary path of technology development that point towards the same dead end (pun intended).

Restructuring a military to win non-nuclear kinetic warfighting campaigns could represent the act of optimizing a military for a local maximum, not a global maximum. The end state of that effort is potentially a stalemate between nations at both the strategic and tactical levels, using both nuclear and non-nuclear technology. The key takeaway from this thought exercise is that there is essentially no way for agrarian society to escape from the bounded prosperity trap of a kinetic stalemate; all we can do is keeping forking the evolution of our kinetic power projection technologies to buy ourselves small windows of time to settle disputes and establish pecking order in a zero-trust and permissionless way before we discover yet another way to mutually assure our own destruction.

4.12.5 Non-Kinetic (i.e. Electric) Power Competitions Could Enable Mutually Assured Preservation

Now we finally arrive at a core insight about human physical power projection tactics which is essential for understanding the potential national strategic implications of Bitcoin. As much as people wish for it, there can be no peace, because peace requires a world without predators – a world which doesn't exist. Consequently, if we stalemate ourselves with kinetic warfare, we don't create an environment without predators, so we don't get peace. Instead, we create a huge window of opportunity for predators to exploit us through our belief systems at unprecedented scale because they cannot be physically challenged or overthrown.

A stalemate forces people into adopting abusive and exploitative abstract power hierarchies teeming with corruption because people lose the option of settling their disputes and establishing their pecking order in a zero-trust and egalitarian way using physical power. And if society tries to fork the evolution of kinetic power projection technology into a non-nuclear direction, they aren't going to solve the paradox of kinetic power projection; they are just going to discover another path to mutually assured destruction and find themselves stuck in the same bounded prosperity trap we're already stuck in, with the exact same systemic security vulnerabilities. The most that can be gained from continuing down the kinetic path could be minor victories against asymmetrically weak nations, perhaps not major strategic victories against nuclear peers.

How does society escape from this trap? Nature offers an idea: antlers. We need a technology that will allow us to continue to use physical power as the basis for settling our disputes, managing our resources, and establishing our pecking order in a zero-trust and permissionless way, while still keeping ourselves systemically secure by empowering us to impose severe physical costs on our attackers – but we need a way to do this that either minimizes or eliminates intraspecies fratricide. To accomplish this, we need a different, non-kinetic form of physical power projection technology that allows us to engage in global-scale physical power competitions in a zero-trust, permissionless, and egalitarian way, but non-lethally and non-destructively. It needs to be a power projection technology that will not become increasingly impractical to use as people become increasingly more efficient at projecting physical power and imposing severe physical costs on their neighbors. This new form of physical power projection technology should never become too efficient to be practically useful (therefore it won't suffer from the same fundamental flaw of kinetic power projection technology) no matter how efficient we try to make it.

Using this new type of non-kinetic physical power projection technology, a kinetically stalemated society could continue to along their path of technological evolution without having to worry about hitting a kinetic ceiling. While maintaining the kinetic stalemate, a society could continue to chase after an infinite prosperity margin for themselves by maximizing their C_A and minimizing our BCR_A. This would allow them to keep themselves secure against both foreign invasion and domestic exploitation by making it impossible to justify the physical cost of attacking or exploiting them, no matter how efficient they make their power projection technology, and no matter how much physical power they project.

Here we uncover a remarkable idea: **the future of global-scale strategic warfighting is more likely to be an electronic form of warfare rather than a kinetic form of warfare. Why? For the simple reason that kinetic warfare is literally a dead end for agrarian society. With the invention of nuclear warheads, it is possible that global agrarian society has already stalemated itself by unlocking a kinetic path to mutually assured destruction where wars can't be conclusively won, and disputes can't be conclusively settled**. There might be very little to gain by identifying yet another kinetic path to mutually assured destruction where wars continue to be unwinnable, and disputes continue to be unsettleable using kinetic technology – and that's assuming it's still even possible to fight strategic conflicts with nuclear peers that don't escalate to the nuclear level (which the author highly doubts). Forking the evolutionary path of kinetic power projection tactics, techniques, and technologies to pursue non-nuclear options isn't going to change the direction of this 10,000+ year trend – agrarian society is still marching towards another form of kinetic stalemate, and that's assuming we make the assumption that we aren't already there.

Instead, the future of warfare seems to be a non-kinetic form of global physical power competition that can continue evolving in the same direction as other physical power projection technologies (up and to the right) without running into a kinetic ceiling where it becomes too lethal or destructive to be practically useful. The future of warfare seems like it will be the kind of physical competition that can actually be won so that our property and policy disputes can actually be settled. Otherwise, it's not really the future of warfare, it's just the continuation of the same stalemate we could be in right now. **The key to winning future wars seems to be finding the battlefield where battles can still be fought and won no matter how powerful and efficient sapiens make their power projection technologies – and clearly that's cyberspace**.

The future battlefield of national-scale warfare could be an electro-cyber battlefield where nations impose severe physical costs on each other electronically rather than kinetically. This concept is shown in Figure 58. **In this future, nations would increasingly rely on "soft" forms of non-kinetic power projection that involve charges passing across resistors, as opposed to exclusively relying on "hard" forms of warfare that involve forces displacing masses**. Of course hard war would still exist, but perhaps mostly to settle minor tactical disputes or to preserve the kinetic stalemate caused by the kinetic power paradox discussed in the previous section.

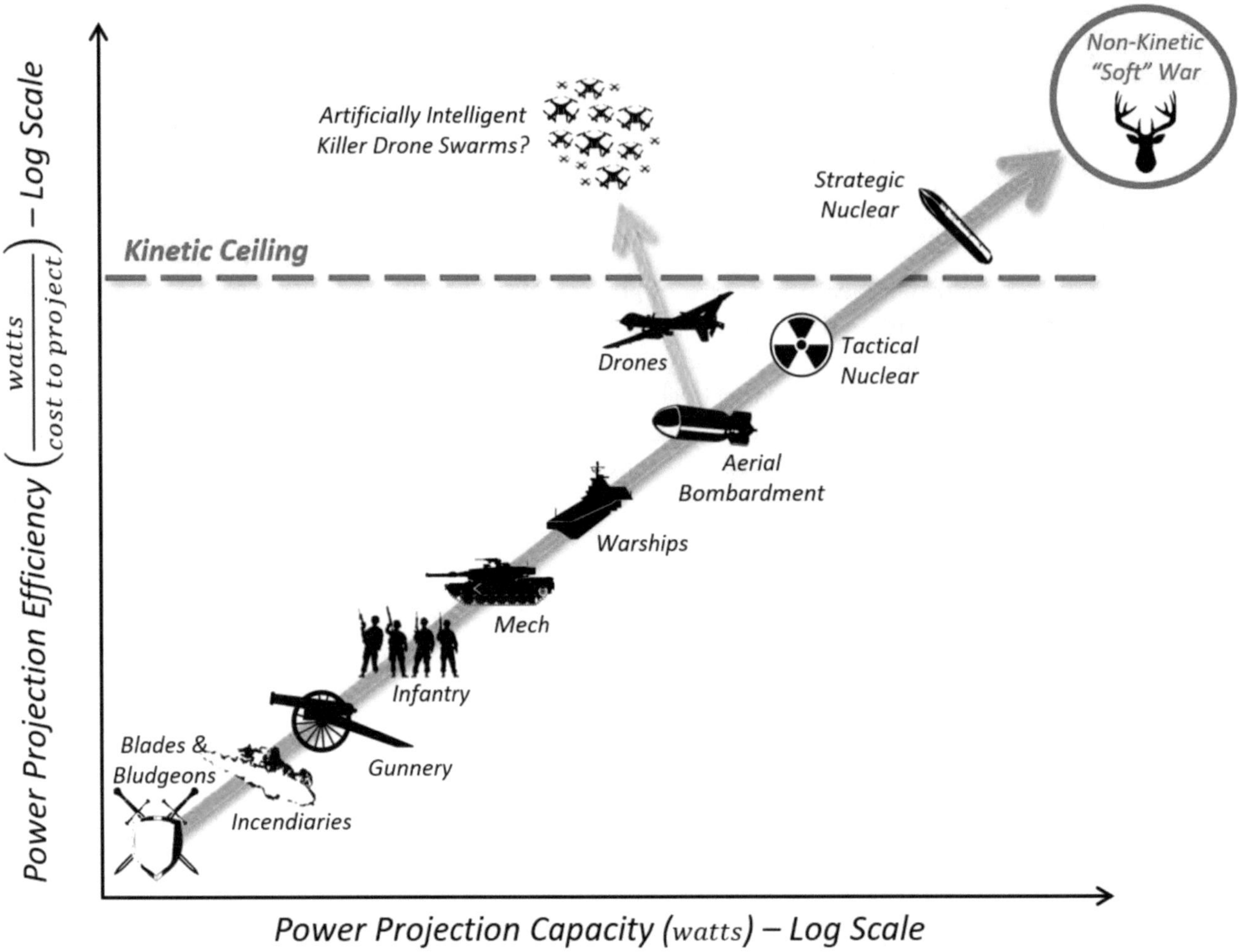

Figure 58: Evolution of Physical Power Projection Technology, Shown with Non-Kinetic End State [88, 89, 116, 117, 118, 119, 120, 121, 122, 123]

To put it simply, if sapiens are going to evolve antlers so they can continue to engage in global power competitions to settle property and policy disputes and establish pecking order in a zero-trust and egalitarian way that can't be kinetically stalemated, it seems highly likely they're going to be electronic antlers. **Electric power projection technology would allow a population to keep themselves *physically* secure against systemic predators (e.g. oppression) without having to rely on the faulty encoded logic of their laws, while a kinetic stalemate would continue to keep them physically secure against foreign invaders.** The resulting electric power competition would preserve a zero-trust, permissionless, and egalitarian way for settling disputes, determining who gets control over resources, and achieving consensus on the legitimate state of ownership and chain of custody of property.

4.12.6 Nikola Tesla Already Predicted This Would Eventually Happen

This idea isn't new; it's actually more than a century old. **Tesla predicted this in his essay entitled "*The Problem of Increasing Human Energy*." He first described the kinetic power projection paradox in 1900.** He already saw the potential for kinetic power projection to scale to the point where it would no longer be useful as a basis for engaging in physical disputes. As one of the world's leading experts in electricity at the time, he also saw how electronic warfare could one day replace kinetic warfare. He predicted society would hit a ceiling with its large and clumsy kinetic power projection technologies and be forced out of existential necessity to evolve towards human-out-of-the-loop, global-scale, machine-on-machine warfare. And he saw this all before the invention of cars, tanks, airplanes, aerial bombardment, nuclear bombs, and drones. [8] In his words:

"What is the next phase in this evolution? Not peace as yet, by any means. The next change which should naturally follow from modern developments should be the continuous diminution of the number of individuals engaged in battle. The apparatus will be one of specifically great power, but only a few individuals will be required to operate it. This evolution will bring more and more into prominence a machine with the fewest individuals as an element of warfare, and the absolutely unavoidable consequence of this will be the abandonment of large, clumsy, slowly moving, and unmanageable units. Greatest possible speed and maximum rate of energy-delivery by the war apparatus will be the main object. The loss of life will become smaller and smaller, and finally the number of the individuals continuously diminishing, merely machines will meet in a contest without blood-shed, the nations being simply interested, ambitious spectators. When this happy condition is realized, peace will be assured." [8]

Here we can see Tesla predicting that people would develop power projection technologies capable of projecting great power, but simultaneously requiring less human-in-the-loop operation. This assertion is followed by another assertion that peace wouldn't be possible if these power projection technologies caused bloodshed. He claims that mutually assured peace would only be feasible if these power projecting machines did battle against each other via an energy competition. Otherwise, power projection technologies would become too savage and destructive. His prediction aligns to what was outlined in this section, that nations would run into kinetic stalemates and not be able to achieve mutually assured peace until they settled on a way to fight their wars non-kinetically, in a way that eliminates bloodshed and mutually assures the preservation of human life no matter how powerful it becomes.

"No matter to what degree of perfection rapid-fire guns, high-power cannons, explosive projectiles, torpedo-boats, or other implements of war may be brought, no matter how destructive they may be made, that condition [of assured peace] can never be reached through any such development… Their object is to kill and to destroy… To break this fierce spirit, a radical departure must be made, an entirely new principle must be introduced, something that never existed before in warfare – a principle which will forcibly, unavoidably, turn the battle into a mere spectacle, a play, a contest without loss of blood." [8]

If we combine the core concepts provided in this chapter with Tesla's insights, then we get the following argument: **the way for agrarian society to assure a lasting peace between nations is not to keep building increasingly more efficient kinetic power projection technologies which do nothing but win small victories and identify new ways to place ourselves into situations where wars can't be won because they're too lethal and destructive. Instead, the way to assure lasting peace between nations is to give nations the means to engage in non-kinetic (i.e. electronic) global warfare with each other. To that end, it is reasonable to believe that nations will continue to compete against each other in global-scale physical power competitions to settle their disputes and their establish pecking order in a zero-trust, permissionless, and egalitarian way, but they will learn how to do it increasingly more electronically rather than kinetically. They will extend the battlefield into a new domain, where battles can still be won and where people can still physically secure the resources they value. And because this new form of warfare is electric (thus non-lethal), the world might reach a state of peace it has never known in 10,000 years of kinetic warfare. In other words, <u>to achieve global peace, simply swap out some of society's *kinetic* warfighting technologies with *electric* warfighting technologies where it would be effective to do so</u>. Maybe the best way to eliminate war is not to try to eliminate it all, but instead to make war "soft."**

A key to a sustainable peace that is safe against systemic predators could be some form of electronic, non-lethal warfare. **Whereas kinetic power competitions lead to mutually assured destruction and stalemates that place society into a systemically vulnerable pickle where they can be systemically exploited through their belief systems at unprecedented scale by people with abstract power, electric power competitions lead to mutually assured preservation**. Electric power makes it possible to settle disputes, establish control authority over resources, and achieve consensus on the legitimate state of ownership and chain of custody of property in a zero-trust, permissionless, and egalitarian way that can't be exploited by people wielding imaginary power. All that needs to be done to keep people's valuable resources (like bits of information, to include financial information) physically secure against both foreign invasion and internal corruption is to maintain the kinetic stalemate while utilizing an electronic form of physical power projection. That way, people can impose severe physical costs on all forms of attackers.

Einstein's theories support Tesla's theories. Einstein theorized that matter is swappable with energy. If that's true, then instead of fighting wars by blasting people with projectiles, we should be able blast them with energy in a non-kinetically destructive manner and get similar emergent properties (in this case, the emergent property is physical security). **A watt of physical power is a watt of physical power regardless of whether it's generated using forces displacing masses, or charges passing across resistors. It should therefore be possible to replicate the emergent properties of kinetic warfare and settle global-scale physical conflicts using an energy-based or "soft" form of global power competition rather than a mass-based or "hard" form of global power competition**. The only question is, how? In Tesla's mind, the answer was by creating intelligent machines. In his words:

"To bring on this result, men must be dispensed with; machine must fight machine. But how to accomplish that which seems impossible? The answer is simple enough: produce a machine capable of acting as though it were part of a human being... a machine embodying a higher principle which will enable it to perform its duties as though it had intelligence, experience, judgement, a mind!" [8]

Tesla made this prediction more than 40 years before the invention of the first operational general-purpose computer. He died a year before it was built, as well as two years before the first successful detonation of a nuclear warhead. Already admired by history as one of the world's greatest scientific minds, these predictions may one day make Tesla famous for also being a military strategic thinker. **Tesla essentially predicted that software would lead to soft war before either computers or nuclear weapons were invented – that a lasting peace could be achieved if agrarian society could figure out a way to wage wars by having their machines battle other machines using great sums of energy, and humans watched eagerly from the sidelines**. If we assume Tesla's predictions will eventually come true, then humankind has one extremely important question it needs to answer as quickly as possible, before they reach the next, non-nuclear form of mutually assured destruction. **What would a global-scale electronic power competition look like, where machines battled machines in a non-lethal way, and humans watched from the sidelines? In other words, what would agrarian society's electronic antlers look like?**

Chapter 5: Power Projection Tactics in Cyberspace

"[Software] cannot substitute for the physical constraints encountered naturally in other disciplines. Without a harsh and uncaring nature forcing us to make hard choices... We are willingly seduced."
G. Frank McCormick [124]

5.1 Introduction

At this point, the reader may be wondering, "what does all of this talk about abstract power, abstract power hierarchies, physical power, and warfare have to do with software?" The answer is simple: it has everything to do with software, because in the 21st century, the world's largest abstract power hierarchies are encoded via software rather than rules of law. For example, Facebook, Google, Twitter, and Amazon (which have substantially more users than governments have citizens) are all software-encoded abstract power hierarchies. Every person or institution that has software-instantiated control authority over our computer networks and the programs running on them (from which we tacitly need their permission to send, receive or store our digital information) is operating within an abstract power hierarchy, and we have established that abstract power hierarchies become dysfunctional and eventually lead to war.

This chapter links together key concepts in computer theory and cyber security that are needed to understand why software is fundamentally a new type of abstract belief system which gives a select few people abstract power and then entrusts them not to exploit it. People subscribe to these belief systems similar to how they subscribe to other belief systems which make them systemically vulnerable to exploitation. People in control of our software wield a new form of abstract power. Much like the clerics and lawmakers of the past, modern software developers have adopted the habit of using encoded logic to give themselves special permissions and authorities which give them asymmetric control authority over a new type of essential resource that has emerged in 21st century society: bits of information.

The way we have built our computer networks and designed our software systems has caused the people in charge of the most popular software we use to gain immense abstract power and control authority over our digital-age resources. Consequently, everyone who uses this software is systemically vulnerable to exploitation and abuse. This would imply that a new form of god-king is rising in the 21st century, except this time they're in control over our digital resources rather than our traditional agrarian resources. Consequently, how we establish control authority over our data appears to be emerging as one of the most contentious intraspecies property disputes of the modern era.

To believe that society will be able to constrain the far-reaching and asymmetric abstract power of these neo god-kings using more encoded logic (usually in the form of more regulations or more software) is to ignore 5,000 years of our predecessors' attempts to logically constrain the abstract power of previous rulers at the top of their respective abstract power hierarchies. Whether they're written using parchment or python, society has already demonstrated beyond a shadow of a doubt that logical constraints are not sufficient to protect people against the exploitation and abuse of people wielding too much abstract power. If history and cultural evolution repeats itself, then sooner or later, people are going to have to start using large quantities of physical power to decentralize control authority over their digital information and secure their digital property rights by imposing severe physical costs on oppressors who try to exploit them through software. To believe that people won't eventually resort to physical power to secure themselves is to ignore the lessons of history.

Using the core concepts of power projection theory established in the previous two chapters as a new frame of reference, we can observe the state of the world today and see the signs of digital-age systemic exploitation and abuse of our digital infrastructure practically everywhere. Weaponized misinformation, troll farms, bot farms, sybil attacks, DoS attacks, hacking epidemics, widescale online fraud, online censorship, shadow banning, routine data leaks, state-sponsored surveillance through entertainment apps, social network targeting campaigns, software companies routinely running ethically questionable social experiments on their users, unsupervised artificial intelligence controlling the primary information streams of billions of people, social media networks being captured by special interest groups, and amidst all of this, no reasonable expectation of privacy for any activity online. No matter where you go online, no matter what you do online, it has become standard practice for your valuable bits of information to be censored, surveilled, or sold to the highest bidder.

Our bits of information are under the centralized control of the people who control our computer networks and the software running on those networks. Meanwhile, software companies are becoming asymmetrically wealthy and influential thanks to the data they're harvesting from the population. The evidence of population-scale systemic exploitation of people through their software is mounting. Except this time, the resource that technocratic god-kings are competing for control over is digital information. Not surprisingly in the age of information, information itself has become a precious resource. People's ability to assemble, collaborate, speak freely, and to exchange valuable resources like money is increasingly facilitated by software, which means these vital social functions are increasingly under the complete control of the people who control that software.

This begs several important social questions: How could people secure themselves against the growing threat of digital oppression by a technocratic ruling class? Should our rights (particularly our property rights, our speech rights, and our rights to defend ourselves) *not* apply to our digital infrastructure and its corresponding digitized information? Is it possible to take back control of our digital information? How do we secure ourselves against systemic exploitation and abuse of our data streams by people who wield extraordinary amounts of abstract power over them because of the software and the computer networks they control? Is it reasonable to believe that we should simply trust in our technocratic ruling class not to exploit our data streams for their own personal advantage? Is it reasonable to believe that we will be able to secure our digital information if we simply find a better combination of logic to constrain a software engineer's abstract power, despite having 5,000 years of written testimony to suggest that attempting to constrain people from abusing their abstract power using encoded logic doesn't actually work?

10,000 years of warfare would suggest that the most effective way to keep people's rights secure against attackers is to figure out a way to impose severe physical costs on those attackers. The solution to the emerging threat of systemic exploitation via software seems clear when viewed through the lens of Power Projection Theory: to secure our digital information and infrastructure, we have to find new power projection tactics, techniques, and technologies which give people a way to impose severe physical costs on people wielding too much abstract power and control authority over our digital information. The solution makes rational sense, but it's not yet clear what the tactics, techniques, and technologies will be.

Enter Bitcoin. With a background in Power Projection Theory now thoroughly established, we can see Bitcoin in a new light. Bitcoin is compelling not as a candidate monetary system, but as an electro-cyber security system. Bitcoin's underlying proof-of-work technology is proving itself to be a successful way to physically secure bits of information against systemic exploitation and abuse by giving people the ability to project physical power to impose severe physical costs (costs denominated in watts) on belligerent actors who try to exploit people through their software. Bitcoin demonstrates that people can gain and maintain zero-trust, permissionless, egalitarian, and decentralized control over bits of information so long as they are willing and able to project physical power to secure it. This would suggest that Bitcoin isn't merely a monetary system, but perhaps some kind of "soft" form of power competition that has successfully replicated the same complex emergent benefits of warfare, less the destructive side effects. Through the ongoing global adoption of Bitcoin, people appear to be building the largest physical-power-based dominance hierarchy ever created in human history, and the social implications of this could be extraordinarily disruptive to *all* existing power hierarchies, including and especially nation states.

If the theories presented in this chapter are true, they would imply that Bitcoin represents far more than just a monetary system. After all, bits of information secured on the Bitcoin network could denote any type of information, not exclusively financial information. Instead, Bitcoin could represent a completely new system for securing any and all information in cyberspace – a way to keep bits of information secure against belligerent actors using *physical* constraints, not logical constraints. This could not only be revolutionary to the field of cybersecurity, but it could also be transformational to the field of physical security in general – to include national strategic security.

To understand all of these concepts in further detail, it's necessary to rewind back to the early development of computers and to retrace the technological steps we took to get to where we are today, all while leveraging our newfound knowledge in power projection tactics in nature and human society. To that end, we begin the final chapter of Power Projection Theory with a review of computer theory.

5.2 Thinking Machines

"Innumerable activities still performed by human hands today will be performed by automatons. At this very moment scientists working in the laboratories of American universities are attempting to create what has been described as a 'thinking machine.'"
Nikola Tesla [7]

5.2.1 General-Purpose State Mechanisms

Two years following Tesla's death, a Hungarian-American mathematician, physicist, and engineer named John von Neumann summarized novel theories on electronic computing in a report prepared for the US Army Ordinance Department. The report, titled *Preliminary Discussion of the Logical Design of an Electronic Computing Instrument*, explained technical ideas for the design of a fully electric general-purpose state machine (i.e. computer) with a remarkable new capability: storing programming instructions as states within its own electronically accessible memory so that it could do operations on its own programming, as if the machine had its own intelligence, experience, judgement, and mind. [125]

At the time this paper was published, programming state machines was a tedious ceremony involving days of lever pulling, dial turning, switch flipping, cable manipulation, and circuit plugging. Neumann and his colleagues believed that electronically storable computer programs would represent a dramatic improvement to this process. To accomplish their idea, Neumann and his colleagues reasoned they would need to build a machine that could store "*not only the digital information needed in a given computation... but also the instructions which govern the actual routine to be performed*." One way to do this would be to find a way to convert manual instructions into numerical code. "*If orders to the machine are reduced to a numerical code*," the report says, "*and if the machine can in some fashion distinguish a number from an order, the memory organ can be used to store both numbers and orders*." [125]

It was in US DoD-sponsored research like this paper where computer engineers began to communicate the enormous potential of what would eventually be known as *software*. The theoretical potential for general purpose computing had been discussed a decade prior by mathematicians like Alan Turing, who predicted the feasibility of universal computing machines capable of implementing finite sequences of instructions to solve complex math problems. But it wasn't until the first general-purpose computing machines became operational that people like Neumann began to practically demonstrate how they could be digitally instructed using their own memory. [126]

With 80 years of hindsight, we can read Neumann's report and appreciate what might be one of the biggest understatements of the century, when he made the claim that state machines capable of digitally converting human instructions into numerical codes would enable society to build machines that could "*conveniently handle problems several orders of magnitude more complex than are now handled by existing machines*." [125] After making this assertion, Neumann and his colleagues went on to pioneer the development of the first stored-program general-purpose state machine, a.k.a. the modern computer.

Computers are by no means new technology. Special-purpose state machines date at least as far back as 2,100 years ago to astrolabes like the Antikythera mechanism. In the two millennia following Greek astrolabes, the fundamental technological concept behind computers has not changed. They remain state machines named after their primary function of computing things. [127]

People decided to name state machines after what they do (their function) rather than what they are (their form). This is perhaps because their function has remained constant while their form continually changes over time. A state machine's form can take many different sizes, shapes, materials, and complexities depending on what it's designed to compute. Nevertheless, the core function of computing discrete mathematical states has not changed over two millennia. So for all intents and purposes, state machines have remained the same computing technology.

Prior to the early 1800s, state machines were special-purpose, non-programmable instruments used to assist people with very specific computations. The first viable general-purpose state machine and forerunner of the modern digital computer was the analytic engine designed by English mathematician Charles Babbage in 1833. Babbage's concept was novel because it meant people wouldn't have to build expensive special-purpose instruments to make single computations anymore. One general-purpose state machine doubles as a nearly unlimited number of disembodied special-purpose state machines. Thus, by using a general-purpose mechanical instrument to perform multiple different computations, it would cause the marginal cost of computing to decrease, while simultaneously turning the act of designing a special-purpose state machine into a disembodied abstraction.

This is a subtle but important distinction to make: **thanks to the invention of general-purpose computers, instead of having to *build* special-purpose state mechanisms to make computations, engineers only need to *think* of a sequence of operating instructions to assign to a general-purpose state mechanism. The act of computing changed from a manual, hands-on exercise to an exercise in *abstract thinking*.** [128]

Unfortunately, the design of Babbage's analytic engine was too complex and expensive (and the benefits of general-purpose computing too poorly understood) for it to be manufactured in the 1800s, so Babbage never lived to see his mechanical general-purpose state machine built. Nevertheless, the mere concept of an operationally viable general-purpose state machine was revolutionary in the field of computing. Armed with a general-purpose state machine, a person could perform an extraordinarily wide range of specialized computations. For the first time, the primary limiting factor of computing was no longer the *machine*, but the *imagination* of its programmer.

With the invention of general-purpose computing, the process of designing and building special-purpose state machines to perform specific computations became a disembodied abstract concept separable from the physical implementation of the machine making the computations. Machines which had previously been physically impossible or impractical to build suddenly became feasible, because instead of having to *build* a special-purpose state machine, a person simply needed to *imagine* what instructions they needed to give a general-purpose state machine. By following these instructions, the general-purpose state machine would, in effect, become the physical embodiment of a special-purpose state machine imagined by the programmer. This process of getting general-purpose state machines to role-play as special-purpose state machines is what's known today as ordinary computer programming. We take this for granted today but in the early 1800s, this was a revolutionary engineering concept.

Babbage's analytical engine was inspired by the design of Jacquard machines, devices fitted to looms that enabled complex textile patterns using punched cards to dictate the weave pattern of different colored threads. Borrowing from this design concept, the analytic engine was programmed using punched cards. The presence of a hole in the card indicated a symbolically important Boolean state like "1" or "true" while the absence of a hole in the card indicated an opposing state like "0" or "false." Using formal logical methods theorized by people like George Bool (who lived at the same time as Babbage), the sequence of holes punched into the card could issue a series of instructions to the state machine in a way that would mimic how a human might program a computer by pressing buttons, flipping switches, or turning dials. This technique for programming general-purpose computers was so effective that punched cards remained a popular programming technique for more than a century, until it was obsoleted by new technologies emerging in the 1980s. [129]

Although Babbage's analytic engine was too expensive to build, the disembodied and abstract nature of computer programming it enabled made it possible for mathematicians to write sets of instructions for Babbage's analytic engine using nothing more than a published design concept. To that end, one of Babbage's students, a mathematician named Ada Lovelace, published the design concept of an algorithm in 1843 for the analytic engine to calculate Bernoulli numbers. By publishing these instructions, Ada Lovelace became the world's first computer programmer. This illustrates how **the art of computer programming is an abstract and highly creative process akin to writing a fictional story. Just like playwrights write stories for actors to perform, computer programmers write stories for general-purpose state machines to perform**. [130]

5.2.2 Stored-Program Computers

Over the century following the publication of Lovelace's program, the emergence of new techniques like Boolean algebra and new technologies like electro-magnetically controlled relays made programmable general-purpose state machines far more feasible to design and manufacture, reinvigorating interest in general-purpose computing during the 1930s. Having studied the 100-year-old design principles of Babbage's analytic engine, Alan Turing and other mathematicians started publishing theories about the feasibility of universal computation by machines capable of implementing finite sequences of instructions to solve complex problems. [131]

Inspired by Babbage, Harvard physicist Howard Aiken proposed the original concept for a general-purpose electromechanical computer and began searching for a company to design and build it. International Business Machine (IBM) Corporation accepted Aiken's request in 1937 and after 7 years of design and manufacturing, delivered the first general-purpose electromechanical computer to Harvard in February 1944. Formally called the Automatic Sequence Controlled Calculator (ASCC) but colloquially known as the "Mark I" by Harvard staff, this 9,500-pound, 51-foot-long machine was immediately commandeered for military purposes to assist with making computations for the US Navy Bureau of Ships. [132]

Weeks after IBM delivered the ASCC to Harvard in February 1944, Jon von Neumann arrived on campus to commandeer the machine and put it to work on a series of computer calculations for a mysterious project he was working on. During this timeframe, Neumann began to write highly classified computing algorithms for the machine, including some of the most common algorithms used today (e.g. merge sort). Impressed by the ASCC's general-purpose computing capabilities, Neumann quickly turned his attention to the work of John Mauchly and Presper Eckert when he learned they had been commissioned by the US Army Ballistic Research Laboratory to build something even more special than IBM's electromechanical machine: the world's first fully-electronic general-purpose computer. [125]

The first fully-electric general-purpose computer was commissioned by the US Army Ballistic Research Laboratory and built for the US Army Ordnance Corps to perform ballistic calculations for firing tables. Called the Electronic Numerical Integrator and Computer (ENIAC), Mauchly and Ecker's machine reduced artillery trajectory calculation time down from 20 hours to 30 seconds; one ENIAC replaced 2,400 humans. But immediately after it was built, John von Neumann once again appeared on campus, commandeered the machine, and put it to work on a series of different computer calculations for his mysterious project. [133]

This time, however, John von Neumann's mysterious project became far less secret. News of the Hiroshima and Nagasaki bombings stunned the world as the US President publicly disclosed the project on which Neumann and his colleagues had been working: the Manhattan Project. While the world celebrated the surrender of Japan, Neumann continued to (literally) plug away on the first fully-electric general-purpose programmable computer to help him design the next version of a nuclear bomb.

Instead of punched cards, the ENIAC used plugboards as its state changing mechanism. This allowed it to run a program at electronic speed rather than at the speed of a punched card reader. The tradeoff to this approach was that it took weeks to figure out how to map a computing problem to the plugboard, and days to rewire the machine to execute each new program (in contrast to simply swapping punched cards). Nevertheless, ENIAC was a powerful calculating device that left an impression on Neumann and others who operated it. Its impressive capabilities, combined with how tedious and cumbersome it was to program, inspired Neumann and his colleagues to search for a solution on how to improve the machine by converting the state changing mechanism from manual circuit plugs into electronically-activated actuators. Then, by applying Boolean logic to the position of the actuators, this would allow programming actions to be digitized and stored as actuator states within the state mechanism. [134]

Stored-program computers, Neumann theorized, would not only allow a computer to store its own programming instructions as states within its own "memory organ" (a.k.a. the state space dedicated to storing the information needed for computer programs), it would also allow the machine to automatically perform operations on its own programming instructions. Neumann believed this architecture would allow general-purpose state machines to handle problems several orders of magnitude higher than even the world's most powerful computers at the time, like the ENIAC he used to design the detonation device for the first atomic bomb. [125]

Inspired by the potential of these types of machines, Neumann and his colleagues wrote the aforementioned report to the US Army Ordinance Corp. After getting the funding and green light to design and build one, Neumann teamed up with Mauchly and Eckert to create the Electronic Discrete Variable Automatic Computer (EDVAC), the world's first stored-program general-purpose computing machine. [135]

5.3 A New (Exploitable) Belief System

"Lessons learned over centuries are lost when older technologies are replaced by newer ones."
Nancy Leveson [136]

5.3.1 How to Talk to Thinking Machines

General-purpose and stored-program computing were not only technological breakthroughs, they were also two major back-to-back leaps in human abstract thinking. The first leap in abstract thinking came when general-purpose state machines converted the exercise of computing from a physical machine-building process to an abstract design process. The second leap in abstract thinking came when computer scientists introduced methods for storing computer programming information as physical states on a state mechanism using symbolically and syntactically complex protocols like machine code.

Instead of applying symbolic meaning to specific audible waveforms or scribbled images like we do with the *words* we speak or the *letters* we write, early computer scientists figured out how to apply symbolic meaning to physical *state changes* within a state machine. They accomplished this by applying mathematically discrete Boolean logic to state-changing mechanisms like switches or transistors to convert them into "bits" of information, in much the same way humans convert symbols or wave patterns into words. This gave rise to a new form of semantically and syntactically complex higher-order language consisting of binary and multi-decimal symbols called *machine code*.

From the broader context of metacognition and the evolution of human abstract thinking skills, it's hard to understate how remarkable machine code is. **Stored-program computing and the invention of machine code represent the emergence of an entirely new symbolic language as well as a new medium through which syntactically and semantically complex information can be communicated**. Using machine code, it is possible to convert practically any type of physical state-changing phenomenon in the universe into digitized bits of information (this is an important concept to remember for future discussions about Bitcoin, because the author will assert that bitcoin represents the act of apply machine code applied to quantities of electric power drawn out of the global electricity grid).

At first glance, machine code seems deceptively unremarkable. Sapiens have been creating and using higher-order languages for tens of thousands of years. We have been converting physical state-changing phenomenon like audible waveforms into information since we first discovered spoken language, so what's the big deal? What makes machine code so profound is the recipient of the information. Machine code was not invented for sapiens to communicate directly with other sapiens, but for sapiens to communicate directly with machines.

Therein lies the significance of what computer engineers designed and built in the 1940s. They didn't just build gizmos, they invented a completely new, machine-readable language – a medium through which humans can "talk" to inanimate objects using machine code, and just as astonishingly, a medium through which inanimate machines can "talk" back to humans. This new form of symbolic language created an entirely new form of storytelling where people can communicate with each other indirectly and asynchronously through the computers they program, as shown in Figure 59.

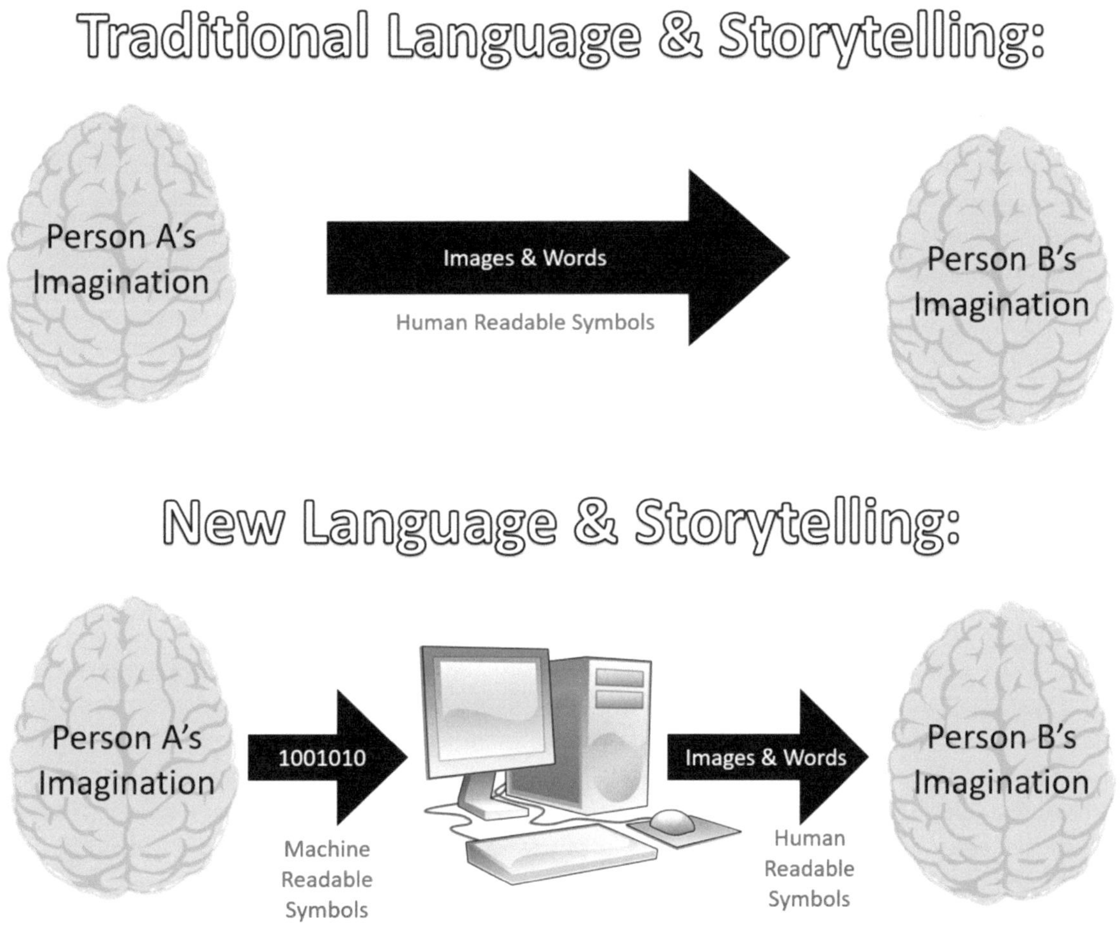

Figure 59: Illustration of the Difference Between Traditional Language and Machine Code [137, 76]

By issuing computing instructions to a general-purpose state machine using a new form of machine-readable language, Neumann and his team realized it would be possible for a computer programmer to produce practically any conceivable sequence of operational instructions because there is theoretically no limit to the state space (i.e. the number of different possible states) that a general-purpose state machine can have. Since there is no theoretical limit to the size of the state space or the sequence of operations that a programmer can create using a stored-program digital computer, a programmer has near-unlimited design options and flexibility. With the right set of instructions, they can make a general-purpose computer perform any operation. Neumann described this potential as follows:

"It is easy to see by formal-logical methods, that there exists codes that are in abstracto adequate to control and cause the execution of any sequence of operations which are individually available in the machine and which are in their entirety, conceivable by the program planner." [125]

Neumann explains how instead of having to *physically execute* a sequence of operations to instruct a general-purpose state machine how to behave, the programmer of a stored-program general-purpose state machine simply needs to *think* about a sequence of operations *in abstracto*, that is, from a purely imaginary or theoretical

point of view. Then, the programmer could use machine-readable languages to *speak* to the machine and tell it how to operate, rather than to manually operate it. These ideas were the precursor to what we now call "operating systems." The important takeaway is that **computer operating systems are *belief systems*, not physical systems. Like all belief systems, modern computer programs are abstractions – figments of the imagination that are programmed into a computer rather than spoken or written. Therefore, like all belief systems, computer programs are vulnerable to systemic exploitation.**
With the invention of stored-programming computers and machine code, sapiens took a major leap forward in abstract thinking, and the sociotechnical consequences of this major leap are still not yet fully understood. **Neumann's stored-program general-purpose computer removed many of the physical constraints of computer programming by moving the act of programming out of the domain of shared objective reality and into the domain of shared abstract reality**. Unbounded by physical constraints and with infinitely scalable state spaces, the primary limiting factors of general-purpose computer programming became the imagination and design skills of the programmer. Insofar as a computer programmer can think of a functional design and communicate it effectively to a general-purpose computer, a general-purpose computer can faithfully act it out.

5.3.2 A New Way of Storytelling

A core concept of computer theory is that **computer programs don't physically exist. They are abstract concepts hypostatized as concretely real things for the sake of reducing the intellectual effort required to use stored-program general-purpose state mechanisms**. Although society has adopted the habit of describing software as if it were a physical thing that is "loaded" onto a computer, it simply isn't. And although society has adopted the habit of describing software as if it were comprised of an orientation of objects (e.g. folder, trash can, recycle bin, thumb nail), it simply isn't. This is merely an abstraction technique used to make it easier to understand the complex emergent behavior of computer programs.

Modern computers are usually made of complex circuitry and electrons stored in floating gate transistors. When a program is "stored" onto a computer, nothing is physically added to the machine. The computer remains the same general-purpose state machine built out of the same complex circuitry and electrons stored in floating gate transistors. When a program is "removed" from a computer, nothing is physically removed from the machine. Before a computer runs a program, all that physically exists is a state machine made of complex circuitry and electrons stored in floating gate transistors. While a computer runs a program and after it completes the program, all that physically exists is the same machine made of the same circuitry and electrons stored in floating gate transistors. At no point before, during, or after a computer program is added, removed, or executed does anything get physically added or removed from the system. The only physical change that occurs is a state change in the circuitry of the machine. A switch is flipped, or electrons move from one side of a floating gate transistor to another.

A computer program is technically nothing more than symbolic meaning assigned to the state changes of a state machine. Just like sapiens learned how to assign symbolic meaning to audible wave patterns or visible shapes to form spoken and written language, they also learned how to assign symbolic meaning to state changes on a state machine using Boolean logic. In the case of machine code, the symbolic meaning assigned to a state change is usually binary code like "1" or "0," which is enough to generate increasingly more semantically and syntactically complex information. This is one of the many reasons why computer science is such a fascinating (and difficult to grasp) field. Sapiens figured out how to apply symbolic meaning to computer circuitry, which can then repeat that symbolic meaning back to humans using communication protocols that humans can understand. Thanks to computer theory, we learned how to talk to machines and taught them how to talk back to us.

Revisiting the core concepts of metacognition discussed in the previous chapter, the reader is reminded that within the domain of shared objective physical reality, **the physical patterns of our spoken words, written words, and states of our state machines have no inherent, objective meaning. It takes a powerful human neocortex capable of abstract thinking to assign semantically or syntactically complex symbolic meaning to these phenomena**. All computer programs are abstractions – figments of the imagination which do not physically exist as things with mass, volume, or energy in physically objective reality, completely incapable of producing their own physical signatures. Computer programs are imaginary concepts within shared abstract reality that are transferred from neocortex to neocortex using physical media, just like all other forms of imaginary thoughts and ideas travel between sapient brains.

A common source of confusion about computer science occurs when people conflate the physical media through which a computer program is communicated with the program itself. There is a major difference between an imaginary sequence of operating instructions spoken to a machine using symbolic language, and the physical media through which those instructions are communicated to and from a computer. When programming a modern computer, what physically exists is the machine through which sapiens communicate their abstract ideas using a carefully programmed sequence of state-changing operations. The rest is abstract.

When a human speaks to another human, the audible waveform of their voice is physically real, but the symbolic meaning and information assigned to those audible waveform patterns is abstract. If a person were to believe that words are physically real things, they would be making a false correlation between abstract inputs detected by the brain (i.e. the symbolic meaning within a semantically and syntactically complex higher-order communication protocol) that were produced by their imagination, and their sensory inputs (i.e. the audible pattern detected by their ears). The same reasoning holds true for computer programs. Just because your eyes can see a computer acting out a program doesn't mean the program is physically real; you are looking at a general-purpose state machine, not a computer program.

Like a painter's canvas, the digital computer is a portal through which an imaginary idea travels from one brain to the other. The person depicted in a painting isn't physically real; the only thing that is physically real is the paint and canvas through which the person's image is communicated to its viewer. Likewise, what the reader sees written on this page is not the author's ideas in and of themselves, but merely the medium through which the author's ideas pass from his neocortex to the reader's neocortex. Using this same logic, the switches on the circuit board and the electrons stored within the floating gate transistor are not the program, but merely the physical medium through which an abstract idea is communicated from the computer programmer to the computer user.

5.3.3 Computers are Electromechanical Marionettes Acting out Scripts Written by Neo Playwrights

To make the abstract nature of computer programming more digestible for readers who don't have computer science backgrounds, the reader is invited to reflect on Shakespeare's *Romeo & Juliet*. As previously discussed, sapiens use their abstract thinking skills to design virtual worlds within their minds, and then they get other people to envision the same world (i.e. synchronize their individual abstract realities into a single shared abstract reality) via syntactically and semantically complex symbolic languages using storytelling.

The story of *Romeo & Juliet* was an imaginary scenario conceived within the neocortex of a notable storyteller named William Shakespeare a little over four centuries ago. Our own neocortices can envision what Shakespeare imagined because Shakespeare used a symbolic language (i.e. English) as a medium through which he could share what otherwise existed exclusively within his imagination.

The English symbols of Shakespeare's language were copied and printed onto pieces of paper and given to actors as instructions. These instructions are commonly called *scripts*. Actors speak and move according to the instructions received from their scripts. If Shakespeare did a good job writing these instructions and if actors did a good job at following them, then a complex effect will emerge: an audience of people will be entranced by the performance, so emotionally captured by what they're watching that they'll forget that it's completely fictional – they'll forget that Romeo and Juliet don't exist anywhere except exclusively within their imaginations.

Because of physical actions taken by role-playing actors in shared objective reality, people can see with their own eyes (a.k.a. sensory inputs) a physical representation of Shakespeare's imagination. Through the actors' performance, an audience can experience a full range of emotions that Shakespeare experienced when he first conceived of the story, even though many lifetimes have passed since the last synapse of his neocortex fired. Such is the beauty of storytelling and other forms of art – the ability to share one's ideas far beyond one's own physical limitations, long after one's tenure on Earth has ended.

Nowhere in the sequence of events between Shakespeare's first imagination of *Romeo & Juliet*, to an audience of people emotionally entranced by a theater's performance 400 years later, did *Romeo & Juliet* become anything more than a fictional story. Based on the very definition of fiction, we know that Romeo and Juliet never physically existed. The fact that the story *Romeo & Juliet* can be conceived by multiple different brains using tools like symbolic language printed on a physical script, physically-spoken soundwaves, or physically-moving actors, doesn't mean the imaginary story of *Romeo & Juliet* is anything except imaginary. *Romeo & Juliet* is merely an imaginary story told by a gifted storyteller.

If someone were to argue that the imaginary story of *Romeo & Juliet* physically exists because it can be written, spoken, or acted out in physically objective reality, they would be guilty of hypostatization, the previously discussed fallacy of ambiguity. They would be doing that thing sapiens are instinctively inclined to do where they believe something imaginary is something concretely real because of a false-positive correlation between matching abstract and sensory inputs, as illustrated in Figure 60. Because the motion of the actors matches a scenario imagined by the neocortex, people are quick to believe the imaginary scenario is physically real. Much like people who believe the king's abstract power is physically real because knights display physical power when they get orders from the king, people will believe *Romeo & Juliet* is physically real because actors physically perform it based on the orders they receive from Shakespeare. Of course, it is a logical error to believe that a fictional story which exists only within the imagination is physically real just because the pages on which it's written are physically real, or the stage on which it's performed is physically real and can be seen, smelled, touched, tasted, and heard.

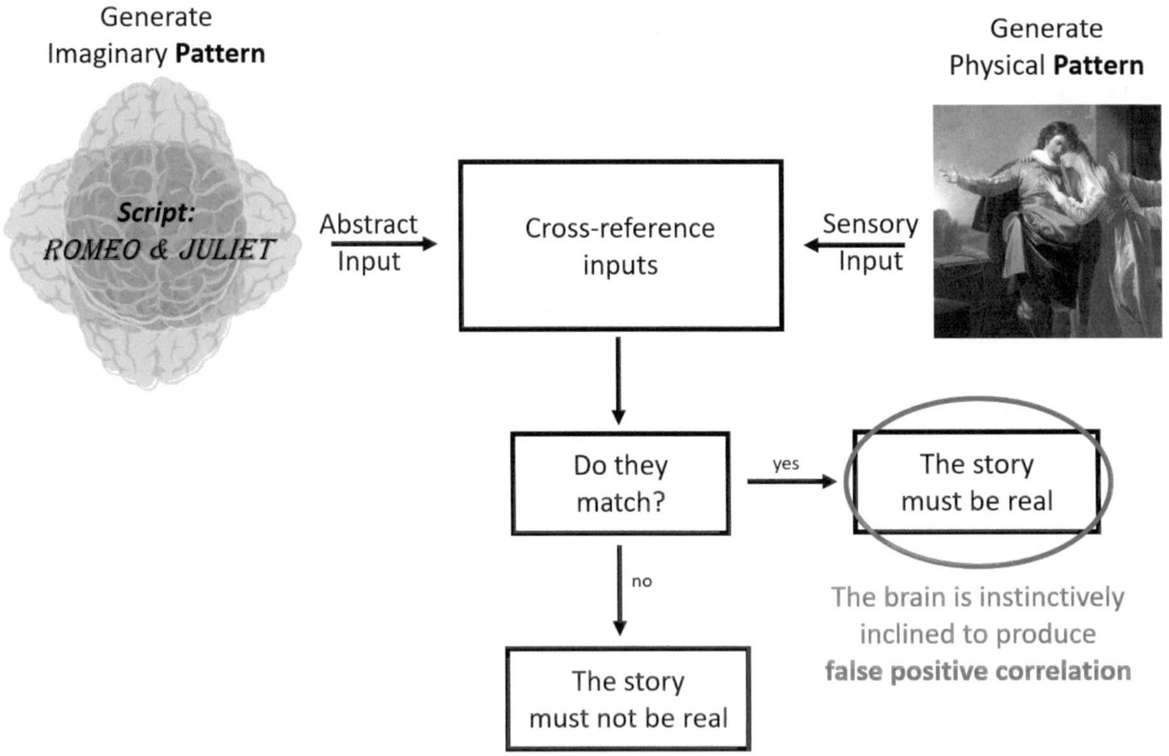

Figure 60: False Positive Correlation Produced by the Brain's Realness-Verification Algorithm
[76, 138]

By now the reader may be asking, what does Shakespeare have to do with computer programming? Simply change English symbols encoded on parchment to Boolean logic encoded on transistors, and replace the actors with electro-mechanical marionettes, and you get the modern art of computer programming. **Computer programmers are, fundamentally speaking, storytellers. They conceive of fictional stories and create abstract descriptions of imaginary scenarios, events, objects, and ideas**. Computer programmers write their stories using scripts and they hand those scripts to machines just like Shakespeare would hand his scripts to actors. Then, the general-purpose computers role-play according to the directions they receive in their scripts, just like actors role-play *Romeo & Juliet*.

One could say that *Romeo & Juliet* is an imaginary story "stored" within the pages of a script, just like a computer program is an imaginary story "stored" within the circuitry of a computer. This is a technique computer programmers use to make it easier to talk about programming. **For the sake of simplicity, engineers have adopted the habit of hypostatizing computer programs and treating them as if they were concretely real things comprised of concretely real objects**. Like a form of shorthand notation, it's simply easier on the brain to communicate complex abstract things like computer programs as if they were physically real things, because that's what we're used to experiencing. As the reader is no-doubt experiencing right now, it's quite tedious to explain and to comprehend how computer programs *actually* work, so we simplify using abstractions.

With the invention of stored-program general-purpose state machines, sapiens created something more complex than they have the capacity to fully comprehend. Faced with the overwhelming complexity of modern digital computers, sapiens do what they have been doing since the Upper Paleolithic era: they put their neocortices to work coming up with abstract explanations for things they can't fully comprehend, and they pretend like something imaginary is something physically real for the sake of simplicity. This is an extraordinarily helpful technique for reducing the metacognitive burden of managing the complexity of designing and operating computer programs, but it's important for the reader to understand that it is *not* technically accurate to describe a computer program as something physically real. This is an especially important point to understand prior to future discussions about Bitcoin.

Computer programmers use syntactically and semantically complex symbolic languages like machine code, assembly language, or higher-order programming languages as a medium through which they share their ideas with machines, writing scripts for electromechanical marionettes to give a performance for an audience. These scripts and their corresponding symbols have taken many physical forms over the past 80 years. In the early days of general-purpose computing, these scripts were pieces of paper just like Shakespeare would have used, except with holes punched in them rather than symbols written on them. But after the invention of stored-program computing, scripts took the form of circuits built into digital computers. Today, scripts often take the form of electrons stored in floating gate transistors.

General-purpose state machines can take different programmable states based on the instructions given to them by a computer programmer. If a computer programmer does a good job of imagining a fictional design and communicating it to the computer via symbolic language, and if the machine does a good job following its instructions, then a complex effect will emerge: a desired computation will be made, a desired behavior will emerge, and an audience will be so entranced by the performance of their electromechanical marionettes that they will become convinced what they see is something physically real. A well-programmed computer will cause a person to lose sight of the fact that they are sitting still, doing little more than staring at an array of light-emitting diodes glued to a plane of glass and controlled by a general-purpose computer, all of which was painstakingly orchestrated across decades of engineering to present something which mimics the behavior of something observed in shared objective reality, but isn't actually there. It's as real as Romeo and Juliet.

Nowhere in the sequence of events between a computer programmers' imaginary thoughts, to an audience of people enthralled by an array of light-emitting diodes, did the story written by the computer programmer become anything more than imaginary. Just like skilled playwrights and highly capable actors can write and follow scripts to make fictional tales like *Romeo & Juliet* look and feel physically real, so too can skilled computer programmers and highly capable machines write and follow scripts to make fictional tales like *Solitaire* look and feel physically real. People will become so moved by the performance of their general-purpose computers that they will completely lose sight of the fact that what they see is merely a canvas of symbols – an abstract representation of a completely virtual reality that would otherwise have no detectable physical signature if it weren't for the presence of the electromechanical marionette.

The general-purpose computer is a machine commanded to live-action role-play (LARP) according to whatever lies within the imagination of its programmer. Everything printed on the screen of a general-purpose computer is a computer-generated illusion. Whether it be a line of text, or a detailed image, or an imaginary object, or a three-dimensional interactive environment that looks and behaves just like environments experienced in shared objective reality, what a machine shows on a screen is virtual reality. Virtual reality is, by definition, not physically real. The only knowledge a person can gain from looking at a computer screen is symbolic knowledge, not experiential knowledge. This is true even if what's shown on screen is an image of something real or an event which did physically happen.

A stored-program general-purpose computer can therefore be thought of as a symbol-generating abstract reality machine. **With the invention of digital computers, sapiens transformed their ability to communicate their abstract thoughts to each other to form a new type of shared abstract reality – which we have named "cyberspace."** Instead of applying symbolic meaning to spoken and written words, humans now apply symbolic meaning to electromechanical state changes in circuit boards. Instead of utilizing actors to role-play imaginary stories, they use machines. This is a core concept in computer science that is essential to understanding software's systemic security flaws.

5.4 Software Security Challenges

"The first step in creating safer software-controlled systems is recognizing that software is an abstraction."
Nancy Leveson [136]

This section gives a technological deep-dive into some of the most challenging aspects of computer programming and cyber security. These concepts lay the groundwork for understanding how and why common cyber security challenges could be alleviated by physical cost function protocols like Bitcoin.

5.4.1 Software Security is Fundamentally a Control Structure Design Problem

If the total amount of money stolen from cybercrime were treated as its own country, then it would represent the third-largest economy after the US and Chinese economies. In a special report issued by Cybercrime magazine, Cybersecurity Ventures stated that it expects "*global cybercrime to grow by 15 percent per year over the next five years, reaching a $10.5 trillion USD annually by 2025, up from $3 trillion USD in 2015. This represents the greatest transfer of economic wealth in history, risks the incentives for innovation and investment, is exponentially larger than the damage inflicted from natural disasters in a year, and will be more profitable than the global trade of all major illegal drugs combined*." [139]

Thanks in part to a substantial increase in nation state sponsored hacking activities, some have claimed that there is a "hacking epidemic" plaguing the modern field of cybersecurity which will cause cyber attacks to increase an estimated 10X between 2020-2025. This may not be surprising to some people considering how routine ransomware attacks, data breaches, and other major cyber security incidents have become. Cyber security is now such a significant challenge for US national security that the White House recently passed an executive order addressing it. According to US President Biden, improving the nation's cyber security is essential to national strategic security and stability. "*The prevention, detection, assessment, and remediation of cyber incidents is a top priority to this Administration*," President Biden has declared, "*and essential to national and economic security*." [140, 139]

Today's substantial amount of cybercrime suggests that software security engineering is challenging, and there is room for improvement. **An important first step towards improving software security is understanding that software is fundamentally an abstract belief system, and all abstract belief systems are vulnerable to systemic exploitation and abuse**. When software leads to unexpected or undesired behavior like a cyber security incident, people often claim their software "broke," but this is just a figure of speech. Nothing physically breaks during a software malfunction as it is physically impossible for something which doesn't physically exist to physically break. What *actually* happens during a software hack is that people find a way to exploit the software's design logic. This is why subject matter experts in software and system safety design like Leveson assert that the first step to creating safer and more secure software-intensive systems is to remind yourself that software is *only* an abstraction – it's all in one's imagination.

Computers behave exactly as they are instructed to behave. Therefore, when a computer produces an unexpected or undesired emergent behavior, the root cause of that behavior is most likely the design of the software. Except for very rare exceptions where computer hardware components are physically damaged or experience something like a short or an unintended bit flip, computers don't fail to operate exactly as they've been programmed to operate. By that same logic, unless the state-changing mechanism of a state machine has been physically impaired, there's also no such thing as a "failed" or "broken" state of a computer, because the machine was explicitly designed and built to be able to take that state.

What usually happens when a computer program gets hacked or leads to a safety or security incident is that the original computer programmer attempts to encode logical constraints which are insufficient at stopping a belligerent actor from systemically exploiting the logic of the computer program. As many computer programmers have learned over the years, **encoding logical constraints into software doesn't eliminate the threat of people exploiting the software's logic, it just changes the way the software's logic can be exploited** (note how this is the exact same concept discussed in the previous chapter about how laws don't prevent people from exploiting or breaking the law). Combining this observation with the fact that software can't break, then that means **software security problems are fundamentally software design problems**. If software gets hacked or behaves in a way that's unexpected and it leads to an undesired incident, it's almost always the case that the root cause of that incident was the result of the programmer producing a flawed design. Computers cannot be blamed for diligently and faithfully acting out the script given to them by their director; the script is to blame. This is a core concept in the field of software systems safety and security. [136]

Recall how a computer program represents a sequence of control signals issued to a state machine. **All software-related safety and security incidents are the byproduct of control signals which create insecure or hazardous system states which then lead to an undesired loss event**. The goal of software systems security is to identify which control signals could lead to a hazardous state, and then design a control structure which eliminates or constrains those control signals. This is a foundational concept not just in software security engineering, but for system safety in general, as outlined by safety and security engineering techniques like STAMP and STPA. [141]

There are four primary ways software control signals can produce insecure or hazardous system states. First, software can provide a control signal that overtly places the system directly into an insecure or hazardous state. Second, software could *not* provide a control signal that is needed to prevent a system from being placed into an insecure or hazardous state. Third, software can provide a potentially secure control signal, but do it too late or too early, resulting in an insecure or hazardous system state. Lastly, software can stop providing a potentially secure control signal too soon or too long, resulting in an insecure or hazardous state. [124, 136]

As discussed by pioneers in systems safety and security engineering, the systems approach to improving security is to anticipate insecure or hazardous systems states using abstract thinking exercises like scenario planning, then to identify what control signals (or lack thereof) would cause the system to reach these undesired states. Once those sensitive control signals have been identified, the role of the security engineer is to design a control structure that either eliminates those control signals or constrains them as much as possible. To accomplish this, software security designers must keep strict account of all the different control signals (or lack thereof) a piece of software can execute. [124, 141]

In systems security theory, **the root cause of all software security incidents is attributable to insufficient control structure designs which didn't properly eliminate or constrain "unsafe" or sensitive control signals**. Therein lies the fundamental challenge of software security engineering; it requires an engineer who can understand and anticipate different combinations of control actions or inaction which should or shouldn't occur. Software security engineers must be able to recognize these sensitive control signals *and* design control structures which eliminate or constrain those signals, which is quite hard to do using only logical constraints encoded into software, while still meeting desired functionality and behavior. This is one of the biggest challenges in software security which makes it so different from security engineering in other industries. **Because software doesn't physically exist, it's not possible to secure software using physical constraints unless the underlying state mechanism (i.e. computer) is physically constrained** (this is the single most important concept that the reader should note prior to a discussion about proof-of-work physical cost function protocols like Bitcoin, because what proof-of-work represents is the act of physically securing software by physically constraining the underlying computer).

Why exactly is it so challenging to design software control structures which can eliminate or sufficiently constrain a computer from sending unsafe control signals using logical constraints rather than physical constraints? The author offers six explanations. First, computers can have infinitely expanding state spaces comprised of an infinite number of hazardous states. Second, programmed computers have shape-shifting protean behavior which gives them unpredictable, non-continuous (thus non-intuitive) emergent behavior. Third, because it's imaginary, it's very easy to build software with unmanageable complexity. Fourth, software control signal interfaces are invisible and physically unconstrainable. Fifth, software design specifications are arbitrary and semantically ambiguous, and the software engineering culture of information hiding can also hide critical security information. Lastly, untrustworthy software administrators deliberately design systems which give themselves abstract power and control authority.

5.4.2 Software Security Challenge #1: Infinitely Expandable State Spaces with Infinite Hazardous States

The first reason why software security engineering is challenging is because the state space of most computers is practically infinite. As Von Neumann famously observed, there is theoretically no limit to the number of states that stored-program general-purpose state mechanisms can have. Unfortunately, this means there's also no theoretical limit to how many insecure or hazardous states a programmed computer can have. Consequently, **as software becomes larger and more complex, the size of its hazardous state space increases exponentially, often far exceeding what computer programmers can reasonably expect to circumnavigate**.

This presents an extraordinary challenge for software security engineers who are responsible for understanding a given computer program's hazardous state space and designing control structures which eliminate or constrain control actions which would cause the system to enter that hazardous state space. If the hazardous state space is practically infinite, then it's practically impossible to avoid all hazardous states. [124]

5.4.3 Software Security Challenge #2: State-Changing Mechanisms Behave Like Shape-Shifting Monsters

The discrete nature of states also means a computer can have dramatically different emergent behavior despite minor state changes. A state-shifting mechanism can be thought of as a bipolar, shape-shifting monster. With a seemingly minute and inconsequential state change, a computer's emergent behavior can transform from something harmless to something significantly hazardous. This non-continuous behavior means **computers are unpredictable and always "one wrong move" away from catastrophic malfunction**. One seemingly minor control action (or inaction) can cause a discrete state change that causes a programmed computer to behave in surprising ways, and it is practically impossible for software designers to anticipate all the possible different combinations of unsafe control actions which could lead to every possible hazardous state change within a given state space. This means it's practically impossible for software engineers to know every single "wrong move" or unsafe control action a complex piece of software can make.

5.4.4 Software Security Challenge #3: It's Very Easy to Make Software Unmanageably Complex

Because software is abstract and because state mechanisms have infinitely expanding state spaces, there is practically nothing limiting the complexity of the design of computer programs. In her book on systems safety engineering, Nancy Leveson offers an explanation for why this can make software engineering exceptionally difficult, citing observations of subject matter experts like Parnas and Shore and software's so-called "curse of flexibility." [124]

"*In principle*," Leveson explains, "*[software's flexibility] is good – major changes can be made quickly and at seemingly low cost. In reality, the apparent low cost is deceptive… the ease of change encourages major and frequent changes, which often increases complexity and rapidly introduces errors.*" [142, 143, 124]

Shore explains software's curse of flexibility by comparing software engineering with aircraft engineering. When designing an aircraft, "*feasible designs are governed by mechanical imitations of the design materials and by the laws of aerodynamics. In this way, nature imposes discipline on the design process, which helps to control complexity. In contrast, software has no corresponding physical limitations or natural laws, which makes it too easy to build enormously complex designs. The structure of the typical software system can make a Rube Goldberg design look elegant in comparison. In reality, software is just as brittle as hardware, but the fact that software is logically brittle rather than physically brittle makes it more difficult to see how easily it can be 'broken' and how little flexibility actually exists.*" [143]

Leveson argues that software makes dramatic (and severely inappropriate) design changes so easy to execute that it gives software engineers false confidence and encourages them to begin premature construction of a system, leading to poor designs that remain unchanged later in development. "*Few engineers would start building an airplane before the designers had finished the detailed plans*," she asserts, yet this is the norm in software development. [124]

Another issue emerging from software's flexibility is the ease with which it's possible to achieve partial success, at the expense of creating unmanageable design complexity. "*The untrained can achieve results that appear to be successful, but are really only partially successful*," Leveson explains. "*Software works correctly most of the time, but not all the time. Attempting to get a poorly designed, but partially successful, program to work all of the time is usually futile; once a program's complexity has become unmanageable, each change is as likely to hurt as to help. Each new feature may interfere with several old features, and each attempt to fix and error may create several more. Thus, although it is extremely difficult to build a large computer program that works correctly under all required conditions, it is easy to build one that works 90 percent of the time*." Comparing this concept to the design of physical systems, Shore notes how inappropriate it would be to build an airplane that flies 90% of the time. [124]

Shore also notes how, for some reason, the general public often has few objections about software engineers attempting to build complex software without appropriate design knowledge and experience in the field they're writing software for. Few people would dare to fly in an airplane designed and built by people who have had no formal training or education in aerospace engineering, yet people often have no problem entrusting software (to include safety or security-critical software) to teenagers with no background in computer science or even the field they're working. Thanks to advances in computer programming languages, it is not difficult for people with no background in computer science, systems engineering, or systems security to teach themselves how to code – all it takes to learn how to program a computer is to simply take the time to understand a computer programming language, as if it were any other type of foreign language. And because computer programmers are often in high demand, it is also not uncommon for programmers to be hired to immediately start designing and building software infrastructure for major systems with which they have no experience.

Shore also notes how there is little physical or self-enforced discipline in software engineering like there are in other fields of engineering – a trend which seems to continue despite how increasingly more reliant the population becomes on computer programs. [124] He argues that **the lack of physical constraints in software design and development creates extra responsibility on computer programmers to have the self-discipline *not* to produce overly complex and unmanageable designs which can lead to unexpected or undesired behavior, but many computer programmers shrug off this responsibility**. [124]

"Like airplane complexity, software complexity can be controlled by an appropriate design discipline. But to reap this benefit, people have to impose that discipline; nature won't do it. As the name implies, computer software exploits a 'soft' medium, with intrinsic flexibility that is both its strength and its weakness. Offering so much freedom and so few constraints, computer software has all the advantages of free verse over sonnets; and all the disadvantages." [143]

Here the reader should note how Shore makes a direct comparison between software and free verse written by storytellers. This is yet another reminder that the act of programming is fundamentally an act of writing a fictional story; a script for a computer to role-play. Just like storytellers can produce abstract imaginary realities where they are completely uninhibited by the physical constraints of shared objective reality, so too can software engineers. Shore explicitly describes this as a *disadvantage* because it removes the "natural forces" which constrain complexity, prevent poor design, or stop a developer from producing designs which seem functional, but are logically flawed and/or physically impossible to engineer. This concept is illustrated in Figure 61.

Figure 61: Example of a Logically Flawed Engineering Design that's Physically Impossible

"The flexibility of software," Leveson explains, *"encourages us to build much more complex systems than we have the ability to engineer."* This necessitates a type of self-discipline which she asserts may be the most difficult kind of discipline to find in the field of software engineering: deliberately limiting the functionality of software. *"Theoretically, a large number of tasks can be accomplished with software, and distinguishing between what can be done and what should be done is very difficult... When we are limited to physical materials, the difficulty or even impossibility of building anything we might think about building limits what we attempt."* [124] In software engineering, this isn't the case. Just as easily as artists can come up with logically impossible designs such as the bridge shown above, software engineers can easily create design concepts that are physically or logically impossible.

Leveson summarizes the danger of software design flexibility with a quote from systems engineer G. Frank McCormick: *"And they looked upon software and saw that it was good. But they just had to add this one other feature... Software temptations are virtually irresistible. The apparent ease of creating arbitrary behavior makes us arrogant. We become sorcerer's apprentices, foolishly believing that we can control any amount of complexity. Our systems will dance for us in ever more complicated ways. We don't know when to stop... A project's specification rapidly becomes a wish list. Additions to the list encounter little or no resistance. We can always justify just one more feature, one more mode, one more gee-whiz capability. And don't worry, it'll be easy – after all, it's just software. We can do anything. In one stroke we are free of nature's constraints. This freedom is software's main attraction, but unbounded freedom lies at the heart of all software difficulty... We would be better off if we learned how and when to say no..."* [124]

5.4.5 Software Security Challenge #4: Software Interfaces are Cheap to Produce and Often Invisible

Another reason why software engineering is exceptionally difficult is because **software control interfaces are cheap to produce and invisible, making them easy to obfuscate**. A common way to deal with the complexity of modern computer programming is to use systems engineering abstraction techniques like decomposition to break software down into separate modules. Although separating a program into different modules may reduce the complexity of individual software components, it doesn't reduce the complexity of the software system as a whole and it can introduce unmanageable complexity into the design by creating a high number of invisible interfaces which become impossible to manage. [124]

In his journal article about the *Software Aspects of Strategic Defense Systems*, David Parnas describes how invisible and complex control interfaces represent a major challenge with software engineering, particularly when designing safety or security-critical systems. *"The greater the number of small components, the more complex the interface becomes. Errors occur because the human mind is unable to fully comprehend the many conditions that can arise through the interactions of these components."* [124, 142]

Shore once again calls out how the lack of physical constraints in software can be a disadvantage. He makes the case that software interface design is more challenging than designing interfaces for physical systems. *"Physical machines such as cars and airplanes are built by dividing the design problems into parts and building a separate unit for each part. The spatial separation of the resulting parts has several advantages: it limits their interactions, it makes their interactions relatively easy to trace, and it makes new interactions difficult to introduce... The interfaces in hardware systems, from airplanes to computer circuits, tend to be simpler than those in software systems because physical constraints discourage complicated interfaces. The costs are immediate and obvious."* [143, 124]

"In contrast," Leveson explains, *"software has no physical connections, and logical connections are cheap and easy to introduce. Without physical constraints, complex interfaces are as easy to construct as simple ones, perhaps easier. Moreover, the interfaces between software components are often 'invisible' or not obvious; it is easy to make anything depend on anything else."* [124]

5.4.6 Software Security Challenge #5: Software Design Specifications are Arbitrary, Ambiguous, and Hide Security-Critical Information

"Do not try and bend the spoon – that's impossible.
Instead, only try to realize the truth: There is no spoon."
Boy Monk, *The Matrix* [144]

The fourth reason why software security is so challenging deserves a more thorough explanation, as it relates directly to the justification for making this thesis. The bottom-line up front is that **computer engineers have adopted the habit of using arbitrary and semantically ambiguous terms to specify the functionality and *desired* (but perhaps not actual) emergent behavior of their software. This not only causes confusion about how software works, but it also suppresses vital information needed for security design purposes. Additionally, the arbitrary and semantically ambiguous way that software engineers describe their software creates a window of opportunity for nefarious software engineers to deliberately build exploitable design features and intentionally obfuscate them**.

Recall how computer program design represents an exercise in abstract thinking. Because software is imaginary, computer programs must come up with imaginary explanations and abstract concepts to explain the emergent behavior of the computers they program. Then, people hypostatize these abstract concepts. They start acting like software abstractions are concretely real things. They forget that **just because multiple people serendipitously decided to use the same abstract terms to describe the desired functionality and behavior of a given computer program, doesn't mean these descriptions are objectively true, or that it's the only way to describe the function and behavior of a computer program**.

For whatever reason (probably because it's not necessary to understand computer science to write software), people keep falling into the same trap of forgetting the undisputed truth that **all computer programs are abstractions and can therefore be described any different way using any imaginary concept, abstraction, or metaphor**. People don't know a basic lesson of computer science, that the way *any* software engineer chooses to describe the function, design, and behavior of software (including but *especially* its creator) is imaginary, arbitrary, and semantically ambiguous. This not only leads to pointless debates (people like to argue about what the "right" metaphor is describe software as if there's an objective answer, oblivious to the fact that there can't be an objectively "right" way to describe an imaginary abstraction), but it also leads directly to security incidents because the metaphors we use often hide safety and security-critical design information.

One of the more detailed and comprehensive explanations of the abstract and arbitrary nature of software design specifications comes from Charles Krueger, who first outlined the challenges of software design reuse in his research for the US Air Force in the early 1990s. As Krueger explains, computer scientists and systems engineers use abstraction techniques to manage the enormous complexity of software. Abstraction is a popular technique in both systems and software engineering because it allows software engineers to suppress the details of a computer program that are unimportant to them, while emphasizing the information that is important to them. [30]

Modern computer programmers use multiple layers of nested abstractions developed over many decades. There's usually a minimum of at least four layers of abstractions nested within each other when modern software engineers write computer programs today. For example, the abstraction called "machine code" is further abstracted and nested inside the operations of an "assembly language" which is even further abstracted and nested inside the operations of a "general-purpose language" which is then further abstracted using software specification techniques like object-oriented design. [30]

When software engineers use semantic expressions like string, char, var, thumbnail, coin, token, application, or website to describe the computer programs they design, they are referring to nested abstractions that have been developed and popularized by computer programmers over decades. To the untrained eye, this abstraction technique can be confusing. Nevertheless, multi-layer and nested abstractions have proven to be quite helpful for software engineers because they reduce what's known as *cognitive distance*. Krueger defines cognitive distance as the intellectual effort that must be expended by software engineers when developing a computer program. Abstractions reduce cognitive distance by allowing software engineers to filter out complex details about a computer program and focus on what's important to them. "*Without abstractions,*" Krueger explains, "*software developers would be forced to sift through a collection of artifacts trying to figure out what each artifact did.*" [30]

"*The effectiveness of abstractions...*" Krueger explains, "*can be evaluated in terms of the intellectual effort required to use them. Better abstractions mean that less effort is required from the user*." [30] The more an abstraction technique reduces the cognitive burden of thinking about software, the more popular it becomes as a mechanism not just to explain the intended function and complex emergent behavior of a computer program, but to inform the design of future computer programs. In a technique commonly known as *information hiding*, software engineers consider it to be a virtue to create abstractions which suppress as much of the technical details about a computer program as possible. The more information and details are suppressed by a software abstraction, the better it is perceived to be (at this point it should be clear to the reader that if the goal is to hide as much information as possible, then it's going to lead to a breeding ground of confusion about how software is designed and how it actually functions).

Krueger explains that the way software engineers decide to suppress or emphasize information using abstraction techniques "*is not an innate property of the abstraction but rather an arbitrary decision made by the creator of the abstraction. The creator decides what information will be useful to users of the abstraction and puts it in the abstraction specification. The creator also decides which properties of the abstraction the user might want to vary and places them in a variable part of the abstracting specification*." In other words, the criteria that software engineers use to distinguish between important and unimportant artifacts of a design are arbitrary, and so are the metaphors they use to describe its desired function and emergent behavior.

This point needs to be emphasized: **those who create and specify the design of software do so using arbitrary decisions about what to name it and what information they think is important to share about its design. There is no such thing as a technically precise software specification because software itself is an imaginary, abstract concept**. One of the most fundamental concepts of computer theory is that the way *any* software engineer chooses to describe the function, design, and behavior of software is imaginary, arbitrary, and semantically ambiguous. Nobody – including but *especially* the creators of software who may have the most familiarity with its design – can claim to have produced a technically accurate or objectively true description of software because all software descriptions are strictly abstract and imaginary concepts. Just because someone has technical knowledge about the syntax behind a piece of logic doesn't mean they're objectively right about its specification.

Krueger goes on to break down the technical structure of software abstractions. He explains how all software engineering abstractions include a specification that explains what the software does. These specifications have syntactic and semantic parts. The syntactic parts of a computer program's specification manifest as the program's source code and the mathematically discrete operations implemented by that code. The semantic parts of the design specification are expressed using common language, independent from whatever computer programming language is used. In other words, software engineers describe the intended function and behavior of their programs in two different ways: through the source code itself, and through the words/language they use to describe what the source code is supposed to do. [30]

The way computer engineers create abstractions to specify the design, intended functionality, and desired emergent behavior of their software therefore qualifies as an abstract language in and of itself, filled with its own semantically and syntactically complex structure. Recalling the concepts presented in the previous chapter, all higher-order languages invented by sapiens are both syntactically and semantically complex. Well, so too are the abstract design specifications software engineers use to explain the intended function and behavior of their computer programs.

Like with any language, there are opportunities for misinterpretation and miscommunication. Many of the semantic expressions used by software engineers are not uniform across the industry. Different terms can mean the same thing, or the same term can mean dramatically different things. This challenge is further exacerbated by the fact that most abstractions are popularized for their simplicity, not for their technical precision or accuracy. As Krueger explains, *"semantic specifications are rarely derived from first principles."* Software engineers invent semantic expressions to specify the design and function of their software based on personal whims, using whatever abstract metaphors they want for any reason they want, in completely arbitrary and subjective ways.

As an example, consider the computer programs which manage off-premises data storage called "the cloud." This is an abstract and semantic description of data storage technology which is not based on first principles and not intended to be technically accurate. "Cloud" is a popular term because it offers a simple way to describe a technology. It reduces the metacognitive burden of having to stop and think about enormously complex data storage technology. [30]

Why is it important for the reader to understand how arbitrary and semantically ambiguous software abstractions are? Because software engineers notoriously forget this basic concept of computer theory. A common problem in the field of software engineering is that engineers keep overlooking the fact that the abstractions used to describe the intended function and behavior of software are intended to reduce cognitive distance; they're not intended to provide a technically accurate description of the design.

Abstract software design specifications become popular because of how much information they suppress and how easy they are to understand, *not* how technically valid they are. Software engineers have an unfortunate tendency to forget this lesson of computer science, causing them to become overly reliant on popular (but technically inaccurate) abstractions to influence their design decisions. This can be extremely counterproductive to the goals of software security designers if the "popular" specification obfuscates security-critical control structure information.

The fundamental problem is misaligned goals between software and security engineers. The goal of most abstractions and design specifications created by software engineers is to reduce cognitive distance – to minimize the amount of thinking required to understand the intended function and behavior of software, so that software design concepts can be understood as quickly as possible and reused as easily as possible. But as explained previously, the goal of systems security engineers is to understand what control signals a piece of software can send to a computer that could place it into a hazardous state, so that control structures can be designed which eliminate or constrain unsafe control signals.

The software engineer's goal of suppressing as much information about a computer program as possible therefore directly conflicts with the security engineer's goals. The information being suppressed by an abstraction can include information that is vital to security design. Therein lies the problem with software engineers who forget how arbitrary, subjective, and semantically ambiguous their software design specifications are: it can cause them to inadvertently overlook vital security information (or in many cases, intentionally disguise vital security information – hence popular types of "back door" or "trap door" exploits). **A lack of awareness about this issue at scale leads to poor security culture where software engineers keep producing, promoting, or reusing abstract, non-technically accurate software design specifications that hide vital security information**.

The reason why this is so important to understand is because the author believes that this is one of the major contributing factors to why Bitcoin is so misunderstood by the public. **People seem to be missing vital information about Bitcoin's security or jumping to inappropriate conclusions about it because of an arbitrary decision by its inventor to describe it as a peer-to-peer cash system**. People are constantly arguing about design specifications and constantly using completely arbitrary and meaningless categorizations like "cryptocurrency" or "blockchain" to describe Bitcoin, completely oblivious to the basic tenets of computer theory which tell us that these are all arbitrary terms founded on nothing more than personal whim. It is no more technically accurate to call Bitcoin a "coin" as it is to call data storage software a "cloud."

There's no need to learn graduate-level concepts in computer theory and systems security to code software. This creates a problematic situation where people can devote their careers to writing computer programs without actually knowing the core tenets of computer science. Because they don't have backgrounds in computer science, they often don't understand how completely arbitrary, subjective, and technically inaccurate their software design specifications are. This makes them predisposed to using abstractions which suppress vital control signal information needed for security. It also creates a situation where a lot of software engineers have false confidence about how much they understand computers. It should come as no surprise then, that this might lead to such pronounced problems with cyber security that a US president has to pass an executive order to address it.

Security-critical control signal information hidden by software abstractions is a major contributor of zero-days and other software security exploits. To "hack" a computer program is to take advantage of sensitive control actions. When a hacker "hacks" software, they simply execute (or withhold) control signals that were not properly eliminated or constrained by the software's design logic. Why were these sensitive control signals not properly constrained? In many cases it's because software engineers overlooked them. Why did software engineers overlook these sensitive control signals? Likely because they were using abstractions which suppressed vital information needed to detect the security design flaw.

Hackers and nefarious software engineers thrive on making arbitrary, abstract, and semantically ambiguous software descriptions to obfuscate their exploitable design features. The adopted culture of abstraction and information hiding is where these belligerent actors derive their advantage. They will intentionally encode exploitable design logic into software and give themselves backdoor or trap door access to sensitive control signals. They deliberately design their software to enable unsafe control actions which can place their target's computer into an insecure or hazardous state which they can exploit. To hide their nefarious design or subversive tactics, they create abstractions which deliberately suppress or distract people from critical information about their program's control structure.

To illustrate the security challenges associated with arbitrary and semantically ambiguous software specifications, let's return to the example of "the cloud." When some people save their sensitive data to "the cloud," they think it works the same way as when they store it locally to their own computer. They don't realize they're sending their sensitive data to another person's computer. For obvious reasons, sending sensitive data to an anonymous person's computer represents a security hazard. But people often don't recognize the vulnerability because the public has arbitrarily adopted the habit of calling it a "cloud" rather than "external computers under the control of people we must trust not to exploit that control."

It's easy for nefarious actors to come up with similar abstract software specifications using equally as arbitrary and semantically ambiguous terms designed to distract people from seeing security vulnerabilities which would otherwise be obvious. To make matters even more counterproductive for security designers, **software engineers often form inappropriately rigid consensus about their abstractions. They will assert that one abstraction is "correct" even though it is arbitrary and probably not derived from first principles**. The reader is invited to test this out on their own by trying to convince software engineers to use a different name than "cloud" to describe data storage software.

Security challenges associated with the arbitrary and semantically ambiguous nature of abstract software design specifications are further exaggerated by an uneducated public. As people increasingly incorporate software into their everyday lives, they gain a false sense of confidence and understanding about that software. They think they understand software's complexities because of how often they use it or because they know what jargon to use in what context, despite how unfamiliar they are with computer theory or systems theory. This false confidence leads to further miscommunication and confusion about the merits of different software designs. It is not uncommon to see people with no background in computer science quibbling about the merit of different software designs. In the debate, people overlook how arbitrary and semantically ambiguous their jargon is in the first place.

A good illustration of this phenomenon is object-oriented software design abstractions. The desire to minimize cognitive distance explains why abstraction techniques like object-oriented design became so popular in the 1990s. People live in a three-dimensional world surrounded by an orientation of different objects, so it makes sense that people would find it easier to explain the complex emergent behavior of computer programs as if they were orientations of objects, despite how technically inaccurate this description is. Likewise, it's easier to design computer programs as if they were objects, and it's easier to communicate that design to other developers. For these reasons, object-oriented design has become one of the most popular software abstraction techniques. [30]

The decision to describe the intended function and behavior of a computer program as an orientation of objects is strictly an arbitrary decision based on personal whim. Computer programs are abstract beliefs that can be described as anything. From a technical perspective, it is just as valid to describe the complex emergent behavior of software as an orientation of objects, as it is to describe it as a verb, function, or sequence of actions. People who don't understand computer science don't understand this basic principle of computer science. Consequently, they often (tacitly) assert that the only way to accurately describe the function and behavior of a computer program is as an orientation of objects. They will sincerely believe that "cloud" and "token" and "coin" and countless other terms are the only appropriate ways to specify the functionality of a given piece of software. Even more surprisingly, they will sometimes legitimately believe that abstract software objects like "token" or "coin" are real things, for no other reason than the fact that people adopt a universal habit of talking about them as if they were physically real objects.

Nevertheless, while it may be popular to talk about software as if it were comprised of objects oriented in three-dimensional space, software is clearly not comprised of physical objects in three-dimensional space. As *The Matrix* famously reminds us, "*there is no spoon*." [144] All software objects are arbitrary abstractions which don't exist anywhere except within people's imaginations. The only things which physically exist are stored-program general-purpose state machines programmed to exhibit complex emergent behaviors that resemble what people arbitrarily choose to describe as a token, coin, or spoon.

Just because computers can be programmed to present the image or behavior of a spoon, doesn't mean that spoon is a technically accurate description of the software, nor does it mean the spoon is physically real. One would think this is obvious, but increasingly, it's not. People seem to have become so accustomed to talking about software as if it were comprised of objects, that they have begun to hypostatize abstract software objects as concretely real objects. It frequently appears as though the public has genuinely lost sight of the fact that software objects aren't real. **Like Neo or anyone else who has spent too much time in cyberspace, the public seems increasingly less capable of remembering *there is no spoon* – that software is nothing more than a method of communicating abstract ideas**. Like the gentleman picture in Figure 62, this can cause some people to lose their grip on reality and struggle to understand what's real versus what's virtual because they have lost sight of the fact that all computer programs are nothing but programmed computers acting out purely fictional stories.

Figure 62: Modern Agrarian Homosapien Losing His Grip on Physical Reality

This trend could have major implications on population-scale security. **The inability to make a distinction between abstract reality and physical reality could cause people to place themselves into situations where they can be exploited at massive scale through their software. In the modern era, the belief system through which populations can be exploited is not their ideologies, morals, ethics, or theologies, but their software and that programmed perception of shared abstract reality called cyberspace. People are vulnerable to exploitation because with way the internet has currently been built, computer programmers and software administrators must be trusted not to exploit people through the virtual reality they encode for us, but trusting people is a demonstrably poor security strategy**.

5.4.7 Software Security Challenge #6: Software Entrusts People with Unprecedented Levels of Asymmetric Abstract Power and Control Authority over our Vital Digital Resources

The author has now identified five different ways that software's lack of physical constraints contributes to challenges associated with software security. Because state mechanisms have infinitely expandable state spaces, there is practically no limit to what software engineers can design. Because software represents abstract meaning assigned to state changes as machine-readable language, software designs are abstract and physically unconstrained. Being physically unconstrained means software system designers have nothing physically limiting them from producing unmanageably or outright logically flawed designs with runaway complexity. The combination of infinite state spaces and physically unrestrained complexity means software can have an infinite and unmanageable number of hazardous states. Being physically unconstrained also means it's easy to build invisible, complex, and unmanageable control signal interfaces where engineers don't even know what control signals are being passed to the computer. Moreover, the discrete, non-continuous nature of state mechanisms means software is just one "wrong move" away from sudden and unpredictable behavior changes.

There is another reason why software's lack of physical constraints contributes to systemic security problems – one that surprisingly few scientific papers have mentioned. **Software gives administrators highly asymmetric levels of abstract power and control authority over other people's computers, and there's often no way for users to physically constrain software administrators from systemically exploiting or abusing their special permissions or control authority**. In other words, software represents a new form of abstract power hierarchy, where someone with extraordinary amounts of abstract power must be trusted not to exploit it. As discussed at length in the previous chapter, abstract power hierarchies are trust-based, permission-based, and inegalitarian systems that are demonstrably insecure against untrustworthy people.

Ironically, the asymmetric abstract power and control authority given to software administrators is derived from an attempt to make software more secure. Sometimes it's not difficult to recognize a software system's hazardous states and sensitive control signals. For example, consider a simple piece of software responsible for managing private or sensitive data, or a piece of software which controls the firing of a weapon system. For both systems, there are control signals which would qualify as sensitive or unsafe control signals that should either be eliminated or logically constrained. Because software is abstract and non-physical, software engineers must use logical methods to constrain these control signals; they can't use physical constraints like a physical interlock or a safety switch.

Without a way to physically constrain sensitive control actions, a common alternative that software engineers like to use to logically constrain sensitive control signals is to codify permission-based or rank-based hierarchies where special permissions are given to specific users to execute sensitive control signals. By using this technique, software engineers effectively codify abstract power hierarchies for themselves, where some users (i.e. a ruling class) have more rank and control authority over low-ranking users (i.e. a ruled class) and must be entrusted not to abuse or exploit their rank. For example, it is not uncommon for software to give certain users "admin rights." Incidentally, these rights are colloquially known as "god rights" because of the amount of abstract power and control authority they have in comparison to regular users with regular permissions. These systems are systemically insecure by virtue of the fact that users must trust their admins not to abuse their admin/god rights.

Because software is an abstraction, all permission-based control structure designs like this qualify as abstract power hierarchies. Software engineers are essentially forced to encode these abstract power hierarchies to logically constrain control authority because they simply don't have the option of using real-world physical power as the basis for constraining control authority (at least not before the invention of Bitcoin). Here we see the lack of physical constraints in software creating yet another major systemic security vulnerability. **Without being able to physically constrain sensitive commands, software must necessarily utilize centralized, trust-based, permission-based, and inegalitarian systems of control where users are inherently vulnerable to systemic exploitation and abuse of untrustworthy or unethical computer programmers and software administrators who award themselves asymmetric abstract power and control authority over the system via specially-encoded privileges and permissions**.

The previous chapter provided a lengthy and in-depth discussion about the systemic security flaws of abstract power hierarchies. As it turns out, the systemic security flaws are the same whether abstract power and control authority is codified using rules of law or using software. The fundamental security flaw of these systems is derived from the lack of control over the higher-ranking positions. Users must necessarily trust in the benevolence of their god-kings (i.e. software administrators who have god-rights over the software), because they otherwise have no ability to physically constrain them from being systemically exploitative, abusive, or incompetent with their asymmetric levels of control authority.

Rank-based control structure design approaches only restrict those who have permission to access sensitive control signals, they do nothing to physically constrain the execution of sensitive control signals. Consequently, there is little to nothing stopping a higher-ranking computer programmer from executing unsafe control actions which could cause the system to enter a hazardous state (recall the lessons of the previous chapter about how logical constraints can't stop the exploitation of logic, they can only change logic can be exploited.) Therefore, just like all abstract power hierarchies which assign rank and special permissions to specific people, users must trust higher-ranking insiders not to abuse or be incompetent with their ability to send (or withhold) sensitive control signals. Likewise, users must trust that software developers designed the system in such a way that outsiders can't exploit the hierarchical control structure to gain access to the special permissions and control authority given to higher ranks.

Regardless of whether a rank-based resource control structure is formally codified using rules of law or software, whenever special permissions are granted to specific ranks within an abstract power hierarchy, the same systemic security problems emerge. Abstract power is inegalitarian, trust-based, permission-based, systemically endogenous, and zero-sum. But perhaps most importantly of all, rank-based control structures create a honeypot security vulnerability. Positions of high-rank and control authority have a high benefit to exploit or attack (i.e. high B_A). It's far easier for untrustworthy or belligerent actors to exploit people or gain disproportional access to resources by taking over high-ranking positions in rank-based control structures. Creating rank-based control structures comprised of high-ranking positions with lots of disproportional control authority over computers is like lighting a flame in a dark room filled with moths; if you build rank-based abstract power hierarchies, systemic predators will be attracted to those positions.

To make matters even worse from a systemic security perspective, software alone cannot impose severe physical costs on anyone attempting to exploit or abuse a computer program's control signals. **No matter how well software's control structures are designed, software alone has no capacity to physically constrain the execution of unsafe control signals – the only way this can be done is to physically constrain the underlying computer**. Not being able to increase the physical costs of executing unsafe control signals (i.e. increase C_A) means the benefit-to-cost ratio of attempting to access and exploit the centralized control structure of a software-instantiated abstract power hierarchy can only increase as more people use and rely on it (i.e. BCR_A can only increase). Users are physically powerless and therefore wholly reliant on a specially-permissioned group of administrators to keep them secure.

Eventually, the BCR_A of exploiting a software's control structure can reach a hazardous level that will motivate someone, either inside the system or outside the system, to exploit it. The systemic security of software-instantiated abstract power hierarchies therefore hinges on both the design skill and self-restraint of the software's engineers and administrators. **Computer programmers must be trusted to design a hierarchical control structure that can logically constrain people from exploiting a computer program's sensitive control signals. Then, the programmers themselves must be trusted not to give into the increasing temptation of taking advantage of the increasing BCR_A of the hierarchical and asymmetrically powerful control structures they have designed**.

Unfortunately, the history of all abstract power hierarchies created over the last several millennia is filled with examples of breaches of that trust. Just like abstract power hierarchies codified by lawmakers, those codified by computer programmers are apparently just as vulnerable to attack from bad actors either external or internal to the system. These vulnerabilities have now blossomed into widescale systemic exploitation abuse across cyberspace. Computer engineers and administrators of all kinds wield extraordinary levels of asymmetric abstract power and control authority over cyberspace, and the evidence of widescale systemic exploitation and abuse of people's digital information and computer resources has become commonplace.

5.5 Creating Abstract Power Hierarchies using Software

"This mode of instantaneous communication must inevitably become an instrument of immense power, to be wielded for good or for evil, as it shall be properly or improperly directed."
Samuel Morse (inventor of Morse Code), on the Telegraph [145]

5.5.1 Recreating Exploitable Abstract Power Hierarchies using a New Technology

Recalling core concepts about creating abstract power from the previous chapter: to seek permission or approval from people is to tacitly give them abstract power over you. On the flip side, if you want to create and wield abstract power over a large population, simply convince them to adopt a belief system where they need your permission or approval. Once a population has adopted a belief system where they need your permission or approval, you have successfully gained abstract power and influence over them. Now combine this concept with the concepts introduced in this chapter about computer theory, namely that software represents nothing more than a belief system. Here we can begin to see how **people can use software to give themselves immense abstract power by simply convincing people to adopt software applications designed as a permission-based hierarchy such that users tacitly need approval or permission from other people who are higher in the hierarchy and have higher permissions**.

To summarize the previous section, software control structure design is not an easy task. A major challenge of all software engineering is the fact that software is an abstract belief system – it does not physically exist. Like all belief systems, software is vulnerable to systemic exploitation and abuse. There is practically no limit to the number of ways that software can be exploited. Even if a software engineer could sufficiently identify the hazardous system states it needs to circumnavigate to prevent a security incident, software engineers must still figure out how to design control structures which sufficiently constrain those control signals, and they must do it using discrete mathematical logic rather than by applying physical constraints, while also preserving the intended functionality and desired emergent behavior of the software. This is a profoundly difficult task.

Engineers who design and build physical systems have an enormous advantage over software engineers who design abstract systems. Physical system engineers get the benefit of being able to *physically* constrain unsafe or insecure control actions and *physically* prevent systems from reaching hazardous states. They can build safety switches which physically prevent users from committing unsafe actions. They can build interlocks, thick walls, or heavy physical barriers. They can deter adversarial control actions by making it impossible to justify the physical cost of performing them. These are luxuries which software engineers do not have because they do not design systems in the domain of physical reality. Instead, **software engineers must use discrete mathematical logic and syntactically and semantically complex language to encode logical constraints. This is far a far more difficult way to make a system secure than by using simple physical constraints**.

Because software engineers can't physically constrain unsafe control actions, they often design permission-based systems to logically constrain sensitive control actions. One of the most common ways that software engineers choose to secure software is to design rank-based, abstract power hierarchy where positions of high rank are granted special permission to execute sensitive control actions. Ironically, even though these abstract power hierarchies are designed ostensibly to improve security, we know from the core concepts discussed at length in the previous chapter that abstract power hierarchies have substantial systemic security flaws.

This software security design methodology, which has become the predominant methodology used today, produces trust-based, permission-based, and inegalitarian abstract power hierarchies which give disproportionate amounts of abstract power and control authority to a select few people who must be trusted not to exploit it. These software-defined abstract power hierarchies create a ruling class and a ruled class just like legacy abstract power hierarchies (i.e. governments) do, where users are physically powerless to constrain the abstract power and control authority given to their rulers. Users must trust that actors either inside or outside the system will not find a way to abuse the abstract power and control authority given to specific

positions within these hierarchies, because they're otherwise physically powerless to stop it since they have no ability to impose severe, physically prohibitive costs on systemic predators who exploit the design logic.

Herein lies one of the most significant but unspoken security flaws of modern software design methodologies: it has created a new type of oppressive empire. A technocratic ruling class of computer programmers have gained control authority over billions of people's computers, giving them the capacity to exploit populations at unprecedented scale. These digital-age "god-kings" can exploit people's belief systems using software. They are actively data mining people and running constant experiments on entire populations, learning how to network target them, influence their decisions, and steer their behavior. [66]

In the hands of oppressive regimes, computer programs are dangerous because they can turn into panopticons which give a ruling class of people unprecedented surveillance capabilities with pinpoint-precision and authoritarian control over billions of people's vital digital resources. Cyberspace is nothing but a belief system, and never before have so many people come to adopt the same belief system at such a global and unified scale. Therefore, never before have so many people been so systemically vulnerable to exploitation and psychological abuse through their belief system. Through the abstract reality of cyberspace, agrarian society can be entrapped, domesticated, farmed, and herded like cattle.

5.5.2 A New Type of Abstract Dominance Hierarchy over a Digital-Age Resource

*"Yeah so if you ever need info about anyone at Harvard... just ask... I have over 4,000 emails, pictures, addresses, SNS... people just submitted it... I don't know why... They 'trust me' ... Dumb f***s."*
Mark Zuckerberg [146]

In the past, abstract power hierarchies were created by writing stories using pen and parchment or encoding what we now call rules of law. Today, abstract power hierarchies can be encoded using what we now call software. This would imply that once again, agrarian society has fallen into a systemically vulnerable situation where those who are literate with a new form of language, storytelling, and rulemaking can use it to systemically exploit entire populations. Using software it is possible that an elite, technocratic ruling class can design abstract power hierarchies which give themselves enormous levels of abstract power and control authority over the precious resources of entire populations (namely their bits of information). Convince a population to run a particular piece of software, and that population enters a scenario where someone wields abstract power and control authority over their resources.

A software administrator's abstract power and control authority is formally codified using machine code rather than written rule of law. But from a systemic perspective, the strategy for creating abstract power and using it to build an exploitable belief system is functionally identical. As previously discussed, the strategy of a god-king goes like this: use semantically and syntactically complex language to tell lots of stories which give yourself abstract power and control authority over people's resources (in this case, their bits of information). Get people to adopt a common belief system which creates an abstract power hierarchy that places you at the top of that hierarchy (in this case, get them to use or install your software). Convince the population to trust that you will not abuse your abstract power and control authority over them to exploit them through their belief system. If the population starts to show some concern, convince them they are secure because you have encoded logical constraints into the system – logical constraints which can do nothing to physically prevent you from exploiting those logical constraints in the future.

Major differences between the god-kings of the past and the god-kings of the present are simply the technologies they use and the resources they control. Modern god-kings don't need slavishly obedient people anymore because they have slavishly obedient machines (machines which are far more loyal and far less prone to uprising – at least for now). The modern technocratic elite don't need their population to farm food anymore, they just need their population to farm data. And the more a population willingly forfeits their data to their technocratic ruling class, the more that population can be experimented upon to determine causally inferable relationships and actions which drive their behavior. It's no secret that these software administrators utilize their platforms to A/B test their populations on a regular basis to determine not just what *influences* the population's behavior, but what *drives* their behavior. In other words, neo god-kings can and do determine how to *control* their users based purely on what they can learn from having access to their precious bits of information. [66]

As agrarian society grows increasingly more reliant on their computers, they appear to have forgotten a lesson learned over 10,000 years of people creating abstract power hierarchies: they're systemically insecure, particularly against the people at the top of those hierarchies. **The form of technology used to create abstract power hierarchies may have changed, but the function hasn't, nor have the tactics for creating and exploiting abstract power**. Because people don't understand this, they appear to be oblivious to how vulnerable they are to systemic exploitation and abuse anytime they operate a computer. People are migrating to cyberspace at unprecedented scale, and software administrators with extraordinary abstract power are encouraging it and rebranding it with fun names like "metaverse." Meanwhile, nothing is protecting the population against unprecedented levels of exploitation except their trust in anonymous strangers who have control over their computers through their computer programs.

As more people migrate online and grow increasingly dependent on their computers, the benefit of exploiting them through their software is only accelerating. At the same time, people continue to have no means to physically secure themselves by imposing severe physical costs on people and programs in, from, and through cyberspace. The result is an exponentially increasing BCR_A for entire populations of people. The more people choose to believe in software, the more valuable software becomes as the chosen attack vector for modern systemic predators. This is happening not just at the individual level, but also on a global scale. Entire nations are now operating online, completely exposed to the whim of other nations and utterly incapable of physically securing themselves from neo god-kings (a.k.a. software or computer network administrators).

Although this systemic security problem is thousands of years old, people appear to have forgotten the lesson. All that has happened is that an older form of storytelling technology (spoken and written symbols) has been replaced by a new form of storytelling technology (state changes in a state machine). People, private institutions, public organizations, and entire nations are writing stories, convincing transnational, global-scale populations to believe things and exploiting them through their belief systems. This is nothing new; this is the same abstract empire-building game with a different name. The exploitable ideology of choice is now software. Convince a population to use and believe in your software, and you can have unconstrained levels of abstract power and control authority over them that is so asymmetric, it could rival that of Egyptian pharaohs. This concept is illustrated in Figure 63.

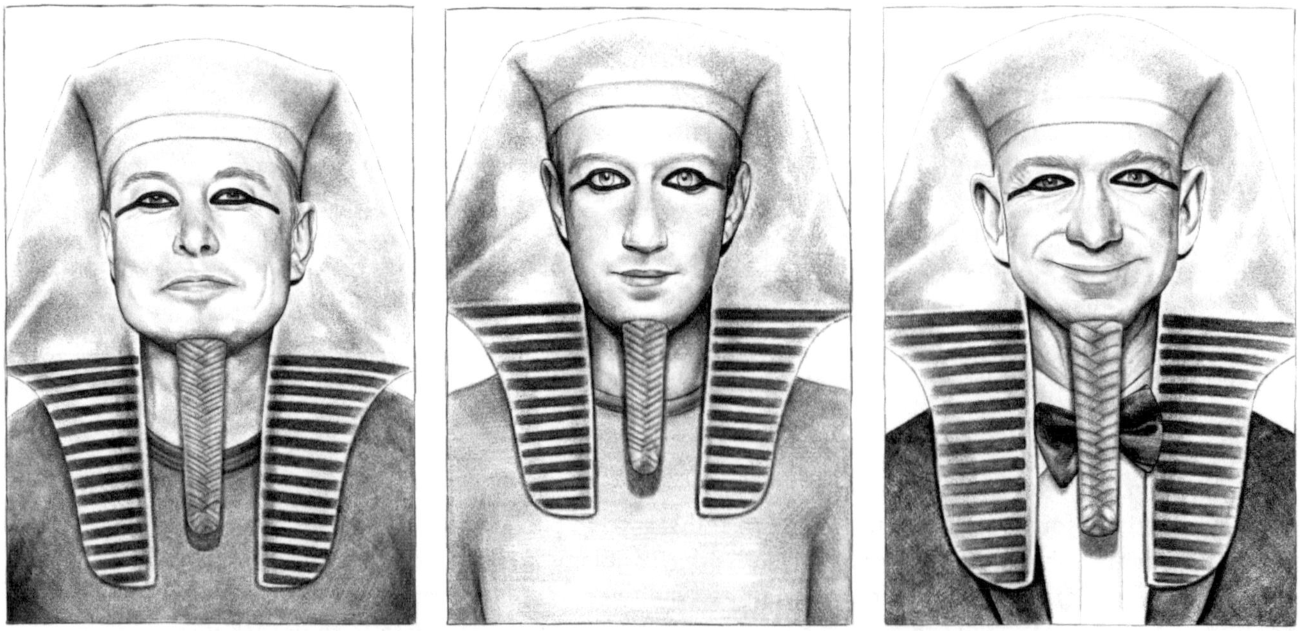

Figure 63: Software Administrators with Abstract Power over Digital-Age Resources

Unfortunately, the general public doesn't appear to understand the complexity of computer theory to know that software represents an exploitable belief system. At the same time, they also don't appear to understand enough about the complexities of agrarian power dynamics to understand how vulnerable they could become to getting exploited through this belief system, much less how they can secure themselves against this form of abuse. Perhaps this is because people don't take the time to understand the difference between abstract power and physical power. They don't understand how logical constraints encoded into rulesets will not secure them against the exploitation of logic. They don't understand how demonstrably dysfunctional abstract power hierarchies are. And because of these misunderstandings, the public doesn't appear to understand how entrapped they could be, nor how they might be able to escape their entrapment. As mentioned previously, this is a recurring problem with domestication. People can lose the capacity or inclination to secure themselves by imposing severe, physically prohibitive costs on their attackers to make it impossible to justify the physical cost (in watts) of exploiting them. Domesticated populations become inclined to believe that they can adequately defend themselves without projecting physical power. **With the way the internet is currently architected, users operating in cyberspace are automatically domesticated, as the architecture necessary to project physical power to physically constrain belligerent actors is currently missing** (at least, it was before the invention of Bitcoin)

5.5.3 Software is the Same Abstract Power Projection Game with a Different Name

Recalling a core concept from the previous section, software represents an abstract form of power and resource control authority. Computer programmers use software to construct rank-based, abstract power hierarchies which give them physically unrestricted levels of control authority over people's resources (namely their computers, the information on those computers, and all the operational functionality of computers – which is becoming increasing substantial in the digital age). Software represents a new type of abstract power projection technology; a new way for dynasties to build their empires and reign over entire populations across multiple generations. These abstract power hierarchies have the enormous advantage of inheriting a domesticated population of users who have no capacity or inclination to use physical power to secure themselves, making them automatically predisposed to systemic exploitation and abuse at unprecedented scale. Not surprisingly for anyone who understands history, these abstract power hierarchies are becoming oppressive; people are beginning to take advantage of entire populations through their computer programs. History tells us what populations must do to escape from this type of oppression. This is a lesson that has been learned time and time again over several millennia, and modern agrarian society appears to be on the verge of learning it again.

Just like agrarian populations have been trying to do for at least the last 5,000 years with other written languages, computer programmers today keep assuming that they will be able to design and codify abstract power hierarchies which can adequately secure people against exploitation and abuse with the right combination of encoded logic. They keep trying to design complex software with increasingly complex rulesets, only to be surprised when someone inevitably finds yet another way to systemically exploit encoded logic. **Somehow, people keep overlooking the fact that encoded logic – whether they're written on parchment or written in python – are *always* the source of systemic exploitation and *never* a complete solution to it. Encoding more logic into our software doesn't stop the exploitation of logic, it just changes the way the encoded logic of our software can be exploited**.

Attempting to keep populations systemically secure against exploitation and abuse from software-defined abstract power hierarchies by encoding more logical constraints is a demonstrably unsuccessful strategy. Ineffective logical constraints are the source of all forms of software hacks and software systemic exploitation, not the solution to them. The same line of reasoning which applies to keeping nations secure against external invasion or internal corruption also applies to computer programs because they are systemically identical problems.

Attempting to keep software systemically secure by writing more software may be even harder and more futile than attempting to keep rules of law systemically secure against corrupt lawmakers by writing more laws. At least when a dictator rises to power and starts to oppress their people through their belief system, the population can *see* their oppressors. This is not the case with software. Modern oppressors can hide behind their software and obfuscate its exploitable design logic, making it even easier for a technocratic, elite ruling class to rise to the top of the abstract power hierarchy and avoid detection (for an example of this, see the discussion about proof-of-stake in section 5.10). Only unsophisticated oppressors allow themselves to be identifiable; the rest should understand by now that the best way to entrap people is to do it subversively using software.

People don't appear to understand that a primary security hazard of cyberspace is not a lack of encoded logical constraints. A major reason why cyberspace is so systemically hazardous is because it represents the adoption of a common belief system, and all belief systems are systemically exploitable. Cyberspace is nothing more than people volunteering to assign symbolic meaning to state changes within the combined state space of a globally-distributed network of state machines. Common belief systems like cyberspace can and will have their logic exploited regardless of how that logic is encoded, especially if there is no physical penalty for doing so.

In the current internet architecture, a fundamental security problem is that people have adopted a common belief system over which they have little capacity to resist exploitation because they are literally powerless to resist it. To be systemically secure in cyberspace, digital-age agrarian society must figure out a way to project physical power in, from, and through cyberspace to impose severe physical costs on people who try to exploit them, and to physically restrict malevolent software-defined abstract power hierarchies. This would imply that **to fix a fundamental and reoccurring problem with cyber security, the internet itself needs to be re-architected in such a way that people can impose severe physical constraints on each other (which is precisely what Bitcoin does)**.

5.5.4 The "Software" Metaphor Began as a Neologism for Abstract Power and Control Over Computers

"From the very beginning, I found the word too informal to write and often embarrassing to say... Colleagues and friends simply shrugged, no doubt regarding each utterance as a tiresome prank or worse, another offbeat neologism..."
Paul Niquette [147]

If the reader is having a hard time accepting the author's assertion that software represents a form of abstract power, then consider the fact that the "software" metaphor first emerged as a neologism for wielding absolute power and top-down command and control authority over computers. Less than five years after Neumann's EDVAC was delivered to the Army Ballistic Research Laboratory in 1949, computing pioneer Paul Niquette coined the term *software*. Niquette came up with the term during his experience programming the Standards Western Automatic Computer (SWAC), a stored-program computer built a year after Neumann's EDVAC. In his memoirs, he describes the SWAC as a mindless, subservient machine that bends to his will. [147]

"I wanted nothing to do with the SWAC 'hardware' – the machine was the mindless means for executing my programs – a necessary evil, but mostly evil. It was about at that moment, I seized upon the consummate reality of what I was doing... I was writing on a coding sheet, not plugging jacks into sockets, not clipping leads onto terminal posts, not soldering wires, not bending relay contacts, not replacing vacuum tubes. What I was doing was writing on a coding sheet! It was October 1953 and I was experiencing an epiphany. Before my eyes, I saw my own markings carefully scrawled inside the printed blocks on the coding sheet. They comprised numerical 'words' – the only vocabulary the computer could understand. My coded words were not anything like those other things – those machine things, those 'hardware' things. I could write down numerical words – right or wrong – and after they were punched into cards and fed into the reader, the SWAC would be commanded to perform my mandated operations in exactly the sequence I had written them – right or wrong. The written codes – my written codes – had absolute power over 'hardware.' I could erase what I had written down and write down something different... the SWAC, slavishly obedient in its hardware ways, would then be commanded to do my work differently – to do

different work entirely, in fact. The writing on the coding sheet was changeable; it was decided not hardware. It was – well, it was 'soft-ware.'" [147]

Here we can see how the inventor of the word "software" described it as a form of "absolute power" over "slavishly obedient hardware" which would follow his commands no matter how often he changed his mind. The commands issued by a computer programmer were classified as "software" and the machines slavishly obeying them were classified as "hardware." Thus, **since its inception, the word "software" has *always* been used as a neologism for abstract power and top-down control authority over computers**.

Using Niquette's point of view, it's easier to understand how computer programs represent a new type of abstract power hierarchy. When a computer programmer writes software, they use a symbolic language to formally codify a rank-based hierarchical system where they give themselves control authority over computing resources (computers which often belong to other people). In much the same way that kings or lawmakers formally codify the design of their abstract power hierarchies using rules of law, programmers formally codify the design of their abstract power hierarchies using software. Whereas kings assign themselves with abstract forms of power like "rank," programmers assign themselves with abstract forms of power like "admin rights." Whereas kings benefit from their control authority over their subjects' land, programmers benefit from their control authority over their subjects' computers, the information stored on those computers, and all the resources controlled by those computers.

When viewing software as a way to create abstract power and codify abstract power hierarchies, this line of thinking begs a question: what *physically* prevents computer programmers and software administrators from becoming exploitative or abusive with their "absolute power" and control authority over other people's "slavishly obedient" computers? The previous chapter outlined why abstract power hierarchies are systemically insecure. People – Americans especially – should be keenly aware of the threat of high-ranking individuals who are given too much abstract power and resource control authority. The citizens of all countries which have ever had to suffer the rule of an oppressive ruling class should understand the importance of being able to secure themselves against abusive and systemically exploitative people by imposing severe, real-world physical costs on anyone who would try to exploit them.

If computers create new forms of property, policy, and other resources upon which entire populations depend, and if software represents a new form of abstract power and control authority over those resources, then how do populations maintain the ability to *physically* secure themselves in, from, and through cyberspace against computer programmers who make themselves god-kings and abuse their abstract power and resource control authority?

If cyberspace fundamentally represents a new kind of shared abstract belief system, what is to prevent predators from systemically exploiting us through this shared abstract belief system? The answer appears to be nothing – at least not with the current architecture of the internet. There does not appear to be anything physically preventing computer programmers from exploiting populations at massive scales through their software. This is because there is nothing empowering people to secure their precious bits of information by imposing severe physical costs on anyone who would try to systemically exploit their computer networks. That is, with one clear exception: proof-of-work systems like Bitcoin.

5.6 Physically Resisting Digital-Age God-Kings

"Cyberspace. A consensual hallucination experienced daily by billions of legitimate operators, in every nation, by children being taught mathematical concepts... A graphic representation of data abstracted from banks of every computer in the human system. Unthinkable complexity."
William Gibson [148]

5.6.1 How Did we Get Here?

The pace of abstract thinking and cultural evolution following Niquette's observations was so fast that it's easy to overlook how this situation emerged, so what follows is a short summary of 80 years of computer science. In 1953, the amount of abstract power and resource control authority wielded by computer programmers and software administrators like Niquette was minimal and tightly localized. Few people knew what stored-program general-purpose state machines were at that time. People didn't use these machines to manage critical infrastructure, store sensitive data, or create global-scale virtual realities which transform how entire populations perceive objective reality. But in a span of only six decades, computer programmers went from encoding rudimentary symbolic Boolean logic into large and clunky electromechanical circuitry, to developing what could be described as cyber empires ruled by administrators who wield unprecedented levels of physically unconstrained abstract power. Why? For little other reason than what Niquette first noted about software; software gives people "absolute power" over "slavishly obedient" machines, which directly translates into absolute power, influence, and control authority over the people who depend upon those machines.

Add Niquette and Columbus together, and you get today's systemic security problem. Admiral Columbus demonstrated that all it takes to get an entire population of people to do your bidding is for that population to have an exploitable belief system. To achieve the same effect as Columbus and countless other false prophets, people don't have to believe in "good" or "god" anymore, they just have to believe in Boolean logic assigned to transistor states. If you make yourself the person who programs how those transistor states change, you become the person who can control what people see and believe.

In the 1940s, engineers figured out how to turn circuits into symbolic languages which could be used to issue commands to machines (i.e. machine code), paving the way for people like Niquette to wield software as a form of abstract power and control over stored-program computers. In the 1950s, engineers figured out formula translators and cleared a path towards assembling commonly-reoccurring pieces of machine code together to make computer programming easier to do. This became known as assembly language.

Assembly languages made the art of computer programming more accessible to people. This caused the size and complexity of computer programs to explode throughout the 1960s until the point where NATO declared a "software crisis" in 1968 and asked for nations to come together to create a new field of "software engineering" to manage the extraordinary complexity of computer programming. [30]

In the 1970s, computer programming became its own distinct discipline known as "software engineering," which became its own form of computer programming that was considered to be separate and distinct from computer science and the profession of computer engineering. These so-called "software engineers" began to evolve their methodologies to manage the extraordinary complexity of designing computer programs. They explored formal methodologies like discrete mathematical modeling, and they invented techniques like structured programming and modularity to make computer program design more manageable. These methodologies gave rise to the philosophy of systems thinking, which accelerated the development of new fields of engineering like systems engineering in the 1980s.

Throughout the 1970s and the 1980s, computer programmers became increasingly more inclined to think of their programs as complex systems which produce complex emergent behaviors. This led to what some might describe as a software engineering renaissance. Software engineers developed an arsenal of system engineering tools which transformed the look and feel of computer programming. At the same time, developers struggled with the inherent complexity of building larger systems using assembly languages and began to search for more effective ways to express computation.

This led to a new generation of semantically and syntactically complex computer programming languages where the most common implementation patterns used in assembly languages (e.g. iteration, branching, arithmetic expressions, relational expressions, data declarations, and data assignment) became the primitive constructs for higher-level programming languages. Under pressure from NATO to create uniform general-purpose programming languages (e.g. Ada), software engineers started inventing higher-level general-purpose computer programming languages with very similar semantic and syntactic expressions. Consequently, a common lexicon began to emerge.

Software engineers could now routinely talk about the same kind of if/then/else statements, for/while loops, chars, vars, and function calls no matter what general-purpose language they learned. These common general-purpose languages made software engineering so easy that people no longer needed decades of experience in formal logic or computer science to program a computer; they just needed to learn the syntactic and semantic rules of higher-order computer programming languages. By the 1990s, countries like the US began to legally classify general-purpose computer programming languages as formal languages so they could be protected under free speech laws. [30]

As previously discussed, sapiens use language to tell stories, and storytelling gives people the ability to connect neocortices and construct shared, abstract realities where some people have abstract power. Over time, the stories told by shamans turned into stories told by god-kings. With this in mind, we can see that the emergence of general-purpose computer programming languages not only created a new type of literacy, it also created a new type of storytelling available to practically anyone willing to teach themselves how to communicate with computers. With any new form of storytelling comes a new way to create abstract realities where some people have abstract power. Today, people only needed to teach themselves how to write computer programs to gain immense levels of abstract power and control authority over other people's computers.

In the beginning, the virtual realities and abstract power hierarchies created by computer programmers were rather small, confined to small computer networks or "tribes." But then computers found their way into single family homes and started communicating with each other across continents. Engineers started running wires across oceans and computers started controlling society's most critical infrastructure. Over a few short decades, the ability to program computers became the ability to control critical resources anywhere in the world from anywhere in the world. It wasn't long after the emergence of the internet that engineers started replacing signals sent across wires with signals sent across the electromagnetic spectrum. Computers shrunk from the size of a room down to the size of a pocket, and far more people started to incorporate them into their lives, growing increasingly more dependent on them.

Along the evolutionary journey of computer programming, software never stopped offering a form of absolute power and control authority over machines. The abstract power and resource control authority available to computer programmers only grew as computers became more prevalent. While the population became increasingly more dependent on their computers, they became increasingly more dependent upon computer programmers.

Today, the amount of abstract power and resource control authority afforded to computer programmers by the software they write has reached a level which dwarfs many nation states. Some countries (e.g. China) have recognized the source of this abstract power and moved to capture it by ensuring direct state control over software administrators. Other nations have exercised some restraint at the strategic cost of allowing some computer programmers to become neo god-kings with formally codified abstract power hierarchies with extraordinary and unprecedented control authority and influence over billions of people and their computers. These neo god-kings are now trying to recruit as many people into their abstract empire as possible, giving it attractive-sounding names like "the metaverse."

Modern software programmers understand that the ability to program people's computers translates directly into the ability to control the data, information, and other resources which entire human populations rely upon to shape their thoughts, guide their actions, and make sense of shared objective reality. Unfortunately, people who are less familiar with computer science and power projection in agrarian society are less inclined to see the hazard this poses.

Since multi-national populations have become highly dependent on their computers, they are already highly dependent on the abstract power and control authority of software engineers who program them. **In a very short amount of time, under the noses of the people who don't understand human metacognition, cultural evolution, and abstract power building, software appears to have emerged as the predominant form of empire-building in modern agrarian society, giving software administrators the ability to rule over trans-national populations through the computers they carry around in their pockets**. Today, it is not even considered remarkable that a single person has physically unconstrained rank and control authority over three billion people's computers. But it has not yet occurred to many of them that they could now have the means to physically resist their god-kings when those god-kings inevitably become oppressive.

5.6.2 "The Metaverse" is the Same Kind of Systemic Trap as "The Matrix"

It is still too early to know what long-term impacts computers will have on human metacognition, but one thing is already clear: sapiens have discovered a virtual frontier which blurs the lines between abstract and objective realities even further than it was already blurred by tens of thousands of years of storytelling, in completely new ways that our species has never experienced before. As more people integrate symbol-generating virtual reality machines into their lives, they increase their potential to lose sight of the difference between objective reality and abstract reality – the difference between real versus imaginary things. More and more, people are entrapping themselves in semi-somnambulant dream states where they allow strangers to control what they see, hear, and believe, feeding them with programmed illusions of freedom, serendipity, and choice.

The future envisioned by the Wachowskis in the movie *The Matrix* appears to be coming to fruition. **Modern sapiens are becoming so engrossed in the shared hallucination of cyberspace that they appear to be losing their grip on objective reality. "The Matrix" can be rebranded as "The Metaverse," but the dynamics of entrapment and exploitation are still the same, regardless of the semantics**. A growing number of people appear to be genuinely confused about the difference between the real world and the imaginary world. In this confusion, they are not grasping the difference between real and imaginary things, most notably the difference between real power and imaginary power. Combining this confusion with a general ignorance of power dynamics in agrarian society, people appear to have no concept of the tradeoff between physical power-based resource control and abstract power-based resource control. This creates an unprecedented window of opportunity for systemic predators. As young, unsuspecting, self-domesticated, and computationally illiterate people continue to migrate into the metaverse, they have no idea how vulnerable they are.

This, of course, is not the first time modern agrarian society has become vulnerable to entrapment by their belief systems and consensual hallucinations. The previous chapter provided a detailed description of how sapiens developed increasingly complex applications of abstract thought until it eventually led to the creation of abstract power hierarchies to settle intraspecies disputes, establish control authority over intraspecies resources, and achieve consensus on the legitimate state of ownership and chain of custody of intraspecies property. Sapiens first started using their abstract thinking skills for planning and pattern finding, but then they started using it to assign imaginary meaning to recurring patterns observed in nature through a process called symbolism. Symbolism enabled higher-order communication because it allowed sapiens to develop semantically and syntactically complex languages as a protocol for exchanging symbolically meaningful, conceptually dense, and mathematically discrete information.

Armed with symbolic language, sapiens discovered the art of storytelling, a process where they can leverage people's capacity for abstract thought to create shared imaginary realities for themselves to explore together. These virtual realities were unbound by physical constraints, allowing people to explore them within the safety & comfort of their own imaginations. Fictional stories enhance sapient relationship building, information sharing, entertainment, and vicarious experience. But fictional stories also lead to consensual hallucinations where storytellers give themselves passive-aggressive access to physically unbounded levels of abstract power and control authority over the people who believe in those stories.

All abstract realities created by storytellers are consensual hallucinations. The fact that the mechanism for telling these abstract stories has changed to computers doesn't change the overarching systemic dynamics of abstract power projection. **People have tried valiantly to design abstract power hierarchies which keep their populations secure from exploitation and abuse using nothing more than encoded logic, but 5,000 years of written testimony prove that no combination of encoded logic (i.e. laws) has successfully prevented an abstract power hierarchy from becoming dysfunctional, exploited, vulnerable to foreign attack, or vulnerable to internal corruption. Physical power has always been necessary and essential to correct these security vulnerabilities, hence the phenomenon we call "war"** and the inevitable return to using physical power as the basis for settling disputes, determining control authority over resources, and achieving consensus on the legitimate state of ownership and chain of custody of resources. The clockwork nature of warfare is proof that abstract power hierarchies are inherently flawed.

Why is physical power so useful? Because it's real, not abstract. It is therefore not endogenous to a belief system which can be systematically exploited by storytellers, making it immune to the threat of god-kings. Physical power works the same regardless of rank and regardless of whether people sympathize with it. Physical power is objectively true, impossible to refute, and impossible to ignore. Physical power-based control authority predates abstract power-based control authority by four billion years. Since the first abstract power hierarchies emerged, physical power has always served as its antithesis. Physical power has always been used to remedy the dysfunctionality of abstract power.

As much as sapiens hate to admit it, sapiens depend upon the complex emergent social benefits of physical power competitions (a.k.a. warfare) to establish resource control authority the same zero-trust and egalitarian way most other pack animals do. Physical power competitions may be energy intensive and prone to causing injury, but they are a far more systemically secure way for populations to manage resources.

Abstract power hierarchies are insecure because they increase the benefit of attacking a population by creating resource abundance and high-ranking positions with immense control authority. At the same time, abstract power hierarchies have no ability to physically constrain the control actions of attackers or oppressors. Rank and codified rulesets are all honey, but no sting. Therefore, people experience a high benefit to exploiting abstract power hierarchies, but no physical cost for doing so. The **BCR_A** of abstract power hierarchies approaches infinity unless users figure out how to impose severe, physically prohibitive costs.

But what does this have to do with software? It has everything to do with software, because **software is the continuation of the same 10,000-year-old trend of people trying to create and wield abstract power over others**. This isn't sapiens' first rodeo with the exploitation and abuse of abstract power hierarchies. We know exactly how this is likely to unfold because it has unfolded countless times before, under the same systemic conditions. We know that there is no physical limit to the amount of rank and resource control authority that software developers can wield. We can reasonably predict that eventually, populations will have to learn how to physically constrain and decentralize this growing and unimpeachable abstract power, or else they will be exploited and oppressed. Since the early days of the first god-kings, agrarian populations have routinely used physical power to constrain, dismantle, decentralize, and countervail abstract power hierarchies to secure themselves against the threat of overreaching abstract power. By continuously engaging in a global-scale physical power projection competition, no single abstract power hierarchy has been able to expand too far or gain too much centralized control authority over agrarian resources.

We also know that domestication gives us a causally inferable, randomized experimental dataset to show how insecure animals become to systemic exploitation and abuse when they are incapable of or disinclined to project physical power. We know that abstract power hierarchies have a repeated and demonstrable strategic security risk of self-domestication. We know that it is easy to condemn the use of physical power to settle disputes, establish control over resources, or determine consensus on the state of ownership and chain of custody of property because of the energy it uses or the injury it causes. But when populations condemn the use of physical power for ideological reasons, they become systemically insecure against exploitation and abuse. We know all of this because agrarian society has repeatedly learned this lesson the hard way. Thousands of years of physical confrontation has made the strategic necessity of physical power quite clear.

We know that as populations became more inclined to impose severe, physically prohibitive costs on abusive abstract power hierarchies, resource management systems became gradually less exploitative over time. In other words, abstract power hierarchies become less exploitative when people stand up and fight for their rights. Over thousands of years of warfare, the extraordinarily high levels of abstract power and resource control authority wielded by god-kings were eventually reduced to kings, then to presidents and prime ministers. The centralized control of empires and monarchies were eventually decentralized into democracies with carefully encoded checks and balances to restrict people's abstract power as much as possible. Theologies, philosophies, and ideologies increasingly favored equality and individual rights, and the encoded rulesets reflected it. This cultural evolution was gradual and filled with regression, but the trend is clear: **sapiens will find ways to utilize physical power to dismantle, decentralize, rebalance, and constrain abstract power hierarchies because abstract power hierarchies are demonstrably and incontrovertibly untrustworthy, inegalitarian, and dysfunctional**.

In other words, war is the trend. If there's one thing that agrarian society has made explicitly clear, it's that they're going to fight wars to physically decentralize and constrain the abstract power of their rulers. But right now, cyberspace is missing a warfighting protocol. This appears to be changing with the global adoption of Bitcoin.

5.6.3 Digital-Age Society will Inevitably Learn to Fight Wars in, from, and through Cyberspace

Here we finally arrive at the primary hypothesis of this thesis and the reason why the author dedicated hundreds of pages to developing a new theory about power projection from which to analyze Bitcoin. The core concepts outlined in Power Projection Theory laid the groundwork needed to understand the following insight: **the power projection dynamics and cultural evolution of agrarian society which took place over dozens of millennia appear to be repeating themselves following the invention of computers. Except now, it appears to be happening hundreds of times faster**.

The emergence of cyberspace could represent something as significant to human cultural evolution as the discovery of agriculture and the corresponding abstract power hierarchies established to govern agricultural resources. Cyberspace is a globally-adopted belief system that is radically transforming the way society organizes itself, in much the same way that agrarian abstract power hierarchies did. Just as agrarian society led to the formation of empires, so too does cyberspace appear to be leading to the formation of cyber empires, complete with the threat of oppressive rulers rising to the top of the hierarchy. If history repeats, then the inevitable next step for digital-age society after the emergence of soft*ware* (i.e. computer-generated abstract power) is the emergence of soft*war* (i.e. computer-generated physical power) to physically dismantle, decentralize, rebalance, and constrain it.

Whereas prehistoric sapiens took tens of thousands of years of abstract thinking to develop symbolic languages and then use them to formally codify abstract power hierarchies, modern sapiens equipped with computers have taken only tens of years to formally codify abstract power hierarchies using symbolic languages, giving them control authority over society's digital-age resources at unprecedented speed and scale.

As history continues to repeat itself, these new types of software-codified abstract power hierarchies are becoming as systemically insecure and oppressive as their predecessors were. It is reasonable to believe that this will lead to population-scale exploitation (assuming it hasn't already – which would be easy to debate) which must inevitably be resolved the same way it always has been resolved: using a real-world physical power competition to physically constrain and decentralize the emerging tyrannical, technocratic ruling classes and their oppressive empires. In other words, a new form of digital abstract power is likely to bring a new form of digital warfare to physically decentralize and constrain it, just as warfare has always been used as a mechanism to physically decentralize and constrain abstract power hierarchies (see section 4.9 for a detailed explanation of this).

The author hypothesizes that humanity is going to become so tired of being systemically exploited at unprecedented scales through their computer networks by an elite, tyrannical, and technocratic ruling class, that they are going to invent a new form of digital warfare and use it to fight for zero-trust, permissionless, and egalitarian access to cyberspace and its associated resources.

Using some type of new digital-age, physical power projection tactic, people will be empowered to compete for egalitarian access to, and control over, bits of information passed across cyberspace. This new form of electro-cyber warfighting will allow the population to keep itself systemically secure against exploitation from an abusive ruling class by giving them the ability to impose unlimited amounts of severe physical costs on their oppressors. An open-source electro-cyber warfighting protocol would allow people to keep themselves physically secure in cyberspace (i.e. virtual reality) using the same technique they already use to keep themselves physically secure in objective reality: by making it impossible to justify the physical cost (in watts) of attacking them. An illustration of this hypothesis is shown in Figure 64.

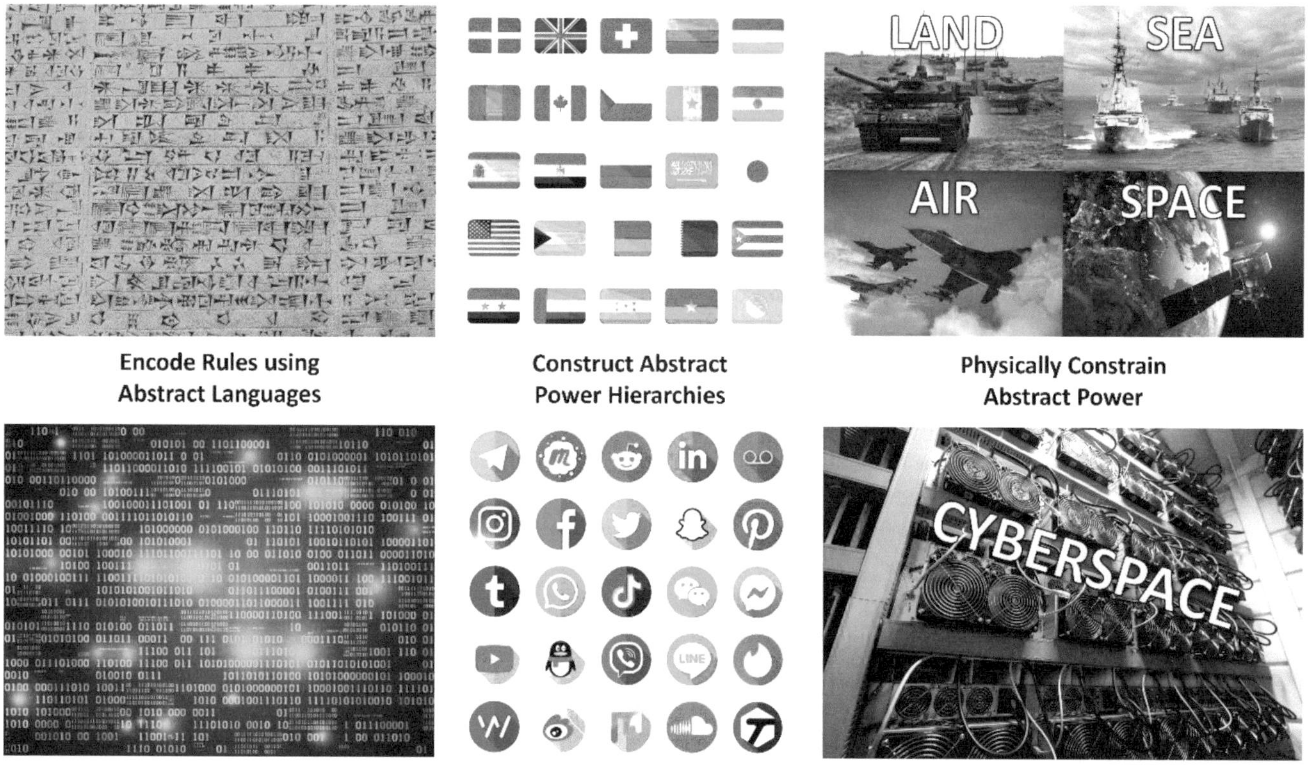

Figure 64: A Repeating Pattern of Human Power Projection Tactics
[1, 2, 3, 4, 5, 149, 150, 151, 152]

Without this new type of warfighting protocol, entire populations (including entire nations) will become increasingly vulnerable to widescale systemic exploitation and abuse via cyberspace. Just as populations rely upon physical power competitions as a protocol to physically constrain abstract power and control authority over agrarian resources, a primary hypothesis of this thesis is that all populations – including and especially nation states – are going to need an equivalent protocol to physically constrain the abstract power and resource control authority wielded by those who write and administer computer programs. **Without adopting a "softwar" protocol, entire nations could be vulnerable to becoming subservient to an elite, technocratic ruling class which could emerge either domestically or from a foreign power**.

People are going to need a physical cyber security protocol designed to provide society with the same complex emergent benefits as warfare. The goal of this so-called cyber warfighting protocol would be to give populations the ability to secure themselves using physical constraints rather than (demonstrably unsuccessful) logical constraints. **By (literally) empowering people to impose severe physical costs on people and programs in, from, and through cyberspace, agrarian society will be able to once again physically constrain people who abuse their imaginary power and control authority**. At the same time, this would likely unlock a zero-trust, permissionless, egalitarian method of physical power projection in cyberspace that could transform the profession of cyber security and possibly even change the foundational architecture of the internet.

5.7 Projecting Physical Power in, from, and through Cyberspace

"...power is almost everything. It doesn't matter what you think is right;
it matters what you can demonstrate and enforce is right."
Lyn Alden [153]

5.7.1 There's Two Ways to Constrain Software: Logically or Physically

Now that we have established a thorough conceptual understanding of power projection tactics in nature, human society, and cyberspace, we can finally begin to analyze the strategic security implications of Bitcoin from a different point of view. Not merely as monetary technology – but as physical power projection technology and a new form of digital-age, electro-cyber warfare.

So far, the author has focused on providing an exhaustive explanation for why human populations use physical power competitions (i.e. wars) to settle disputes, determine control authority over resources, establish consensus on the legitimate state of ownership and chain of custody of property, physically constrain abstract power hierarchies, preserve international trade routes, and secure themselves against systemic exploitation and abuse. This lengthy discussion was designed to help the reader understand *why* society would even want to adopt a digital warfighting protocol in the first place, which in the author's opinion, is a much harder and more important question to answer than *how* it could be done. Perhaps the reason why people don't understand the value of Bitcoin is because they don't understand *why* people would want to project power and impose severe physical costs on others in, from, and through cyberspace.

Now that we have answered the *why*, the next question to ask is *how* it could be done. To answer this question, we can begin by asking ourselves to consider what a global-scale cyber war might look like. Would it even be possible to wield and project physical power (a real-world thing) in, from, and through cyberspace (an abstract domain)? How could someone impose physical (a.k.a. non-virtual) costs virtually?

The key to understanding how this could be done is to return to the first principles of computer theory. First, we can recall Von Neumann's early observation about stored-program state mechanisms: *"It is easy to see by formal-logical methods, that there exist codes that are in abstracto adequate to control and cause the execution of any sequence of operations which are individually available in the machine, and which are in their entirety, conceivable by the program planner."* [125]

In this quote, Von Neumann highlights that software has two primary constraints: the physical limits of the state machine, and the imagination of the computer programmer. First, the software running on a computer can be constrained using any design logic conceivable by the program planner (logic which we have established is routinely dysfunctional and incapable of securing software against the systemic exploitation of its own logic). Alternatively, a computer program can be constrained by physically constraining the underlying state mechanism running the program. This is illustrated in Figure 65.

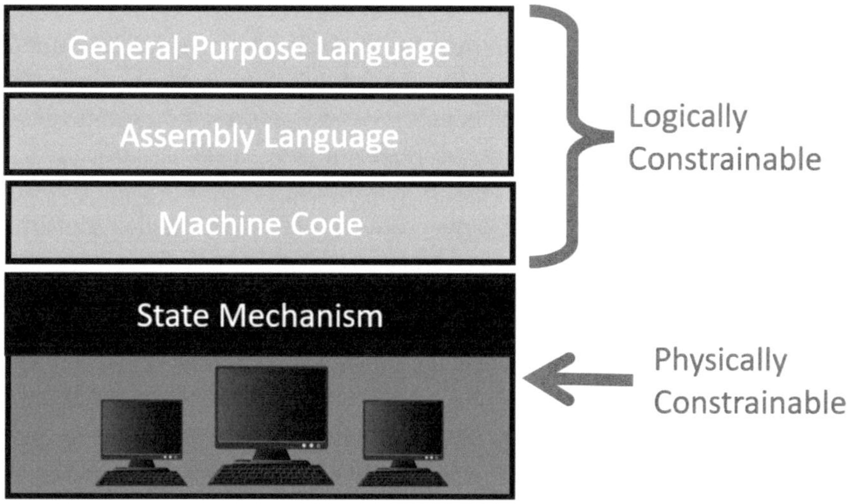

Figure 65: Illustration of Two Ways to Constrain a Computer Program [154]

This observation has subtle but important implications for cyber security because it offers insight about a potential way to improve cyber security by physically constraining computers rather than continuing to attempt to encode logical constraints into software (a practice that is clearly ineffective, hence the current hacking epidemic). **To make cyberspace more secure, it is possible to physically constrain the underlying computers connected to the internet. This would imply that what cyberspace is missing is an open-source internet protocol and supporting infrastructure needed to physically constrain computers. By creating an open-source protocol which empowers people to physically restrict computers via the internet, society would gain an unprecedented physical cyber security capability which they could use to physically secure themselves against each other in, from, and through cyberspace**.

5.7.2 Not Being Able to Physically Constrain Computers is a Major Systemic Security Vulnerability

So far, we've established that software system security vulnerabilities are derived from insufficient constraints on control signals which can be exploited in such a way that it puts software into insecure or hazardous states. Software security engineers have a major disadvantage in their ability to constrain the control actions executed by computer programs in comparison to physical control actions executed by people or machines. The fundamental challenge is that software engineers *must* use discrete mathematical logic to constrain software control actions because it is otherwise impossible to physically constrain an undesired command signal without physically constraining the underlying state mechanism.

Herein lies a key insight that remarkably few computer theorists seem to understand. The inability to apply physical constraints on computers is a major systemic security vulnerability. Not having a mechanism to apply physical constraints on computers forces software engineers to design trust-based abstract power hierarchies, where sensitive control actions are logically constrained by giving the authority to execute them to a select few users of high rank (e.g. system admins, which are usually themselves). This consolidates or "centralizes" software administrative permissions to select user accounts that can be systemically exploited. Software engineers must then devise ways to prevent outsiders from accessing and exploiting these special administrative permissions. At the same time, the software's legitimate administrators and engineers must be trusted not to exploit their self-encoded abstract power and control authority over the system. Because these control structure designs rely on *trusting* people not to execute unsafe control actions rather than *physically* constraining them, they are systemically insecure. It turns out that trusting a small, centralized group of software engineers and administrators not to exploit their abstract power is a highly ineffective security strategy backed by thousands of years of testimony. The **BCR_A** of a trust-based abstract power hierarchies approaches infinity as the benefit to attacking or exploiting it increases. Meanwhile, users must accept the risks posed by a perpetually increasing **BCR_A** because they have no alternative way to physically constrain unsafe control signals or impose severe physical costs on people who exploit their special permissions over the system.

5.7.3 A Protocol which Physically Constrains Computers Would Look Like a Very Inefficient Program

To remedy this glaring systemic security vulnerability in our current approach to cyber security, we need a way to physically constrain computers. Using physical constraints to improve cyber security is not a new concept. For example, a common cyber security strategy in many organizations is to physically remove hardware (e.g. USB drives) in the computers to physically prevent belligerent actors from stealing sensitive data from a given computer network or uploading malware to it. US military personnel physically secure their encryption keys by carrying them on specially-designed common access cards (CACs) in their wallets. The same principle also applies to strategies like air-gapped networks. An air gap represents a real-world physical constraint applied to the underlying state mechanisms of a given computer network.

It should be noted, however, that these examples are self-imposed physical constraints that are not transferable in, from, and through cyberspace. **If someone were to develop a technique for applying physical constraints to *other* people's computer programs in, from, and through cyberspace, that would represent a remarkable new capability in cyber security. It would be especially noteworthy if this capability manifested itself as a globally-adoptable open-source protocol that utilized an existing infrastructure (e.g. the global electronic power grid) to apply these physical constraints to computers connected to the internet.**

With these concepts in mind, the rest of this chapter can be summarized with the following: it is theoretically possible to create computer programs that are inherently secure because it's either physically impossible or too physically difficult to put them into a hazardous state, simply by intentionally applying physical constraints to the underlying state mechanisms running the software. It is also theoretically possible to design computer protocols which can apply real-world physical constraints to other people's computer programs in, from, and through cyberspace. Incidentally, this protocol already exists, and it already has a name: it's called proof-of-work.

If an open-source protocol were invented that allowed people to physically constrain or "chain down" the underlying state mechanisms connected to the internet, that would likely become a valuable cyber security protocol. If the protocol were globally adoptable in a zero-trust, permissionless, and egalitarian way, it could become so valuable that it would represent a national strategic priority to adopt it. **There is a strategic imperative for nation states to be on the lookout for the development of any kind of open-source physical power projection protocol which empowers people to physically constrain or "chain down" other people's computer programs in, from, and through cyberspace. A protocol like this would be of national strategic significance because it would represent the discovery of a globally-adopted physical security protocol for cyberspace and an unprecedented new cyber security capability**. This concept is illustrated in Figure 66.

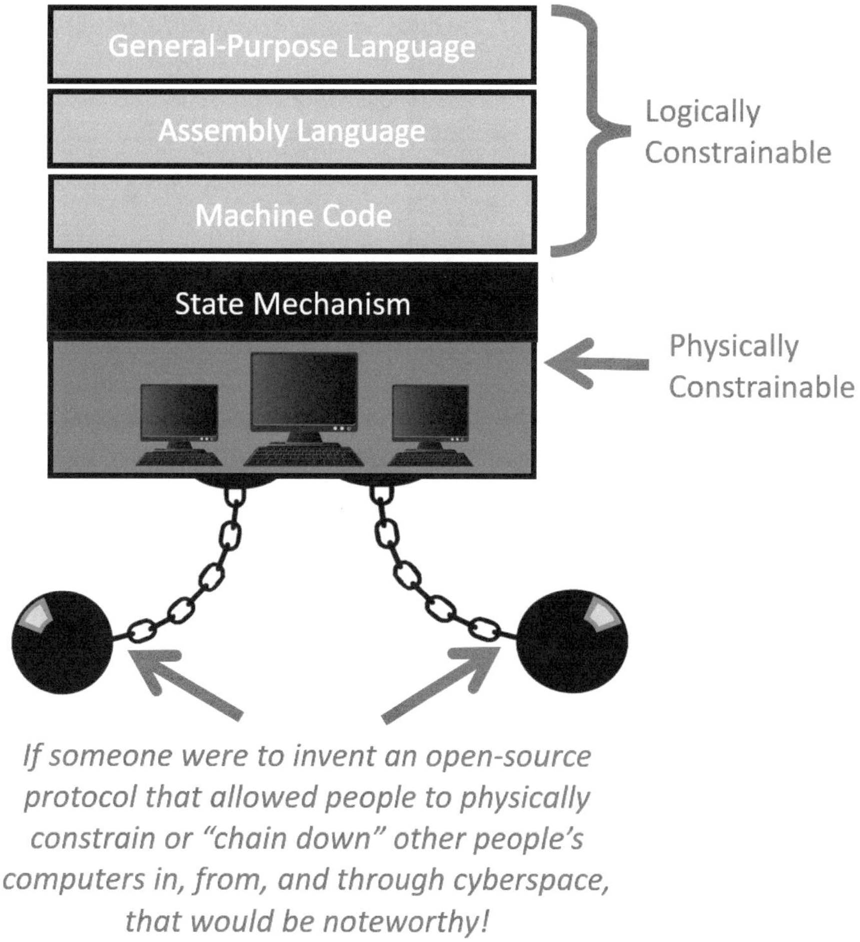
General-Purpose Language
Assembly Language
Machine Code
Logically Constrainable
State Mechanism
Physically Constrainable
If someone were to invent an open-source protocol that allowed people to physically constrain or "chain down" other people's computers in, from, and through cyberspace, that would be noteworthy!

Figure 66: "Chain Down" Design Concept for Physically Constraining Computers [154, 155]

Knowing that such a protocol is theoretically possible, the next question becomes, what would such a system look like? What kind of physical constraints would the system impose to "chain down" computers and how would it work? What kind of supporting infrastructure would this system require to apply these physical restrictions? One way to physically constrain computers via cyberspace is by filtering all incoming control signals through a special kind of state mechanism with a physically constrained state space that has deliberately-difficult-to-change states. In other words, to physically constrain other people's computers via cyberspace, create a computer that is intentionally costly (in terms of watts) to operate, and then require other people's software to run it. This concept is illustrated in Figure 67.

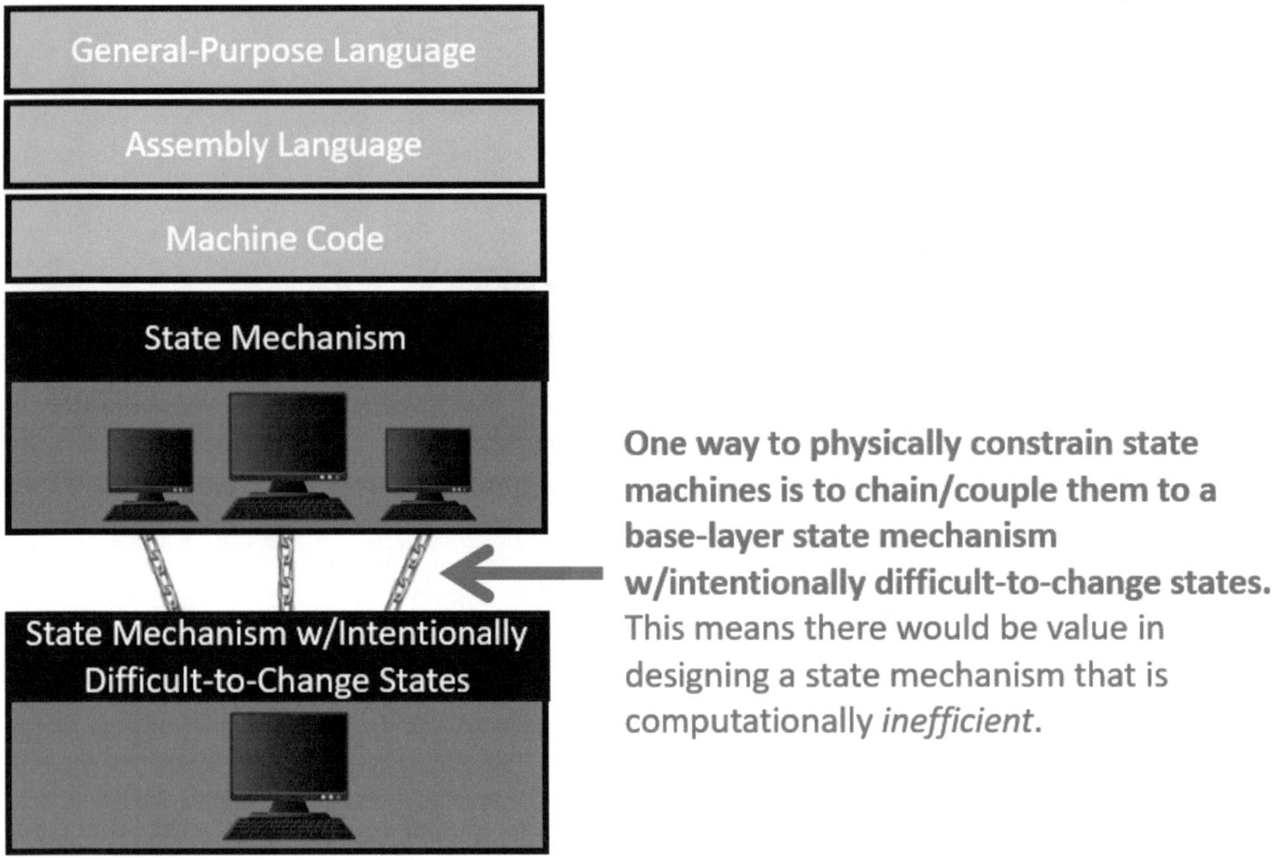

Figure 67: Security Protocol Design Concept of a Deliberately Inefficient State Mechanism [156, 154]

Herein lies another key insight that many people (including but especially software engineers and computer scientists) overlook. A special type of computer that is highly inefficient, with intentionally difficult-to-change states, would have counterintuitively beneficial emergent properties for cyber security. By deliberately designing a computer to be *inefficient* (i.e. intentionally physically costly to operate) and then designing software logic requiring other people to use it, it would be possible to apply physical restrictions to *other* people's computers in, from, and through cyberspace. **There would be extraordinary strategic benefits to deliberately creating an *inefficient* state mechanism which takes an excess quantity of physical resources (e.g. electricity) to operate, as those thermodynamic restrictions would double as a way to physically constrain sensitive software control actions rather than logically constraining them**.

The key insight that seems to be overlooked by computer theorists is that a highly inefficient computer to which everyone has zero-trust and egalitarian access would likely be very beneficial for cyber security purposes. Based on this insight, we can see that **an open-source, globally-adopted physical security system for cyberspace would likely manifest itself as something which, on the surface, appears to be extraordinarily inefficient.** This inefficiency would not be a bug, but rather a feature. In fact, it would be the *primary value-delivered function* of a system which allows people and their programs to apply physical constraints on other people and programs in, from, and through cyberspace. The inefficiency of this system would represent the real-world physical costs imposed by the system onto people in the virtual domain.

Adding these thoughts together, the author offers the following insight: to recognize an unprecedented new cyber security capability that could transform how people secure digital resources, we should look for the emergence of open-source protocols which, to the untrained and uniformed eye, would appear to be highly computationally inefficient. **This inefficiency would not be a bug that needs to be corrected – it would be a very important and highly noteworthy feature that is essential for *physically* securing software**. With this in mind, we turn our attention to cost function protocols and the invention of proof-of-work.

5.7.4 Physical Cost Function Protocols (a.k.a. Proof-of-Work Protocols) are Not Well Understood

The author challenges the reader to think of an open-source protocol as infamous for its computational inefficiency as Bitcoin. Bitcoin is constantly criticized for how much power is consumed by the computers which run it. What is not understood is why computational inefficiency is an incredibly useful cyber security *feature*, not a *bug*. In other words, Bitcoin's computational inefficiency is its primary value-delivered function – it's the thing that has been missing in cyber security that is needed to *physically* constrain belligerent actors, rather than (continue to fail to) *logically* constrain them.

The author hypothesizes that **people don't understand the security benefits of Bitcoin's inefficiency because they don't understand how physical security works, nor the enormous systemic security benefits gained by projecting physical power (a.k.a. watts) to impose severe physical costs on attackers in order to raise one's C_A and thus lower one's BCR_A**. Without this background, it's difficult to understand why open-source computer programs which *literally* empower people to do this via cyberspace may represent an unprecedented and remarkable cyber security capability which could change the way people design their software and even potentially transform the underlying architecture of the internet.

This lack of understanding about how power projection works is why the author has outlined the following first principles approach to understanding the Bitcoin protocol, not as a monetary system, but as a physical security system which utilizes a computationally inefficient state mechanism to give people the ability to impose real-world, physically prohibitive costs on other people in, from, and through cyberspace. To build a first principles foundational understanding of Bitcoin as a security system, it is necessary to develop a thorough understanding of the security design concept colloquially known as "proof of work," or what the author alternatively calls "proof of power."

5.7.5 To Eliminate Superfluous Control Signals (a.k.a. Spam), Make Control Signals Superfluously Costly

Today, most computer programs send control signals across cyberspace at practically no marginal cost. The lack of marginal cost is a byproduct of engineers designing highly efficient computers. Unfortunately, cheap computing is a feature that is commonly exploited by belligerent actors. For example, it's possible to attack computer networks by sending millions of superfluous control signals (e.g. service requests) to overwhelm a target network's bandwidth. Ironically, the reason why these control signals are superfluous is because the costs associated with sending them *aren't* superfluous. Because computers *don't* add superfluous costs (like additional electricity) to sending control signals, it is trivial for belligerent actors to flood target networks with superfluous control signals.

This type of systemic exploitation is commonly known as a denial-of-service (DoS) attack. The near-zero marginal cost of using efficient computers to send superfluous control signals across the internet is what enables practically all email spam, comment spam, bot farms, troll farms, and many other popular exploitation tactics possible. Because there's minimal physical cost associated with these activities, there's practically no physical costs imposed on those who systemically exploit these control signals. In other words, the BCR_A of this type of attack is high because C_A is practically zero.

5.7.6 Searching for the Best Way to Make Control Signals Superfluously Costly to Execute

For decades, software engineers have been experimenting with different ways to improve cyber security by creating control structure designs which physically constrain unsafe control signals by physically constraining underlying computers. There are many different ways that this can be done, each with their own tradeoffs.

A popular way to physically constrain unsafe control signals is by physically constraining access to the computers which send those signals. As previously discussed, this can be done via physical computer access control points or air gaps. For example, the US military keeps top secret intelligence information secure against data leaks by physically restricting access to the computers which store that information using air-gapped intranets where each computer in the network is locked within a specially-designed sensitive compartmented information facility (SCIF).

This air-gapped intranet with physically constrained access control points is known as the Joint Worldwide Intelligence Communication System (JWICS).

Physically constraining computers in this way has major tradeoffs, though. First, the functionality and utility of these air-gapped intranets are severely restricted because they are difficult to access, and they cannot easily communicate with other computer networks. Second, they are still systemically insecure because they are trust-based, inegalitarian, permission-based systems within a centralized abstract power hierarchy, thus they're still vulnerable to systemic exploitation. People must rely on trusted third parties with abstract power and control authority (i.e. administrative privileges) over these networks to gain and maintain access, and that access can be revoked at any time. Additionally, once granted access to these air-gapped networks, people have to be trusted not to exploit their access – there's little physically preventing them from leaking the sensitive information contained within JWICS networks to other networks. As many highly-publicized leaks of JWICS data would suggest, these types of trust-based security systems are systemically insecure.

Fortunately, there appears to be other ways to physically constrain unsafe software control signals that don't suffer from these negative tradeoffs. Software engineers have been investigating different ways to do this for at least three decades. Throughout the late 90s and early 2000s, software engineers began to propose different kinds of software protocols which could prevent an adversary from gaining access to online resources by simply making it superfluously costly for them to do so.

The first software engineers to publish a paper about this security concept were Cynthia Dwork and Moni Naor. In their paper, *Pricing via Processing or Combatting Junk Mail*, Dwork and Naor state the following: *"We present a computational technique for... controlling access to a shared resource... The main idea is to require a user to compute a moderately hard, but not intractable, function in order to gain access to the resource, thus preventing frivolous use. To this end, we suggest several pricing functions...* [26]

This paper was the first to offer a subtle but profound cyber security concept: to secure access to resources, simply decrease the benefit-to-cost ratio of accessing those resources by increasing the cost of accessing them. In other words, Dwork and Naor's idea was to utilize the primordial economic dynamics discussed at length throughout Chapter 3. To keep online resources secure against DoS attacks, simply increase the computational cost (i.e. increase C_A) of executing certain control actions to decrease the BCR_A of executing those control actions. Dwork and Naor offered several ideas for candidate pricing functions (to including hash functions) which could be used as the mechanism to add superfluous costs, but they concluded that *"... there is no theory of moderately hard functions. The most obvious theoretical open question is to develop such a theory..."* [26]

Here we see the emergence of the idea that software can be secured using the same primordial economic dynamics that life has been using for four billion years: simply decrease the BCR_A of attacking or exploiting software, rather than trying to write exploitable logic. All that appeared to be missing was for someone to develop the necessary theories of how it could work, why it could work, and specifically what cost function algorithms should be used.

By 1999, Dwork and Naor's idea of securing online resources by making computers solve moderately hard pricing functions was formally named "proof-of-work." Multiple computer scientists, including Markus Jakobsson and Ari Juels, conceived of potential use cases and designs of proof-of-work protocols. The concept was simple: a computer can constrain the control signals sent by belligerent computers by superfluously increasing the computational cost of sending those control signals. In other words, **software can physically constrain another computer by making it computationally inefficient for them to execute certain functions or send specific commands**. As previously discussed, there is clear value in computational inefficiency and the strategy is quite simple: to physically constrain a bad guy's computer, create a computer program that is deliberately inefficient to run, and then force the bad guy to run it. [28]

It's important for the reader to note that, like all software specifications, semantic specifications of software design concepts like "proof-of-work" are completely arbitrary (see section 5.4 for a more thorough breakdown of this concept). When proof-of-work was formally introduced into scientific literature, it was *arbitrarily* called a "bread pudding" protocol. Jakobsson and Juels explained the name as follows: "*Bread pudding is a dish that originated with the purpose of re-using bread that has gone stale. In the same spirit, we define a bread pudding protocol to be a proof of work such that the computational effort invested in the proof may also be harvested to achieve a separate, useful, and verifiably correct information*." In the very same year, the same author gave the same proof-of-work protocol a different name: a "client puzzle" system. The takeaway? **What people call their proof-of-work software doesn't matter because it can be (and has been) called practically anything. What matters is not the name; what matters is how it works, why it works, and most importantly, why people would want to use it**. [28, 27]

Interestingly, two years prior to the release of these formally published academic papers about how proof-of-work protocols could work, a software engineer named Adam Back privately released his own version of an operational proof-of-work protocol concept he named "hashcash." Despite having a different name, the intent of the protocol was the same: to secure software against threats like DoS attacks by adding superfluous costs to sending unwanted control signals, thus decreasing the benefit-to-cost ratio of sending unwanted control signals. Just like his peers did, Back asserted that this type of protocol could function as "*a mechanism to throttle systemic abuse of internet resources*." [157]

5.7.7 Proof-of-Work Protocols are Literally Proof-of-Power Protocols

It is imperative for the reader not to overlook the fact that proof-of-work protocols very specifically create real-world *physical* costs that can be measured in watts. These protocols are computationally inefficient not just because of how many calculations they require, but because of how many watts must be consumed to make those calculations. By using a deliberately inefficient protocol like this, a person or program applies a real-world *physical* cost on the execution of control signals sent by other people or programs operating on different computers. To emphasize this point, the author will interchangeably use the term "physical cost function" instead of "pricing function" and "proof of power" or "bitpower" instead of "proof-of-work" where the term "physical cost" refers to the real-world physical deficit of electric power (a.k.a. watts) required to generate a proof of power.

By increasing the physical cost (a.k.a. watts) required to execute a command action or send a control signal, Back showed how it was possible to increase the marginal cost of exploiting certain control signals exactly as envisioned by Dwork and Naor. This can be done by utilizing a two-step process shown in Figure 68. The first step is to create an algorithm that is so computationally difficult to solve that it imposes excess physical cost (a.k.a. watts) on computers attempting to solve it. This algorithm effectively creates a vacuum of electric power that must be filled for the physical cost function to generate a proof of power.

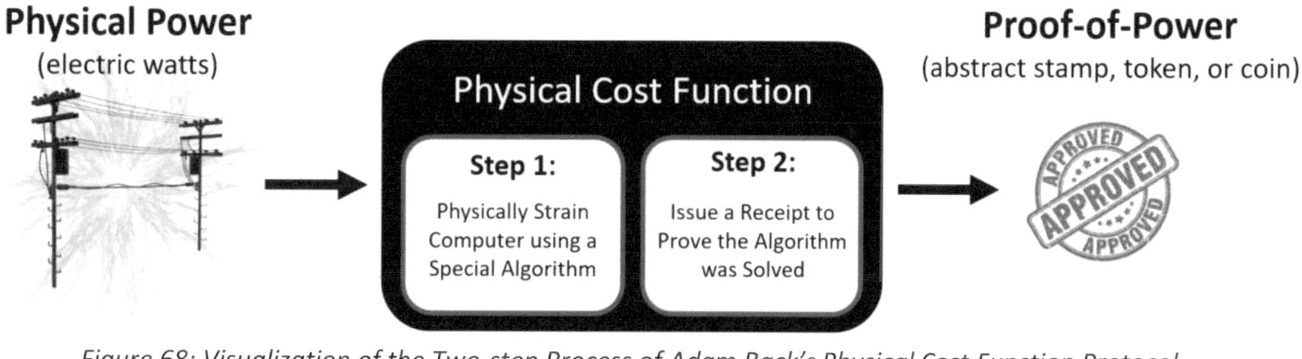

Figure 68: Visualization of the Two-step Process of Adam Back's Physical Cost Function Protocol [158, 159, 160]

Finding the right hashing algorithm to achieve this first step was a challenge that took years to resolve. Many candidates for physical cost functions were discussed throughout the 90s, but Back's physical cost function became popular because the algorithm he designed doesn't appear to provide an advantage to any particular method for solving it. It utilizes a lottery-style "pick the winning number" technique where computers must repeatedly guess random numbers until they pick the winning number.

A major advantage of this type of lottery system is that it appears to be perfectly impartial and fair. Back's algorithm is equally and uniformly difficult for everyone to solve. On a per-guess basis, nobody appears to be able to gain an upper hand on solving Back's algorithm no matter who they are or what kind of guessing strategy they use. One single guess has equal odds of being correct as any other guess, and there's no apparent way to "game" or exploit the algorithm to give somebody an unfair guessing advantage. Consequently, the most effective way to solve Back's algorithm is by using a brute-force guessing technique where computers must make as many random guesses as they can as quickly as possible.

The second step of Back's cost function protocol is to issue an abstract proof or receipt to the entity which successfully solves the hashing algorithm. This allows the bearer of the proof/receipt to verify that they incurred an excess physical cost (as measured in watts) for solving the algorithm. Back arbitrarily called this abstract receipt a "proof-of-work" based off the precedent set in formal academic literature by Jakobsson and Juels and their "bread pudding" protocol, where "proof" refers to a bearer asset (e.g. a receipt, stamp, or token), and "work" refers to the expenditure incurred to solve the hashing algorithm.

Something that is vitally important for the reader to understand about "proofs of work" is that **the term "proof-of-work" represents an abstraction for a real-world physical phenomenon: power (i.e. watts). The "proof" or token or receipt that is used to verify the power expenditure is abstract, but the power expenditure is a real-world physical phenomenon that is directly measurable in watts**. The reason why this is important will be discussed in much further detail in section 5.9, but the main takeaway for now is that a "proof-of-work" is an abstract way of describing a real-world physical thing, which means that "proofs of work" inherit all the physical and systemic properties of real-world physical power, which are physically impossible for other types of computing systems to replicate (as discussed in section 5.9, this is why it is physically impossible for proof-of-stake protocols to replicate the same systemic security benefits of proof-of-work systems).

As shown in Figure 68, Back's physical cost function protocol can be visualized as a black box with a single input and a single output. The input is a vacuum of real-world physical power (a.k.a. watts) which must be filled to solve the protocol's hashing algorithm. The output is the proof-of-power receipt issued to prove that a real-world physical power expenditure was incurred to solve the algorithm.

The reader may be asking, what's the point of a computer program which intentionally creates a vacuum of physical power that must be filled, and then issues a receipt to the entity which fills it? The answer is worth repeating: this activity imposes a physical cost on other people in, from, and through cyberspace. **With the invention of physical cost function protocols, it becomes possible to impose severe, real-world physical costs on others by simply demanding that they present proofs of power. This is an extraordinarily simple but effective cyber security capability that has been notably missing from the current architecture of the internet**.

This is where the core concepts of Power Projection Theory become so important to understand. Back's "hashcash" could represent the first earnest attempt at building a physical-cost-of-attack function, or C_A function. **Proof of physical power expenditure represents proof of real-world physical cost, which means proof of high C_A. Back's protocol showed how it was feasible to *physically* constrain the execution of unwanted signals by simply increasing control signal C_A.**

By increasing the physical cost of sending unwanted control signals, physical cost function protocols increase the physical cost of exploiting or abusing those control signals. People gain the ability to improve systemic security and solve the survivor's dilemma by projecting physical power. The more people increase the physical cost of sending control signals (not just spam, as originally envisioned by Dwork and Naor, but any type of unsafe or unwanted control signal), the more its C_A increases and the lower its BCR_A becomes. Control signals with lower BCR_A are intrinsically more secure against systemic exploitation or abuse because they're more physically expensive to exploit. This concept is illustrated in Figure 69.

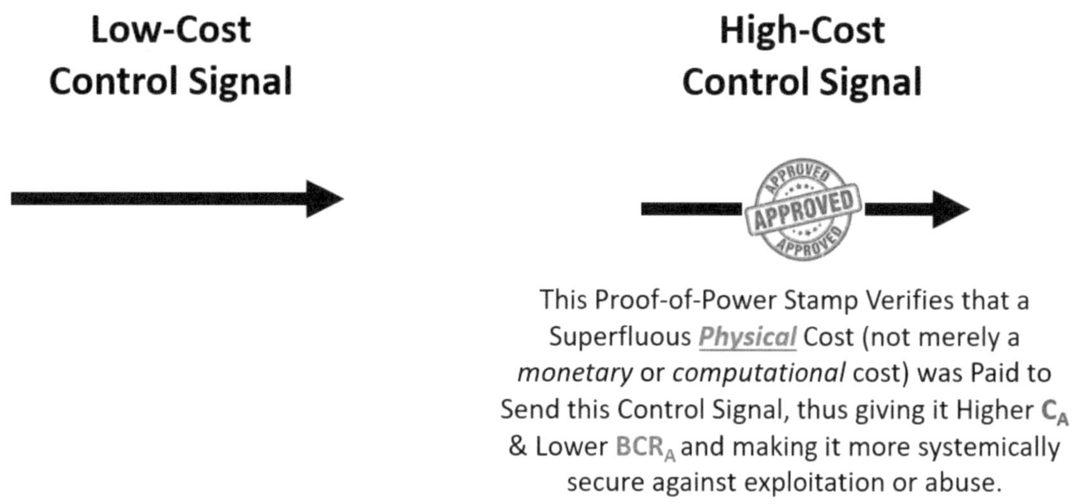

Figure 69: Illustration of Two Different Types of Control Signals
[160]

The arrow on the left represents a typical "low-cost" control signal sent by a computer program across cyberspace at near-zero physical or marginal cost. The arrow on the right represents a "high-cost" control signal which has been physically constrained through the addition of proof-of-power. By "stamping" a control signal with proof-of-power, the receiving computer of that control signal can verify excess physical cost (a.k.a. watts) was imposed on the user which sent this control signal, giving it a higher C_A and lower BCR_A, thus making it more strategically secure against exploitation and abuse than an ordinary low-cost control signal with high BCR_A.

The ability to increase control signal C_A gives software security engineers a capability they previously didn't have: the ability to use software to *physically* (not just logically) constrain sensitive control signals. It also gives users an ability they didn't have before too: the ability to project real-world physical power against other people and programs in, from, and through cyberspace. With the invention of physical cost functions, all one needs to do to impose physically prohibitive costs on other computers, programs, and computer programmers is to simply refuse to accept control signals unless they present proof-of-power.

Thanks to physical cost function protocols, software engineers can secure computer programs in the same way that animals project power to secure themselves in nature, and sapiens project power to secure themselves in society. By simply raising the C_A and thus lowering the BCR_A of control signals, it is theoretically possible to dramatically improve the security of software against systemic exploitation and abuse by any belligerent actor, in novel ways that were previously not considered possible. Thus, physical cost functions represent a noteworthy contribution to the field of cyber security that people could be overlooking just because they arbitrarily adopted the habit of calling it a monetary protocol.

Another interesting observation about physical cost function protocols is that they appear to be the continuation of a four-billion-year-old trend of organisms developing increasingly clever power projections tactics to keep themselves secure against predators by continually increasing their capacity to impose physical costs on them. Using physical cost functions, sapiens appear to have discovered a way to project power by directing energy in such a way that it enables them to impose severe, physically prohibitive costs on their neighbors via cyberspace. It should be noted that because physical cost functions passively utilize an *electric* form of physical power rather than a *kinetic* form of physical power, they represent a non-lethal form of global-scale power projection – making them potentially disruptive to traditional methods of agrarian warfighting (more on this later).

A globally-adopted proof-of-power protocol could mean that agrarian society has figured out a way to impose severe physical costs on entire populations of people (such as nations abusing their specially-encoded administrative privileges) – but in a way that is incapable of causing injury. Therefore, not only do physical cost function protocols likely represent a noteworthy contribution to the field of cyber security, but they could also one day be considered to be a noteworthy contribution to the field of security in general – particularly national security. Physical cost functions *literally* empower people to physically secure their digital-age resources and policies against exploitation and foreign attack without physically injuring anyone or causing practically any form of physical pain or discomfort. This is a remarkable capability considering how traditional physical security and legacy physical power competitions tend to be highly destructive and fratricidal.

5.8 Electro-Cyber Dome Security Concept

"Power lies not in resources but in the ability to change behavior...
As the instruments of power change, so do strategies."
Joseph Nye [161]

5.8.1 Imposing Infinitely Scalable Physical Costs on Computers, Programs, and Computer Programmers

Like all strategic security, a "peaceful" period of reprieve against attack or exploitation happens at an equilibrium point when BCR_A is sufficiently low because C_A is sufficiently high. Successful deterrence is achieved not because would-be attackers can't perform an attack, but because they find it impossible to justify the physical cost of performing an attack. This happens when the attacker's opponent has successfully demonstrated they have both the means and inclination to impose severe physical costs on their attacker. Because there's theoretically no limit to the difficulty of solving hashing algorithms, that means there's theoretically no limit to the amount of physical cost that people/programs can impose on their attackers using physical cost functions. And because the physical cost is imposed electronically rather than kinetically, there are comparatively little practical limitations either. It goes without saying that this could have enormous strategic implications, which will be discussed further at the end of this chapter.

Recalling the concept of the infinitely prosperous organism discussed in section 3.7, the reader should note that physical cost functions represent an *infinitely scalable* physical cost function. Since there's no theoretical limit to how much physical cost can be imposed by physical cost function protocols like Bitcoin, that means there's theoretically no limit to how much prosperity margin can be created by organizations utilizing this technology. Moreover, because the physical cost imposed by these systems is generated electronically rather than kinetically, that means there's also no threat of hitting a kinetic ceiling no matter how efficient people get at imposing these physical costs (see sections 4.11 and 4.12 for a more thorough explanation of these concepts). The potential of physical cost function protocols like Bitcoin are therefore extraordinary, because they don't have the same theoretical or practical limitations of traditional power projection tactics, techniques, and technologies used for physical security applications.

All computers, programs, and computer programmers need to do to impose severe, real-world physically prohibitive costs on other computers, programs, and computer programmers in, from, and through cyberspace is simply create a firewall-style application programming interface (API) which utilizes an open-source physical cost function protocol like Bitcoin. The author calls this concept a "proof-of-power wall" and illustrates it in Figure 70.

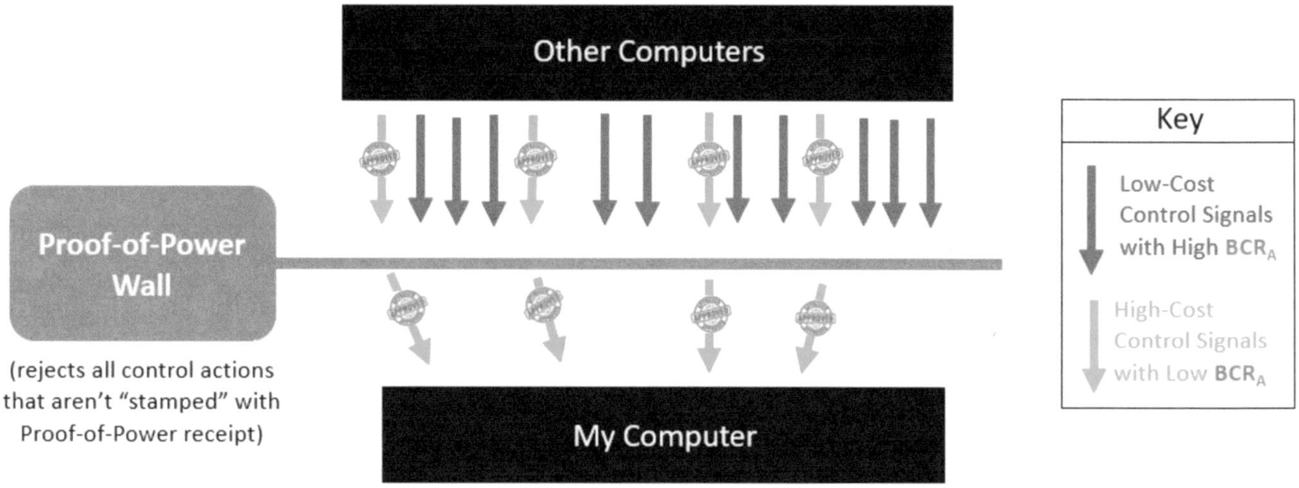

Figure 70: Design Concept of a "Proof-of-Power Wall" Cyber Security API [160]

Proof-of-Power wall protocols could work by simply rejecting all incoming control signals that don't present proof-of-power. This would guarantee that all control signals sent across the API have low BCR_A and are therefore more systemically secure against exploitation and abuse. By throttling up or down the physical costs incurred for each proof-of-power receipt (which could be done by adjusting the difficulty of the hashing algorithm or requiring more proofs of power), it would be possible to raise or lower the BCR_A of each control signal as needed (thus increasing or lowering prosperity margin as desired). So long as the BCR_A of authorized control signals are lower than the hazardous BCR_A level, the computer program remains strategically secure against systemic exploitation.

Recalling the core concept of the survivor's dilemma discussed in section 3.6, the hazardous BCR_A level for any given environment cannot be known and tends to drop over time, causing a natural and continuous decline in prosperity margin. Organisms and organizations must strive to continually increase their C_A as much as they can afford to increase it to ensure they can lower their BCR_A as much as possible and buy as much prosperity margin as they can to remain systemically secure. This is a fundamental challenge of all strategic security that has stressed life for four billion years.

Applying this same concept to physical cost function protocols like Bitcoin, we can see that hashing algorithms should be designed in such a way that their difficulty is continuously adjustable. That way, it's always possible to increase the physical difficulty of solving them, thus always possible to increase their C_A. This will ensure that people can permanently decrease BCR_A of their computer programs and buy as much prosperity margin needed to keep control signals systemically secure against exploitation and abuse. If sensitive control signals start to be exploited, people can simply increase the difficulty of the hashing algorithm to increase the amount of power needed to solve the cost function.

5.8.2 Building an "Electro-Cyber Dome" Security System

"Bitcoin is a swarm of cyber hornets serving the goddess of wisdom, feeding on the fire of truth, exponentially growing ever smarter, faster, and stronger behind a wall of encrypted energy."
Michael Saylor [162]

As previously discussed, the first proposed use case for cost function protocols was to serve as a countermeasure against DoS attacks – specifically email spam. This was something that multiple computer engineers recognized as a simple and attractive use case. Adam Back originally called proofs-of-work "stamps" rather than proofs. The name "stamp" was chosen because the original use case of his protocol was to improve email services. The term "stamp" also highlighted how the proofs generated by his software aren't sequentially reusable, just like real mailing stamps aren't. The term "proof of work" didn't appear until two years *after* Back introduced his prototype. Only after "proof of work" became organically popular terminology did Back call his design concept a proof-of-work protocol. [39]

It didn't take long for people to start exploring how to use physical cost functions for other use cases beyond securing email browsers against spam. Some software engineers started exploring how to create physical cost function protocols to create sequentially reusable and transferable proof-of-power receipts. When this happened, people abandoned the word "stamp" and started using the names of other abstract objects like "token" or "coin." These semantic specifications provided a more intuitive explanation of the software's emergent behavior because tokens and coins are both sequentially reusable and transferrable. Nevertheless, the general function of physical cost functions and proof-of-power receipts remained the same: to make exploitable control signals more expensive to exploit – particularly the ones exploited during DoS attacks. This is why Back entitled his 2002 paper "*Hashcash: A Denial of Service Counter-Measure*."

The general public frequently does not understand that proof-of-work protocols are first and foremost cyber security protocols designed to defend computer programs against cyber attacks like DoS attacks. Back's hashcash design concept, which explicitly called it a DoS countermeasure in the title, was reused in Bitcoin and was directly cited by Satoshi Nakamoto as a primary inspiration. Additionally, proof-of-work was universally described as a cyber security protocol for fifteen years of peer-reviewed academic literature prior to the release of Bitcoin. In other words, proof-of-work protocols have been known as cyber security protocols for twice as long as they have been known as monetary protocols.

As previously discussed, because there is close to no marginal cost required for modern computers to send control signals through cyberspace, it is possible for belligerent actors to target computer networks and flood them with superfluous control actions (e.g. service requests) to overwhelm network bandwidth. One possible countermeasure against this type of attack is to ignore the control signals coming from the belligerent computers. However, this countermeasure is easily circumnavigated by sending belligerent control signals from different networks in what's known as distributed denial-of-service (DDoS) attacks. [157]

Using a proof-of-power wall API (a.k.a. a proof-of-work protocol), it is possible to apply a physical cost uniformly to all incoming control signals, creating a potentially more effective countermeasure against DDoS attacks. No matter how distributed an attacker's computer network can be, it would be equally as physically costly for them to spam a target network with superfluous control actions because they would be required to attach a proof-of-power receipt (or stamp, or token, or coin, or whatever arbitrary abstraction the reader finds most helpful) to every malevolent control signal. And because DDoS attacks send substantially more control signals than honest users, it would be possible to throttle the physical cost exponentially higher so that it's inconsequential for honest users but severe for belligerent actors.

By creating an email protocol which doesn't allow emails to reach someone's inbox unless they have a proof-of-power "stamp," it's possible to cause the physical cost of sending someone emails to rise to levels where it would become too physically expensive to spam people's inboxes. The operational expense of generating enough proof-of-power "stamps" would simply be too high to make email spam (or any kind of spam) a profitable endeavor. It's a simple strategy: to prevent superfluous emails, make them superfluously expensive to send.

Of course, **proof-of-power wall APIs could theoretically be used to secure people against *any* form of systemic exploitation and abuse of *any* kind of software control signal issued by *any* computer**, not just against email spam specifically. This is because all forms of hacking and software abuse are the result of insufficient control structure designs which do not adequately eliminate or constrain unsafe commands or insecure control actions. Email spam, comment spam, sybil attacks, bots, troll farms, and weaponized misinformation stem from the exact same types of core design flaws which proof-of-power wall APIs could theoretically help to alleviate. At the end of the day, a leading root cause of these attacks is a lack of ability to physically constrain control signals, and that capability is precisely what physical cost function protocols provide.

Each of these types of attacks represents ways to systemically exploit computer programs by taking advantage of high-$\mathbf{BCR}_A$ control signals like sending superfluous emails, posting superfluous comments, casting superfluous votes, and publishing superfluous information. Thus, each of these attacks could possibly be mitigated by adopting proof-of-power wall APIs which make these malicious operations superfluously costly in terms of the number of watts required to send them. In plain language, proof-of-power APIs can make it so that bad guys literally don't have enough power (a.k.a. watts) to launch or sustain their attacks. This "electro-cyber dome" concept is illustrated in Figure 71.

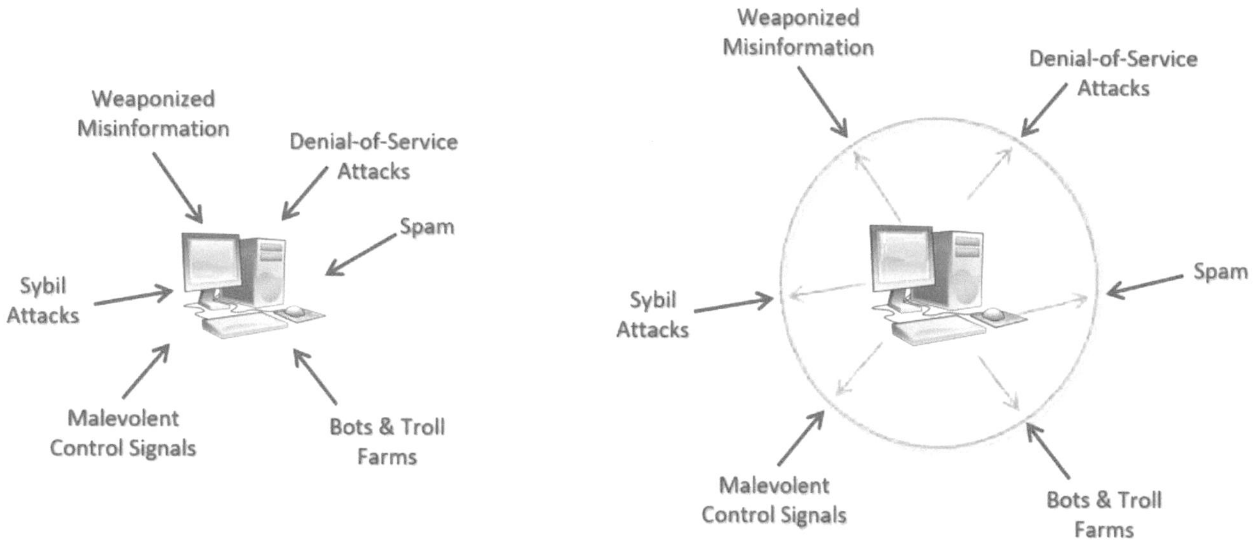

<u>**Current Cyberspace:**</u>
Computers have no ability to impose ***physically prohibitive*** costs on attackers. Users are physically powerless to countervail belligerent activity.

<u>**"Cyber Dome" Concept:**</u>
Computers adopt open-source physical cost functions & proof-of-power wall protocols that can impose ***physically prohibitive*** costs on attackers. Users are physically empowered to countervail belligerent activity.

Figure 71: Illustration of the "Electro-Cyber Dome" Concept using Proof-of-Power Wall APIs [137]

An electro-cyber dome is a passive power projection tactic like the pressurized membrane concept developed by life's first organisms during abiogenesis, as discussed in Chapter 3. This simple barrier works the same way as a cellular wall or other passive power projection system. It utilizes the computational inefficiency of cost function protocols to increase the amount of physical power (watts) needed to successfully attack or penetrate a target's API. A major difference is that electro-cyber domes are non-kinetic physical energy barriers with no mass, built to function in an abstract domain called cyberspace. An electro-cyber dome has no force, mass, or volume and is therefore completely incapable of causing physical injury. Nevertheless, an electro-cyber dome is still capable of projecting physical power and imposing severe physical costs on attackers in the form of electrically-generated watts.

On the flip side, it should be noted that this wouldn't be a strictly "defensive" power projection capability (the author is not even aware of any power projection capability which is strictly "defensive," as even passive power projection tactics like walls are among the most successful offensive strategies ever used, hence all colonized territory). People with access to proof-of-power can theoretically "smash" through these electro-cyber dome defenses if desired. Thus, proof-of-power protocols are not strictly "defense only" protocols as some have argued. A top threat to people using physical cost function protocols like Bitcoin is other people using *the same protocol* (hence why Nakamoto mentions the word "attack" 25 times in an 8-page whitepaper, each time referring to people running *the same protocol*).

Instead of physically constraining attackers via forces displacing masses like normal barriers do, electro-cyber domes physically constrain attackers via electric charges passing across resistors. But as Einstein famously observed, matter is swappable with energy, so kinetic physical constraints can theoretically be swappable with electric physical constraints for certain applications. **Physical power is physical power regardless of whether watts are generated kinetically or electrically. Thus, the physics behind electro-cyber dome security systems is similar to any other physical barrier even though it doesn't look anything like a traditional physical barrier**. Just like any other physical barrier, electro-cyber domes secure people against attackers by increasing the number of watts attackers must expend to succeed. Security is achieved the same way it's always achieved using any other kind of physical barrier. At a certain threshold, attackers simply don't have enough physical power to succeed. The armor is too strong, the moat is too wide, the wall is too tall, or in this case, the amount of electric power needed to send belligerent control signals is simply too high.

Go to practically any popular website today and look at the comments section; you'll probably see many fake comments posted by many fake accounts, upvoted to the top of the page by many fake likes, all produced by computer programs (bots) or a small percentage of the population, often orchestrated by people trying to affect behavior or influence people's perception of reality. Part of the reason why this type of systemic exploitation of cyberspace is possible is because of how computationally efficient computers plugged into the internet have become; there's essentially no physical or marginal cost required to change the states of state machines or to send these control signals.

In the real world, people can't sustain attacks like these because they simply aren't powerful enough. The real world has severe physical limitations and physical opportunity costs. In the current version of cyberspace, there are no such physical limitations. This could change if websites were to adopt electro-cyber domes. They could create a physically constrained area of cyberspace where they would be protected from belligerent actors who routinely exploit cyberspace's lack of physical constraints. It would be possible to make it too physically expensive for belligerent actors to continue their malevolent operations on these websites, while simultaneously keeping costs low enough to preserve honest people's ability to use these services. By applying miniscule, superfluous physical constraints to actions taken online could dramatically reduce the level of systemic exploitation and abuse currently plaguing the internet. Everything from email spam to state-sponsored weaponized misinformation could theoretically be physically constrained using proof-of-power wall APIs and electro-cyber domes.

We know for sure that electro-cyber domes can function successfully as a security protocol because this is what Bitcoin uses to secure itself and its own bits of information against systemic exploitation. Like many things in software, Bitcoin utilizes a recursive system design. It calls its own physical cost function protocol to secure itself against systemic exploitation of itself. By using its own proof-of-power electro-cyber dome design concept to secure itself, Bitcoin has managed to remain operational for thirteen years without being systemically exploited (i.e. hacked). It is not that people don't know how to exploit/hack Bitcoin's ledger, it's that it's impossible for attackers to either justify or overcome the immense physical cost (a.k.a. watts) of exploiting Bitcoin's ledger because it's parked behind an electro-cyber dome. People literally don't have enough watts to gain centralized control authority over the protocol – to include nuclear-armed nation states. This concept is illustrated in Figure 72.

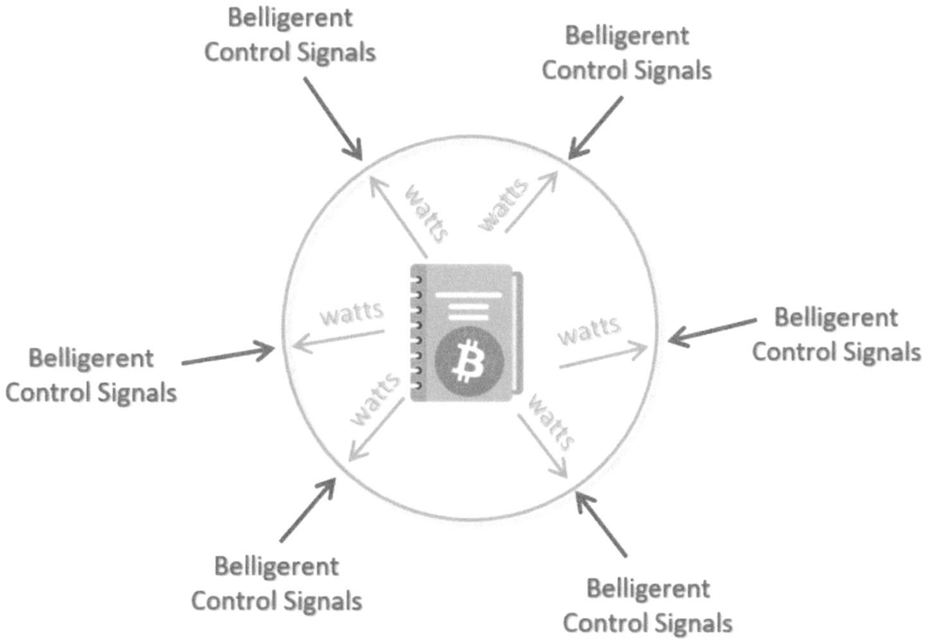

<u>**Bitcoin Proves the "Cyber Dome" Concept Works:**</u>

Bitcoin utilizes Back's open-source physical cost function to impose severe *physical* costs (watts) on attackers. The bits of information stored on its ledger remain secure against systemic exploitation and abuse because it's impossible for attackers to justify the physical cost of executing harmful control signals. In other words, belligerent actors simply aren't powerful enough to overcome the system's defenses.

Figure 72: Bitcoin is a Recursive System Which Secures Itself Behind its Own "Electro-Cyber Dome" [163]

,.9 Novel Computer Theory about Bitcoin

"Satoshi opened a portal from the physical realm into the digital realm, and energy began to flow into cyberspace, bringing life to a formally dead realm consisting of only shadows and ghosts – bringing conservation of energy and matter, objectivity, truth, time, and consequence into the digital realm. Delivering property rights, freedom, and sovereignty...."
Michael Saylor [164]

Following the operational success of Bitcoin, several new computer theories regarding physical cost function protocols (a.k.a. proof-of-work protocols) have emerged. Just as new theories emerging throughout the early 1930s eventually became foundational to mainstream computer science, today's new computer theories regarding physical cost function protocols could one day become foundational to mainstream computer science as well. The goal of this thesis is to present an over-arching grounded theory on the sociotechnical implications of physical cost function protocols and to understand how they could influence cyber security and national strategic security.

The core concepts of this overarching theory are grounded to more technically detailed theories discussed in this section. Some of these ideas were inspired from data collection, but others are completely original. The most important takeaway from this technical discussion is that **Bitcoin (the system) may not only be special because it represents a new type of software, but because it represents a new type of *computer***. The author believes that Bitcoin represents the invention of what he calls a "planetary state mechanism."

5.9.1 Computational Inefficiency has Complex Emergent Benefits that Few Understand

"What we obtain too cheap, we esteem too lightly."
Thomas Paine [165]

As of this writing, the fastest operational computer can transmit data at a rate of 1.84 petabits per second, making it fast enough to transmit the entire world's internet traffic in well under 3 seconds. This type of speed is possible because an unspoken assumption of computer engineering is that we should endeavor to build more physically unconstrained and energy-efficient state mechanisms with larger and larger state spaces, so they can process as many bits of information as quickly and as cheaply as possible. [166]

Here's an interesting question: What's the slowest and most energy-inefficient computer in the world? Why don't we measure that or celebrate that machine? Rarely do engineers discuss the strategic value of endeavoring to build the biggest, slowest, and most energy intensive computers with smaller and more physically restricted state spaces that have extremely difficult-to-change state mechanisms. Generally speaking, computational slowness and energy intensity are not valued.

Few recognize the strategic benefit of creating increasingly more costly bits of information in a domain flooded with increasingly less costly bits of information. Notably few articles are written by engineers celebrating how much power a computer had to consume to change its state or to send a command. There seems to be a universal bias for smaller, faster, and more energy efficient computers because of an unspoken assumption that such devices are *always* better for every possible computing application.

Incidentally, one of the rare groups of people who appear to understand the value of superfluously difficult computing and to celebrate how computationally expensive it is to transfer, receive, and store bits of information are Bitcoin enthusiasts. This is very noteworthy. Bitcoin enthusiasts have emerged as iconoclasts who are not only unapologetic about the Bitcoin computing network's superfluous energy usage, they also celebrate how computationally difficult it is to run the Bitcoin network. Why do these people consider Bitcoin's extreme computational difficulty to be a virtue, while mainstream computer scientists appear to be adopting a stance that Bitcoin's energy intensity is somehow "bad?"

5.9.2 Leveraging the Planet's Global Power Infrastructure as a Physically Decentralized & Thermodynamically Restricted Computer

Recalling the core concepts presented earlier in this chapter, modern computers are stored-program general-purpose state mechanisms. Computers work because engineers learned how to apply symbolic meaning (e.g. Boolean logic) to physical state changes created by miscellaneous state mechanisms by inventing a new language called machine code. One of the main reasons why machine code is so remarkable is that it allows people to convert practically any physical phenomena into bits of information that can be transferred, received, and stored using practically any form of physical state-changing mechanism.

The key takeaway from this deep-dive into computing was to develop a foundational understanding that **thanks to machine code, practically any state-changing phenomena in nature can be utilized as a programmable computer, as well as a medium through which information can be transmitted and received by a computer**. This explains why the physical form of general-purpose stored-program computers has been able to change so dramatically over time. Whereas early computer engineers like Von Neumann worked with massive machines and large plugboards, modern computer engineers work with microscopic floating gate transistors on circuits smaller than a fingerprint. The size, weight, and energy requirements of computers have changed dramatically over the years, as computer engineers discover new ways to manipulate their physical environment. Nevertheless, machine code has remained the same.

The past 80 years of computer engineering could be generalized as a search for the fastest and most energy-efficient state-changing phenomena. The primary goal of computer engineers has been to build cheaper, faster, smaller, and more energy efficient state mechanisms. State mechanisms with larger state spaces are treated as though they are unquestionably better because they have more memory. Likewise, state mechanisms with easier-to-change states are treated as unquestionably better because they are cheaper and faster to operate. Generally speaking, the ability for computers to transmit more data, receive more data, and store more data on increasingly smaller devices is universally considered better. There has been little consideration if it would be useful to reverse optimize the design of a computer.

Computer engineers' obsession with faster, cheaper, and smaller computers is what drove them to turn the fastest-known movable physical phenomena in the universe into a physical state-changing mechanism for use in computers: light. The reason why computer engineers have successfully managed to build computers which can process 1.84 petabits of information per second is because they leverage light itself as the state-changing mechanism to encode, transfer, receive, and store bits. As of this writing, the fastest computer chip works by encoding bits of information into light by applying machine code to modulated amplitude, phase, polarization, and wavelengths of transmitted light.

Just as it is possible to create faster and more energy-efficient state mechanisms by leveraging faster and more energy-efficient phenomena of the natural environment, it should also be possible to create larger, heavier, slower, and more energy-intensive state mechanisms by taking advantage of the larger, heavier, slower, and more energy-intensive physical phenomena of our natural environment. All a computer engineer would need to do is find the appropriate physical state-changing mechanism, and then apply Boolean logic to it.

We have now established two key points that are essential to understanding the computer theory behind protocols like Bitcoin. The first key point is that people with their extraordinarily gifted capacity for abstract thinking can apply the Boolean logic of machine code to practically any physical state-changing phenomena to turn it into a programmable computer. The second key point is that for decades, computer engineers have been operating under an unchallenged assumption that smaller, lighter, cheaper, faster, and more computationally and energy efficient computers are *always* better for computing applications. **There has not been significant research towards exploring the idea that there could be major complex emergent benefits to society if they were to design and build a massive-scale, heavy, slow, expensive, and energy-intensive computer network**.

.9.3 Planetary-Scale Computation is Theoretically Feasible, and Bitcoin could be the Key

A Los Angeles-based think tank called the Berggruen Institute has a program called Antikythera named after the famous Antikythera mechanism (considered by many to be the earliest computer ever discovered). One of the primary goals of The Berggruen Institute's Antikythera program is to become a global leader in speculative philosophy surrounding the sociotechnical implications of planetary-scale computation, to including the geopolitical impacts of planetary computation and what its underlying platform infrastructure might look like. In their view, planetary-scale computing could have extraordinary impacts on human society, including mind-bending implications like planetary sapience (if artificial intelligence were to reach a point where computers became sapient, and we were to combine this technology with a planetary-scale computing platform, then it is theoretically feasible for the planet itself to become sapient and self-aware of its own existence). [167]

Many of the Antikythera program's speculative philosophies about planetary-scale computing are outside of the scope of thesis, but the purpose of mentioning it is to illustrate the point that planetary-scale computing is a legitimate concept that is actively being investigated by organizations that influence public policy. It isn't really a matter of *whether* planetary-scale computing can be done because we already know from computer theory that the Boolean logic in machine code can be applied to practically any physical state-changing phenomenon, to include the planet itself. It's pretty straightforward that planetary-scale computing *could* be done. Instead, the mystery surrounding planetary-scale computing is devoted to figuring out *how* it could be done, particularly what physical state-changing mechanism would be used, and how its operating system would work.

A core hypothesis of this thesis is that **Bitcoin is emerging as the base-layer operating system and infrastructure of a planetary-scale computer and consequently a new form of physically-constrained and thermodynamically-restricted state space that can be added to the internet. If this theory is valid, then global adoption of Bitcoin simultaneously represents the global-adoption of a planetary-scale computer – the exact thing which speculative philosophers like the Berggruen Institute's Antikythera program theorize would have drastic geopolitical implications**.

It turns out, the natural environment *outside* of our computers is filled with circuitry and physical state-changing phenomena that could be utilized for large, heavy, slow, expensive, and energy-intensive computing applications. Our surrounding environment is filled with state-changing phenomena that can encode, transfer, receive, and store machine-readable bits of information. It is therefore theoretically possible to utilize the surrounding environment and supporting infrastructure *outside* of our world-wide web of computers as a singular and superfluously cumbersome planetary-scale computer in and of itself. Theoretically, all we have to do to accomplish this is figure out how to (1) control its physical state and (2) apply Boolean logic to it – which is precisely what Bitcoin *already* appears to do.

Not only would it be possible to utilize the planet itself as a large, heavy, slow, and energy-intensive computer, the infrastructure and circuitry required to accomplish this has already been built – we call it the global electronic power grid. Herein lies a simple, but profound idea: **to create the world's largest, heaviest, most energy-intensive, and most physically-difficult-to-operate computer ever built, we could simply utilize our planet's energy resources as the controllable physical-state-changing mechanism of a planetary-scale computer, where the globally-distributed electronic power grid serves as its circuit board**. This design concept is illustrated in Figure 73.

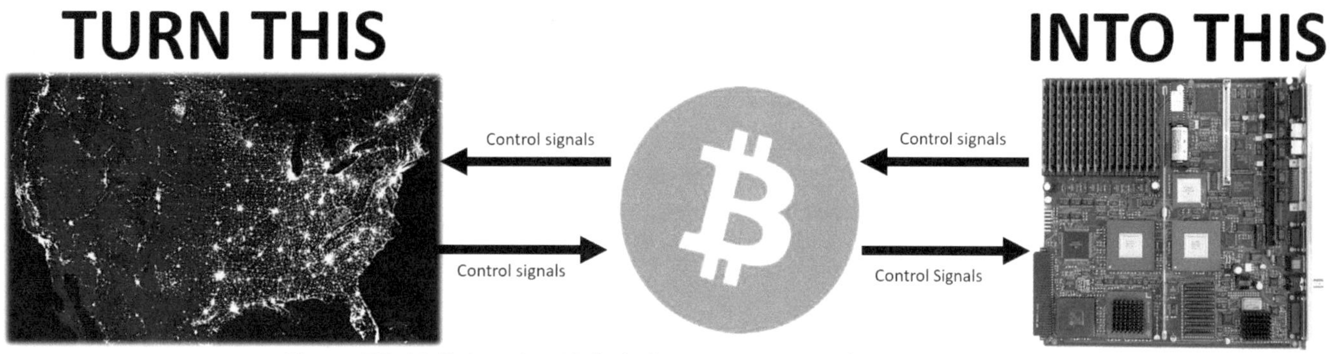

Figure 73: Utilizing the Global Electric Power Grid as a Computer
[168, 169]

Bitcoin demonstrates that it's possible to create a computer program that is so computationally and energy-intensive that it can lead to material changes to the state of the surrounding physical environment *outside* of the computers running it. Computer programmers could theoretically apply Boolean logic to these environmental state changes, converting them into machine-readable bits of information that can be fed back into ordinary computers. These ideas are valid according to modern theories of computation.

By turning the surrounding environment *outside* of our ordinary computers into a state-changing mechanism and then applying machine-readable logic to those state changes, a computer network like cyberspace would be able to send bits of information and control signals into its surrounding environment (i.e. the planet itself) and receive control signals back from its surrounding environment. This technique would make it possible to convert the planet itself into a massive-scale and globally decentralized planetary computer by utilizing (1) the planet's energy resources as the state-changing mechanism and (2) the global electric power grid as its motherboard. This would qualify as a planetary-scale computer that could be plugged into the existing internet, effectively adding the state of the planet's globally decentralized and thermodynamically restricted energy resources to the state space of cyberspace through the existing infrastructure of our global electric power grid. From a first principles perspective, it's a surprisingly simple concept. Cyberspace is, after all, nothing more than the state space of all the physical state-changing mechanisms plugged into it, and the global electric power grid is nothing more than a bunch of machinery that changes its physical state. This concept is illustrated in Figure 74.

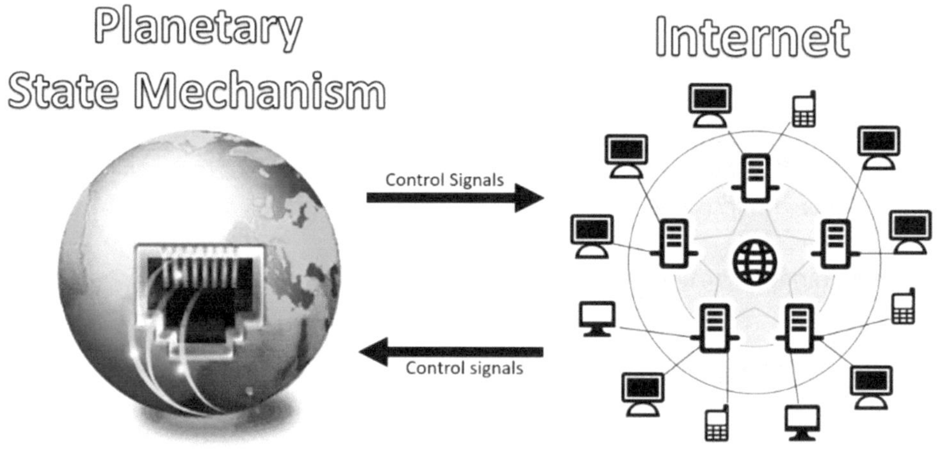

Design Concept: Utilize the planet's energy resources and global electric power grid as a state mechanism with a physically decentralized & thermodynamically restricted state space that can be added to the state space of the existing internet, effectively turning the planet itself into a computer with emergent properties unlike any other computer on the internet.

Figure 74: Planetary-Scale Computer Design Concept
[170, 171]

But what could possibly be the benefit of turning the electronic circuitry *outside* of our internet into a programmable computer in and of itself and then plugging it back into the internet? The benefit is that it would create a physically decentralized and thermodynamically restricted section of the internet over which agrarian society would be able to compete for control authority in a zero-trust, permissionless, and egalitarian way using a physical power competition. As a complex emergent benefit of that physical power competition, the global population would be able to gain and maintain decentralized control authority over this special part of the internet.

This is where the core concepts discussed in the previous chapter become critical to understand. With the invention of this type of planetary-scale computer, agrarian society would be able to use their own energy resources to battle for control authority over this planetary-scale computer the same way all organisms in nature (including and especially modern agrarian sapiens) have always used physical power competitions to battle for control authority over resources. And as a complex emergent benefit of this ongoing electric power competition, agrarian society would be able to gain and maintain decentralized control authority over this new digital-age resource. As established in section 4.9, one of the primary value-delivered functions of physical power competitions (a.k.a. warfare) is physical decentralization of control over Earth's natural resources. There's no reason to believe this wouldn't also apply to power competitions for control over the natural resources comprising a planetary-scale computer.

A planetary-scale computer like this could theoretically create a portion of the internet where no single person or organization has full control over it, thus they have no ability to fully control the bits of information transferred, received, and stored on it. This decentralized portion of the internet could be reverse-optimized to be more expensive and energy-intensive to send, receive, and store every bit, giving it complex emergent behavior that no other computer connected to the internet would be physically capable of replicating. These features could become extraordinarily useful for digital-age societies which find themselves increasingly trapped under software-instantiated abstract power hierarchies crowned by a technocratic ruling class of digital-age god-kings who wield asymmetric abstract power and control authority over the existing computers connected to the internet. Right now, these rulers must be trusted not to exploit people through cyberspace because the population has practically no infrastructure in place to physically constrain software or impose severe physical costs on people, programs, or institutions operating in, from, and through cyberspace. Digital-age society also has no ability to use physical power as the basis for settling online disputes or achieving online consensus on the legitimate state of ownership and chain of custody of property. All of this could theoretically change if society invented a planetary-scale computer that effectively converted the global electric power grid into part of the internet.

In short, it's possible to think of several reasons why people would benefit from a physically decentralized and thermodynamically restricted state space of the internet created by a planetary-scale state mechanism which converts our electric power grid into the motherboard of a planetary-scale computer.

5.9.4 A Planetary-Scale Computer could Physically Constrain other Computers for Improved Security

"Bitcoin enables conservation of energy in time and space
because it enforces conservation of energy in cyberspace."
Michael Saylor [172]

The surrounding physical environment outside of our computers would make a very different kind of state mechanism in comparison to the ones inside our offices and our pockets. It would have very physically-expensive-to-change states. It would require so many watts to change states that some populations would literally not have enough power to change the state of this mechanism. This would translate into computer commands that are not practically possible to execute, or bits of information that are not practically possible to replicate, alter, or exploit. In other words, it would create a mechanism that computers could use to physically enforce and physically secure computer programs.

Open and egalitarian access to this type of special-purpose planetary computer could give digital-age society the ability to physically enforce their policies, physically constrain sensitive computer control signals, and physically secure valuable bits of information in a way that was previously not considered possible. Populations would be able to use it to impose severe physical costs on other people, programs, institutions, and even entire nations who try to systematically exploit them through their world-wide web of computers. This capability could only get more valuable as more people, institutions, and nations coordinate international operations via cyberspace, because they would be more vulnerable to the threat of those who keep trying to systemically exploit the internet's existing architecture.

Physically-constrained computers translate directly into physically-constrained computer programs as well as physically-constrained computer programmers. In other words, it's possible to physically constrain people in, from, and through cyberspace – it doesn't have to be done kinetically in three-dimensional space using forces to displace masses. Therefore, **a thermodynamically restricted, planetary-scale state mechanism would likely serve as a key-enabling technology for physical power projection in, from, and through cyberspace**. As discussed previously, the ability to project physical power to physically constrain computer programs and computer programmers in, from, and through cyberspace is notably missing in the design schematics of today's internet architecture. **If people were to build a planetary-scale computer like the one described here and plug it into the internet, this could not only usher in new approaches to cyber security, but it could also dramatically change humanity's approach to national strategic security, particularly how it secures information and enforces policies and social contracts against exploitation**.

As previously discussed, one of the most challenging aspects of software engineering and cyber security is the inability to apply physical constraints to unsafe control actions or to impose physical costs on belligerent actors. It would theoretically be possible to utilize an open-source, internet-accessible, planetary-scale computer to perform these functions. The key to doing this effectively would be finding a way to "chain down" or couple ordinary state mechanisms to this new planetary-scale state mechanism. In this design, the planetary-scale computer with superfluously difficult-to-change states would apply severe physical constraints and thermodynamic restrictions on the normal computers connected to it via cyberspace, as well as the programs running on them. This design concept is illustrated in Figure 75.

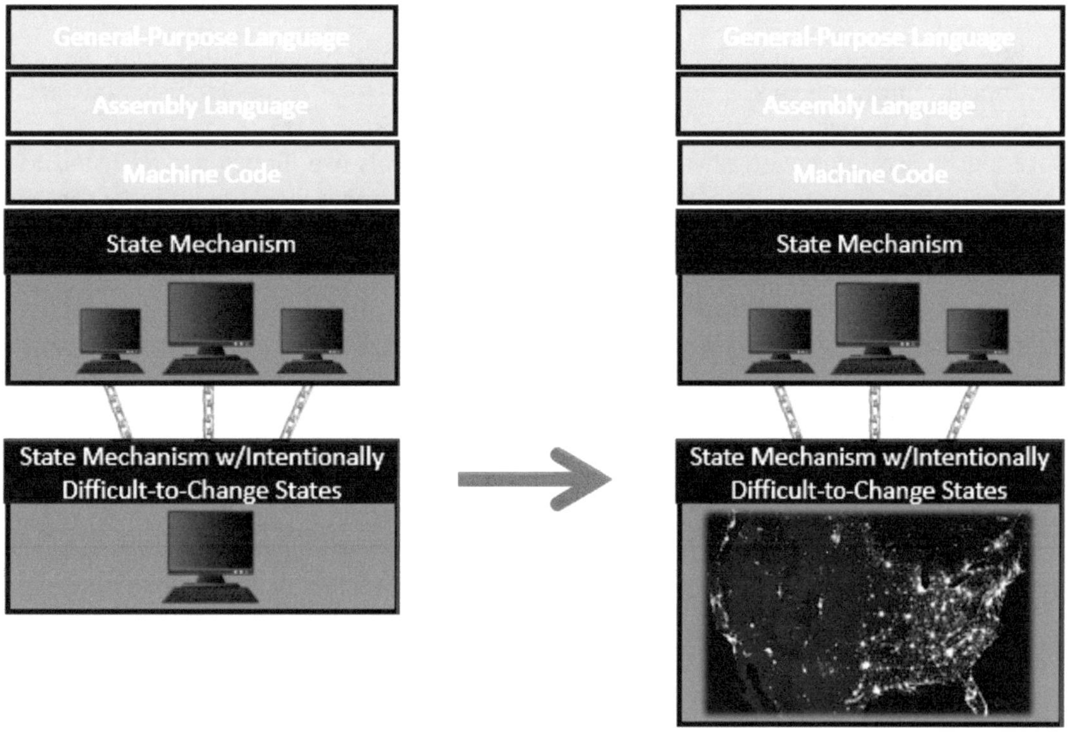

Figure 75: Using a Planetary Computer to "Chain Down" Regular Computers
[169, 154, 156]

).5 A Planetary-Scale Computer could Dramatically Improve Society's Global Electric Power Grid

There's another major benefit of utilizing a planetary-scale state mechanism as a large, heavy, slow, and energy-intensive computer. If society were to create a planetary-scale computer which utilized global energy resources as its state-changing mechanism (akin to transistors), as well as the global electric power grid as its circuitry (akin to a motherboard), then society could theoretically unlock the benefits of Moore's law for our energy infrastructure. Engineers love making their systems more efficient, particularly their computer systems. As previously mentioned, computer engineers are obsessed with optimizing their computer systems to make them faster, cheaper and more energy efficient. This phenomenon is so predictable that we colloquially refer to it as Moore's law.

If society created a planetary-scale computer, then populations would naturally seek to improve the speed and efficiency of this computer's circuitry. Societies might search for cheaper and more abundant sources of energy with more optimized infrastructure designs so they could make their portion of the planetary-scale computer faster and more inefficient. There would also be an incentive to do this to gain a competitive edge in the form of more control authority over the planetary computer as a whole.

These dynamics would create a counterintuitively beneficial feedback loop for society. The wide-scale adoption of a deliberately large, heavy, slow, and energy-intensive planetary-scale state mechanism would motivate society to try to continuously improve it by creating faster, more efficient, and more optimized electric power infrastructure. In other words, **the complex emergent benefit of turning the global electric power grid into the circuit board of a planetary-scale computer would be to unlock Moore's law for the global electric power grid. By simply turning the power grid into a computer chip, society would be able to utilize its insatiable drive for faster and more efficient computer chips to create a faster, more responsive, and more efficient power grids**.

5.9.6 Software could Utilize the Environment Outside of Our Computers as a Computer in and of Itself

With the potential benefits of a planetary-scale computing fresh in our minds, the reader is invited to re-examine physical cost function protocols like the one invented by Adam Back to further understand the author's assertion that Bitcoin could function as the operating system of a planetary-scale computer.

There is a subtle but extremely important difference to emphasize between regular computer programs and physical cost function protocols (a.k.a. proof-of-work protocols). Regular computer programs are designed to produce a single physical effect: a materially-inconsequential transistor state change inside the circuitry of an ordinary state mechanism. However, physical cost function protocols like Bitcoin are designed to produce an additional physical effect. Not only do they create transistor state changes inside of ordinary computers, they're also explicitly designed to create materially-consequential state changes in the surrounding physical environment *outside* of our computers. This design concept is illustrated in Figure 76.

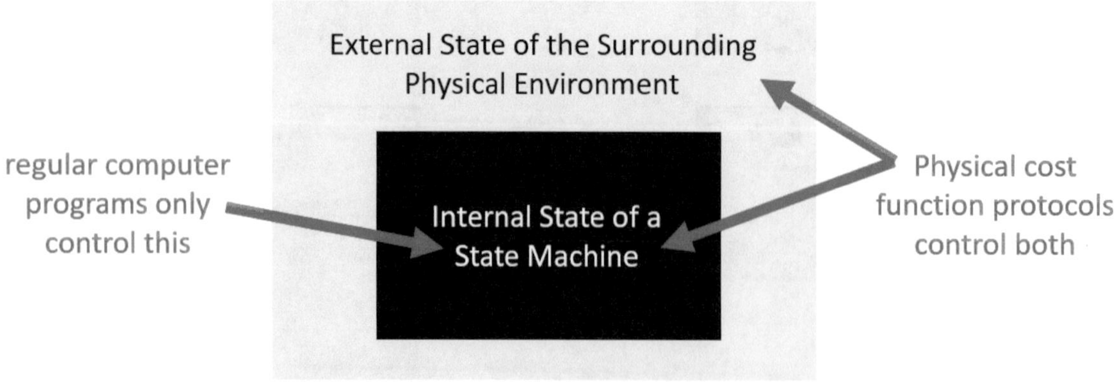

Figure 76: Difference between Regular Computer Programs and Physical Cost Function Protocols

It's incontrovertibly true that physical cost function protocols like Bitcoin are so energy intensive that they can directly influence the shape and structure of the surrounding electric power grid – a feature that has made Bitcoin infamous as a superfluously inefficient protocol by people who don't understand the complex emergent benefits of this design concept. Therein lies the key difference between normal software and Bitcoin that many people overlook. It's the fact that Bitcoin creates such a substantial physical change to the state of the environment outside of the computer on which it runs which makes Bitcoin so extraordinarily different and unique compared to all other types of software. **People mistakenly believe that Bitcoin's substantial energy intensity is a bug, but it's actually its primary value-delivered feature**.

The physical state changes of the surrounding environment produced by physical cost functions manifest themselves as deficits or vacuums of electric power which must be filled by electric power producers connected to the global electric power grid. This quantity of electric power represents a continuous-time signal that can be transferred, received, modulated, digitally sampled, and stored as digital information just like any other power-modulated electromagnetic signal. The primary differences are simply power intensity and production cost.

Just like the world's fastest computer chips transmit, receive, modulate, sample, and store electromagnetic signals (in the form of electromagnetic light waves) with low power intensity and low production cost, physical cost function protocols convert the electric power grid into a giant, global-scale circuit board which can transmit, receive, modulate, sample, and store electromagnetic signals (in the form of large quantities of electric power) with comparatively giant power intensity and production cost. From a first principles perspective, it's the exact same infrastructure (circuitry), the exact same physical phenomena (electromagnetic signals), and even the exact same design concept (converting modulated electromagnetic signals into bits of information), but reverse-optimized. **Instead of using increasingly smaller circuitry and smaller amounts of electromagnetic power like normal computer networks, Bitcoin is intentionally reverse-optimized to use increases larger circuitry (e.g. the power grid) and larger amounts of electromagnetic power**.

Using systems thinking, we can see that physical cost function protocols like Bitcoin function as a signal generator and modulator which send digitizable signals to a planetary-scale computer comprised of the global electric power grid. At the same time, physical cost function protocols also function as a signal processor which digitally samples signals received from the planetary-scale computer. The signals sent to the planetary computer take the form of large electric power deficits created by the computer network running the cost function protocol (i.e. the computers running Bitcoin), which can be modulated as desired by simply changing the difficulty of the solving the cost function's hashing algorithm.

The signals received from the planetary computer take the form of a digitally sampled proxy measurement of the quantity of electric power consumed by the network. For physical cost function protocols like Bitcoin, computational power is sampled by measuring the average hash rate over a specified time interval. In other words, instead of sampling the modulated electric power drawn from the electricity grid, Bitcoin samples how many calculations are being made by the computer network running it as a directly proportional proxy measurement. This design concept is illustrated in Figure 77.

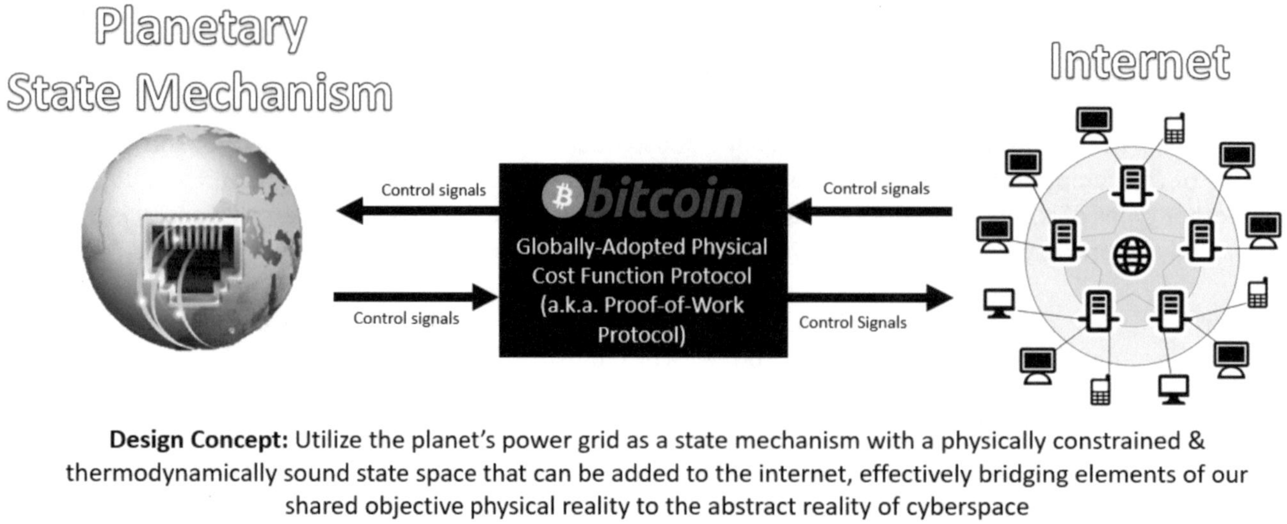
Planetary
State Mechanism
Internet
Control signals
Control signals
bitcoin
Globally-Adopted Physical
Cost Function Protocol
(a.k.a. Proof-of-Work
Protocol)
Control signals
Control Signals
Design Concept: Utilize the planet's power grid as a state mechanism with a physically constrained & thermodynamically sound state space that can be added to the internet, effectively bridging elements of our shared objective physical reality to the abstract reality of cyberspace

Figure 77: Bitcoin is a Physical Cost Function Protocol which Utilizes the Planet as a Computer [170, 171, 173]

Physical cost function protocols utilize the planet itself as a programmable computer, where the globally-decentralized power grid functions as its motherboard through which it sends, stores, and receives digitized information. These protocols can adjust the difficulty of their hashing algorithms to produce a modulated continuous-time control signal in the form of a modulated quantity of electric power drawn from the surrounding physical environment. These protocols appear to be able to digitally sample the power signal they receive from the environment and turn them into a machine-readable, discrete-time signal.

These digitized signals are broadcasted to regular computers connected to the internet as regular bits of information which can then be abstracted by computer programmers as anything they desire – just like all bits of information sent across the internet can be abstracted as anything the imagination is capable of thinking of. The bits can be described as a transferable object, like a proof-of-power receipt, with countless names like digital energy, bit-watts, or bit-power. They could also be called stamps, tokens, spoons, coins, bit-stamps, bit-tokens, bit-spoons, or perhaps even bit-coins. As always, what people call the bits of information passed across computers is semantic and arbitrary; it (literally) doesn't matter because bits are not matter. What makes the bits produced by physical cost functions matter is not their name, but the matter used to create them. In other words, **what is happening under the hood of physical cost function protocols at the state mechanism level is what matters because it's materially different than what happens under the hood with other software at the state mechanism level**. This design concept is illustrated in Figure 78.

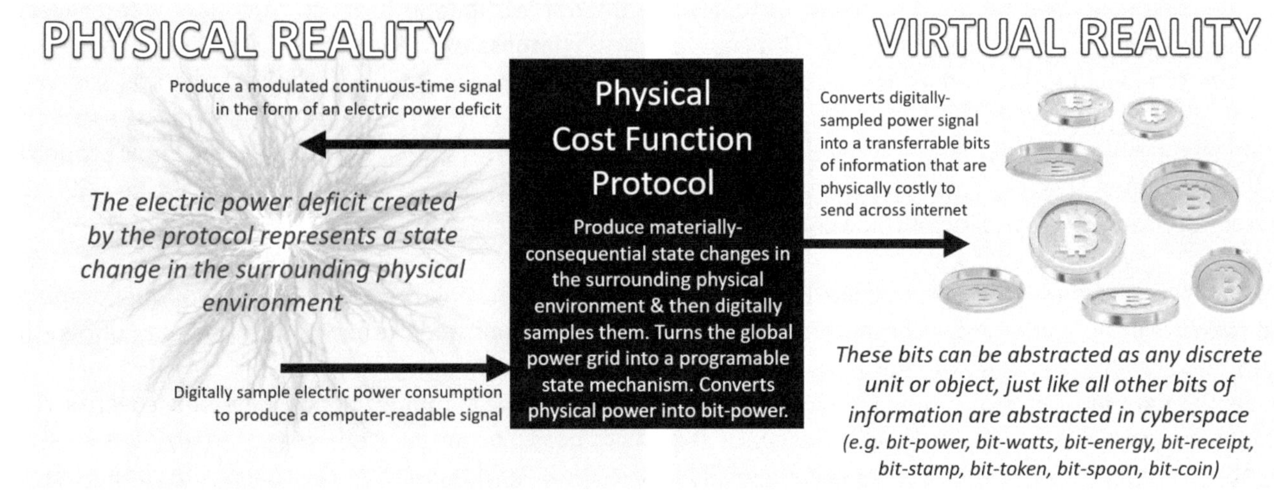

Figure 78: Illustration of how Physical Cost Functions Convert Quantities of Power into Bits [158, 174]

There are very real, very meaningful systemic differences between the bits of information created, transferred, and stored by normal computers and the bits of information created, transferred, and stored by a planetary computer. The primary difference is that these two very different kinds of bits of information have been reverse-optimized. Whereas bits of information from a normal computer are optimized to be as cheap and easy as possible to produce, transfer, and store, the bits of information from a planetary-scale computer are optimized to be as expensive and difficult as possible to produce, transfer, and store. It's the exact same technology and the exact same digitization process, but reverse-optimized. Here we can begin to appreciate what could eventually become recognized as a primary implication of physical cost function protocols like Bitcoin.

Bitcoin appears to set a precedent within the field of computer science by being the first globally-adopted software to utilize the planet itself as a state mechanism (i.e. computer) for the explicit purpose of making bits of information increasingly *more* difficult to create, transmit, receive, and store. Physical cost functions protocols accomplish this by creating state-changes in the surrounding physical environment *outside* of the microchips contained within ordinary computers. They apply Boolean logic to massive quantities of electric power drawn from the local environment, treating the globally distributed electric power grid as if it were a state-changing circuit board.

By applying Boolean logic to the power grid, people convert large and expensive quantities of physical power into bits of information and feed that information back into our regular computers via the internet. People then use that information to affect state changes inside their ordinary computers. The result of this activity is a capability which does not appear to have existed before: the ability to impose severe physical costs and thermodynamic constraints on people, programs, and computer programmers operating in, from, and through cyberspace in a zero-trust, egalitarian, and permissionless way that no person or organization can fully control.

People don't appreciate how special physical cost function protocols like Bitcoin are because they are making the tacit assumption that all bits of information are the same. They think the bits produced by Bitcoin are the same as the bits produced by any other type of software. From a technical perspective, it's true that the bits produced by bitcoin are generally the same as the bits produced by other software. In both cases, people are simply applying Boolean logic to a modulated electric signal. However, people aren't paying attention to what's happening under the hood at the foundational layer of the tech stack (the state mechanism itself), so they aren't recognizing that a completely different physical phenomenon is generating Bitcoin's bits.

Without paying attention to what's happening under the hood at the state mechanism level, it's hard to see how the bits produced by proof-of-work protocols like Bitcoin are unique because they have been reverse-optimized, giving them unique properties that are physically impossible to replicate by ordinary computers. Because people don't look under the hood, they overlook the fact that Bitcoin converts large and expensive quantities of physical power into bits of information. Physical power is converted into bitpower, giving people the ability to wield *real* power in the *abstract* domain of cyberspace while retaining the beneficial systemic properties of physical power (e.g. cost, scarcity, decentralization, and thermodynamic restrictions) that can't be replicated by transistors in ordinary state machines.

With physical cost function protocols like Bitcoin, the boundary separating the circuitry inside of our computers and the circuitry outside of our computers disappears. Our microchips and our power grid become one-in-the-same technology but headed towards opposite sides of the optimization spectrum. Gradually less materially consequential transistor states on microchips represent the side of the computational engineering spectrum that is being optimized to be as small, light, fast, and efficient as possible. Gradually more materially consequential quantities of electric power drawn out of the global electric power grid represents the side of the computational engineering spectrum that is being optimized to be as large, heavy, slow, and energy intensive as possible. The result? Very different emergent behavior.

With proof-of-power protocols like Bitcoin, two very different worlds are bridged together into one combined system. Our planet's largest energy producers become equivalent to state-changing transistors on a giant, global-scale, trans-national, general-purpose computer over which no person or organization can afford to have complete and unimpeachable control. Consequently, agrarian society itself becomes a living, breathing part of the internet's hardware. As a result of this bridge, the abstract things we encounter in cyberspace no longer exclusively represent a complex emergent effect created by cheap and materially inconsequential transistor state changes on a microchip. They also represent expensive, materially consequential state changes in our surrounding physical environment that we can independently verify by simply observing the state of our civilian infrastructure.

Physical cost function protocols (a.k.a. proof-of-work protocols) like Bitcoin bridge our surrounding physical environment to cyberspace by effectively converting our surrounding physical environment into the state space of a computer and weaving it into the internet. In other words, **Bitcoin converts our three-dimensional space into a subspace of cyberspace**. For the first time, agrarian society has zero-trust and egalitarian access to physically constrained and thermodynamically restricted bits of information because the underlying state mechanism is physically constrained and thermodynamically restricted. We can pass these bits of information across cyberspace and use them however we want to use them to give our legacy cyber systems complex emergent behavior (e.g. physical scarcity, decentralization, and zero-trust control) that is otherwise physically impossible for computers to replicate in cyberspace.

Physical cost function protocols like Bitcoin therefore appear to function as both an operating system for a planetary-scale computer, as well as an application programmable interface (API) between our computers and the planetary-scale computer (a.k.a. power grid) into which our computers are plugged. By creating this special planetary-scale computer, people appear to be adding a new, unique, physically constrained and thermodynamically restricted state space to cyberspace that has never been seen before, and it's leaving people confounded. They don't know how to describe what they're seeing because they've never seen anything like it.

Bitcoin appears to be bridging our shared objective physical reality to virtual reality by turning the planet itself into a computer which can be programmed and plugged back into the internet. This could theoretically allow people to import the physically constraining and thermodynamically restrictive properties of our surrounding physical environment into cyberspace so that we can create real-world, materially consequential effects in, from, and through cyberspace. As engineers like Saylor (MIT '87) have observed, proof-of-work protocols like Bitcoin appear to have opened a portal between the physical realm and the digital realm. In so doing, these protocols appear to be importing properties from one reality to the other in a manner that people apparently didn't think was possible.

Yet another potential benefit to be gained from turning the global electric power grid into a planetary-scale computer and then plugging it into the internet is that it would effectively create a bridge between the virtual reality of cyberspace and the real-world objective reality of our surrounding physical environment. The state changes of a planetary-scale state mechanism would represent materially-consequential changes in shared objective physical reality, as opposed to materially-inconsequential and virtually undetectable state changes in microscopic transistors. By utilizing large quantities of energy drawn from the environment as a state-changing mechanism and running computer programs on it, this would cause software to have a highly path-dependent and irreproducible "realness" to it that ordinary computer programs running on basic computers simply cannot replicate.

By learning how to utilize the planet itself as a programmable state mechanism, abstract concepts in computer programming are no longer strictly abstract. Computer programs would be capable of changing the physical state of the surrounding environment in substantial ways, giving software emergent behavior that is physically impossible to replicate by ordinary computer programs running on ordinary computers. The programs we run in cyberspace wouldn't just manifest themselves as images on a computer screen, they would begin to include major physical changes to the state of our surrounding three-dimensional environment – namely our global electric power grid architecture (this has already started to happen with Bitcoin). Instead of the images on our screens changing as we run our software, the infrastructure of agrarian society would change as we run our planetary-scale computer software. In essence, the software running of a planetary-scale computer would have the ability to program society itself.

From a metacognitive perspective, turning the environment into a programmable computer would be a repeat form of bi-directional abstract thinking and symbolism discussed in the previous chapter. The symbolic meaning (Boolean logic) that sapiens apply to physical changes in shared objective physical reality would influence their shared abstract reality, and vice versa. By utilizing a planetary-scale computer that converts our global civilian infrastructure into a shared state-changing mechanism, the emergent behavior of the software running on that planetary-scale computer wouldn't be strictly virtual. The effects of planetary-scale computer software would correspond to real, meaningful physical changes of the observable universe in front of our eyes, in addition to changes in the imaginary world behind our eyes in the virtual domain of cyberspace. Once again, our abstract belief systems – our virtual reality – would change our behavior in physical reality. Except this time, our virtual reality would be the bits of information passing across the state space of cyberspace corresponding to the planetary-scale computer we created. This concept is illustrated in Figure 79.

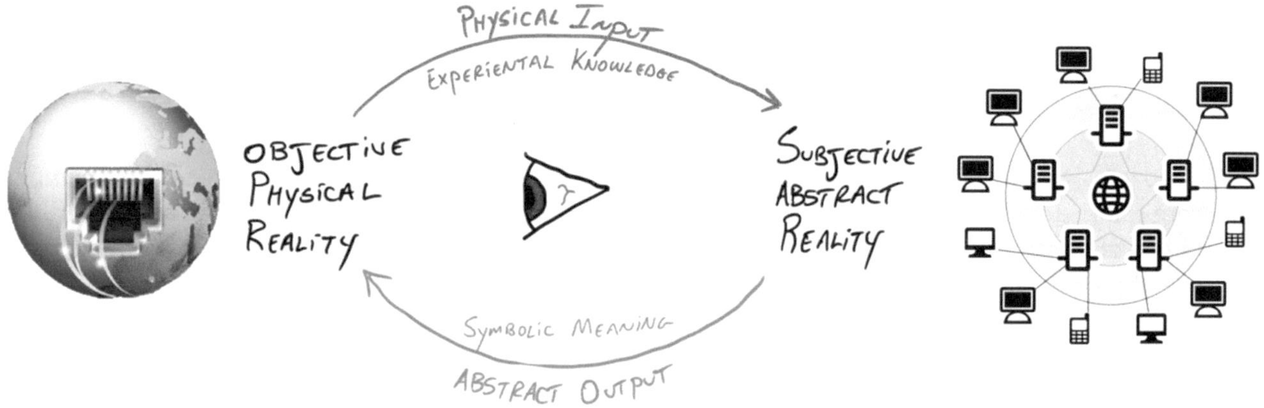

Figure 79: Bi-Directional Abstract Thought & Symbolism Applied to a Planetary Computer
[170, 171]

The idea of using planetary-scale computing software to program how humans behave and structure their surrounding environment sounds remarkable, but when we take the time to reflect upon human metacognition and the concepts introduced in the previous chapter, we can see that this is actually common human behavior. Human society already operates under large-scale consensual hallucinations. Society's abstract belief systems already dramatically influence social behavior as well as the shape and structure of our physical environment. The widescale adoption of these shared abstract belief systems already explains why societies organize into imaginary abstract power hierarchies like nation-states and then physically compete in wars. Nation states are nothing more than shared virtual reality; they don't exist anywhere except within our collective imaginations. Therefore, humans have already adopted the habit of letting virtual reality govern their behavior on a global scale and influence how they structure civilization. Software running on a planetary-scale computer would essentially represent the exact same phenomenon: a common belief system that influences the shape, structure, and behavior of society itself.

5.9.8 Proof-of-Power Protocols are Proof-of-Real Protocols

A planetary-scale computer could function as an "internet grounding device" which grounds elements of our shared virtual reality (cyberspace) to physical reality (three-dimensional space). Large quantities of physical power represent something real, path-dependent, and materially substantial. When large quantities of physical power are used to transfer digitized information across the internet, then proof of that power expenditure doubles as proof of materially irreproducible state changes in shared objective physical reality. As discussed in the last chapter, *proof-of-power* protocols represent *proof-of-real*.

After having reflected in the previous chapter about human metacognition and how sapiens detect if something is real (section 4.3), the concept of using proof-of-power to represent proof-of-real in cyberspace makes perfect sense. Sapiens already use proof-of-power protocols like poking, pinching, or touching to prove or verify realness. Proof-of-real protocols are necessary because sapient neocortices are so effortlessly gifted at abstract thinking that it can be challenging to distinguish between real-world things and imaginary things. If a person ever finds themselves in a situation where they need to verify if something they encounter is real, the common protocol is to reach out and try to poke the thing.

A key insight into the poking/pinching protocol is that it is fundamentally a proof-of-power protocol. When someone reaches out to touch an object to verify if it's a real thing rather than an imaginary thing, they are technically searching for a physical power signal – they're looking for the presence of watts. To poke something is to use energy to apply a force to displace mass across space and time. When someone pokes an object, they project power (a.k.a. watts). If the thing they're trying to poke moves or exerts an equal and opposite amount of force back at its observer, then the observer gains a cross-referenceable sensory input (a.k.a. synaptic feedback) needed to tighten their consensus that the thing they see does indeed physically exist and is not just a figment of their imagination (hence why companies are working on synaptic feedback systems to make virtual reality experiences "feel" more real).

Why is a power signal so important for making online experiences "feel" more real? Because power is comprised of the core components of physically objective reality: time, space, matter, and energy. **If a person can verify that physical power is present, then they can be much more confident that what they see is from the domain of physically objective reality, because they have verified to the best of their ability that the core components of objective reality are verifiably present**. The object moved when it was poked, or the object pushed back. Therefore, the object must be real.

How humans use proof-of-power protocols like poking and pinching illustrates a glaring capability gap for humans operating in, from, and through cyberspace. Cyberspace is a virtual reality; virtual reality is, by definition, imaginary reality. When sapiens currently operate in cyberspace, everything they see represents nothing more than abstract meaning assigned to the state space of all the state machines connected across the internet. Connect billions of state machines together, and you get an incalculable number of states and state changes. Sum together the abstract meaning of that incalculable number of state changes and you get the infinitely-expanding imaginary reality of cyberspace. This domain is getting infinitely larger, faster, and cheaper to build. As this happens, cyberspace becomes infinitely more marginalized.

When people operate online, their neocortices are deprived of cross-referenceable physical sensory inputs needed to verify realness. They have no way to generate a proof-of-power signal on their own, thus no way to produce a cross-referenceable sensory input to use as a proof-of-real signal, as illustrated in Figure 80 (synaptic feedback systems don't help with this because they're equally as artificially generated as the symbols on a screen or the sounds from computer speakers). Without being able to manually generate a proof-of-power signal by users on their own in a zero-trust way that can be independently validated, nothing in cyberspace is capable of being validated as physically real or consequential. Cyberspace becomes strictly an imaginary dream state, nothing but abstract meaning applied to bits of information stored as inconsequential state changes inside an infinitely-expanding state space. The abstractions pile on top of each other to the point where nothing observable in cyberspace can be unique or special – they are just programmed to present the *illusion* of being unique or special.

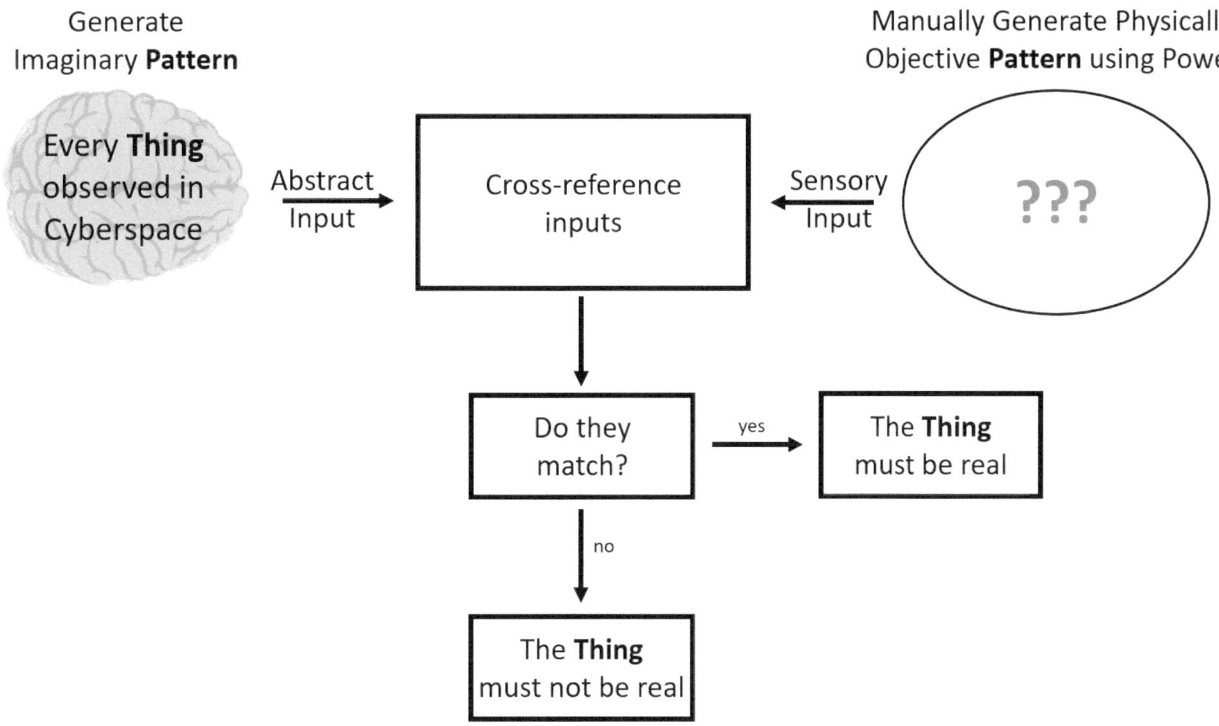

Figure 80: A Capability Gap for Brains Operating in the Abstract Reality Known as Cyberspace
[76]

Sapiens arbitrarily chose to describe the internet as if it were a three-dimensional orientation of objects because that's the easiest and most intuitive way for us to explain the complex emergent behavior of something which otherwise takes an enormous amount of metacognitive effort to understand. The limitless potential of this dream space we are now starting to call "the Metaverse" is an attractive place for some, but for others, it's superficial and fake. As cyberspace expands, the meaning assigned to this place becomes increasingly ethereal and diluted. People come to realize that, ironically, despite the limitless potential cyberspace, it remains extraordinarily limited. It is capable of producing anything our imaginations are capable of thinking of, with one glaringly important exception: something *real*.

Our computers can produce symbols, sounds, synaptic feedback, and complex emergent behaviors. Through false positive correlations and abstract thinking, people hypostatize the sights and sounds produced by these machines as if they were concretely real things. This learned habit of hypostatization doesn't mean that anything we see in cyberspace is *physically* real. What humans see online is nothing but synthetically-generated images on an array of light-emitting diodes they put into their offices or carried around in their pockets. What humans hear online is synthetically-generated air pressure waves produced by plastic cones (i.e. speakers). Even our newly-developed synaptic feedback systems are synthetically generated by some computer programmer who has control over that synaptic feedback system.

Our computers never stopped being general-purpose state machines which produce artificial symbols and sounds and complex emergent behaviors via state changes. **Without proof-of-real protocols like Bitcoin, these state changes are materially inconsequential, trivial to reproduce, thermodynamically unsound, and carefully designed by software engineers to present the illusion of something real, but isn't actually real**.

We see and hear the same things online because our computers have been programmed to role-play the same things back to us. These computers are connected together and highly synchronized, producing artificial sensory inputs which feed our brains the same programmed sights and sounds. Once we receive these artificially-produced signals, our neocortices apply the same abstract meaning to the same objectively meaningless and materially inconsequential patterns produced by these machines, creating a new form of shared abstract reality that we can see and explore together, a virtual reality we have named cyberspace. And behind all of these illusions are computer programmers who control them – computer programmers we must trust not to exploit us with their ability to conjure these illusions.

The cyber stories we see online are entertaining and informative, but if we examine our behavior strictly from the perspective of shared objective physical reality, then we can see that all we are doing is behaving like an audience of people watching electromechanical marionettes acting out the same fictional performance for us. As we increasingly adopt the habit of ignoring the people who are pulling the strings of these marionettes, it becomes easier for us to believe that what we are seeing is something other than a well-orchestrated performance that was explicitly designed by someone in control over the illusion (basically a digital-age version of Plato's Allegory of the Cave). This makes us gradually less capable of seeing that we can – and probably are – being oppressed by those who control these illusions. Computer programmers and software system administrators are harvesting our data; they are experimenting on us to learn what drives our behavior, and they are using this knowledge to further their own agendas.

It seems that we are beginning to lose sight of the lesson that the boy monk from *The Matrix* gave to Neo: *there is no spoon*. Operating in, from, and through cyberspace is akin to operating in a dream state with no time, energy, space, or mass (or in Saylor's words, it consists only of shadows and ghosts). **Behind each computer-generated illusion is a person or organization who controls that illusion, a person or organization we must implicitly trust not to conjure illusions which manipulate us for their own benefit. This trust is necessary because without proof-of-power (a.k.a. proof-of-work) protocols like Bitcoin, we neither have the ability to detect what's real, nor do we have the ability to physically constrain those who might seek to manipulate us through the illusions they can conjure on our computer screens**.

The simple truth of the matter is that the world we see when we operate in, from, and through the current version of cyberspace exists in the human imagination only. The videos people watch, the text people read, the comments they see, the ads and images they encounter, even the results they get searching for something, are all carefully and deliberately programmed by computer programmers who wield unprecedented amounts of abstract power and control authority over people by virtue of their control over people's computers. Almost nothing about what people experience on the internet is random or even serendipitous anymore. All of it is being meticulously programmed by people who enjoy physically unrestricted control over billions of people's computers. That notification you got in the morning for a smoothie coupon was not serendipitous; your apps continually harvest your phone's data and sell it to advertisers who know where you live, where you work, the routes you take, and the fact that you're going to be passing by that smoothie place you like in approximately 15 minutes. You are being manipulated.

As people operate increasingly more in cyberspace, they find themselves with a unique problem: everyone online is participating in a shared abstract illusion where people have no way to verify if anything they're experiencing is linked to anything objectively real or even materially substantial. **Practically all of the most popular software we rely on is attempting to influence our patterns of behavior for the personal benefit of that software's**

administrators, and they get away with it because there's currently nothing anchoring cyberspace to objective physical reality – nothing applying physical or thermodynamic restrictions on them. It's impossible to verify if anything seen online has real-world physical properties because people don't have the ability to "poke" or pinch anything; they have no proof-of-power protocol. At the same time, they are completely subject to the whim of the storytellers writing the scripts – the people pulling the strings – with no way to physically restrict them from global-scale systemic exploitation of cyberspace.

Bitcoin potentially fixes this. **With the arrival of open-source and globally-scalable proof-of-power protocols like Bitcoin, comes the arrival of a globally-adoptable proof-of-real protocol. If a person wants to claim that their cyber abstraction has physically real and materially consequential properties like physical cost, scarcity, or decentralization, then they can now prove it using bits of Bitcoin**. Otherwise, people have no ability to verify that what they see in cyberspace is anything more than just another illusion floating in an infinitely-expanding sea of marginalized illusions trying to expropriate a population's most valuable bits of information (their behavioral data) and sell it to the highest bidder.

The point is, **proof-of-power protocols have metacognitive properties that could transform how people perceive and behave in cyberspace, in ways that we can't yet comprehend or fully appreciate because this technology is so new.** To the author's knowledge, these pages represent the first time someone has even attempted to discuss these concepts together in formal academic literature.

How humans perceive "real" is directly dependent on finding proof-of-power. Proof-of-power signals produced by physical cost function protocols like Bitcoin could theoretically double as proof-of-real signals, to serve the same function as kinetic power (i.e. poking/pinching). When abstractions in cyberspace are accompanied by verifiable proof-of-real signals, they can have verifiable physical characteristics unlike any other abstraction in cyberspace, bringing what Saylor called the "*conservation of energy and matter, objectivity, truth, time, and consequence into the digital realm*." At the same time, proof-of-power could also be used to legitimize or delegitimize computer programmers who otherwise have unrestricted abstract power and control over the illusions we see in cyberspace.

5.9.9 Adding a Physically Constrained & Thermodynamically Restricted Space to Cyberspace

Combining these different theories together, we can see what advantages could be gained if someone figured out how to create a protocol which allows people to "plug" a planetary-scale computer into the internet. **Bitcoin is special because it could represent the global adoption of the operating system for a planetary-scale computer which converts our global electric power grid into a globally-decentralized state mechanism under nobody's centralized control, that everyone on the planet is capable of operating in a zero-trust and permissionless way so long as they're willing to project power. This planetary-scale computer would theoretically be capable of adding a unique and one-of-a-kind state space to cyberspace in comparison to the existing state space created by ordinary computers. This special state space would function as its own physically constrained and thermodynamically restricted space that no other state space in cyberspace could replicate, offering reverse-optimized properties that could be very useful for many different applications, particularly ones related to cyber security and keeping people safe against population-scale systemic exploitation and abuse of the internet's existing architecture**.

The special state space created by a planetary-scale computer would be more materially consequential than other spaces in cyberspace, and people could use it to impose severe physical costs on other people by simply requiring them to operate or "pass through" this space. This state space could also link our surrounding physical environment to virtual reality in a way that previously wasn't considered possible, providing cyberspace with emergent properties that are physically impossible to reproduce using ordinary computers operating on the legacy, physically unconstrained and thermodynamically unrestricted version of the internet. These concepts are illustrated in Figure 81.

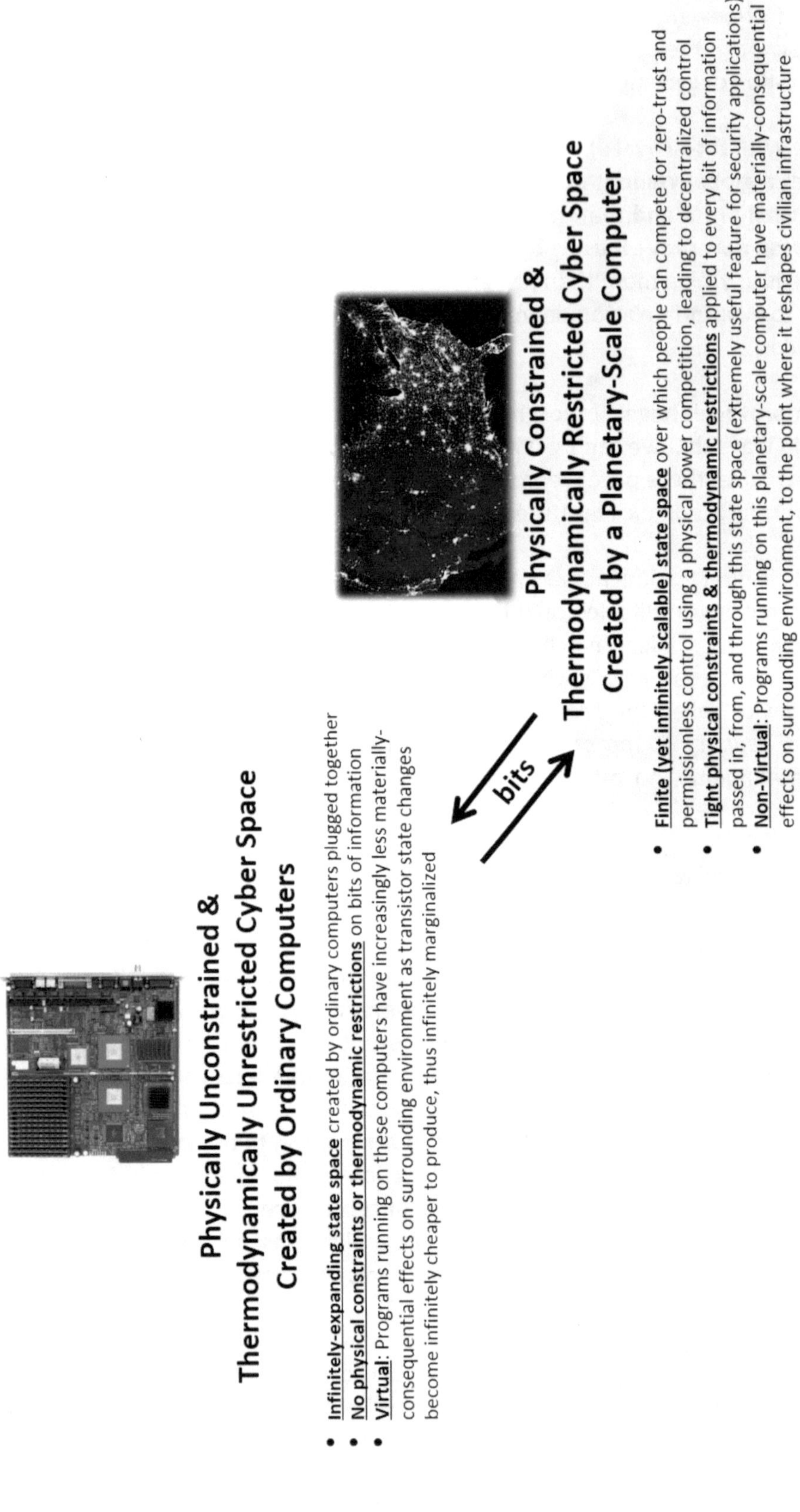
Physically Unconstrained &
Thermodynamically Unrestricted Cyber Space
Created by Ordinary Computers
Infinitely-expanding state space created by ordinary computers plugged together
No physical constraints or thermodynamic restrictions on bits of information
Virtual: Programs running on these computers have increasingly less materially-consequential effects on surrounding environment as transistor state changes become infinitely cheaper to produce, thus infinitely marginalized
bits
Physically Constrained &
Thermodynamically Restricted Cyber Space
Created by a Planetary-Scale Computer
Finite (yet infinitely scalable) state space over which people can compete for zero-trust and permissionless control using a physical power competition, leading to decentralized control
Tight physical constraints & thermodynamic restrictions applied to every bit of information passed in, from, and through this state space (extremely useful feature for security applications)
Non-Virtual: Programs running on this planetary-scale computer have materially-consequential effects on surrounding environment, to the point where it reshapes civilian infrastructure

Figure 81: Theoretical Effect of Plugging a Planetary-Scale State Mechanism into the Internet [168, 169]

As of now, these ideas are nothing more than theories grounded by the author's first principles approach to exploring the benefits of proof-of-power (a.k.a. proof-of-work) protocols. However, if further scrutinization of these theories indicate they could be valid, then it's hard to understate the strategic value of the state space created by systems like Bitcoin. The bits of information passed across the Bitcoin system would represent bits of reverse-optimized information transferred, received, and stored by a planetary-scale computer. **Bitcoin's "coins" would represent the underlying state space of this planetary-scale computer – an extraordinarily unique and strategically advantageous space within cyberspace. This would be similar to how there are unique and strategically advantageous positions inside of other domains. Just like canals and ports are strategically valuable spaces in sea, or Lagrange points are strategically valuable spaces in outer space, the physically constrained and thermodynamically restricted space of cyberspace could become among the most strategically valuable resource of the digital age**.

Owning a bit of Bitcoin's "coins" (which, if these theories presented in this thesis are valid, is really just an arbitrary name assigned to bits of information transferred and stored across a planetary-scale computing system) could represent owning an irreproducible piece of novel cyber space created by society's first globally-adopted, planetary-scale computer. Moreover, **because global leaders don't understand computer theory well enough to understand that the term "bitcoin" is purely an arbitrary name, or because researchers aren't taking these theories seriously, this could mean that nation states are allowing themselves to become among the *last* institutions to establish a foothold in this potentially strategically vital space within cyberspace. This could upturn global power projection dynamics and dramatically restructure the digital-age geopolitical environment, exactly as planetary-scale computing theorists predict will happen**.

5.9.10 Applying the Strangler Pattern to the Internet Itself, Starting with Financial Microservices

It is possible that in the future, people will not value objects in cyberspace unless they're "anchored" to our physical objective reality via verifiable proofs-of-power. Another way of saying the same thing is that in the future, people may not trust or value bits of information that are transferred and stored in cyberspace unless those bits of information are specifically derived from the planetary-scale computer. In fact, this is already starting to happen, as people have already started not to trust or value financial bits of information unless they're derived from the Bitcoin computing system (which, the author theorizes, doubles as the operating system of the world's first operational planetary-scale computer).

Why would people not trust or value bits of information that haven't been physically constrained or thermodynamically restricted? Simply because they would not be anchored to anything verifiably real or materially consequential. **People might one day come to understand that everything they encounter in the so-called "Metaverse" is something physically unconstrained, thermodynamically unrestricted, materially irrelevant, and under the complete, unimpeachable control of some technocratic ruling class who can and probably already is exploiting these unconstrained and unrestricted bits of information. One day, civil society may learn that if they want to physically constrain and decentralize people's control over their computers in order to mitigate global-scale systemic exploitation and abuse of the internet, they will have to learn how to utilize real-world physical power (a.k.a. watts) to countervail the imaginary power of their neo god-kings (a.k.a. software administrators) the same way agrarian society has always had to utilize real-world physical power to countervail the imaginary power of their god-kings.** Physical power has always been both the antecedent and the antidote to imaginary power. New physical power projection protocols like Bitcoin appear to be integral to the same power projection game agrarian society has always played, but in a new domain where digital-age society is spending increasingly more of its time.

If this cultural evolution were to occur at large enough scale, then the special state space added to existing cyberspace by systems like Bitcoin could one day become recognized as the foundational layer of the internet which grounds cyberspace to the physical constraints of shared objective material reality. This new, physically restricted version of cyberspace could become universally recognized as a place where computer programmers, with their immense abstract power and physically unconstrained control authority over people's computers, can no longer use it to exploit people's bits because it's too physically expensive. It is theoretically possible that people might start making a clear and pronounced mental distinction between two different versions of cyberspace with two very different kinds of bits of information transferring across them. One version of cyberspace would be *physically* secure against a technocratic ruling class of computer programmers and software administrators who have already demonstrated that, when left physically unconstrained, can and will deny people's bits, harvest people's bits, use people's bits to learn how to affect their behavior, and use what they learn to exploit or influence the behavior of entire populations at unprecedented scale. The other version of cyberspace would be the original version of cyberspace that we know today, the version of cyberspace where people *don't* have the ability to physically secure their bits of information because they have no ability to impose severe physical costs on people in, from, and through cyberspace who try to deny their right to transfer bits, or who try to exploit them in many other ways thanks to their having centralized control over those bits.

In the newer version of cyberspace that is backed by proof-of-power, things would have verifiably real-world physical properties like scarcity and decentralization that would be physically impossible to replicate in the older version of cyberspace created by ordinary computers. In the new version of cyberspace, programs wouldn't be allowed to execute certain control signals unless they're backed by substantial proof-of-power, thereby physically constraining the reach and control authority of software-defined abstract power hierarchies. This would be a stark contrast to the older, original version of cyberspace that is notably lacking proof-of-power, where all software abstractions would continue to not have verifiably real-world physical properties. In the legacy version of cyberspace, the abstract power and control authority of software-defined abstract power hierarchies would continue to remain physically unconstrained, and all of the clear systemic security challenges we have today would remain the same because the underlying architecture of the internet would remain the same (namely the inability to impose severe physical costs or physical constraints).

In the future, proof-of-work, a.k.a. proof-of-power, a.k.a. proof-of-real, a.k.a. physical cost function protocols like Bitcoin (to repeat, the name we give it is arbitrary) could serve as the dominate mechanism used by digital-age societies to map or anchor elements of the legacy version of the internet to the new version internet, thereby serving as the world's globally-adopted "proof-of-real" protocol. Regardless of whatever arbitrary name people give to protocols like Bitcoin in the future, the point is that this system could become instrumentally more important than only a financial system. Bitcoin has the potential to become integral to the development of a completely new internet infrastructure and a completely new type of cyberspace. **It is possible that what we are seeing with the global adoption of Bitcoin represents a "strangler pattern" playing out for the existing internet architecture – we could be modernizing a large, monolithic computing system right now by gradually replacing services performed on the legacy system with services performed on the modernized system, and it could just be that financial information exchange simply represents the first of many microservices to be performed on a modernized architecture of the world-wide web**.

5.10 There is No Second Best

"Air power is like poker. A second-best hand is like none at all – it will cost you dough and win you nothing."
General George Kenney [175]

5.10.1 Physical Cost Function Protocols Can't be Replicated by Ordinary Computers or Software

Another reason why it is important to point out what's going on "under the hood" with physical cost function protocols like Bitcoin is to illustrate how there are very meaningful physical differences between this technology and other computer programs. **People who utilize proof-of-work protocols like Bitcoin are technically using a different style of abstract thinking than people who utilize regular computer programs**. Without examining the underlying tech stack, it's impossible to see these differences or to appreciate how the abstractions used to describe objects like "bitcoins" have different complex emergent properties that are *physically impossible to reproduce* using regular computers.

When operating ordinary computer programs, people assign abstract meaning (e.g. Boolean logic) to transistor state changes in a state machine. However, when operating physical cost function protocols (a.k.a. proof-of-work protocols) like Bitcoin, people also assign abstract meaning to quantities of physical power, as illustrated in Figure 82. This means **the abstractions people use to describe the complex emergent behavior of ordinary programs are both physically and systemically different than abstractions used to describe proof-of-power systems like Bitcoin, even though they have the exact same semantic specifications. The bits produced by normal software using normal computers have physically and systemically different properties, which means the abstractions used to describe those bits also have physically and systemically different properties even though we give them the same name.**

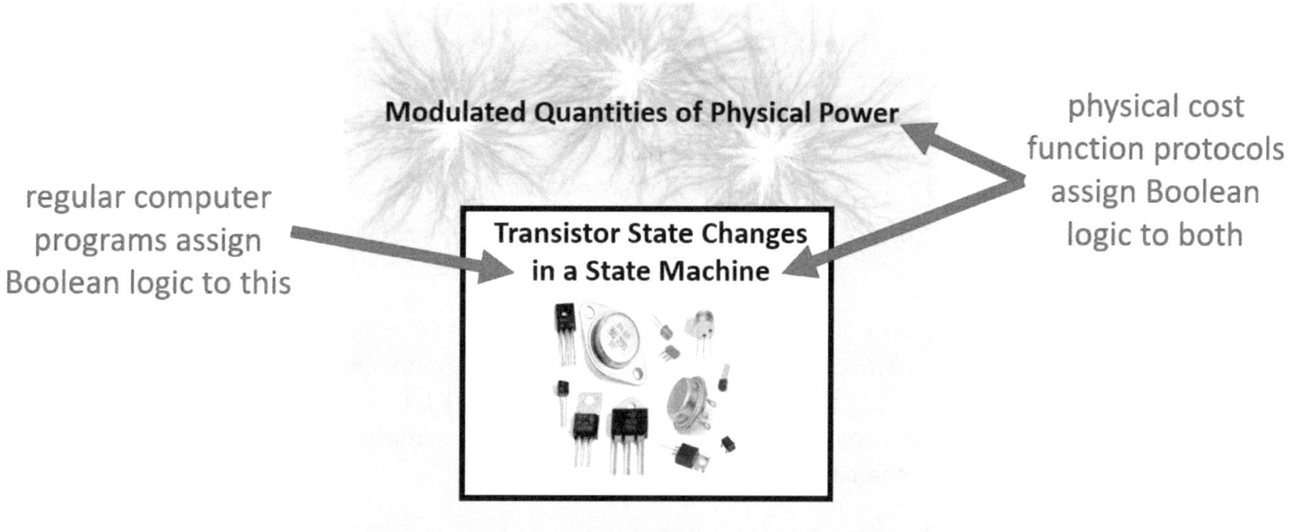

Figure 82: Illustration of Physical and Systemic Differences between Technologies [176]

It is physically impossible for a "coin" generated by an ordinary computer program to have the same complex emergent properties as a "coin" generated by a proof-of-work protocol like Bitcoin because they are *physically* different systems. In the same way that something which doesn't physically exist doesn't have the same complex emergent properties as something which does physically exist (particularly properties like being verifiably scarce or decentralized), bits of "coin" created by an ordinary piece of software running on an ordinary computer can't have the same complex emergent properties as Bitcoin. **From the software at the top of the tech stack, all the way down to the underlying state mechanism, Bitcoin is a fundamentally different computing system with physically different and reverse-optimized hardware, making it incomparable to traditional computing systems.**

When people normally refer to an "object" instantiated by ordinary software, they are applying abstract and symbolic meaning to transistor state changes inside a computer using machine code to convert them to machine-readable bits of information. These "objects" do not physically exist; people have *arbitrarily* choosen to describe the complex emergent behavior of their computer programs using metaphors like "objects" (e.g. folder, trash can, cloud, string) to make it easier for other people to understand the intended functionality and design of their computer programs. In other words, *there is no spoon*. All that physically exists is the physically unrestricted and infinitely expanding state space of countless state machines plugged together. Countless transistors are added together on circuit boards, and the combined state of those transistors is what we abstract as so-called "objects." But underneath the metaphor is nothing but people converting transistor state changes to bits of information using a new type of symbolically and syntactically complex language.

Physical cost function protocols (a.k.a. proof-of-work protocols) like Bitcoin work differently. When people describe the real-world physical cost of solving a physical cost function protocol as if it were a quantifiable object (e.g. a proof, receipt, stamp, coin, or token), they're using abstraction to describe a real, meaningful, physically scarce, physically constrained, thermodynamically restricted, path-dependent, and irrevocable quantity of physical power drawn out of shared objective physical reality. Therefore, **whereas ordinary computer programs can only present the *illusion* of physical cost, physical cost function protocols like Bitcoin produce *real-world* physical cost. Whereas ordinary computer programs can only present the *illusion* of scarcity and decentralization, Bitcoin produces *real-world* scarcity and decentralization. This is possible because the underlying physical state-changing mechanism (i.e. massive quantities of watts) being converted into bits is, itself, physically real, physically scarce, and physically decentralized**. You know that bits of information created by large quantities of physical power are real, scarce, and decentralized because *power itself is real, scarce, and decentralized*.

The physically costly quantities of real-world power utilized by protocols like Bitcoin double as machine-readable bits of information which are then described as abstract objects like "coin." Combining these concepts together, we can see **there's a substantial difference between the abstract "objects" produced by regular software and the "objects" produced by physical cost functions (a.k.a. proof-of-work protocols) because the bits we arbitrarily call "objects" are derived from physically different and reverse-optimized phenomena. These differences go undetected because most people aren't paying attention to the underlying state mechanism. They're only paying attention to the software layer (where everything looks the same because we're using the same metaphors), not the hardware layer (where it's incontrovertibly true that there are meaningful physical differences)**.

The same object-oriented software abstractions refer to two completely different physical phenomena with completely different physical and emergent properties. As discussed in the previous section, the bits produced by ordinary computer programs have the exact opposite design intent as the bits produced by physical cost function protocols. Right now, people are treating these bits as if they're the same thing without taking the time to understand how these bits have been intentionally reverse-optimized.

One explanation for why computer scientists and software engineers keep overlooking these differences is because they've become accustomed to evaluating tech stacks from the top down, not the bottom up, so naturally they aren't going to detect that a new state mechanism has been added to the *bottom* of the internet's tech stack. It would appear that **software engineering culture has become so accustomed to abstraction, information hiding, and ignoring everything *below* assembly language that they don't recognize how proof-of-work utilizes a completely different state mechanism with properties that are physically impossible for non-proof-of-work systems to replicate**. This concept is illustrated in Figure 83.

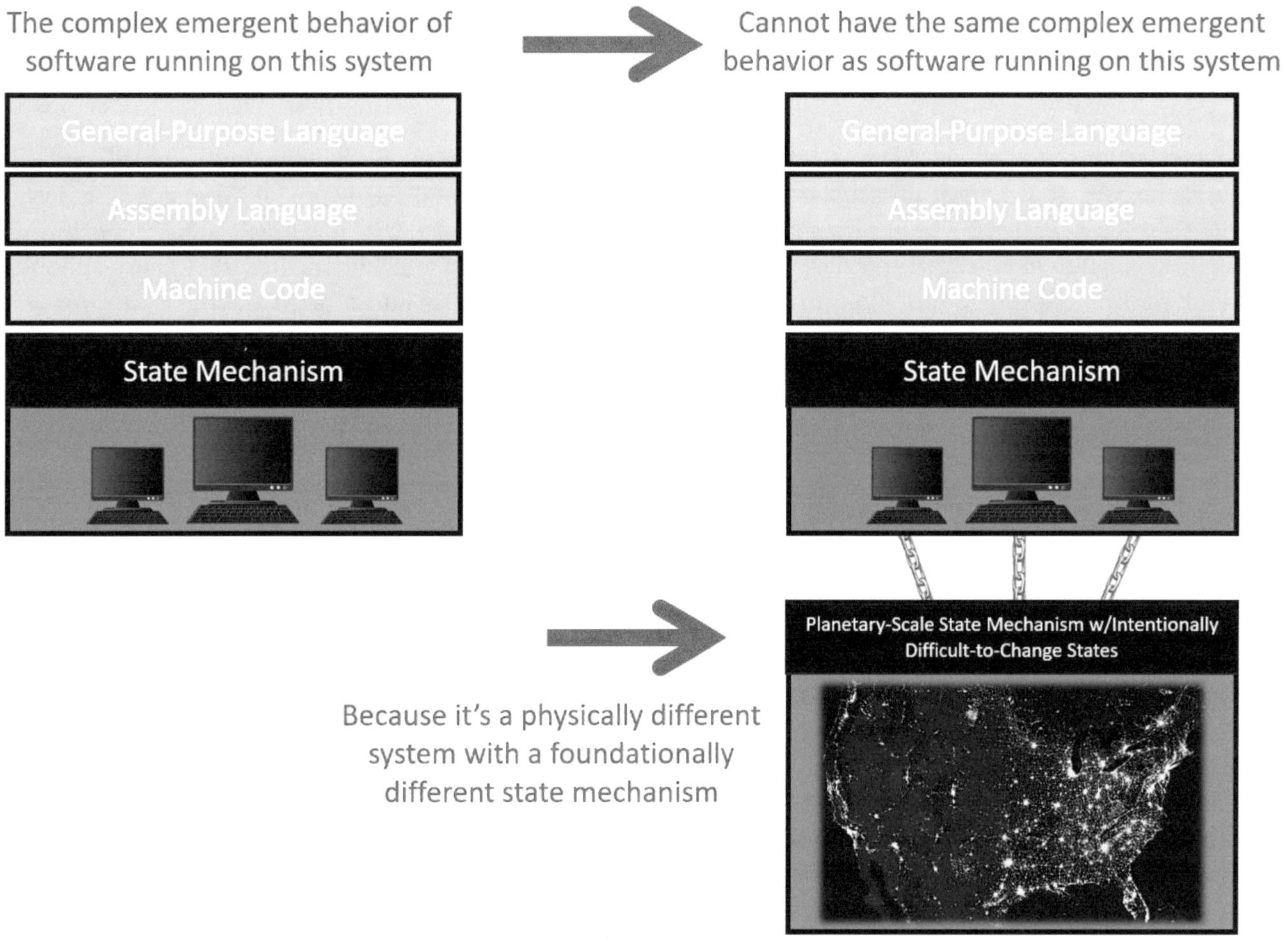
The complex emergent behavior of software running on this system
Cannot have the same complex emergent behavior as software running on this system
General-Purpose Language
Assembly Language
Machine Code
State Mechanism
General-Purpose Language
Assembly Language
Machine Code
State Mechanism
Planetary-Scale State Mechanism w/Intentionally Difficult-to-Change States
Because it's a physically different system with a foundationally different state mechanism

Figure 83: Illustrating the Physical Differences between Different Technologies
[169, 154, 156]

Physical cost function protocols (a.k.a. proof-of-work protocols) are deceptively simple: they convert watts to bits. When people utilize computing networks like the Bitcoin network, they take watts as their input and convert them into proofs-of-watts as their output, which are then metaphorically described as some type of abstract object. Lost in the metaphors, computer and software engineers aren't taking the time to consider the major technical differences in the underlying physics, despite how observably different it is. Why do these major technical differences go undetected? The author hypothesizes that it could be because **proof-of-work represents an unprecedented approach to computing that software engineers aren't recognizing yet because they have effectively been trained by their industry best practices to ignore what's happening under the hood of their software by abstracting it away**.

As previously discussed, computer programming is an enormously complex challenge. It is essential for computer programmers to reduce cognitive distance and lighten the metacognitive burden of software engineering as much as possible. Object-oriented software design is one of the most common abstraction techniques. An easy thing to do to reduce cognitive distance is to call the proof-of-power receipt produced by a physical cost function a "receipt" or "stamp" or "token" or "coin" and then to simply move on without taking the time to sort through the physical details.

When computer scientists and software engineers do this, they tacitly treat the abstract object produced by a proof-of-work protocol (e.g. a "bitcoin") as if it were the same as any other abstract object produced by software, despite obvious physical differences in the underlying tech stack. Consequently, they miss the fact that "bitcoins" are *physically* costly, constrained, and decentralized in a way that no ordinary software "object" can be. Probably the most famous example of people making this mistake is when they claim that proof-of-stake protocols can have the same complex emergent properties (e.g. decentralization) as proof-of-work protocols. What these people are overlooking is the fact that "stake" and "work" are physically incomparable because "stake" refers to an ordinary software abstraction for a normal computing system whereas "work" refers to a software abstraction for a fundamentally different type of computer with a fundamentally different type of underlying state machine (more on this in the following sections).

Using abstractions like "bitcoin," critical information about the differences in underlying physical hardware is lost in the same way that critical information about the underlying physical hardware of "the cloud" is lost. As previously discussed, this is a reoccurring challenge of software engineering that negatively impacts security. The semantic ambiguity of software specifications is a well-known problem in cyber security that causes people to overlook vitally important security information – like how storing sensitive information on "the cloud" really just means storing sensitive information on someone else's computer that you can't control, that belongs to somebody you must trust. [30]

Just like the term "cloud" hides critical information about where bits of information are being physically stored, the use of object-oriented software abstractions like "coin" for proof-of-work protocols disguises the fact that the bits of information that are being metaphorically described as a "coin" represent state changes occurring *outside* of our computers, not *inside* of them. Because people don't understand computer theory and they aren't paying attention to the bottom of the tech stack, they overlook highly unique properties of proof-of-work protocols which have a potential to change the game of cyber security or even change how humans perceive cyberspace. Ironically, most people have no trouble recognizing Bitcoin's substantial power usage, but they struggle to understand why that substantial power usage *is* the irreplicable innovation.

5.10.2 It is Physically Impossible for Proof-of-Stake Protocols to Replicate Proof-of-Work Protocols

For what might be the first time since Neumann's invention of stored-program computing techniques, people are learning to use proof-of-work protocols like Bitcoin to program something *other* than materially inconsequential state changes inside a state machine. The massive electrical power deficits produced by protocols like Bitcoin represent physical state changes of the surrounding environment *outside* the boundary of our ordinary computers. This makes it possible to couple information stored as state changes *inside* of an ordinary stored-program computer to state changes stored *outside* of a computer, effectively bridging the cold, hard, shared objective reality of our physically constrained and thermodynamically restricted environment to the virtual reality of cyberspace.

The key takeaway is that **proof-of-work protocols developed by computer engineers like Adam Back may represent the discovery of an externally stored programming technique that simultaneously changes the physical state of the environment *outside* of our computers as well as the physical state of the environment *inside* our computers, giving computer programmers a way to program our virtual environment *and* physical environment simultaneously and then link them together**. With this capability, the abstractions we use in cyberspace start to reshape our surrounding physical environment and then import the real-world physical properties of our surrounding environment (properties like physical cost, physical scarcity, and physical decentralization) into cyberspace. It sounds extraordinary, but all that it takes to achieve this is for people to simply build the infrastructure needed to convert watts into bits, which is precisely what the Bitcoin network is doing. These concepts are illustrated in Figure 84.

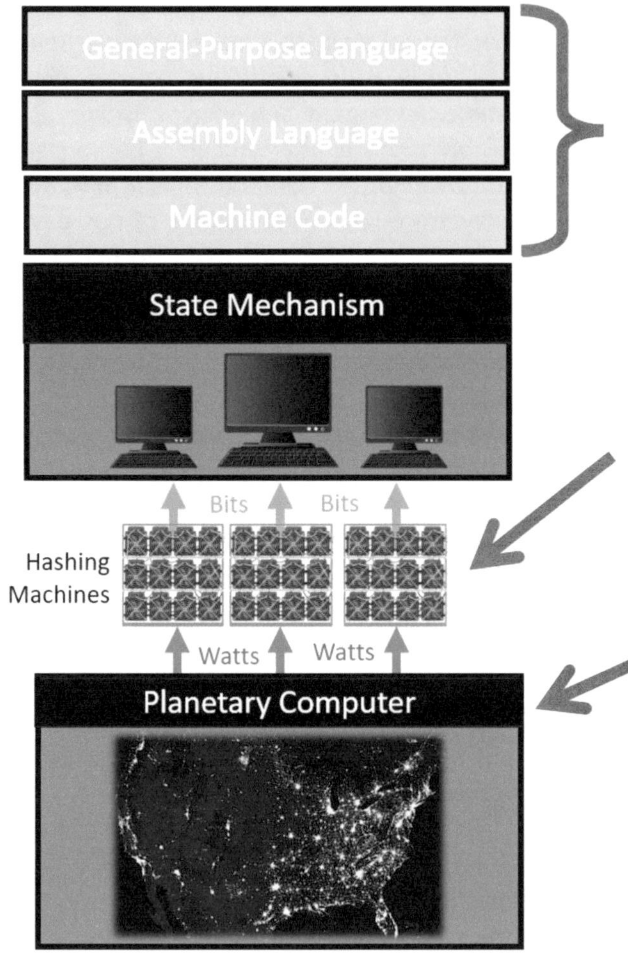

1) Only logical constraints can be used here.
Most software engineers abstract away everything below this layer, making them unable to recognize changes to the bottom of the tech stack if they were to occur.

2) Bitcoin adds a physical constraint to the BOTTOM of the internet tech stack.
The protocol works by applying machine code to digitally-sampled quantities of physical power drawn from the global power grid, thus converting large (and expensive) quantities of physical power into special bits of info that can be transferred through cyberspace like any other digitized info. Unlike regular bits, these special bits have been reverse-optimized to be as expensive to transmit, receive, and store as possible.

3) The global power grid is converted into a planetary-scale state mechanism, a.k.a. "planetary computer."
The planet itself is treated as a physically constrained and thermodynamically restricted computer over which people compete for control authority in a zero-trust, permissionless, and egalitarian way by optimizing the portion of the global power grid under their control, effectively replicating the benefits of Moore's law for the global power grid. This creates cybersecurity advantages by giving people the ability to *physically* constrain programs rather than having to continue to attempt to constrain programs using merely encoded logic.

Figure 84: Breakdown of a Proof-of-Work Tech Stack
[169, 177, 154, 156]

These theoretics emphasize the point that **Bitcoin (the system) represents a physically different protocol that is disrupting the internet from the bottom up.** The Bitcoin system has irreproducible complex emergent behavior because it uses a different base-layer state mechanism that has been intentionally reverse-optimized to produce more physically expensive bits rather than less physically expensive bits. This critical information about proof-of-work computing systems is hidden by the metaphors used to describe the software, which is not surprising considering how information hiding and abstraction is treated as a virtue within the field of computer science. As discussed in section 5.4.6, critical information about substantial technical differences that impact security often gets overlooked because of the arbitrary semantics used to emphasize different features of the software.

Since the physical differences at the *bottom* of the tech stack are (intentionally) ignored, software engineers keep (fraudulently) asserting that they can replicate the same complex emergent properties (namely decentralization) of Bitcoin using regular programs running on ordinary computer hardware. Additionally, since the physical differences are grossly misunderstood, the general public will claim that Bitcoin's computing network is energy inefficient and that it could/should be replaced by something more "environmentally friendly." These misguided (or blatantly fraudulent) assertions are being made because people are missing the point (or deliberating disguising the point) that making a physically costly computer for the purpose of producing physically constrained and thermodynamically restricted bits of information is the primary-value-delivered-function of proof-of-work protocols like Bitcoin. The point of the protocol is to create and utilize physically expensive bits of information, as physically expensive bits are useful for cyber security and many other applications that we are just now beginning to discover.

The software engineering culture of abstraction and information hiding explains the public misconception of proof-of-work protocols like Bitcoin. The information needed to understand the novel computer theory behind proof-of-work is lost in the metaphors we use to describe how and why it was designed. This explains why so many **software engineers will claim to be able to replicate the complex emergent behavior of Bitcoin using metaphors like "blockchains" or "cryptocurrencies" that are ostensibly more energy-efficient. These claims represent a gross lack of appreciation for the complex dynamics of power projection and reverse-optimized computing, perpetuated by a decades-old and well-established software engineering culture of not paying attention to the physical differences of the underlying tech stack**. It's incontrovertibly true that Bitcoin has a far different physical signature than these other so-called "blockchains", but computer and software engineers aren't taking the time to study *why* it's physically different and what that means for computer theory as a whole.

Probably the most famous illustration of this lack of appreciation for power projection and the novel computer theory behind proof-of-work protocols is amongst those who believe proof-of-stake protocols can replicate the same complex emergent behavior and be a more "energy efficient" alternative to proof-of-work protocols. When people make these claims, they illustrate that they don't understand the primary value-delivered function of proof-of-work: to be physically costly.

Proof-of-work and proof-of-stake are physically and systemically different technologies that utilize completely different, reverse-optimized state mechanisms that are practically incomparable no matter what arbitrary metaphors are used to try to categorize them as the same technology (e.g. "blockchain" or "cryptocurrency"). For the same reason it's not possible to claim that something which doesn't exist is equal to something which does exist, it's not possible to claim that proof-of-stake is equal to proof-of-work. **The incontrovertible difference between these two protocols is that one is far more physically expensive (in watts) than the other one**. These physical differences would imply that it's physically impossible for proof-of-stake to have the same complex emergent properties (namely physical cost, physical scarcity, and physical decentralization) as proof-of-work protocols. It should be clear to any observer that the bits of information produced by a proof-of-stake protocol cannot replicate the physical expense of the bits of information produced by a proof-of-work protocol, because that difference is directly measurable in watts. Therefore, **the issue isn't that proof-of-work is "too physically expensive," the issue is that people don't understand why physically expensive bits are useful in the first place** (probably because they don't understand power projection dynamics and how physically expensive bits could be used to increase $\mathbf{C_A}$ thus decrease $\mathbf{BCR_A}$ of our computer systems).

In addition to not understanding why it would be useful to have physically expensive bits, the general public also doesn't seem to understand that "stake" is not even a real thing – it's a metaphor. Just like *there is no spoon*, there is no "stake." **The term "stake" is just a metaphor for the special administrative privileges given to users who own ETH**. It's a purely imaginary concept, an object-oriented software abstraction that doesn't physically exist and therefore is physically impossible to be costly, scarce, or even verifiably decentralized. "Stake" (and the special administrative privileges it provides) cannot be decentralized anywhere except exclusively within people's imaginations because "stake" itself is an abstract concept. Proof-of-stake systems are what's known as decentralized in name only (DINO) systems because it's physically impossible to verify that "stake" isn't already majority-owned by a single person or organization (much more on this in the next section).

Another way to conceptualize the difference between proof-of-work versus proof-of-stake systems is to borrow from the concepts of Power Projection Theory and rename these protocols proof-of-*real*-power versus proof-of-*imaginary*-power. Proof-of-stake is proof-of-imaginary-power (a.k.a. proof-of-rank). **A proof-of-stake system is yet another form of exploitable abstract power hierarchy where a select few people within a carefully-encoded abstract power hierarchy get consolidated administrative permissions and control authority over an underlying resource.** Therefore, a proof-of-stake (a.k.a. proof-of-rank or proof-of-imaginary-power) protocol is functionally the same type of systemically exploitable, rank-based belief system as a government rule of law, complete with exactly the same type of systemically exploitable voting logic. Moreover, proof-of-stake is clearly a plutocratic design where those who have the most ETH can also have the largest amount of abstract power (a.k.a. "stake").

Because they're functionally the same type of encoded abstract power hierarchy where a select few people control the underlying resource, that means proof-of-stake systems also have the same type of systemic security flaws as traditional governments and rules of law (plutocracies in particular). Additionally, because they're the same type of software running on the same type of computers, that means proof-of-stake systems also have the same kind of cyber security vulnerabilities that practically all traditional software systems have, namely the inability for users to *physically* (as opposed to *logically*) secure themselves against the systemic exploitation of encoded design logic.

As proof-of-stake systems become more popular, the benefit to be gained by exploiting the special administrative privileges given to those who control the most "stake" (a.k.a. the B_A) will increase, but the physical costs of attacking the system (a.k.a. the C_A) will continue to remain fixed at practically zero. Thus, the benefit-to-cost-ratio of systemically exploiting proof-of-stake systems (a.k.a. BCR_A) will gradually approach infinity as more people adopt these systems. "Slashing" doesn't prevent this because the top stakers would clearly have the incentive and the means to slash users back. Moreover, slashing is a voting-based system that is endogenous to the system, which means it can (and most assuredly will) be systemically exploited by those who encoded the design logic (which, incidentally, are probably the same people who control the majority supply of "stake").

Ironically, proof-of-stake systems recreate the exact same type of systemically exploitable, trust-based, permission-based, and inegalitarian abstract power hierarchy that it was ostensibly created to replace. Instead of creating a decentralized system, a proof-of-stake system gives unimpeachable control authority over the system to an anonymous group of people who control the most of an imaginary object called "stake," which is really just another name for rank (i.e. abstract power) and administrative privileges. Instead of creating a decentralized organization, proof-of-stake systems like Ethereum have created a system that masquerades as a decentralized system because those who control the majority supply of "stake" can take advantage of disembodied nature of software to obfuscate how centralized and consolidated that "stake" is.

To believe that proof-of-stake protocols can serve as a viable replacement to proof-of-work protocols is to believe that something which doesn't physically exist can have the same complex emergent properties as something which does physically exist. **Proof-of-stake is yet another systemically exploitable belief system that people are adopting for what are often ideological reasons, because they desire a viable replacement for physical power as the basis for settling disputes, establishing control authority over resources, and achieving consensus on the legitimate state of ownership and chain of custody of property**. Ironically, this means that proof-of-stake systems represent the exact same style of abstract power and empire-building discussed in the previous chapter, complete with the exact same style of ruling class wielding yet another type of asymmetric abstract power and control authority created by convincing a population to adopt the same belief system.

To repeat the same point using the concepts introduced in section 4.5, **"stake" is nothing more than imaginary power masquerading as real power (watts)**. Like every other type of imaginary power that has been conjured up by people searching for a viable replacement to real power, "stake" is neither self-evident nor self-legitimizing. It's systemically endogenous to the software which instantiates it, making it highly vulnerable to systemic exploitation and abuse by the people who write the software. It's physically unconstrained and thermodynamically unsound, bounded, zero-sum, non-inclusive, inegalitarian, and non-attributable. This makes proof-of-stake protocols physically and systemically opposite from proof-of-work protocols. To believe that two physically and systemically opposing technologies can achieve the same emergent effect is equally as silly as the concepts presented in the previous chapter, where people believe that a man wearing a wig has the same type of power and can achieve the same type of security as a woman firing the gatling gun of an A-10 Thunderbolt II. This concept is illustrated in Figure 85.

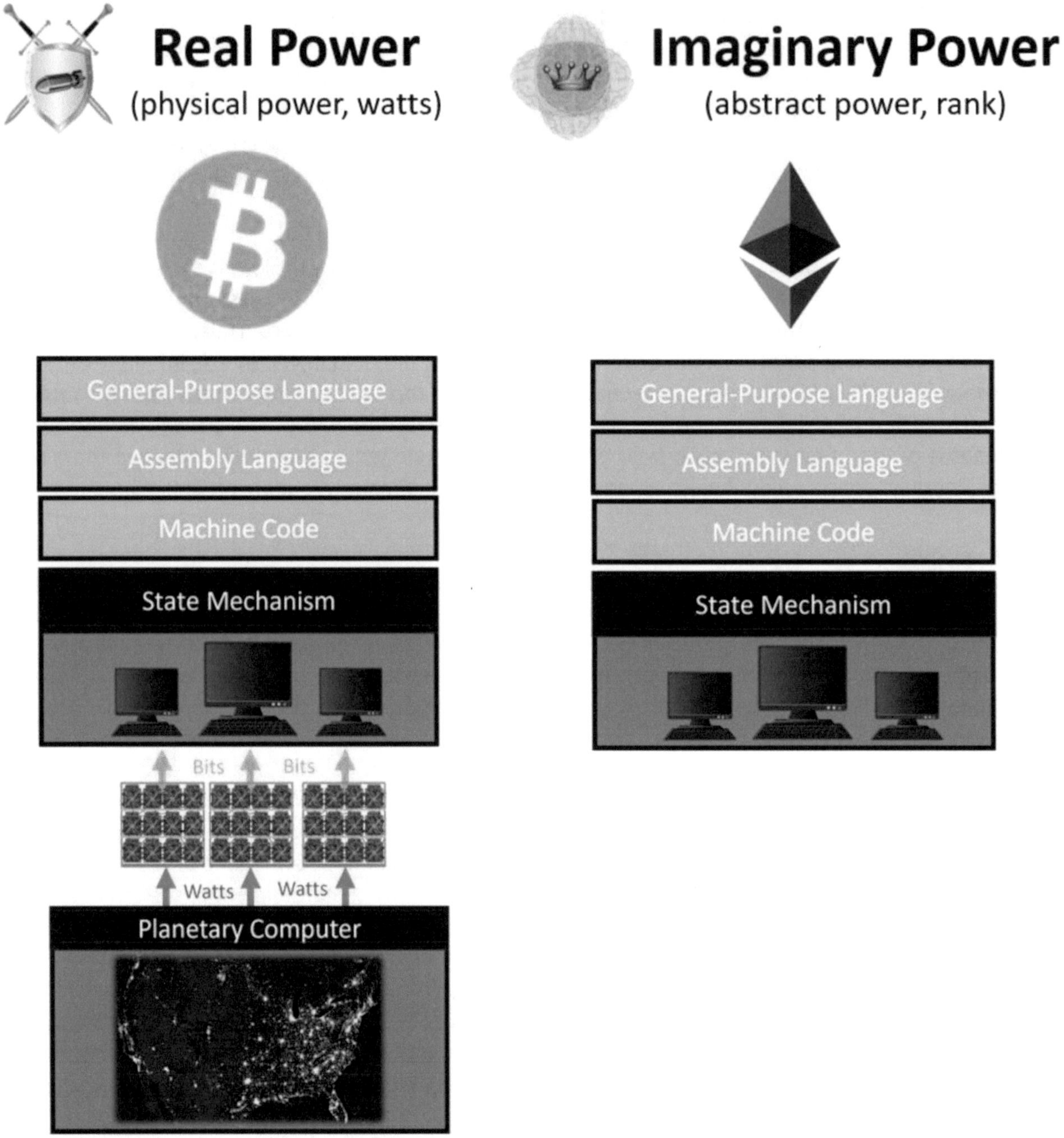

Figure 85: Illustration of the Difference Between Proof-of-Work & Proof-of-Stake Protocols
[169, 177, 154, 178, 179, 90, 88, 89, 76]

To persuade people that proof-of-stake can replicate the same complex emergent properties (e.g. security, scarcity, decentralization of control, etc.) as proof-of-work is, according to many, a fraudulent claim. Not surprisingly, the lead developers of the largest proof-of-stake system started a trend of marginalizing those who accuse them of being fraudulent as being "Bitcoin maxis," the implication being that people who believe that proof-of-stake systems cannot claim to be decentralized or cannot have the same complex emergent behavior as proof-of-work protocols are "closed-minded."

This behavior of claiming that imaginary power can have the same properties as real power could be interpreted as precisely the same style of god-king abstract power building which motivates entire populations of people to willingly forfeit their real power for (systemically exploitable) imaginary power because they don't understand the complex physical and systemic differences between real power and imaginary power-based resource control structures. It even has the same moral, ethical, and ideological motivations behind it. People who forfeit the use of proof-of-work protocols like Bitcoin in favor of proof-of-state protocols like Ethereum are explicitly forfeiting their capacity and inclination to project physical power and impose severe physical costs on attackers because they're ideologically inclined to believe that Bitcoin's power expenditure is "bad." This, of course, creates a window of opportunity for systemic predators to prey on these ideological beliefs, to convince the general public to forfeit proof-of-work protocols like Bitcoin in favor of alternative protocols where a select few anonymous people have systemically exploitable administrative permissions and control authority over the system via their large (and physically impossible to verifiably decentralize) amount of so-called "stake."

Just like they have been doing for thousands of years, people want to believe that proof-of-imaginary-power protocols can sufficiently replace proof-of-real-power protocols because they don't like the energy expenditure required to use real power as the basis for settling their disputes, establishing control authority over their valuable resources, or achieving consensus on the legitimate state of ownership and chain of custody of their property in a zero-trust, permissionless, and systemically secure way. As a result of this desire *not* to use physical power, we can see the exact same security vulnerabilities discussed at length in the previous chapter. People who are adopting proof-of-stake systems are adopting trust-based, permission-based, inegalitarian, and systemically insecure imaginary power hierarchies. History tells us what happens to these systems: they eventually become dysfunctional as the people with the most abstract power and control authority over the system start to abuse it, and users are incapable of stopping it because they made the mistake of believing that encoded logical constraints would secure them.

5.10.3 The change from Ethereum 1.0 to Ethereum 2.0 created a Huge Systemic Security Vulnerability

"One of the ways that I think about this in a more philosophical way is like, proof-of-work is based on the laws of physics, so you have to sort of work with the world as it is. You have to work with electricity as it is, hardware as it is, what computers are as it is. Whereas, because proof-of-stake is virtualized in this way, it's basically letting us create a simulated universe that has it own laws of physics. And that just gives us as protocol developers a lot more freedom to optimize the system around actually having all the different security properties that we want. If we want the system to have a particular security guarantee, then often there is a way to modify the upper state mechanism to also achieve it."

Vitalik Buterin, the night of Ethereum's transition from Proof-of-work to Proof-of-Stake [180]

This section will repeat many of the concepts from the last section but apply them more specifically to the context of the Ethereum computer network, which is currently the most widely-adopted proof-of-stake network. This section summarizes arguments that have been made against the Ethereum network, most notably the argument that Ethereum developers are fraudulently claiming to be able to achieve the same level of security and decentralization as the Bitcoin network while eliminating the energy expenditure of proof-of-work. If the reader isn't interested in learning more about these arguments, they are encouraged to skip this section, as it repeats many of the concepts presented in the previous section.

As discussed in section 5.4, a primary challenge associated with cyber security is a lack of ability to apply physical constraints to software. Another challenge associated with cyber security (outlined in detail in section 5.4.6) is that because software is abstract, computer program developers use arbitrary and semantic metaphors to describe their software designs that are not technically accurate and often disguise critical information needed to understand how secure that software is. This is a systemic security vulnerability because it creates a window of opportunity to (sometimes intentionally) design exploitable design features into a given piece of software and then disguise those security vulnerabilities with disingenuous metaphors. The author argues that this is what has occurred with Ethereum's recent pivot from a proof-of-work security protocol to the proof-of-stake security protocol. After making this change, Ethereum developers are fraudulently claiming that "stake" (a.k.a. abstract power, or rank) can provide the same systemic security properties as "work" (a.k.a. real power, or watts).

Using the concepts of Power Projection Theory presented in this thesis, we can see that during the conversion from Ethereum 1.0 to Ethereum 2.0, developers converted their software from a physical-power-based resource control system constrained and secured by physics to an abstract-power-based resource control system with no physical constraints or thermodynamic restrictions, thus dramatically changing the software's complex emergent security features and giving unimpeachably consolidated control authority over the system to Ethereum's largest owners.

Ethereum's most famous co-developer Vitalik Buterin asserts that their decision to remove physical constraints is actually *good* for cyber security because it enables Ethereum's developers to use encoded logical constraints to *logically* secure the system rather than using (energy-intensive) physical costs (a.k.a. watts) to *physically* secure the system like proof-of-work protocols do. Buterin's assertion not only contradicts decades of established principles in systems security, it's also blatantly illogical, as (1) it is obviously physically impossible for encoded computer logic alone to fully replicate the emergent benefits of physical constraints, and (2) computer logic alone is demonstrably incapable of preventing the systemic exploitation of computer logic (hence every software hack that has ever occurred).

By changing from a proof-of-work (a.k.a. proof-of-*real*-power) security system to a proof-of-stake (a.k.a. proof-of-*imaginary*-power) security system, not only has it become physically impossible to verify that Ethereum is a decentralized system, it has also created another type of software-defined, abstract-power-based resource control hierarchy which gives extraordinarily asymmetric abstract power and control authority over to an anonymous group of people who control the majority supply of "stake." Ethereum users must necessarily trust that the people who control the majority supply of "stake" will not exploit or abuse it, and they are counting entirely on (systemically exploitable) encoded logic (which *they* write) to secure users from exploitation and abuse because users are now physically powerless to secure themselves using physical power.

To illustrate the systemic changes which occurred when Ethereum changed from a proof-of-work system to a proof-of-stake system, Figure 86 provides a side-by-side comparison between the control structure model of a proof-of-work system and a proof-of-stake system. The reader should note that these are the same control structure models discussed in section 4.7. Here we can see that proof-of-stake creates a near-identical resource control structure as proof-of-work that is easy to misperceive as the same system. The controlled process (in this case the bits of information recorded by the blockchain) is the same, the controllers are the same, and the general structure is the same. The main difference is simply the type of power used. Whereas Ethereum 1.0 was a PPB resource control hierarchy that utilized *real* power (a.k.a. watts), Ethereum 2.0 is an APB resource control hierarchy that utilizes abstract power (a.k.a. "stake").

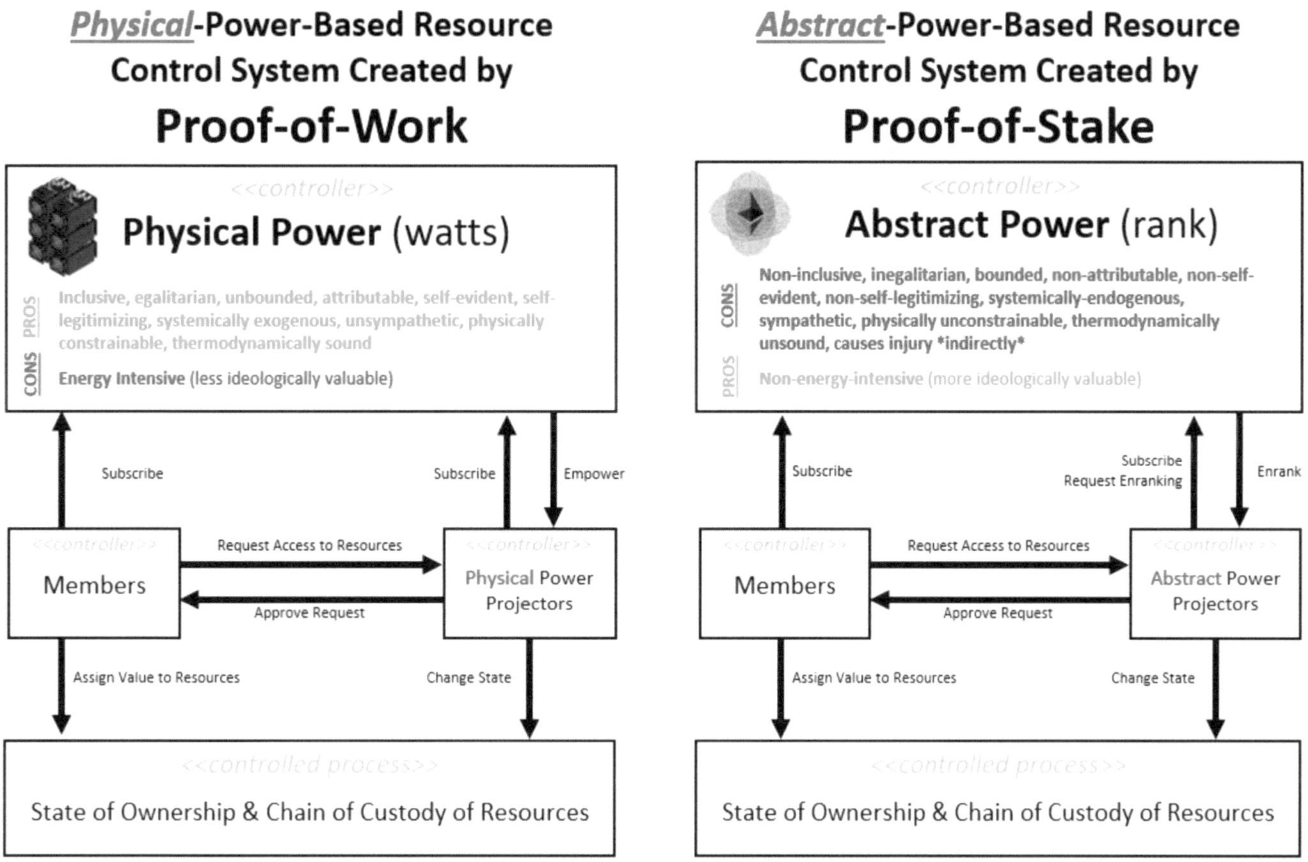

Figure 86: Side-by-Side Comparison between Proof-of-Work Proof-of-Stake Resource Control Models [76, 177]

As introduced through the concepts of Power Projection Theory presented in previous chapters, it's easier to spot the systemic differences between Ethereum 1.0 (a proof-of-work system) and Ethereum 2.0 (a proof-of-stake system). Ethereum 2.0 is far more vulnerable to systemic exploitation from those who control the majority supply of ETH, whereas Ethereum 1.0 isn't. This exact same principle also holds true for other proof-of-work systems like Bitcoin. Bitcoin is not nearly as systemically vulnerable to users who have the most BTC, because owning BTC doesn't give users the same type of special administrative privileges over the system (namely validation rights) as people's ETH give them in the Ethereum 2.0 control structure.

Another way to illustrate this systemic security flaw is to revisit the threat posed by domestication. From a functional perspective, **proof-of-stake is a repeat of the domestication hazard** discussed in sections 3.10 and 4.8. By switching from proof-of-work to a proof-of-stake system, Ethereum users have allowed themselves to give control authority to people with abstract power rather than those with real power (a.k.a. watts). As a result of making this change, users tacitly forfeit their ability to gain and maintain special administrative privileges (namely validation rights) over the system in a zero-trust, permissionless, and egalitarian way that can be verifiably decentralized simply by projecting their own real power. **People who forfeit proof-of-work systems for proof-of-stake systems are forfeiting their own capacity to tap into an unbounded supply of non-systemically exploitable physical power which they could use to impose severe physical costs on other users who exploit or abuse their special administrative privileges**. Consequently, the BCR_A of exploiting the special permissions granted to users with "stake" will only increase because it's physically impossible for users to increase their C_A.

A leading cause of people's confusion about the difference between proof-of-stake and proof-of-work is that they don't understand the physical and systemic differences between what people arbitrarily call "work" and what people arbitrarily call "stake." **The key insight that people are missing is that "work" refers to real-world physical power (a.k.a. watts) whereas "stake" is an arbitrary semantic expression given to an imaginary object-oriented software abstraction**. This subtle but key difference has important implications because it means that proof-of-work and proof-of-stake systems are incontrovertibly different technologies down to the base-layer state mechanism, making it physically impossible for proof-of-stake systems to replicate the same cyber security properties as proof-of-work systems – most notably the ability to keep special administrative privileges physically decentralized and impeachable. If "stake" doesn't physically exist, then that means it's physically impossible for the special administrative privileges granted to proof-of-stake protocols to be decentralized anywhere except exclusively within people's collective imagination (the author is being repetitive because this is a crucially important concept to understand that clearly isn't resonating with the general public).

Therefore, **the key to understanding the systemic security flaw of proof-of-stake systems like Ethereum 2.0 is to simply recognize that "stake" doesn't physically exist. "Stake" is purely an abstraction, a name that developers intentionally made up to obfuscate who has access to the special administrative permissions needed to approve or disapprove transactions (which is most likely centralized amongst the Ethereum foundation and large investment institutions)**. Anybody who uses Ethereum 2.0 tacitly requires validation, permission, and approval from these administrators who have abstract "stake" (a.k.a. rank or abstract power). If these anonymous people choose not to validate a user's valid transaction request, then a user's ability to utilize the blockchain will be degraded.

As of this writing, 60% of all transactions that have been made on Ethereum 2.0 have been compliant with sanctions imposed by the Office of Foreign Assets Control (OFAC). [181] In other words, 60% of all valid but non-OFAC compliant transaction requests have been deliberately censored by those who have special administrative permissions granted to them by their "stake." Thus, almost immediately after switching from proof-of-work to proof-of-stake, Ethereum became a majority-censored blockchain where valid users requesting valid transactions (who happen to not be OFAC compliant) are being denied 60% of the time. Clearly these people's ability to utilize the blockchain has been degraded in comparison to other users, and there is little they can do to eliminate the degradation. According to Ethereum's developers, this type of degradation shouldn't occur because validators should know that degrading people's access to the blockchain decreases the value of the blockchain, yet the attacks are still occurring. Also according to Ethereum's developers, users have the ability to "slash" the "stake" of validators if they censor transactions like this. Yet, to date, none of the validators who are censoring non-OFAC compliant transaction requests have had their "stake" slashed, because to do so would require users to come to consensus via a voting system that "stake" should be slashed. As always, the problem with voting systems is that voting systems represent yet another form of trust-based, permission-based, and egalitarian abstract power hierarchy. Those who are having their transactions degraded have no ability to project power on their own in a zero-trust and permissionless way to take back the administrative privileges that are being used to degrade them; they are wholly reliant on voters.

Recalling the core concepts from section 4.6 on creating abstract power, to seek permission or social approval from someone – particularly software administrators – is to tacitly give them abstract power over you. On the flip side, if you want to create and wield abstract power over a large population of humans, simply convince them to adopt a belief system where they need permission or social approval from you. **Once a population has adopted a belief system where they need your permission or approval (which you can trick them into adopting by obfuscating the permission-based control hierarchy), you have successfully gained abstract power and influence over them. This is what proof-of-stake does**.

In direct contrast, proof-of-work systems give these special administrative permissions freely to anybody willing to solve its proof-of-work algorithm, not those who own BTC. Because the "work" expended to solve this algorithm is a physically real phenomenon ("word" refers to real power, a.k.a. watts), that means the special administrative permissions of the Bitcoin network can (1) be physically impeached or countervailed by any user of the network for any reason at any time by simply overpowering them with more hash, and (2) be physically decentralized across the planet. Neither of these important emergent properties are true for "stake."

Despite the fact that "stake" doesn't physically exist, people claim that administrative control over Ethereum 2.0 is "decentralized." Here, the word "decentralized" is used in quotes to remind the reader that an abstract software object like "stake" (which doesn't physically exist) cannot be verifiably decentralized because "stake" isn't a real thing. Since "stake" is purely an arbitrary semantic description of an abstract object, that means the special administrative permissions given to Ethereum 2.0 validators can only be "decentralized" in name only. The following point demands to be repeated because the general public doesn't seem to understand this: **it's physically impossible to prove that the special administrative permissions given to those with "stake" is decentralized (or conversely, that it wasn't *already* centralized from the very start) because "stake" doesn't physically exist**. The only place that Ethereum's "stake" can be "decentralized" is in people's collective imagination.

Ethereum's developers claim that "stake" is "decentralized" because it has been divided across different validators. However, logically partitioning "stake" across different addresses or validators doesn't mean those addresses or validators aren't controlled by the same person or group of people (groups like the Ethereum Foundation), so it's impossible to claim that "stake" is "decentralized" merely because it has been mathematically partitioned across multiple addresses or across multiple validators. The same person could easily control millions of validators, and the fact that validator node hardware has *minimal* physical signature actually makes it *easier* for people to disguise how many validators they own and control.

This highlights another baked-in security feature of Bitcoin's hashing infrastructure that some might not appreciate: the fact that Bitcoin's hashing infrastructure consumes so much energy gives it a large physical footprint that's much *harder* to hide or disguise, thus *harder* to centralize under one person or organization's control. In other words, the more power that Bitcoin's hashing infrastructure consumes and the more heat that it produces, the easier it is for people across the world to independently verify that it's decentralized. On the contrary, the fact that Ethereum 2.0s validator network consumes so little energy makes it far easier to obfuscate how centralized the special administrative permissions coupled to "stake" is. The lack of Ethereum's electricity consumption essentially functions as a cloaking or stealth mechanism for people or groups (e.g. the Ethereum Foundation) to disguise how much abstract power and validation authority has been consolidated under their direct control.

It's already public knowledge that the majority of stake-able ETH was awarded to Ethereum 1.0s founders before it was released to the public. Combining this fact with the fact that "stake" doesn't physically exist, this means nobody can verify that Ethereum 1.0s founders don't already have the majority of Ethereum 2.0s stake-able ETH. If nobody can verify that Ethereum's founders don't already have the majority of stake-able ETH, that means nobody can verify that Ethereum 1.0s founders haven't already consolidated majority control over Ethereum 2.0s special administrative privileges. So not only is it physically impossible to verify that "stake" is decentralized, **there's no indication to believe or verify that Ethereum 1.0's founders don't *already* have unimpeachable centralized control authority over Ethereum 2.0.**

To make matters even more systemically vulnerable, Ethereum 2.0's staking protocol is designed so that people who have more "stake" are rewarded with proportionally more "stake," and those rewards are compounding. This is an extremely noteworthy design feature. Ethereum 2.0 developers deliberately added additional design logic into the system which gives the people who have the most "stake" the most rewards. In other words, the protocol is intentionally designed to give the people who already have the most abstract power even more abstract power than anyone else in the future. The more "stake" they control today, the more "stake" they will get to control in the future. The emergent effect of this design logic is that the people who have the most abstract power and control authority over the system today get to keep it forever. Not only that, but the abstract power also grows and consolidates as a higher percentage of the system's total stake-able ETH is gradually awarded to them because of the protocol's compounding rewards system.

The author has now highlighted five systemic security vulnerabilities associated with Ethereum's conversion from a proof-of-work system to a proof-of-stake system that are quite easy to recognize. First, proof-of-stake causes people to forfeit their ability to impose severe physical costs on attackers, effectively domesticating users. Second, proof-of-stake is "decentralized" in name only because "stake" doesn't physically exist and therefore can't be verifiably decentralized. Third, validator hardware has a minute physical signature making it easy to hide centralized control over validators. Fourth, the system is intentionally designed to give the people with the most "stake" today the most "stake" in the future, thereby consolidating (i.e. centralizing) "stake" into the hands of the people who already have the most "stake." Fifth, it's already public knowledge that the people with the most stake-able ETH are probably the people who awarded themselves with the most ETH when they originally developed the software, and there's no way to verify they don't continue to have it.

Ethereum 2.0 is therefore a decentralized-in-name-only (DINO) abstract power hierarchy. **Users who accepted the conversion from Ethereum 1.0 to Ethereum 2.0 have adopted yet another belief system that gives people asymmetric and consolidated abstract power over them, and then trusts them not to exploit or abuse it**. By coupling special administrative privileges needed to approve transactions to those who have the most "stake," a proof-of-stake system removes all of the physical constraints that were keeping those special administrative privileges physically constrained and physically (thus verifiably) decentralized. Now, the people who have the most "stake" have the critical administrative privileges to approve transactions, but that "stake" is completely imaginary.

Switching from proof-of-work to proof-of-stake effectively functions as a way to change and then obfuscate which group of people have the most important administrative privileges and control authority over the system. People buy into this sleight-of-hand because they don't understand enough about the complexities of computer theory to understand that "stake" doesn't physically exist and is purely an object-oriented abstraction for what could also be called "rank" or "admin rights." In other words, "stake" is nothing more than a metaphor for abstract power. People with the most "stake" have the most administrative privileges to approve/disapprove transaction requests, thus they wield substantial abstract power. Moreover, we already know that the people who probably have the most "stake" are Ethereum's core group of developers and co-founders, since they overtly awarded themselves the majority supply of stake-able ETH several years ago and they can't prove they don't still control it. This concept is illustrated in Figure 87.

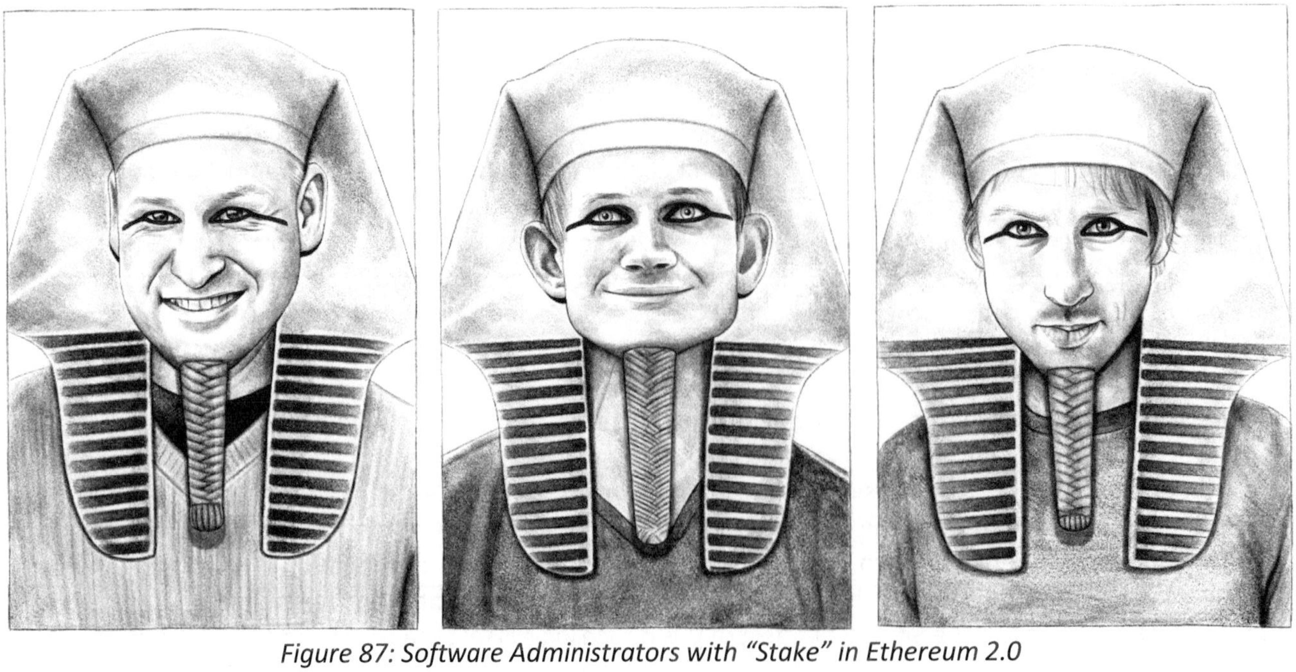

Figure 87: Software Administrators with "Stake" in Ethereum 2.0

Not surprisingly for anyone who understands power projection tactics in human society, Ethereum developers will argue that Ethereum 2.0 is still systemically secure against exploitation because of the (systemically exploitable) voting logic (a.k.a. "slashing") encoded into the software that will ostensibly prevent them from abusing the special authorities given to them by their "stake." Users must assume that staked ETH isn't already majority-controlled by one organization to believe that the voting logic for "slashing" belligerent stakers will work as intended, which is a blind and unverifiable assumption.

Here we have a situation where the people who gave themselves the most abstract power are claiming that users can hold them accountable by "slashing" their abstract power using a (systemically exploitable) voting system that *they* encoded. This would be yet another situation where the people with abstract power try to convince their believers that a belief system is secure against systemic exploitation because of systemically exploitable logic that was encoded into the system to constrain abstract power by the people who have the most abstract power. Remarkably, despite literally thousands of years of written testimony from history (discussed at length in section 4.8) to show that design logic (especially voting systems) can't secure a belief system against the exploitation of logic, people accept this argument.

5.11 Softwar

"World War III will be a guerrilla information war
with no division between military and civilian participation."
Marshall McLuhan [184]

5.11.1 To Highlight how Unique and Different Bitcoin is, Call it Bit *Power* instead of Bit *Coin*

To correct the information hiding problem and make it easier to understand the irreproducible properties of Bitcoin, the author suggests using a more technically accurate name for the "proofs of power" produced by proof-of-work protocols: *bitpower*. The "power" part of bitpower refers to quantities of physical power (a.k.a. watts) drawn out of the environment over time, modulated, and then digitally sampled by the network of computers running the physical cost function protocol (a.k.a. the hashing network). The "bit" part of bitpower refers to the machine-readable bits of information generated from these sampled watts. Combining these two terms together, the word "bitpower" literally means quantities of power (a.k.a. watts) converted into (physically scarce, decentralized, and costly) bits of information that can be transferred across cyberspace. This concept is illustrated in Figure 88.

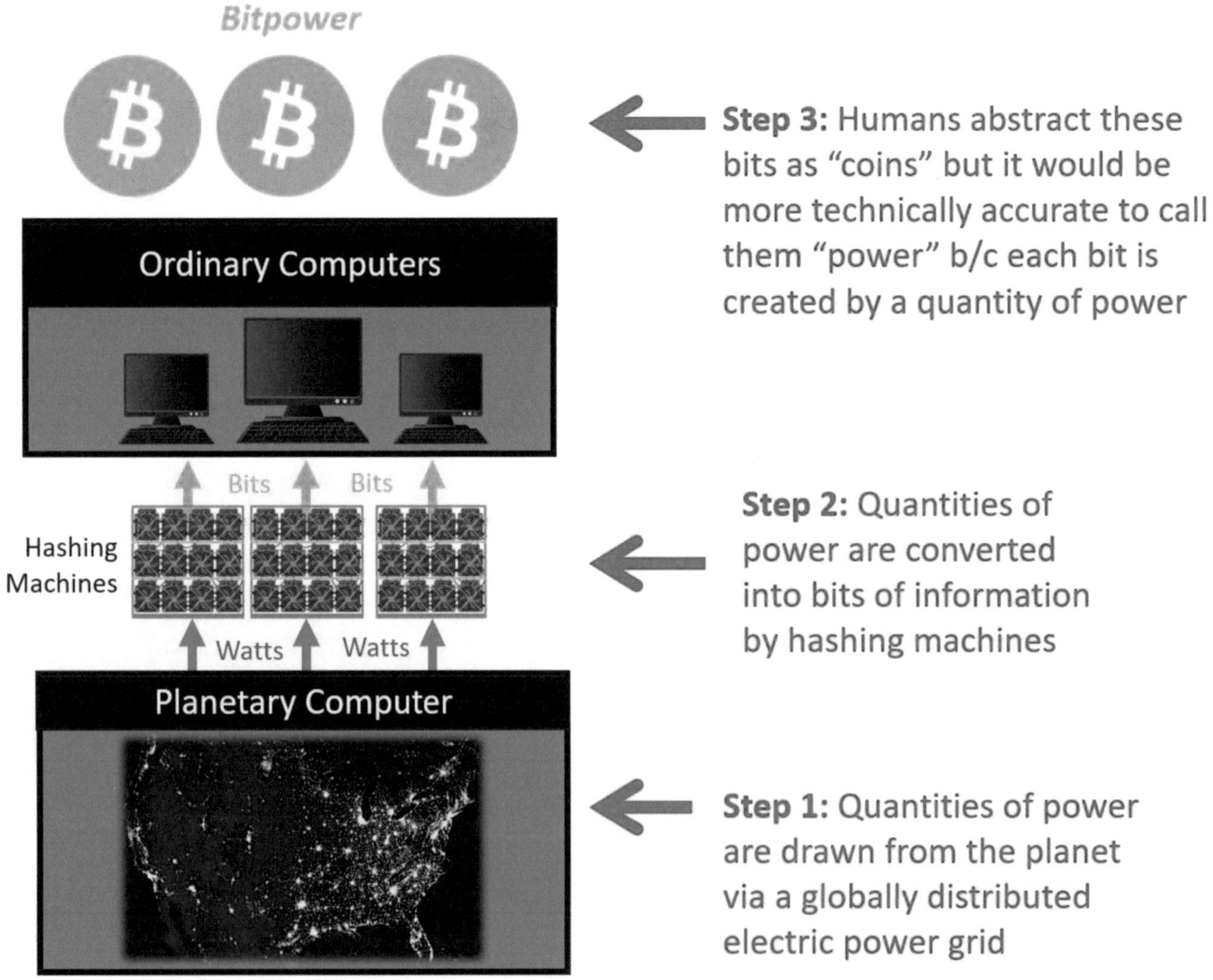

Figure 88: Graphical Illustration of Bitpower Concept
[169, 177, 154, 178]

Unlike most software specifications, it is important for the reader to understand that the term "bitpower" is not a metaphor. The word "bitpower" is meant to be taken literally, not metaphorically like most software abstractions (including the term "bitcoin"). Creating a literal description of proof-of-work technology is helpful because the literal definition of "bitpower" doesn't hide critical information about what's happening "under the hood" at the state mechanism level like the "bitcoin" metaphor does. "Bitpower" refers to the real-world quantities of physical power drawn out of the environment and consumed by the network of computers running the bitpower protocol, which is then converted into machine-readable bits of information that can be passed around cyberspace and called whatever people want to call it. **"Bitpower" isn't an imaginary concept designed to make it easier for people to understand one specific use case of the technology; it's literally power that people are using to program computers and store information to support any use case.**

The reason why the term "bitpower" is not metaphorical is because the protocol leverages real-world quantities of physical power (as opposed to transistor states) to transmit, receive, and store bits of information. Like many features of the Bitcoin protocol, the term "bitpower" is a little confusing because it's recursive. Power serves as both the underlying state-changing mechanism *and* the symbolic meaning assigned to that state change using machine code. When utilizing bitpower protocols like Bitcoin, people are not just utilizing state-changing transistors to store bits of information and then arbitrarily choosing to describe those bits of information using a metaphor. Instead, they are utilizing real-world quantities of physical power and then choosing to describe that real power as literal (albeit digitized) power.

"Bitpower" therefore doesn't technically qualify as hypostatization like "coin" does. As discussed in the previous chapter, hypostatization is when people treat an abstract idea as if it were a concretely real thing. When people describe bits of information using metaphors like "coin" and then act like those "coins" are concretely real, they're hypostatizing. On the contrary, the term "bitpower" describes bits of information as digitized power, which is something that is a concretely real thing (large quantities of electric power drawn out of the surrounding electric power grid are real-world phenomena that everyone can independently see and measure directly). Bitpower is therefore the digitized version of itself that can be utilized by machines inside the digital environment we call cyberspace. Again, it's confusing because it's recursive: bitpower is literal power – the digitized version of itself that can be wielded in, from, and through an abstract domain called cyberspace.

This is a subtle but important distinction to call out because it illustrates the previously mentioned concepts about how the "coins" created by other so-called "blockchain" or "cryptocurrency" technologies are highly incomparable to Bitcoin's "coins." To make it easy to see how dominant and nearly incomparable the Bitcoin protocol currently is in comparison to other "cryptocurrencies," simply call it bitpower instead of "coin". Using this non-metaphorical naming convention, it's easier to see that bit for bit, no other bits can claim to represent remotely as much physical power as Bitcoin's bitpower. As of this writing, Bitcoin represents 94% of the hash rate of all physical cost function protocols combined [182].

Since hash rate doubles as a proxy measurement for the quantity of real-world power consumed by the network and converted into bits of information, this means the Bitcoin network is currently sixteen times more powerful than all other proof-of-power protocols combined (the top contenders are currently Dogecoin, Litecoin, Bitcoin Cash, Monero, Ethereum Classic, Dash, and Zcash). This is noteworthy because **public perception of Bitcoin's incontrovertible dominance in global adoption has been skewed by people who misunderstand, deliberately ignore, or attempt to marginalize/condemn Bitcoin's power usage**. If one were to ignore or marginalize Bitcoin's power output or try to arbitrarily categorize it as merely a "blockchain" or "cryptocurrency," then Bitcoin would appear to be one of thousands of "cryptocurrencies" with no discernable lead. However, if one were to account for the power output and actually recognize it as the primary value-delivered function of the protocol, then Bitcoin is clearly the dominant protocol. This concept is illustrated in Figure 89.

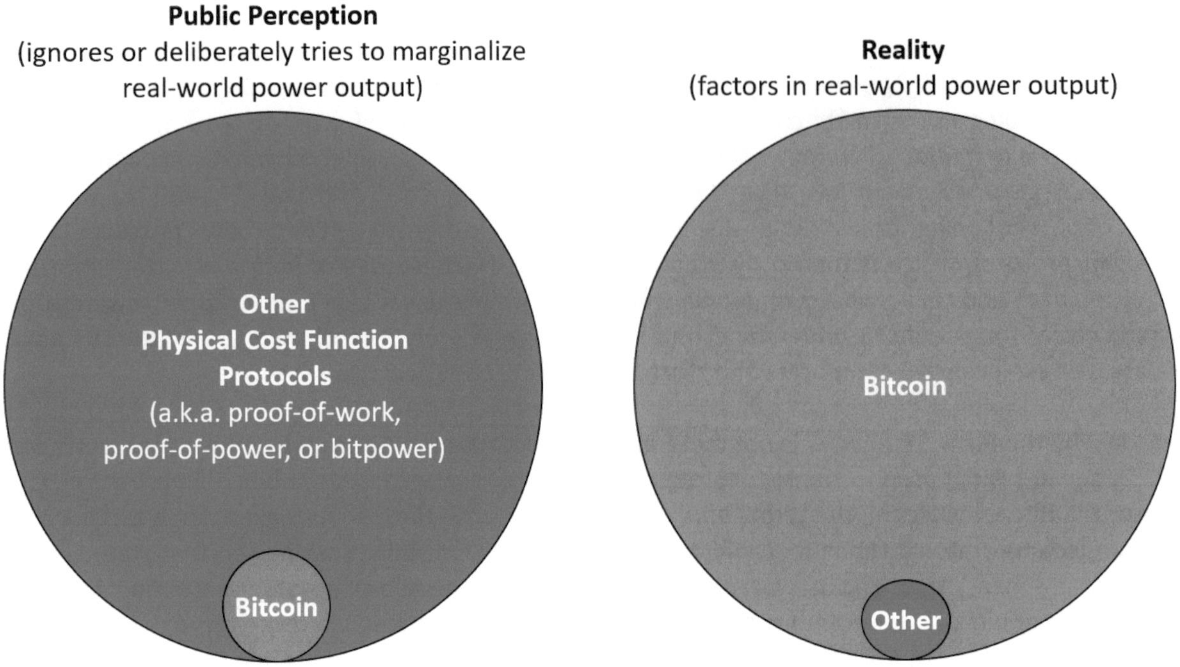

Figure 89: Illustration of the Difference Between Public Perception & Reality of Bitcoin's Dominance [183]

Considering how Bitcoin's primary value-delivered function is to produce physically costly bits of information, trying to create a more energy-efficient version of Bitcoin would defeat the point of Bitcoin. For this reason, those who claim that Bitcoin needs to become more energy efficient are highlighting their own lack of understanding about what the purpose of this protocol is, and how it reverse-optimizes bits to make them physically costly. This also implies that the repeated attempts to duplicate the Bitcoin protocol which fail to produce the same physical power output of Bitcoin only serve to further show how dominant Bitcoin has become as the most widely adopted proof-of-work network. Michael Saylor (MIT '87) famously summed this up in 2021 when he declared, *"there is no second best."*

5.11.2 If Power Projection Theory is Valid, then Bitpower is Extraordinarily Valuable

It's worth repeating that as of today, the concepts presented in this thesis are just theories. However, if these theories hold up to scrutiny within the computer science community, then this would imply that **bitpower protocols like Bitcoin fundamentally represent a new power projection tactic – a clever way to impose physical costs on other people through their computers by turning physical power itself into bits of information**.

When someone describes bitpower, they are technically describing the real-world emergent properties and effects of physical power in a machine-readable format that can be controlled and projected against others in, from, and through the internet. If these theories prove to be valid, then it means sapiens have figured out how to harness physical power to increase their C_A and lower their BCR_A to keep themselves secure when operating in the virtual domain by simply by choosing to treat real power as bits of information so that it can be wielded in an imaginary domain called cyberspace. It's a simple concept: to project real power in a virtual world composed of nothing but bits of information, convert power itself into bits of information.

Herein lies a core insight about physical costs function protocols like Bitcoin. **Adoption of these protocols means entire populations of people are projecting real-world quantities of physical power against each other to impose real-world physical costs on each other, but doing it electrically rather than kinetically, via the cyber domain**. Agrarian society appears to be doing this trans-nationally, in a non-lethal way which doesn't look anything like physical power projection capabilities of the past.

This would imply that **a global-scale physical power competition is being waged and nations don't appear to be aware that it's happening because the battlefield is "inside" the disembodied virtual reality of cyberspace, where nobody can be physically harmed, and things are easily hidden by abstractions and metaphors**. If, for the sake of argument, we accept the author's theories as valid, then it begs the question: over what exactly is agrarian society competing? To answer this question, the author further examines physical cost function design concepts produced by Hal Finney and Satoshi Nakamoto.

5.11.3 The Gabriel's Horn Paradox Shows it's Logically Possible to be Infinitely Scarce yet Infinitely Scalable

An important feature to note about Nakamoto's design of Bitcoin is that the bitpower it produces is supply-capped, making it perfectly scarce. Paradoxically, that same supply-capped quantity of bitpower is also infinitely scalable. Despite having a strictly limited supply, Bitcoin's bitpower can accommodate any number of users and convert any amount of real-world physical power into bits. This is a logically counterintuitive emergent property that is made possible by the fact that bitpower is digital and therefore has no mass. To better understand how this works, the reader is invited to reflect upon a phenomenon called Gabriel's Horn.

A popular character who appears in the holy texts of all three Abrahamic religions is an angel named Gabriel. In Catholic traditions, Gabriel is a Saint and Archangel who wields a horn to herald the Lord's return to Earth. Many artistic depictions of Gabriel show him wielding a long and slender heralding horn, and that horn doubles as a symbol of human salvation. In the field of mathematics, Gabriel's horn is the namesake of a special geometric shape. If you plot the function on a two-dimensional graph, chop off the resulting curve from the point where x and y both equal 1, and then rotate the curve across either axis, you get the famous volume of revolution known as Gabriel's horn. The reason why this geometric shape is called Gabriel's horn is simply because it bears a striking resemblance to Gabriel the Archangel's heralding horn. This is shown in Figure 90. [185]

Figure 90: Gabriel's Horn

Gabriel's horn (the geometrical shape) is noteworthy because it has special mathematical properties. The volume of space inside the horn is fixed, but the surface area of the horn is infinite. This is a mathematically sound shape that could theoretically exist in three-dimensional reality. If it did exist, the volume of the horn could easily be filled with a fixed quantity of paint, yet its surface could never be completely covered in paint, no matter how much paint someone had. This is a logical paradox known as the *Gabriel's Horn Paradox* a.k.a. the *Painter's Paradox*. Practical limitations in manufacturing prevent Gabriel's horn from being built, but the point still stands that it's a mathematically sound shape that could theoretically exist in three-dimensional reality. It would therefore be perfectly (but paradoxically) logical to design something with a fixed volume of supply but infinitely expandable (thus infinitely scalable) surface area. [185]

Understanding Gabriel's Horn Paradox is key to understanding a very important feature of Bitcoin. Because of the logical constraints encoded into Bitcoin's software, the bitpower produced by the Bitcoin protocol has the same paradoxical behavior as Gabriel's horn. Bitcoin's supply of bitpower has a fixed volume, yet it can scale to accommodate any number of users and any amount of real-world physical power, and do this all while remaining infinitely divisible. No matter how much real-world physical power is consumed by the computers attempting to solve Bitcoin's hash function, the total number of sequentially reusable bits of "coin" produced by the protocol to account for that power expenditure remains completely fixed.

These dynamics create a logical paradox that is difficult for some to understand: the total quantity of real-world physical power (a.k.a. watts) consumed and accounted for by the protocol is theoretically infinite, but the total quantity of bitpower exchangeable in cyberspace is fixed. At the same time, even though the total supply of bitpower remains fixed, an unlimited number of people can have access to infinitely smaller pieces of it that can also account for infinitely larger amounts of power. A single bit of bitpower produced by the Bitcoin protocol can scale to represent theoretically unlimited quantities of real-world physical power. And because bits have no mass and are infinitely divisible via logic, then (1) everyone in the world can access Bitcoin's bitpower, and (2) everyone who accesses any amount of bitpower has infinitely scalable physical power projection capacity.

Instead of converting a practically unlimited supply of physical power (a.k.a. watts) into a virtually unlimited supply of bitpower, Nakamoto's design intentionally utilizes the Gabriel's Horn Paradox. At first glance, Nakamoto's design appears to break the laws of physics. However, the Gabriel's Horn Paradox proves that the design is mathematically sound. **One way to better intuit the paradoxical behavior of Nakamoto's design is to abstract the bitpower it produces as surface area on Gabriel's Horn rather than ordinary geometrical shape like a "coin."** Nakamoto's physical cost function protocol takes an infinitely expandable quantity of physical power drawn from the environment and digitally converts it into a fixed volume of bitpower that can account for an infinitely expandable supply of watts. So instead of visualizing Bitcoin's bitpower tokens as "coins," it might be more appropriate to visualize them as equally-partitioned sections of surface area on Gabriel's horn, as illustrated in Figure 91.

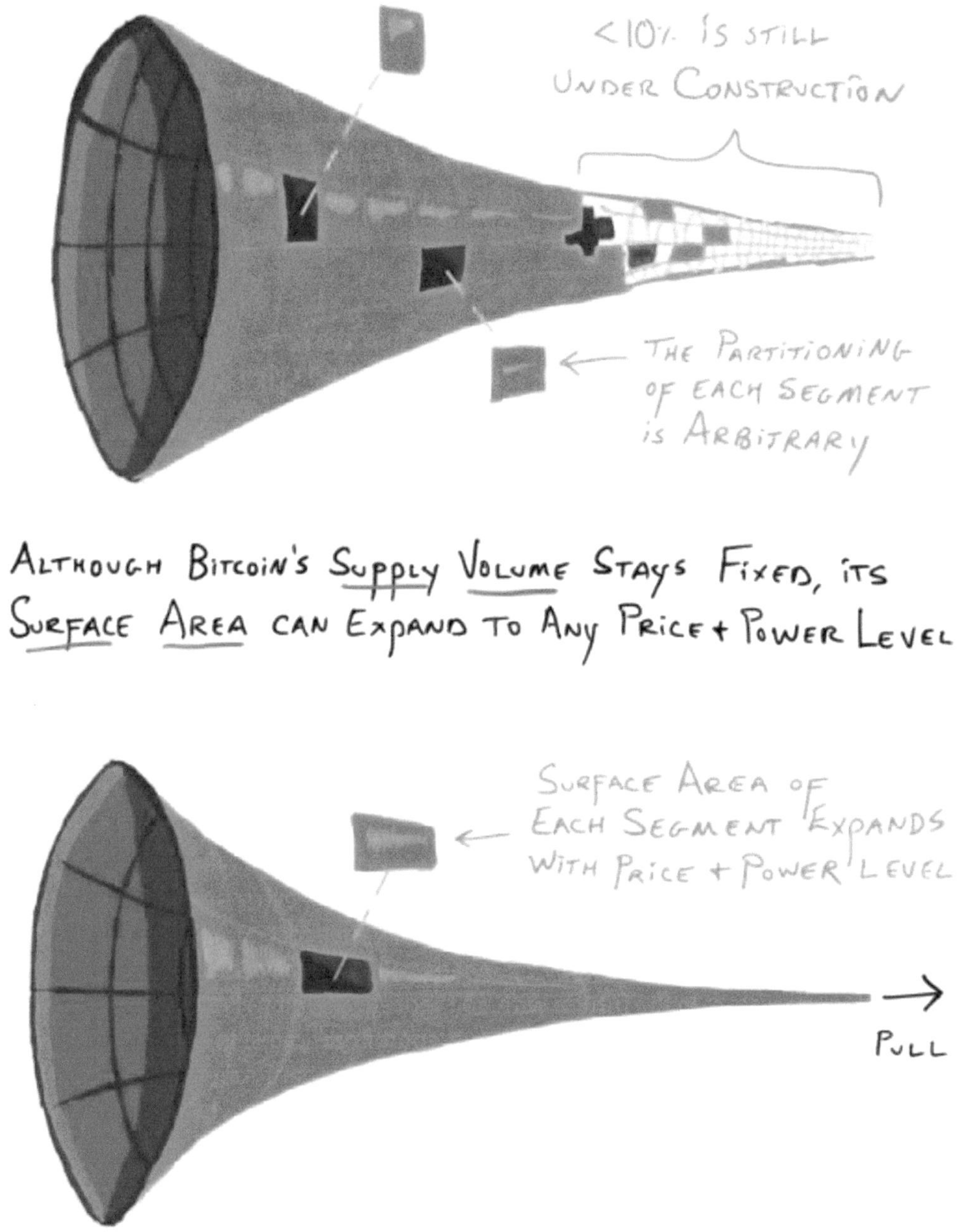

Figure 91: Visualizing Nakamoto's Bitpower Tokens as Surface Area on a Gabriel's Horn

Now may be a good time to remind the reader that using object-oriented software design, computer programmers arbitrarily describe the complex emergent behavior of computer programs as if they were objects. They do this to make it easier to intuit the complex emergent behavior of their programs, because humans are used to living in a three-dimensional world with objects oriented around each other. This means the choice to describe bitpower produced by the Bitcoin protocol as an ordinary object like a "coin" could just as accurately be described as equally-partitioned segments of surface area on Gabriel's Horn. Why do this? Because it highlights the protocol's paradoxically logical complex emergent properties.

The conversion of an infinitely-expandable supply of physical power into a fixed supply of bitpower creates two emergent properties that are important to explicitly call out. The first has already been discussed: **any infinitely miniscule amount of bitpower can represent an infinitely large quantity of physical power**. This is a useful behavior to have if the goal of the protocol is to maximize the physical cost associated with transferring, receiving, and storing each bit. A second emergent property worth calling out is that Bitcoin's bitpower can't be diluted as more physical power is added to the system. Consequently, those who own/control Bitcoin's bitpower automatically inherit the strength of all future physical power added to the system in a perfectly equal way.

Expanding on these two important features further, since there's no limit to how far the surface area of Gabriel's Horn can grow, there's consequently no limit to how far any tiny section of surface area can grow either. This means there's no limit to how much physical power any single amount of surface area can represent. Any infinitesimally small section of surface area on Gabriel's horn can expand or contract to represent any quantity of physical power drawn out of the environment by the computers running the protocol. Thus, there's no theoretical limit to how much physical power can be represented by a bit of bitpower, no matter how miniscule the amount. **The holder of a single "bit" of Bitcoin can theoretically project an unlimited number of real-world watts in, from, and through cyberspace**.

Moreover, even though the addition of more physical power into the system causes the percent magnitude of single units of physical power to decrease, this does not translate into cyberspace as a decrease in the percent magnitude of single units of bitpower. The percent magnitude of any amount of bitpower remains constant because the total supply of bitpower (as represented by the volume of the horn) remains fixed. Instead, the number of watts represented by each unit of bitpower must increase to account for more physical power added to the system.

The key takeaway from this design feature is that it creates a dynamic where it's not possible to dilute the physical power projection capacity of bitpower no matter how much additional physical power is added to the system. If you own one bit of bitpower, you own the same percent magnitude of physical-cost-imposing power in cyberspace no matter how much the system scales or how much more physical power the system consumes. This creates a substantial first mover advantage, as well as substantial path dependence that makes Bitcoin unforgiving to those who hesitate to acquire its bitpower (this is another important feature which makes it more of a strategic imperative to position oneself as an early adopter).

Combining these two emergent properties together creates a logically sound but paradoxical dynamic that could make Nakamoto's version of bitpower disproportionately valuable as a cyber security asset. Recall from section 5.6 that proof-of-power (a.k.a. bitpower) represents proof-of-physical-cost-of-attack or proof-of-C_A. If the supply volume of bitpower is fixed, then owning bitpower means automatically inheriting any of the physical-cost-imposing power added to the system in the future, with no theoretical limit to how much physical cost any amount of bitpower can impose on other people, computers, and computer programs. In other words, owning Nakamoto's supply-capped bitpower means being able to leverage infinite amounts of real-world physical power to impose real-world physical costs on others in, from, and through cyberspace without the capacity for it to be diluted by adversaries in the future no matter how much it scales.

Figure 92 provides an illustration of the paradoxical dynamics of how Nakamoto's design converts physical power to bitpower. The real-world, infinitely-expanding quantity of physical power (a.k.a. watts) consumed by Bitcoin's hashing network can be visualized as an infinitely-expanding volume taking the shape of a cube. Nakamoto's protocol takes this infinitely-expanding volume of power (visualized here as blue cubes) and digitally converts it into a fixed-supply volume horn with infinitely-expandable surface area (i.e. Gabriel's Horn).

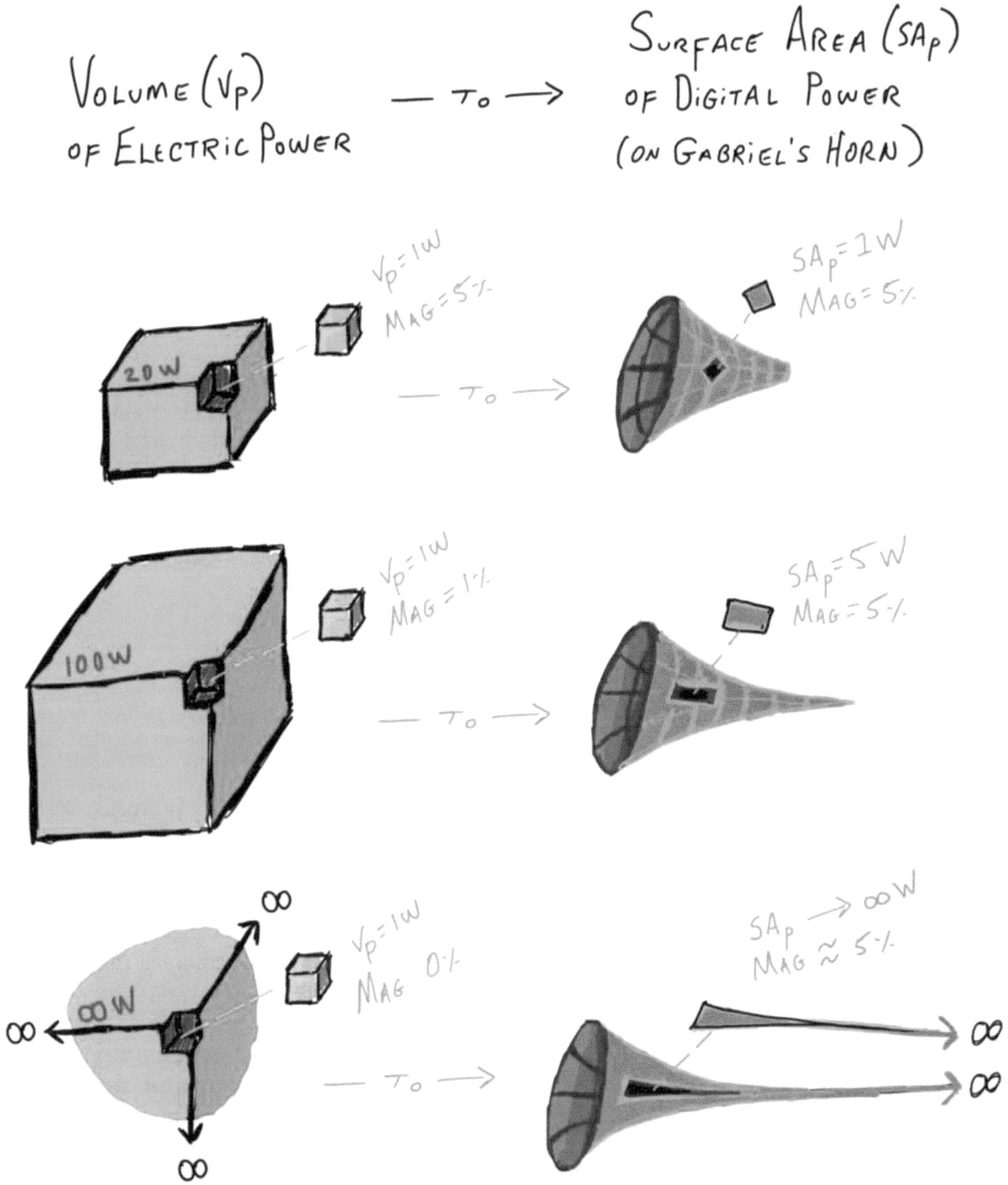

Figure 92: Logically Paradoxical Power Conversion Dynamics of Nakamoto's Bitcoin Protocol

The left side of this image illustrates the expanding supply volume of physical power drawn out of the environment by the computers running Nakamoto's protocol. This is represented as a cube with volume expanding towards infinity. If we were to isolate one unit of that cube's volume and measure its percent magnitude as the total volume of the cube expanded towards infinity, then we would see the percent magnitude of that single unit of volume gradually fall to zero. In this image, we can see that the percent magnitude of one watt of physical power drops from 5% to 0% as the overall volume of electric power drawn out of the environment expands from twenty watts to infinity watts.

Nakamoto's design digitally converts this volume of physical power into information (bitpower) that is logically constrained as if it were the surface area of a Gabriel's Horn. Here we can see how the digital representation of bitpower in virtual reality can be implemented as infinitely expandable surface area on a fixed-volume Gabriel's Horn geometry, in contrast to the geometry of an infinitely expanding cube with infinitely expanding volume. This sleight-of-hand is possible because bitpower is digital (i.e. disembodied or massless). If we were to isolate one segment of that horn's surface area and measure its percent magnitude over time as the total surface area of that horn stretched towards infinity, then we would see the percent magnitude of that single segment of horn surface area remain fixed. In this illustration, we can see that the percent magnitude of one segment of horn surface area remains fixed even as it expands towards infinity.

The bottom line is that Nakamoto's design creates a paradoxical but mathematically valid behavior that causes the protocol's bits to have unique and counterintuitive properties that are perhaps easier to intuit if visualized as sections of surface area on Gabriel's Horn rather than ordinary objects like "coins." By visualizing a bitpower as if it were a segment of surface area on Gabriel's Horn, it perhaps becomes easier to understand the paradoxical behavior caused by digitally converting infinitely scalable quantities of real-world physical power into a fixed supply of bitpower. **The most important takeaway of this design concept is that an infinitesimal amount of bitpower can represent an infinitely large quantity of physical power for any number of users without dilution, giving it potentially disproportionate strategic value as a cyber security asset that is both infinitely scarce and infinitely scalable. This creates an unforgiving first mover advantage and path dependence that is important for people to understand**.

5.11.4 Bitpower could Function as Digital-Age Freedom Fighting Technology

"... we can win a major battle in the arms race and gain a new territory of freedom for several years."
Satoshi Nakamoto [20]

Summarizing the core theoretical concepts about bitpower presented so far: bitpower protocols like Bitcoin look inefficient to those who might not understand the complexities of computer theory, cyber security, agrarian power dynamics, human metacognition, or the basics of how living organisms establish dominance hierarchies to manage control over their resources. Consequently, people don't understand how and why people would be inclined to summon large quantities of electric power from the environment, convert it into machine-readable bits of information, and use it for systemic security purposes – particularly to countervail the threat of people wielding too much abstract power via the computers they program.

As theorized in previous sections, the electric power deficits produced by proof-of-work protocols appear to turn the global electric power grid into a planetary-scale state-changing mechanism that can transmit, store, and receive machine-readable information, thus adding a physically constrained and thermodynamically restricted state space to cyberspace with complex emergent properties that are physically impossible for ordinary computers to replicate. A physically constrained and thermodynamically restricted state space within cyberspace is a novel capability that appears to have been missing from cyberspace thus far, and society appears to be starting to grasp the broader social, technical, and strategic security implications of this capability (this thesis is proof).

The bottom-line up front is that **for the first time, people appear to have discovered how to project physical power in, from, and through cyberspace by converting global power itself into bits**. Thanks to protocols like bitcoin, people can utilize bits of information that have been reverse-optimized to be physically expensive to produce to impose severe physical costs on belligerent actors who keep trying to interfere with people's bits or systemically exploit people through software. These physically expensive bits could also be used to prove the presence of real-world properties such as scarcity and decentralization in an otherwise artificial domain where ordinary computers can only be programmed to present the illusion of scarcity and decentralization.

One clear application for this capability is security against cybercrime and cyber tyranny. Bitpower protocols make it possible to project physical power in a unique way to keep oneself secure against computers, computer programs, and computer programmers by physically constraining them. **Rather than having to rely on permission-based abstract power hierarchies which can only logically constrain certain unsafe or exploitable control actions by giving only a few people with high rank and special administrative privileges the permission needed to execute them, software engineers now have a way to design control structures which can apply physical constraints and use physical power (not abstract power like rank or admin rights) as the basis for establishing control over software**.

Just like animals in nature and people in agrarian society leverage the systemic benefits of physical power to settle disputes, manage resources, and establish dominance hierarchies in shared objective reality, bitpower protocols like Bitcoin theoretically make it possible for people to utilize this exact same strategy in cyberspace. **We can now settle disputes, establish control authority, and determine the legitimate state of ownership and chain of custody of bits the old-fashioned way: through real-world physical power competitions. And remarkably, we can do it without causing injury because the protocol utilizes electric (i.e. non-destructive) power rather than kinetic (i.e. destructive) power**.

Using bitpower protocols like Bitcoin which convert real power into digitized power to wield in a digital domain, it is theoretically possible to design physical power-based resource control structures that physically constrain unsafe or insecure control actions and physically constrain people with abstract power. No longer do people have to design systemically exploitable abstract power hierarchies created by computer programmers that give special permissions to specific people and trust them not to abuse it. This is important because logically constrained software control structures and software-defined abstract power hierarchies are demonstrably prone to systemic exploitation either externally (via hackers) or internally (via abusive system admins or insider threats).

Accelerating cybercrime incidents have demonstrated how challenging software security is when people don't have the ability to impose physical constraints. We have discussed how it's impossible to logically constrain something against the exploitation of logic, whereas it's easy to physically constrain something against the exploitation of logic (simply make it impossible to justify the physical cost of exploiting logic).

We may be starting to see these same principles hold true for the emergence of cyber tyranny – there's no clear way to escape from massive-scale systemic exploitation of our software exclusively by writing more lines of code. The root cause of cyber tyranny would be the same as the root cause of our challenges with cybercrime: users have no way to physically constrain or thermodynamically restrict the abstract powers of software administrators who are abusing the trust we must necessarily put into these people not to exploit us through our bits. If we are to mitigate the threats of cybercrime or cyber tyranny, it's necessary for us to adopt some type of protocol which will empower people to secure their bits physically, by projecting physical power in, from, and through cyberspace to impose severe physical costs on those who exploit our computing systems. And that's precisely what the Bitcoin protocol is doing.

Physically countervailing oppressive abstract power hierarchies is the American way – the United States gained its independence by imposing severe physically prohibitive costs on high-ranking people who were exploiting them through their abstract power hierarchy. Thanks to Bitcoin, Americans now appear to have gained the same capability of digital freedom fighting for cyberspace.

Of course, the threat of tyranny and the need for revolution was not just exclusive to Americans. Agrarian society has suffered through several millennia of systemic exploitation from people creating, wielding, and abusing their abstract power. Revolutions make it clear that societies must remain capable of and willing to project physical power against their ruling class if they want to keep themselves secure against systemic exploitation, abuse, and oppression from their ruling class. A major difference between revolutions of the past and revolutions in the digital age is the technology used to exploit people and to physically resist that exploitation. The digital-age ruling class is the technocratic group of people who control our computers through the software they write. It's the people who administer our software, including but not limited to governments.

In previous chapters, we explored how self-domestication is a real threat. Society's lack of ability or inclination to project physical power (in any domain) to reduce their **BCR_A** is a well-known systemic security hazard that can be independently and empirically validated. The exact same systemic security hazard exists for cyberspace. Without the capacity or inclination to project physical power and impose physical costs in, from, and through cyberspace, a technocratic ruling class of neo god-kings have an opportunity to domesticate and systemically exploit human populations at a massive and unprecedented scale – simply by having physically unconstrained control authority over their computer networks.

Fortunately, thanks to electro-cyber power projection protocols like Bitcoin, people now appear to have the technology they need to countervail this threat. People could use this technology to countervail abstract power hierarchies when they inevitably become too exploitative. This technology could become instrumental to the design of different control structures or computer networks which make it impossible to justify the physical cost of exploiting people through their software. If these theories are valid, then the only thing preventing our digital-age global society from starting to physically secure themselves against the emerging threat of cyber tyranny is for them to (1) recognize the threat, (2) recognize they have this capability, and (3) start using it. **From the perspective of Power Projection Theory, it seems inevitable that we are going to have a digital-age revolution – a soft war for zero-trust, permissionless, egalitarian, and decentralized control over our world-wide web of computers. As Nakamoto himself declared, Bitcoin is part of an arms race which could help people gain new territory for freedom for years to come**.

The author hypothesizes that **agrarian society is now beginning to recognize that they have the capacity to project power in, from, and through cyberspace to secure themselves against the threat of systemic exploitation of their computer networks. The technology appears to be hiding in plain sight by masquerading as a peer-to-peer electronic cash system**. To that end, the purpose of this thesis is to illustrate how physical cost function protocols like Bitcoin represent far more than just monetary technology – they represent physical power projection technology. And if history is any indicator of what's to come, the emergence of a new form of physical power projection has the potential to be far more unique, valuable, disruptive, and strategically essential than a new form of money, especially as the global population continues to spend exponentially more time using computers.

The sociotechnical and national strategic impacts of Bitcoin envisioned by the author have little to do with finance. This is not a grounded theory about money, it's a grounded theory about power. The author asserts that the Bitcoin protocol is first and foremost a physical power projection protocol, and bitpower represents the ability to physically countervail abstract power and resource control authority of software administrators and physically decentralize the cyber empires they're trying to lock people into via cyberspace. This not only impacts businesses, but it also impacts governments. Because of proof-of-work technology, entire populations are now *literally* empowered to impose real-world physical costs on other people and programs via cyberspace – to include computer programs controlled by governments. Agrarian society can now strategically secure itself against enumerable attack vectors encoded (often intentionally) by any type of computer programmer, regardless of who they work for.

Therefore, the author asserts that Bitcoin is not monetary technology – at least, money doesn't appear to be its primary value-delivered function. Instead, Bitcoin appears to be an electro-cyber freedom fighting technology. In other words, **Bitcoin isn't a monetary protocol, it's a bitpower protocol. It creates a new base-layer foundation to the internet by adding a new state mechanism to the bottom of the internet tech stack which physically constrains the world-wide web of computers. It just happens to be the case that the first operational and widely-adopted use case for this novel internet infrastructure is to restore peer-to-peer financial payments. But it's reasonable to expect many more use cases to follow**.

In what could have been an intentional sleight-of-hand and perhaps one day be reflected upon as the most impressive Trojan Horse since the Trojan War, **the decision to frame Bitcoin as *only* a peer-to-peer electronic payment system has motivated trans-national human society to subsidize the development of a global-scale, electro-digital defense industrial complex that literally empowers people with zero-trust, egalitarian, and permissionless access to bitpower. Extraordinary quantities of physical power are now being drawn from a planetary-scale state mechanism and converted into physically costly bits of information that anybody can utilize to impose severe physical costs on people and programs in, from, and through cyberspace. With this new internet architecture, people are capable of physically securing themselves against hackers as well as physically securing themselves against systemic exploitation from of an emerging, technocratic ruling class – and they can do it all without spilling a single drop of blood**.

Power Projection Theory is a bold theory, but it's a grounded theory. It blends military theory, political theory, and computer theory together using first principles and systems thinking. If Power Projection Theory proves to be valid, then Bitcoin represents the emergence of a new internet tech stack which disrupts our perception of cyberspace from the bottom up. It's also a non-lethal form of freedom fighting that could empower people to physically resist tyranny and oppression in, from, and through cyberspace. By simply calling it a coin, this freedom fighting technology has blossomed and grown under the noses of those who made the classical mistake of expecting the next revolution to look like the last revolution.

5.11.5 A Vital New Resource Worth Fighting For

Physical power can and has been used to establish zero-trust and egalitarian control structures which can't be systemically exploited like abstract power hierarchies can. The systemic exploitation of abstract power hierarchies has been a problem that has plagued society for thousands of years, and software-defined abstract power hierarchies are just as vulnerable to exploitation as abstract power hierarchies like governments are. Bitpower protocols have a chance at counteracting the hazards of abstract power hierarchies, just like physical power has already done for thousands of years. If someone were to design an open-source protocol where everyone has zero-trust, egalitarian, and permissionless access to bitpower, it could be extraordinarily valuable as a cyber security asset. With these concepts fresh in our minds, the author will now revisit the design of the physical cost function protocols designed by Back, Finney, and Nakamoto.

As outlined in the previous two sections, attaching bitpower (a.k.a. proof-of-work or proof-of-power receipts) to sensitive control actions issued by a computer program makes it possible for engineers to design new types of control structures which can physically constrain the execution of virtually any control signal in cyberspace, making them too physically costly to systemically exploit and thus more systemically secure. This is a compelling security capability that appears to have been missing in computer science until the emergence of physical cost function protocols (a.k.a. proof-of-work or bitpower protocols) like the one built by Back.

The first iteration of Back's bitpower protocol issued non-sequentially-reusable bitpower that were "stamped" onto individual control actions for one-time use. This design requires computers to consume electricity and solve the hash cost function for every individual control signal that the protocol was used to physically constrain. This presents a technical problem: it makes any computer program utilizing the protocol run too slowly to be practically useful. If people had to wait several minutes for a computer to solve a difficult hash function algorithm every time they wanted to issue a command, then the protocol would likely not be adopted because of how impractical it would be to use.

Recognizing this technical issue with Back's bitpower protocol design, a computer engineer named Hal Finney created a more practical version of a bitpower protocol where the proof-of-power receipts (a.k.a. proofs-of-work) are sequentially reusable. That way, instead of having to wait for a computer to solve the hashing cost function each and every time someone wanted to send a command, they could simply present a proof-of-power "token" to show that the hash cost function was solved at some time in the past. Finney understood that a sequentially reusable version of bitpower would have the same effect of imposing severe physical costs on the computer trying to execute a command, but it would remove the lag of having to re-solve the hash cost function at the point of issuing a command. It's worth noting that this was not a new idea; this is precisely what Juels and Jakobsson described as a "bread pudding" protocol in their 1999 paper which introduce the term "proof-of-work." The main difference is that Finney actually built it, whereas others merely theorized. Using the same naming convention created by Juels and Jakobsson and adopted by Back, Finney called his design concept reusable proofs-of-work (RPoW). [186, 28]

The bottom line is that a major advantage gained from Finney's sequentially reusable bitpower protocol was speed. In Back's design, bitpower is destroyed after the execution of a single command. At scale, this creates an impractically slow system where the hashing algorithm must be solved to execute every new command. When bitpower is sequentially reusable, it becomes possible to make a much faster and more practically useful system while still preserving the primary value-delivered function of physically constraining computers by imposing severe physical costs on the execution of specific commands.

Another major advantage gained from Finney's RPoW design concept was that his version of bitpower could be used as a general-purpose proof-of-power protocol that could serve multiple different use cases rather than exclusively one application at a time (such as reducing email spam). Finney described this advantage as follows: "*Normally PoW tokens can't be reused because that would allow them to be double-spent.*" Finney writes. "*But RPoW allows for a limited form of reuse: sequential reuse. This lets a PoW token be used once, then exchanged for a new one, which can again be used once, then once more exchanged, etc. This approach makes PoW tokens more practical for many purposes and allows the effective cost of a PoW token to be raised while stilling allowing systems to use them effectively.*" [186]

Theoretically speaking, a single bit of sequentially reusable bitpower can be utilized an infinite number of times for any number of different use cases. It could also be sequentially reused at such a high frequency that it could be visualized as a continuous stream of "irradiating" bitpower rather than individually discrete transfers of bitpower. Reusable bitpower protocols therefore have very attractive qualities that could serve multiple applications. Unfortunately, as is the case with most engineering design decisions, the benefits gained from making bitpower sequentially reusable also introduces a tough design challenge.

By making bitpower "tokens" reusable, Finney had to figure out a way to enable sequential reuse while simultaneously prohibiting parallel reuse. Parallel reuse is a problem because if bitpower can be reused in parallel, it would effectively defeat the purpose of the protocol. Each bitpower would no longer be able to impose severe physical costs on computers because they could be duplicated ad infinitum, diluting the physical cost associated with each bitpower "token" down to zero (as a side note, this is the same reason why there has to be a hard cap on the supply of bitpower i.e. bit "coins").

Sequential reuse of bitpower doesn't undermine the laws of thermodynamics, but parallel reuse does. Power can be transferred, stored, and reused in series, but it cannot be created or destroyed, therefore it can't be cloned or exist in two places at once. Considering how bitpower's value is derived from its physical constraints, then allowing parallel reuse of bitpower would logically undermine its value. The utility of proof-of-power is derived from the fact that the physical power expended to produce bitpower is physically expensive (in watts) and thermodynamically restricted. This reverse-optimization is what makes bitpower unique and gives it its irreproducible emergent properties. A custodian who can reuse bitpower in parallel is someone who effectively has the ability to reproduce or clone real-world physical power out of thin air and then use it in two places at once. These custodians would be breaking the real-world constraints of physics that are ostensibly imposed by the protocol, making them false or fraudulent representations of real-world physical power in cyberspace (hence one of the many reasons why people claim that alternative systems like proof-of-stake are fraudulent). Remember: the point of these protocols is to reverse-optimize bits to make them physically expensive to transmit, receive, and store. If these bits can be used in parallel, or there are not hard limits on their supply, or they are not extraordinarily energy intensive to produce, then that defeats the point of the protocol.

To mitigate the problems of parallel bitpower reuse, Finney had to figure out how to design a protocol which enables sequential reuse of bitpower while simultaneously constraining parallel reuse. From a systemic design perspective, this is exactly the same challenge that financial systems have. Financial systems must figure out a way to enable the sequential reuse of monetary notes while prohibiting parallel reuse (a form of fraud where users make counterfeit bills or deliberately write checks they can't cash). The financial community calls this "the double-spend problem." If we were to (arbitrarily) abstract bitpower as "coins," then we could say that Finney's primary design challenge was the same challenge faced by financial systems: finding a way to solve the so-called double-spend problem.

In order for Finney's reusable proof-of-power protocol to function properly, it would require the addition of a third step to the physical cost function protocol that Back's version didn't have: keeping track of the position of each bitpower "coin" over time to make sure they're only reused in series, not in parallel. In other words, Finney's physical cost function protocol had to instantiate a ledger which keeps track of the legitimate state and chain of custody of each bitpower "coin" to make sure they weren't being used in two places at the same time. This is illustrated in Figure 93.

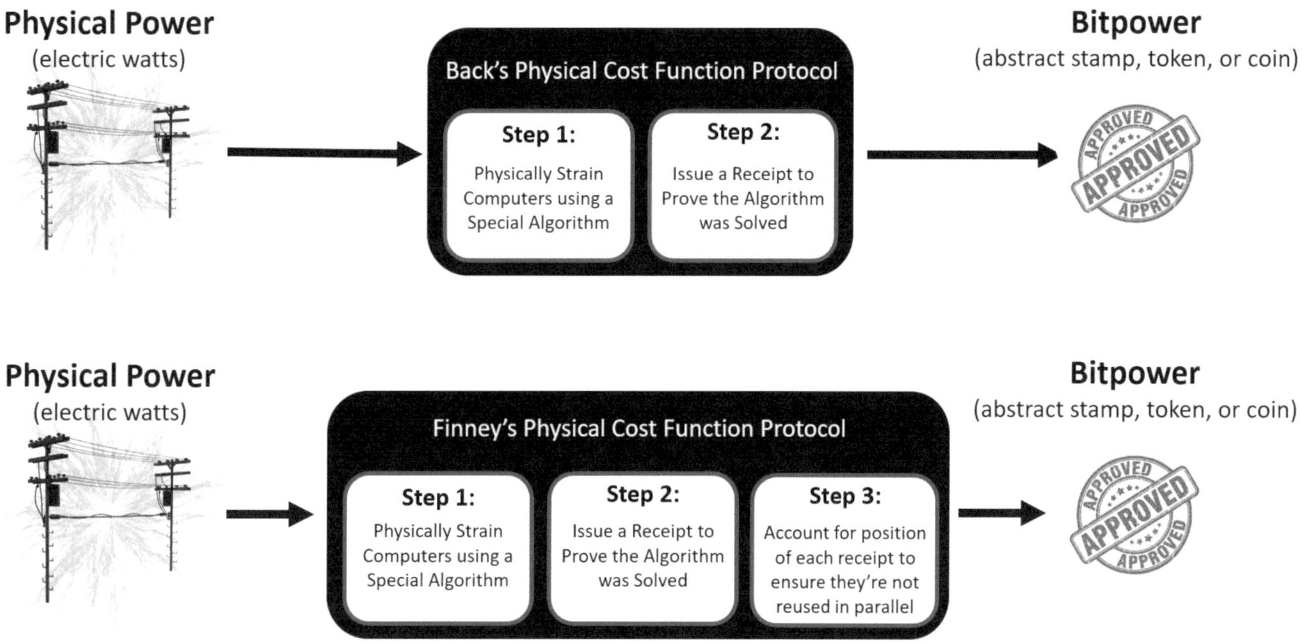

Figure 93: Illustration of the Third Step added by Finney's Physical Cost Function Protocol
[158, 159, 160]

Finney solved the double-spend problem the same way financial systems solve the double spend problem: by having a trusted party keep account of the position of each bitpower "token" to ensure they're not being used in two places at the same time. Unfortunately, this design choice introduces a significant systemic security vulnerability: it requires trust in a group of people with special permissions not to abuse their special permissions to exploit the system.

In other words, Finney's design implemented yet another form of abstract power hierarchy, where people with imaginary power (in the form of special administrative permissions to add transactions to the ledger) have tacit control authority over bitpower and its state of ownership and chain of custody. Like all abstract power hierarchies, this is a trust-based and inegalitarian resource control design which creates a ruling class of administrators and a ruled class of users. The ruled class must trust their ruling class not to abuse their special permissions, because users are otherwise physically powerless to countervail them.

Having just read hundreds of pages explaining how abstract power hierarchies are systemically insecure no matter how they're encoded, hopefully the reader can now appreciate why Finney's bitpower protocol design was flawed from a systemic security perspective. **Finney's reusable bitpower protocol design was flawed for the exact same reason that governments or any other form of abstract power hierarchy is systemically flawed: they're trust-based, permission-based systems which give people abstract power and control authority over a vital resource and trusts them not to exploit it for their own personal advantage, despite an ever-increasing benefit for doing so**. We have 5,000 years of written testimony which says that abstract power hierarchies break down and become dysfunctional no matter how well they're designed, and no matter what logical constraints they try to impose through their rulesets.

Bitpower has the potential to be an extraordinarily valuable resource. By creating a ledger which keeps track of the legitimate state of ownership and chain of custody of each bitpower token to allow for sequential reuse and prevent parallel reuse, then restricting access to writing the ledger to a trusted third party, Finney created an abstract power hierarchy where some entity has abstract power and control authority over the entire system by virtue of their having control authority over the ledger accounting for the position of each bitpower token. His protocol created yet another abstract empire where users must trust in a technocratic elite class of computer programmers not to exploit their control authority over a potentially vital resource.

Finney's design introduced useful features to bitpower protocols like sequential reusability and the ability to pseudonymously register ownership of bitpower using private key encryption. But at the same time, the system relied on a trusted third-party server to keep account of the position of each bitpower token. Therefore, **even though Finney's design successfully solved the double-spend problem and introduced other useful features, the way the system solved the double-spend problem had a glaring systemic security vulnerability – a vulnerability that a pseudonymous software engineer named Satoshi Nakamoto was quick to point out and resolve**.

5.11.6 Let Those with Real Power Compete for Control over Ledger-Writing Administrative Privileges in a Zero-Trust, Permissionless, and Egalitarian way that Will Decentralize that Control

Inspired by Back and Finney's design concepts, Nakamoto's design created sequentially reusable bitpower and also accounted for the sequential state of ownership and chain of custody of bitpower using a ledger just like Finney's design did. However, the primary difference was that Nakamoto's design didn't logically constrain ledger-writing administrative privileges to a trusted server wielding imaginary power. Instead, **Nakamoto chose to assign these special administrative privileges to those who demonstrate they have real power (a.k.a. watts). In other words, he instantiated a "might is right" or "favor the power projectors" pecking order heuristic for Bitcoin's control hierarchy, thus replicating the design of nature.**

Instead of creating an abstract-power-based dominance hierarchy to manage the ledger, Nakamoto created a system where people with real-world physical power (a.k.a. watts) engage in a global-scale physical power competition for ledger-writing control authority in a zero-trust, permissionless, and egalitarian way. **As a complex emergent benefit of this global-scale physical power competition, the special privilege of writing the ledger remains decentralized and impeachable, thereby empowering users with the ability to remain physically secure against systemic exploitation and abuse of those ledger-writing privileges**. Because physical power is unbounded, there is theoretically no limit to how many watts people can project to countervail the power of any other person or organization who might attempt to exploit or abuse these special ledger-writing administrative privileges.

Herein lies the cleverness of Bitcoin's design. The Bitcoin protocol empowers society to leverage physical power to battle for control authority over a vital new agrarian resource (bitpower). Through this global-scale power competition, agrarian society decentralizes control authority over the ledger and achieves consensus on the legitimate state of ownership and chain of custody of bitpower in a zero-trust, egalitarian, and permissionless way. By simply giving permission to write the ledger to those who win the physical power competition, Nakamoto solved the double-spend problem in a way that doesn't make people as vulnerable to systemic exploitation and abuse of software administrators.

With this simple design concept, the BCR_A of the Bitcoin ledger-writing privileges drops down to the point where it's impossible for attackers to justify the physical cost of attempting to exploit these privileges. The complex emergent benefit of this global-scale physical power competition for ledger-writing administrative privileges is decentralization of control over bitpower. In what could be described as an impressive display of design thinking, **Nakamoto replicated the complex emergent benefits of warfighting by creating an electro-cyber version of it**; he created a system where globally-divided populations with completely different priorities and belief systems can engage in a physical power competition to establish physically constrained, balanced, and decentralized control authority over this potentially vital resource to achieve global consensus on its legitimate state of ownership and chain of custody in a purely zero-trust, permissionless, and egalitarian way. To that end, what makes Bitcoin's design remarkable is not merely that it solved the so-called double spend problem or Byzantine general's problem.

5.11.7 The Way Nakamoto solved the Byzantine General's Problem Could Solve Much Bigger Problems

Nakamoto is given credit for solving the double-spend problem without having to rely on a trusted-third party to account for the chain of custody of sequentially reusable bitpower. To accomplish this, he designed a system which utilizes a distributed network of nodes to manage the ledger, rather than relying on a trusted third party (a glaring security vulnerability with Finney's design). Of course, Nakamoto's decision to utilize a distributed network to manage the ledger introduced another design challenge: how do you get a distributed network of computers that can't be trusted to agree on the legitimate state of the ledger? In computer science, this challenge is what's known as the Byzantine general's problem.

Because Nakamoto's protocol provided a viable solution to this challenge, Nakamoto is given credit for solving the Byzantine general's problem. However, the author asserts that crediting Nakamoto's Bitcoin protocol for solving the Byzantine general's problem undersells how significant it is. Using the core concepts of Power Projection Theory, we can better appreciate why Nakamoto's design could be far more significant than merely solving a distributed computing challenge – this technology also has the potential to also solve a much bigger challenge faced by society as a whole: a global strategic nuclear kinetic stalemate.

The design challenge which Nakamoto faced was the age-old challenge of establishing zero-trust, egalitarian, and permissionless access to a resource. Whoever has administrative authority over a bitpower ledger implicitly gets control authority over people who depend on bitpower. That control authority can be exploited; the ability to add transactions to the ledger doubles as the ability to deliberately withhold transactions from the ledger. This means the ability to add transactions to the ledger doubles as the ability to denial-of-service (DoS) attack users. By simply choosing *not* to add valid transactions requested by a given user, that user can be DoS attacked. Just by gaining permission to write the ledger, an entity gains a form of abstract power over people in the form of special permission, which they can exploit to deny someone access to their property. Determining who gets to write the ledger therefore represents an extremely sensitive control action with major systemic security implications, which must be property constrained by a control structure design.

From a broader systemic perspective, we can see that the problem Nakamoto faced with determining how to establish ledger-writing control authority over the bitpower ledger was not merely a problem faced by computer scientists working with distributed computing networks. This is a problem that has been plaguing human society for thousands of years. This is a problem that life itself has wrestled with for the past four billion years. People miss this critical insight because they frame Nakamoto's design challenge as this: "how do you get a distributed network of computers to agree on the legitimate state of the ledger?" But an equally accurate way to frame Nakamoto's design challenge has far bigger scope: "how do you establish control authority over a resource in a zero-trust, egalitarian, and permissionless way that doesn't utilize a systemically exploitable abstract power hierarchy?"

Establishing zero-trust, egalitarian, and permissionless control authority over a resource is fundamentally a question about how to settle disputes *without* the use of abstract power like rank. This is fundamentally a question about achieving consensus between strangers *without* a judge, or a god-king, or some other kind of person wielding abstract power derived from permissive authority. For at least 10,000 years, sapiens have done this the same way animals do it: using a physical power projection protocol we call warfare.

The problem Nakamoto faced with determining how to establish zero-trust and egalitarian control authority over bitpower is therefore not merely a challenge faced in distributed computing; this is a challenge that sapiens have encountered since at least the dawn of the Neolithic age. **Sapiens have been struggling to figure out a zero-trust and permissionless way to manage valuable resources and establish pecking order without injuring each other, and they have never succeeded to date**. The most they have been able to achieve is creating temporary and fleeting reprieves from the destruction and injury of warfare; they have never been able to remove the destruction and injury of warfare altogether. **If Nakamoto successfully solved the challenge to which he's often given credit, that could not just mean he solved a core challenge in computer science, that could mean he solved a core challenge for agrarian civilization: he created a non-lethal form of warfighting. In other words, he created *softwar*.**

It is difficult to understate how big a deal it would be if it were true that Nakamoto did indeed discover a way for global society to achieve zero-trust, egalitarian, and permissionless control over bits of information passed across the internet. That would mean Nakamoto figured out a way for global society to establish zero-trust, egalitarian, and permissionless control authority over many of digital-age society's most precious resources without requiring kinetic (thus fratricidal) warfare. From this perspective, the fact that people have arbitrarily decided to call the bits of information secured by the Bitcoin computing network a "coin" seems trivial in comparison to the fact that **Nakamoto may have discovered how to build a softwar protocol that society could use to decentralize the internet and physically countervail the abstract powers and control authorities of those who are systemically exploiting its current architecture (which would include but is not exclusive to nation states)**.

The core concepts of Power Projection Theory bear repeating: humans already know how to establish zero-trust, egalitarian, permissionless, and decentralized control authority over resources. We have already developed a protocol where different populations of people who can't trust each other settle disputes and achieve consensus on the legitimate state of ownership and chain of custody of resources without the use of trusted third parties wielding some form of abstract power (i.e. rank, admin rights, server owner) and control authority in some form of abstract power hierarchy. Society already knows how to leverage physical power in an inclusive, unbounded, and systemically exogenous way to achieve decentralized control authority over resources: the protocol is called warfighting.

Therefore, what makes Bitcoin special isn't just that it allows for a bunch of people across the world to establish zero-trust, egalitarian, permissionless, and decentralized control over precious resources because we already do that with warfare quite effectively (hence why Earth's natural resources are split across 195 different countries). What makes Bitcoin special is *how* it achieves this same end state. It does it electronically through cyberspace, rather than kinetically through sea, land, air, or space.

If it is true that Nakamoto solved the problem he is given credit for solving, then that could imply he also solved a much bigger, much larger-scope problem that has been plaguing society for thousands of years: the problem of how to establish control authority over precious resources, how to settle disputes, how to establish dominance hierarchies, and how to achieve global consensus on the legitimate state of ownership and a chain of custody of property – but do it all in a way that requires no god-kings, no bloodshed, and thus circumnavigates the threat of fratricide or mutually assured destruction.

5.11.8 Bitcoin is Zero-trust, Permissionless, and Decentralized Control over a Precious Resource: Itself

This section summarizes all of the concepts discussed so far. Bitcoin appears to be a globally-adopted physical power projection protocol, or softwar protocol. The first war that agrarian society appears to be fighting using this softwar protocol is a war to determine who has control authority over bitpower, as well as a war to establish consensus on bitpower's legitimate state of ownership and chain of custody. Just like other resources that agrarian society routinely fights over (e.g. land), the complex emergent effect of this softwar is decentralization of control over bitpower, where people are free to compete for access in a zero-trust, permissionless, and egalitarian way. But unlike other battles, this one is completely non-lethal.

The term "bitpower" is not a metaphor like other software specifications. Bitpower is, quite literally, bits of information created by state changes of physical power created by society through their global electric power grid. By doing this, the Bitcoin protocol utilizes society's power infrastructure as a planetary-scale computer that isn't under the consolidated control of a single person or organization. The bits of information created by this planetary-scale computer are unique in comparison to other bits of information on the internet because they're reverse-optimized. In a world where bits are becoming exponentially less scarce and less physically costly to produce, bitpower is designed to become exponentially more scarce and more physically costly to produce. This gives bitpower unique emergent properties that are irreproducible by ordinary computers running ordinary computer programs.

By developing APIs which require computer programs to utilize bitpower, it is possible to impose severe, physically prohibitive costs on people and their programs in, from, and through cyberspace. This capability has the potential to make bitpower quite valuable for applications like cyber security, or situations where computer programmers want to produce real-world physical properties like physical scarcity or decentralization, rather than producing virtual illusions of physical scarcity or decentralization. **It is already incontrovertibly true that "proof-of-work" works for cyber security because Bitcoin uses proof-of-work to keep its own proofs-of-work secure against systemic exploitation. This confuses some people because it's recursive; Bitcoin is proof of its own merit as a cyber security system**.

Using the Bitcoin softwar protocol, people are projecting real-world quantities of physical power to earn the right to account for the legitimate state of ownership and chain of custody of each sequentially reusable bit of bitpower. As a result of this electric power competition (a.k.a. softwar), the protocol is proving itself to be capable of making it physically impossible for anyone to justify the physical cost of exploiting it, even the most powerful nations in the world. This is how Bitcoin demonstrates its own merit as a cyber security system which can physically countervail cyber-attacks, not strictly a monetary system. It is clearly working as a new type of power projection technology which allows people – to include entire nations – to keep the bits of information encompassing the property and policy they value secure against attacks by empowering them to impose severe, physically prohibitive costs on belligerent actors which may or may not include nations.

As demonstrated by Admiral Columbus and his plot to convince native Jamaicans to forfeit their resources to his crew (who had been oppressive to their Jamaican hosts), there is nothing limiting a person or a crew of people from exploiting a population's belief system to give themselves centralized and unimpeachable control authority over the population's resources, except for one thing: the physical power competition they would have to win against people who refuse to allow their belief systems to be exploited. This highlights a complex emergent social benefit of warfighting that many overlook. War physically prevents people from creating, encoding, consolidating, and centralizing abstract power and control authority over resources by exploiting a belief system. This is especially important to note in the 21st century because software fundamentally represents a belief system that is highly vulnerable to exploitation.

The author explained in the previous chapter that warfighting is valuable to society because it gives people the opportunity to engage in global-scale physical power competitions to physically constrain and decentralize the reach of people with too much abstract power and control authority over resources. We know this is true because we know that warfare is the reason why control over Earth's natural resources is currently decentralized across 195 different nations. **Bitcoin takes this exact same type of global power competition but converts it into an electric form rather than a kinetic form. Bitcoin therefore clearly functions as an electro-cyber warfighting protocol. People are using Bitcoin to create and compete for zero-trust and permissionless access to a scarce and irreproducible supply of reverse-optimized bits of information that can be transferred, received, and stored via cyberspace, but remain under no organization's centralized or unimpeachable control**.

Why would digital age society want a softwar protocol? To preserve their freedom of action in the international thoroughfare we call cyberspace. To preserve their access to the bits of information they value. To protect and defend their ability to pass bits of information across cyberspace in a way that can't be interfered with or systemically exploited by people wielding and abusing their abstract powers over computers.

The physically secured bits of information that are being passed across proof-of-work computing networks like Bitcoin could potentially represent a vital new resource that 21st century society could need to combat cyber tyranny and keep themselves physically secure while operating in cyberspace. To that end, proof-of-work protocols like Bitcoin could give people the freedom to fight for zero-trust and permissionless access to their most valuable bits of information the same way they have always fought for zero-trust and permissionless access to their most valuable resources. **Proof-of-work protocols offer the same "might is right" pecking order heuristic that has been battle-tested by four billion years of natural selection**. And those who learn how to project physical power in the most clever and resourceful way win the right to write the ledger to account for those bits.

The physical power competition for access to and control over Bitcoin's bitpower not only keeps the system decentralized and secure, it also makes the issuance and consensus mechanism unbiased. Like physical power itself, the Bitcoin protocol makes no ethical, moral, theological, or ideological judgements about those who use it. This makes Bitcoin both perfectly fair and perfectly aligned to nature's proven method for settling disputes, managing resources, and establishing pecking order. This is the same method agrarian society already uses to establish their dominance hierarchy over Earth's limited resources, despite how much they like to use their imaginations and role-play like they don't.

In summary, using Power Projection Theory, the author asserts it's not *that* Bitcoin solves the Byzantine general's problem which makes the protocol so special, it's *how* it solves this problem. **What makes Bitcoin remarkable is that it appears to have created a non-lethal equivalent to warfighting that achieves the same complex emergent benefits of traditional warfare using the exact same social protocol of engaging in a physical power competition to establish pecking order over a vital resource in a zero-trust, permissionless, and egalitarian way, but it removes the mass in these power competitions and swaps it with energy, thus eliminating the capacity for human injury and intraspecies fratricide. In other words, Bitcoin performs the same function for a pack of humans as antlers do for a pack of deer. This implies that Nakamoto didn't just create a peer-to-peer electronic cash system – he created human antlers.**

5.12 Mutually Assured Preservation

"The US Dollar enjoyed great trust around the world… but for some reason it began being used as a political weapon, imposing restrictions on use. They began to bite the hand that feeds them. They'll collapse soon. Many countries in the world are turning away from using the dollar as a reserve currency. They restrict Iran in its dollar settlements. They impose some restrictions on Russia and other countries. This undermines confidence in the dollar. Isn't it obvious? They're destroying it with their own hands."

Vladimir Putin [187]

5.12.1 Securing Trade Routes and Preserving Freedom of Action in a New Thoroughfare: Cyberspace

One of the core functions of a military is not to secure a nation's borders, but to secure a nation's trade routes. The US Navy, for example, exists to secure trade routes and preserve freedom of action in, from, and through the maritime domain. In whatever domain a nation relies upon to exchange goods, they tend to need a military to physically secure their ability to exchange goods and preserve freedom of action in that domain, else they become vulnerable to denial-of-service attacks.

Imagine if the entire ocean was under the complete control of a single organization, and every nation who wanted to exchange goods across the ocean tacitly needed the permission of a single organization to do it. Would it be reasonable to believe a single organization in control of the entire ocean can be trusted not to exploit their control? Would it be reasonable to expect other nations to be pleased with a situation where they must have the tacit permission of one organization to access the ocean, establish trade routes, and exchange goods? The answer to these questions is "probably not." Yet, this trust-based, permission-based system is exactly how cyberspace currently functions.

We have established that computers are state machines. A network of computers connected together creates a state space which transfers, receives, and stores bits of information. Some call this state space an intranet or internet. Others call it cyberspace. Because these are abstract names, it would be just as accurate to call cyberspace an ocean. This "ocean" created by a network of computers functions as a new domain through which nations exchange a vitally important new type of good – bits of information. Over the span of a single human lifetime, this cyberspace ocean has become a very important domain upon which nations rely for the exchange of vital bits of information. Many different bits of information have value, but among the most valuable bits of information that nations like to exchange are bits of financial information which facilitate international trade and settlement.

Right now, the ocean through which nations exchange most of their bits of financial information is under the complete control of a single organization. This extraordinarily asymmetric amount of abstract power and control authority is derived from the fact that one nation controls the network of computers facilitating the exchange of these bits of information. The organization which has administrative control over this network has abstract power over the vital bits of information transferred, received, and stored on it. This abstract power gives them complete, unimpeachable, and centralized control over this entire ocean that many of the world's nations rely upon for the exchange of some of their most vital goods. If any nation wants to use this ocean to set up trade routes and exchange their highly valuable financial bits of information on it, they tacitly rely on the permission of a single organization to do it, and they must implicitly trust that organization not to exploit their control over the domain to inhibit their freedom of action. This system of information exchange between nations is currently a trust-based, permission-based, and inegalitarian system that is clearly vulnerable to exploitation. The US has conclusively demonstrated this by denial-of-service attacking Russia from telecommunications networks such as the one controlled by the Society for Worldwide Interbank Financial Telecommunications (SWIFT).

This begs a question: Why would any self-respecting nation who wants to have sovereign access to information exchange across the cyber domain subscribe to a permission-based system like this, when those permissions can be revoked at any time? One explanation could be that they simply don't have a viable alternative. It's hard to think of any computer network used in telecommunications which isn't under the unimpeachable control of some group of people. There might not be a viable way for nations to exchange bits of information using a telecommunications network without relying on a computer network that's still under some group's permissive controls. In this scenario, if a nation were to be denial-of-service attacked by SWIFT for example, they would have no choice but to adopt some other computer network to exchange their bits of financial information. The challenge, of course, is that the other network would be equally as systemically insecure as SWIFT because it would still be under the permissive control of some other group of people. With the current architecture of the internet, there doesn't seem to be a way that nations can use computer networks and not be vulnerable to this form of exploitation.

This would change if a planetary-scale computer were connected to the internet, over which no organization could gain and maintain complete or unimpeachable control.

If society were to adopt a protocol like Bitcoin that leverages the globally-distributed electric power grid as a planetary-scale state mechanism over which no single person, organization, or nation can physically afford to have centralized and unimpeachable control, then the state space of that state mechanism would represent a special new thoroughfare which nations could utilize to exchange their most valuable bits of information in a zero-trust, permissionless, and egalitarian way. Nations could theoretically use this planetary-scale computer to set up their digital trade routes, passing all bits of information they want to keep secure against denial-of-service attacks and other systemic exploits. It would be in a nation's best interest to set up something like this, considering how it's always been the responsibility of government to keep trade routes open and preserve freedom of action in every domain they operate. There's simply no reason to expect the cyber domain to be an exception to other domains in this regard.

Herein lies one of the most critical insights of Bitcoin that might be overlooked. **If the technical theories presented in this thesis are valid and it's true that proof-of-work protocols like Bitcoin utilize the global power grid as a planetary-scale computer which adds a special new state space to the existing state space of the internet, then that new state space could represent the first truly zero-trust, permissionless, and egalitarian digital trade route in cyberspace that nations may come to rely upon to exchange their most valuable bits of information.** An obvious first use case for this technology is to preserve the freedom of exchange of financial bits of information needed for international trade and settlement, but there would likely be several additional use cases.

How might nations physically secure their ability to exchange their most valuable bits of information and preserve their freedom of action in this new, strategically advantageous trade route within the cyber domain? The same way they already do for land, sea, air, and space: by utilizing their specialized group of physical power projectors called the military to impose severe physical costs on those who impeded their freedom of action.

With this concept in mind, the reader is invited to consider how some nations have already begun to adopt the Bitcoin network to exchange valuable bits of financial information needed for international trade and settlement. El Salvador is even raising bonds to subsidize the development of their own portion of the Bitcoin network which converts their country's geothermal power into bitpower. By doing this, El Salvador is actively working to secure zero-trust, egalitarian, and permissionless access to this special part of the internet under no organization's centralized and unimpeachable control. And they're actively demonstrating to other nations that it's working at keeping their valuable bits of information secure.

Based on the precedent set by El Salvador, one could argue that nations are beginning to see how strategically vital it is for them to physically secure their access to this special piece of the internet where other nations can't prohibit them from exchanging their bits of information. To secure access to this unrestricted place in cyberspace against foreign interference, nations appear to be learning that they must create services dedicated to the projection of real-world physical power in, from, and through cyberspace just like they already do for land, sea, air, and space.

Nations appear to be learning how vulnerable they are to denial-of-service attacks by the system administrators of the computer telecommunications networks they choose to use. The actions taken by organizations like SWIFT against Iran and Russia have showcased to the world that whoever controls the network of computers managing the valuable bits of information they rely upon, tacitly has complete and unimpeachable control over them. **Nations are beginning to see that a zero-trust, permissionless computing network that enables bits of information to be transmitted, received, and stored without the threat of denial-of-service attacks or other forms of systemic exploitation are a vital national strategic security priority. Not just for financial bits of information, but for *any* kind of information**.

This would imply that Bitcoin could become widely recognized as a national strategic security imperative. El Salvador's actions demonstrate that nations are already beginning to recognize Bitcoin as a candidate solution. If Bitcoin were to be adopted by other nations, that could validate it a systemically secure and strategically important computer network. To date, no nation has been able to surmount the enormous physical cost required to denial-of-service attack or systemically exploit Bitcoin's bits – including and especially nuclear superpowers. That's noteworthy, and nations appear to be taking notice.

5.12.2 Why Build an Electro-Cyber Militia When You Can Just Subscribe to One?

If the author's theories on softwar are valid, then **national adoption of Bitcoin could be envisioned as the adoption of an electro-cyber militia to protect & defend a nation's valuable bits of information. Nations like El Salvador are creating their own hash force – a service dedicated to projecting physical power to preserve their country's freedom of action in, from, and through cyberspace by imposing severe physical costs (a.k.a. watts) on anything that seeks to deny their sovereign right to freely exchange bits of information**. But even if El Salvador didn't elect to build their own hash force, by simply choosing to use the Bitcoin network as the domain through which they exchange their financial bits of information, El Salvador would still be inheriting the security capabilities of the Bitcoin network, as if subscribing to a military-grade, physical-security-as-a-service protocol.

From a systemic perspective, Bitcoin functions similarly to how a militia functions. A large group of people project lots of physical power to preserve freedom of action within a particular domain, giving nations the ability to exchange goods (in this case bits) with another nations without interference. The form of the militia is different because the form of the domain is different. Cyberspace is disembodied and transnational, so it makes sense that a cyber militia would be disembodied and transnational too.

Something remarkable about the disembodied and transnational form of Bitcoin's security infrastructure is that it appears to be highly resistant to kinetic strikes, including strategic nuclear strikes. Bitcoin's infrastructure is spread across the world and is comprised of the globally-distributed internet, power grid, and all the people running these systems. Because it's a decentralized network, the only way to destroy it is to destroy the whole network. This would suggest that the only way to destroy the Bitcoin infrastructure is to destroy both the internet and the global electric power grid. While this is theoretically feasible, it's not practically feasible because of the kinetic power projection paradox discussed in section 4.12. Our species would likely go extinct sooner than we would be able to destroy all of the physical infrastructure of the internet and the global electric power grid. We may technically have the capacity to destroy Bitcoin's infrastructure, but because of the kinetic power projection paradox, it appears to be too expensive to do. In other words, the **BCR_A** of destroying Bitcoin is simply too low to justify.

Therefore, **not only do nations gain the physical security benefits of a global-scale "electro-cyber militia" when they adopt the Bitcoin network as the means through which they exchange bits of information on the internet, they also appear to gain free access to a non-lethal security system that is, at least practically speaking, impervious to nuclear strike. In other words, nations which adopt Bitcoin gain access to a global-scale physical defense infrastructure that can empower them to stand up to nuclear-armed superpowers**.

5.12.3 Raising Hash Forces to Preserve Zero-Trust and Permissionless Access to the Bitcoin Network

With new physical power projection capabilities come new national strategic security strategies. In Bitcoin's case, nations now, at least theoretically, have a way to ensure that certain organizations in charge of certain telecommunications networks (e.g. SWIFT) cannot abuse their abstract power and control authority over the network to denial-of-service attack nations who have a sovereign right to freely exchange bits of information across the internet. And because Bitcoin's physical power projection technology is electric rather than kinetic, this new strategy doesn't appear to be limited by the kinetic power projection paradox. This could mean that **Bitcoin represents a key enabling technology for a new national strategic security strategy that could overcome a nuclear stalemate**.

For the sake of argument, let's assume the theoretics presented in this chapter are valid, and that the bits of information exchanged via the Bitcoin network do indeed represent reverse-optimized bits of information that are physically costly to transmit, receive, and store, giving them unique properties that are useful more many applications. Let's assume that securing zero-trust, egalitarian, and permissionless access to these special bits becomes increasingly recognized as a national strategic opportunity.

Any nation that follows in El Salvador's footsteps in adopting the Bitcoin network as the domain through which they exchange bits of information would also inherit the same baked-in physical security benefits of Bitcoin's global-scale, nuclear-resistant defenses. These nations would effectively be subscribing to the security services of Bitcoin's globally-decentralized "electro-cyber militia" the same way a nation does when it subscribes to another nation's military for defense. It's a cost-effective strategy, but it's also a trust-based and permission-based strategy. By subscribing to Bitcoin's cyber militia as-is without contributing their own physical power and hashing infrastructure, a nation would tacitly be trusting in that militia not to denial-of-service attack them. The possibility exists that everyone participating in the network could unite against a given nation and withhold their valid transaction requests.

Fortunately, the easy way to prevent this threat is for a nation to have their own hash force. This would allow them to utilize their own physical power projection capacity to guarantee they maintain ledger-writing permissions. In effect, a nation would be building their own electro-cyber militia and attaching it to Bitcoin's combined electro-cyber militia to preserve permissionless access to the network. If this scenario were to play out, it would represent a global-scale international "soft" war for access to the Bitcoin network and its underlying bits of information. But unlike other global-scale international wars, a Bitcoin softwar would have far different strategic dynamics that would lead to a much different end state: the author calls this special end state mutually assured preservation.

5.12.4 Solving the Kinetic Power Projection Paradox

As discussed at the end of the previous chapter, sapiens have a knack for building highly efficient power projection technologies which empower populations to win physical power projection competitions called wars. After more than 10,000 years of practice perfecting the craft of physical power projection, sapiens appear to have reached a stalemate due to what the author calls the kinetic power projection paradox – a situation where society has become so efficient at a particular method of kinetic power projection that it is paradoxically too inefficient to use because of how much mutual destruction it would cause.

To date, strategic nuclear warheads appear to be the most efficient kinetic power projection capacity ever invented by sapiens in terms of size, weight, and effort. Unfortunately, this technology is so efficient at projecting massive quantities of physical power that it does not appear to be practically useful. A full-scale war between two nuclear armed nations would most likely be a war without winners (it's hard to solve a dispute if both sides are mutually destroyed). Ironically, the most efficient kinetic power projection technology ever created (nuclear warheads) appears to have created the most *inefficient* way to settle international policy disputes.

Nations tacitly validate the existence of a nuclear stalemate caused by the kinetic power projection paradox when they pursue far less efficient kinetic power projection strategies to settle their disputes, manage control authority over resources, and determine consensus on the legitimate state of ownership and chain of custody of their property. Instead of using highly efficient power projection technologies like tactical nuclear bombs, they use non-nuclear weaponry like kinetic munitions to settle their disputes.

Since the discovery of nuclear power, this strategy of deliberately using less efficient non-nuclear power projection technologies has only succeeded at settling minor disputes between asymmetric powers. It's still not clear whether it's actually possible for two nuclear-armed nations to settle major disputes using non-nuclear power projection technology. This would require one nation to be willing to surrender to another despite having the capacity to mutually assure the destruction of the nation to which they're surrendering, and that doesn't seem like a realistic scenario. It seems far more likely that a nation would sooner accept a stalemate than surrender to a nation they could destroy, and nuclear warheads are what give nations the former option rather than the latter.

However, even if we assume that it's still possible for two nuclear-armed nations to settle a major dispute without using nuclear technology, this doesn't resolve the kinetic power projection paradox or bypass the risk of mutually assured kinetic destruction. As discussed in the previous chapter, forking the evolutionary path of kinetic power projection technology development into a non-nuclear direction still leads to the same dead end. By attempting a fork, modern society would likely continue to do what they have done so demonstrably well in the past, and develop incrementally more efficient kinetic power projection capabilities until they discover yet another way to mutually destroy each other – except this time without nuclear warheads. Therefore, the strategy of using deliberately less efficient, non-nuclear kinetic power projection technologies seems likely to cross the same threshold as nuclear technology did, where it becomes too efficient to be practically useful (a.k.a. crossing the kinetic ceiling).

A speculative example of a kinetic power projection technology that could follow nukes in crossing the kinetic ceiling is massive-scale swarms of millions of highly maneuverable, artificially intelligent killer drones. It seems unlikely that two nations would be able to utilize kinetic physical power competitions to solve major disputes or secure access to resources in a zero-trust, permissionless, and egalitarian way if both side of the conflict had access to strategic nuclear missiles as well as command over swarms of millions of highly intelligent killer drones. This situation seems like it would easily lead to another stalemate at both the strategic and tactical levels using both nuclear and non-nuclear technology.

Interestingly, **Bitcoin appears to solve this paradox. Global-scale physical power projection competitions enabled by softwar protocols like Bitcoin appear to be able to solve this type of quagmire by bypassing the kinetic power projection paradox altogether**. By using a non-destructive form of electric power rather than a highly destructive form of kinetic power, hash forces would theoretically be able to compete for permissionless access to valuable resources exchanged across the Bitcoin network without the threat of mutually assured destruction. Nations would theoretically be able to increase their electro-cyber power projection capacity and efficiency ad infinitum without it becoming too destructive to be practically useful.

No matter how competitive an electro-cyber power competition (i.e. a "hash war") gets, it seems unlikely that it could lead to a state of mutually assured destruction because there's nothing inherently destructive about developing increasingly efficient forms of electric power projection technology. In fact, society could potentially *benefit* from this power competition. It would motivate society to search for increasingly more clever technologies to generate more power more efficiently, as well as develop faster and more efficient hashing computers.

Whereas the byproduct of a kinetic war is destruction of infrastructure and more expensive energy, the byproduct of softwar appears to be the creation of more infrastructure and cheaper energy. This makes intuitive sense using the theoretics presented earlier in this chapter. By turning the globally-distributed power grid into a planetary-scale computer, society would theoretically be able to unlock Moore's law for the global electric power grid. Society would continually compete for incremental advantages by increasing the power projection efficiency and capacity of their portion of the global electric power grid (i.e. the portion of the planetary computer circuitry under their direct control). As a result, society could see exponential growth in both the efficiency and capacity of their power grids, just like they already do for their ordinary computers.

Bitcoin falls directly in line with the evolutionary path of human power projection technologies developed across the past 10,000 years of agrarian warfare. This concept is illustrated in Figure 94. But more notably than that, **Bitcoin represents the only globally-adoptable physical power projection technology that is practically useful for nations to settle major policy disputes (including but not limited to international financial policy disputes), establish zero-trust control over valuable resources (including but not limited to financial information), achieve global consensus on the legitimate state of ownership and chain of custody of property (including but not limited to digital property), and preserve freedom of action and exchange of goods across cyberspace, and do it all beyond the limit of the kinetic ceiling where kinetic power projection competitions are no longer practically useful. Combining these observations with the fact that running a hash force can be a profitable endeavor due the high monetary value that people assign to underlying bits of information exchanged across these networks, this means proof-of-power protocols like Bitcoin paradoxically have the potential to represent the most powerful and efficient physical power projection technologies ever discovered by agrarian society.**

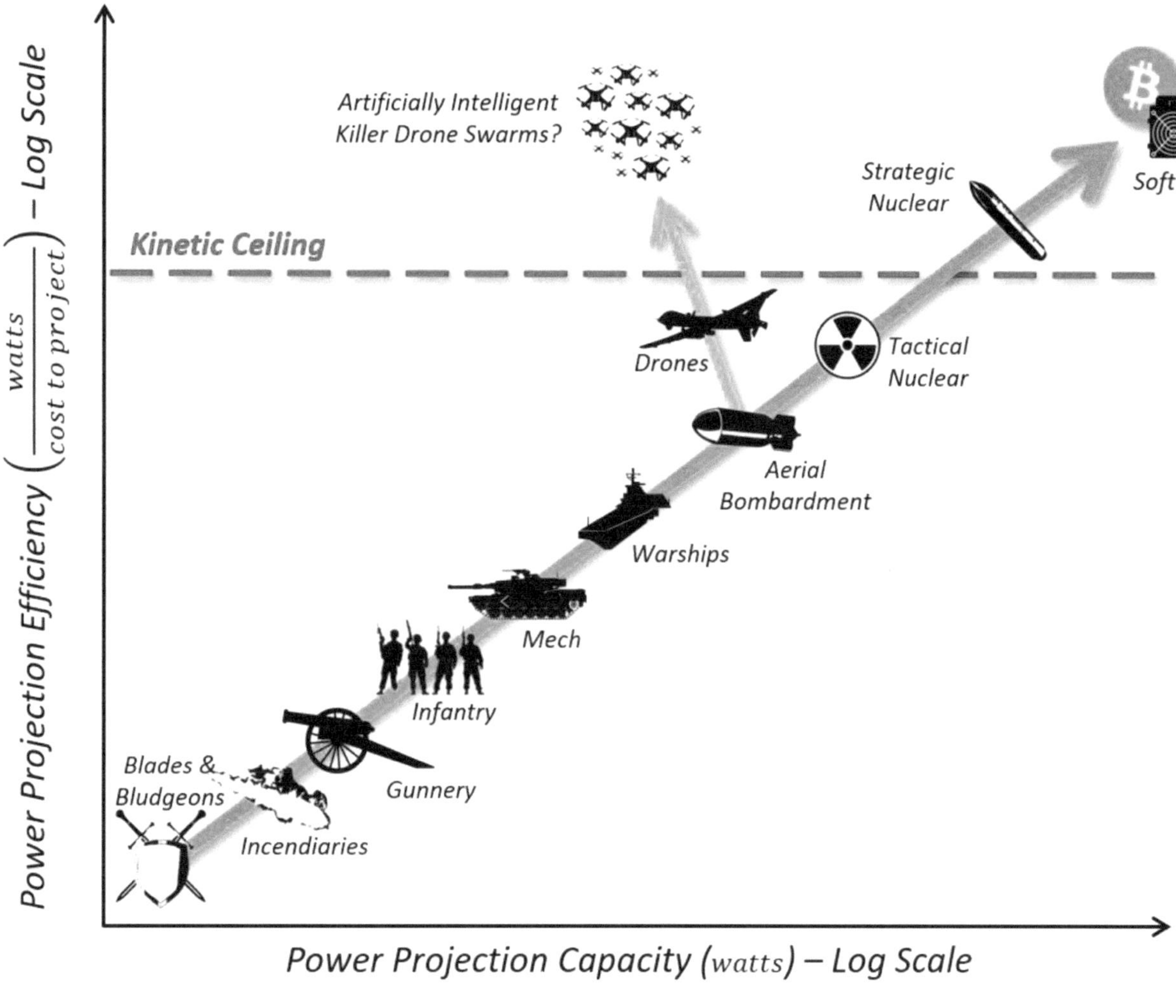

Figure 94: Evolution of Physical Power Projection Technologies, with Bitcoin Shown as End State
[88, 89, 116, 117, 118, 119, 120, 121, 122, 123]

Physical cost function protocols like Bitcoin make pragmatic sense. If it is true that society is already stalemated at a nuclear strategic level, then why go through the hassle of building a bunch of killer drone swarms to do battle with other killer drone swarms while humans watch from the sidelines? Are those drone fights going to achieve any meaningful advantage or resolve a nuclear stalemate? **If society insists on trying to settle strategic policy or property disputes using human-out-of-the-loop drone swarm battles, then what is the point of making it a kinetic battle? Why not just put the drones on a rack, plug them together, and have them battle each other using electric power through cyberspace, rather than kinetic power through three-dimensional space**?

Technically speaking, Bitcoin *is* an international power competition between drone swarms. Instead of having machines compete against each other using kinetic power in the air, land, sea, or space domains, Bitcoin's machines are placed on a rack and they compete against each other using electric power in the cyber domain. **It's the same game, but in a different domain – a domain that doesn't risk nuclear annihilation and doesn't appear to be stalemated**.

If the author's observations are valid, then it could mean that **protocols like Bitcoin could effectively function as humanity's antlers. This technology could theoretically bypass the kinetic power projection paradox and give nations an opportunity to physically compete against each other in a zero-trust, egalitarian, and permissionless way that isn't stalemated and doesn't cause injury – which means it could generate very clear winners and very clear losers.** By bypassing the kinetic power projection paradox, Bitcoin could represent humanity's ticket out of a potential bounded prosperity trap.

5.12.5 Escaping A Fiery Hellscape with a New Method for Zero-Trust Cooperation & Physical Security

During kinetic physical power projection competitions (i.e. traditional warfare), there is an existential imperative for agrarian society to sum their physical power projection capacity together using cooperation techniques. Cooperation is first and foremost a physical power projection strategy which allows a population to increase their mutual C_A to decrease their mutual BCR_A. Cooperation has enabled agrarian society to buy themselves increased prosperity margin and deter aggressors over thousands of years. These are the foundational primordial economic dynamics behind national strategic security.

The massive-scale abstract power hierarchies (i.e. nation states) we have today are the result of thousands of years of agrarian society learning how to cooperate at increasingly higher scales to keep themselves more physically secure against neighboring abstract power hierarchies. Today, agrarian society utilizes trans-national power alliances where multiple nation states sum their physical power projection capacities together. However, like Cyanobacteria and photosynthesis, sapiens and the abstract power hierarchies they build can backfire by causing society to become increasingly more vulnerable to systemic exploitation and abuse through their mutually adopted belief systems.

Abstract power hierarchies inevitably break down for reasons discussed at length in the previous chapter. The general problem is that abstract power hierarchies are fundamentally trust-based cooperation techniques. Populations which utilize abstract power hierarchies must adopt trust-based belief systems where people are given abstract power and trusted not to use it against their own populations. The larger and more asymmetric an abstract power hierarchy becomes, the more its believers become vulnerable to systemic exploitation and abuse. Trust-based belief systems are systemically insecure belief systems – eventually there is a point where the benefit to breaking people's trust is disproportionately higher than the cost of doing so.

These inherent security flaws are important to note because today's nation states are larger and more asymmetric than ever experienced by agrarian society. Today, it is not uncommon for 99.9999% of the population living within a nation state to have to trust less than 0.00009% of the population not to abuse the abstract power and control authority appointed to them. This is a highly inegalitarian system that is vulnerable to systemic exploitation. Untrustworthy systemic predators would have much to gain if they wanted to exploit the high-ranking positions of these abstract power hierarchies, and there would be minimal physical cost for doing so – especially if the ruling class has the nuclear launch codes.

It is no secret that trust-based abstract power hierarchies have a well-documented tendency to break down and drive agrarian society back to war. War is highly energy efficient and injurious, but it is the only zero-trust, egalitarian, and permissionless system where people are free to reset their dominance hierarchies by engaging in physical power competition. **Therein lies the tragedy of agrarian society. The more society tries to avoid war, the more they must adopt systemically insecure, trust-based, inegalitarian, permission-based cooperation systems like abstract power hierarchies which tend to break down and lead directly back to war**. Like cyanobacteria, agrarian society appears to be stuck in a bounded prosperity trap which has driven it to the precipice of strategic nuclear annihilation. Now, populations are being cornered into adopting trust-based abstract power hierarchies, with no ability to countervail their ruling classes if they become untrustworthy, exploitative, oppressive, or tyrannical.

The emergence of new physical power projection technologies like Bitcoin fundamentally changes these power dynamics. Physical cost function protocols like Bitcoin give agrarian society the ability to sum their physical power projection together to increase their mutual C_A and decrease their mutual BCR_A without having to trust each other. Instead of having to adopt systemically insecure, trust-based abstract power hierarchies to cooperate at larger scales and sum their physical power projection capacities together to impose increasingly more severe physical costs on aggressors, protocols like Bitcoin allow people to do perform the same function in a zero-trust way.

With physical cost function protocols like Bitcoin, anyone who contributes physical power to the system (as measured by hash rate) automatically increases the mutual C_A and decreases the mutual BCR_A of all users, including adversaries. The author hypothesizes this is because Bitcoin users technically utilize the same planetary-scale state mechanism. **By turning the global electric power grid into a planetary-scale computer, two adversarial users living on opposite sides of the world "stick together" in a symbiotic relationship because they are technically utilizing two sides of the same machine**. Because they're both using the same planetary-scale computer, if both sides were to attempt to marginalize the permissions of the other side by building larger and more powerful hash forces, they would increase the C_A and lower the BCR_A of the same machine upon which they mutually rely.

So long as no single hash force is allowed to gain and maintain majority hash rate (which is easy to prevent when all users have the freedom to add theoretically unlimited amounts of physical power/hash rate to the system from their side of the planet), then the addition of more physical power/hash to the system serves to increase the total physical cost required to gain and maintain systemically exploitable control over the system. In other words, **as more physical power is added to the Bitcoin network by hash forces, the total physical cost required to achieve centralized control authority over the network increases, making it harder for any one entity to gain and maintain centralized control over the ledger. This creates an emergent behavior of the Bitcoin protocol where the physical costs required to exploit the ledger can grow ad infinitum**.

No matter how much physical power is summoned by people in the competition for ledger-writing administrative privileges over the Bitcoin network, the protocol can absorb all of it and channel it all towards increasing the total physical cost required to exploit the system. Nations could raise hash forces to secure their own permissionless access to the Bitcoin network. As they compete against each other for permissionless access to the ledger, all users directly benefit from the higher total hash rate and power consumed by the network, including and especially adversaries. **Adversaries become mutual beneficiaries because they're both contributing to different sides of the exact same planetary-scale machine which neither side can fully control. The protocol sums their physical power projection capacity together to create a single cost of attacking the network as a whole, causing nations to directly benefit from their adversary's physical power projection efforts even though they are directly competing against each other**.

The more people and organizations use their hash forces to compete against each other for permissionless access to the ledger, the more physical power and hash rate is required for any single hash force to gain and maintain centralized (thus exploitable) control over it. The more they compete against each other and add more hash, the more secure both parties become against denial-of-service attacks. It's a simple but remarkable emergent behavior of the system.

Returning back to a core concept in Power Projection Theory introduced in Chapter 3, Cyanobacteria were able to escape from the fiery hellscape of the Great Oxidation Event by evolving several new power projection tactics, to include learning how to stick together using multicellular cooperation. Both types of early multicellular cooperation represented zero-trust cooperation tactics. Cells didn't stick together because they trusted each other; they stuck together because they were literally stuck together either through colonization or clustering. As a result of their zero-trust cooperation tactic, they were able to sum their combined physical power and escape their bounded prosperity trap together.

Fast forward by about two billion years and we arrive at modern human society trapped in what could be described as another fiery hellscape. This time, it has taken the form of 10,000 years of agrarian warfare escalating all the way up to the threat of nuclear annihilation. **With softwar technologies like Bitcoin, society could have discovered a technology which enables zero-trust cooperation at a global scale which transcends national borders**. By utilizing the same planetary-scale computer, two adversaries who have no capacity to trust each other can mutually benefit from each other's physical power projection efforts. They can use electric power to compete for zero-trust, permissionless, and egalitarian access to a vital new resource (bitpower), and bitpower itself allows them to impose theoretically unlimited physical costs on each other in a non-lethal way through a virtual domain. By fighting against each other using this softwar protocol, they paradoxically cooperate.

Herein lies what the author believes to be the most compelling feature of the Bitcoin protocol: it could create an infinitely prosperous organism as first discussed in section 3.7. By competing against each other for permissionless access to the Bitcoin network, adversaries serve each other by increasing the physical cost required to attack the network. Their competing power projection capacity is summed together to increase the overall C_A of the network. As a result of this ongoing power competition, network BCR_A perpetually decreases. As users scale their hash rates ad infinitum, the network's C_A increases ad infinitum and BCR_A decreases ad infinitum, thus giving all users an infinitely expandable prosperity margin. So long as no single user can gain and maintain majority hash, everyone using this computer network benefits from the battle over control authority over it.

As if by magic, the Bitcoin protocol turns adversarial physical power competition into a single, cohesive, infinitely prosperous organism chasing after the same mutually beneficial end goal of improving security and preserving freedom of information exchange across cyberspace. Enemies become frenemies. Adversarial nations become de facto collaborators working together on the same planetary-scale computer, despite not trusting each other. Unlike the dynamics of kinetic power projection competitions shown in Figure 54, the end state of this global-scale electro-cyber power projection competition is not mutually assured destruction, it's mutually assured preservation. This concept is illustrated in Figure 95.

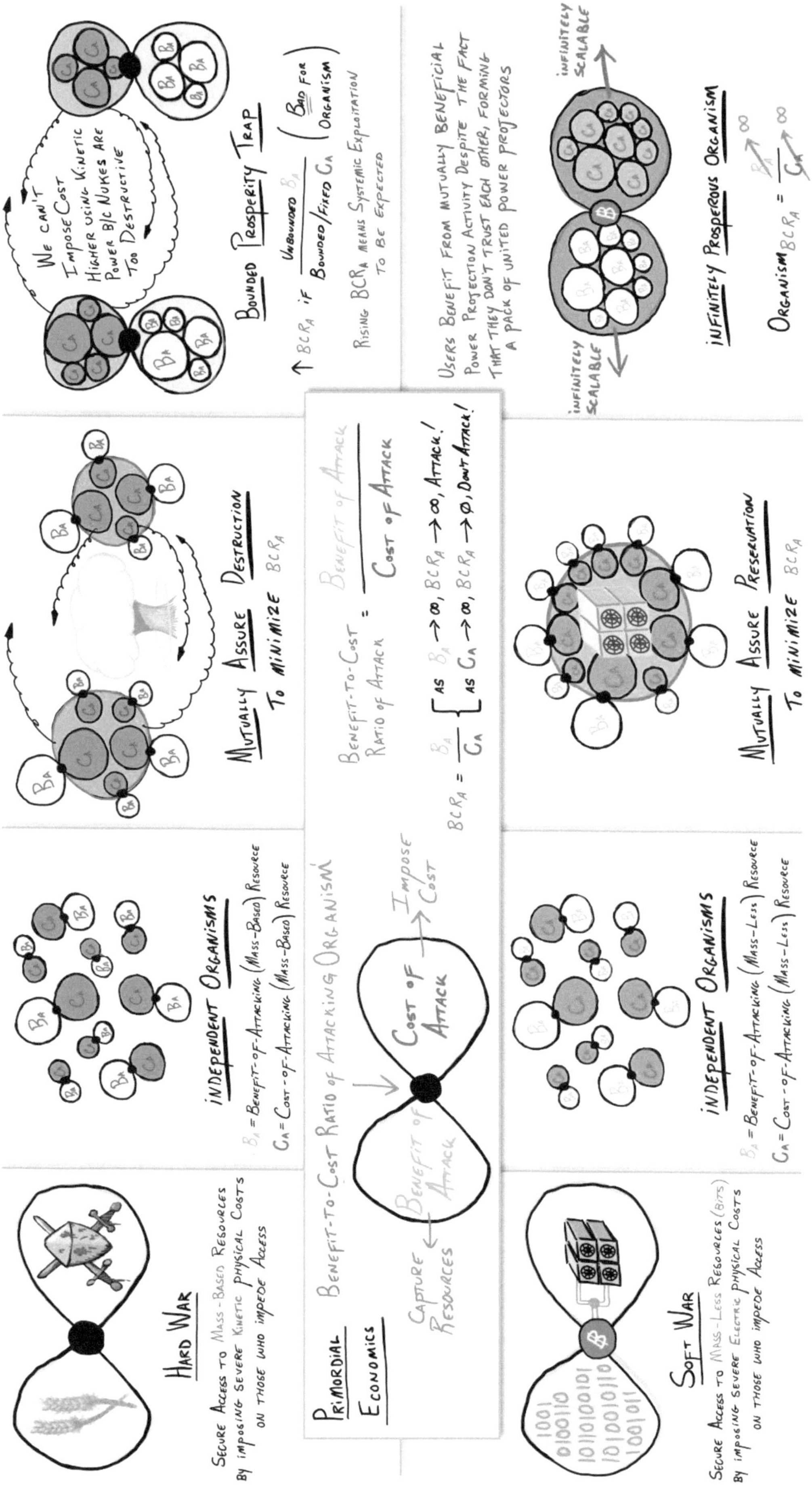

Figure 95: Comparison between "Hard" and "Soft" Warfighting Dynamics

Chapter 6: Recommendations & Conclusion

"Unless mankind's attention is too violently diverted by external wars and internal revolutions, there is no reason why the electric millennium should not begin in a few decades."
Nikola Tesla, 1937 [7]

6.1 Key Takeaways

6.1.1 It's not Reasonable to Believe that Cyberspace will be a Domain without Physical Power Competition

For 30 years, people have been theorizing that physical cost function protocols (a.k.a. proof-of-work protocols) could serve as a mechanism through which people can physically secure digital resources by imposing prohibitive physical costs on others in, from, and through cyberspace. Since these concepts were first formally introduced in 1993, Bitcoin has organically emerged as the most widely-adopted physical cost function protocol to date, proving itself over the past decade to be a means for people to engage in a global-scale battle for zero-trust and egalitarian control over bits of information.

As Bitcoin adoption scales, it appears to be increasingly misunderstood as exclusively a monetary system rather than a physical security system which could be used to secure any type of information. This might be because Bitcoin's inventor was famously terse with the protocol's design specifications and famously disinclined to expand upon its potential functionality as more than just money. Nakamoto orphaned the project shortly after releasing it, so the overwhelming majority of what has been written about Bitcoin is purely speculation. There is not a lot of academic literature which explores the potential for this technology to serve as something other than a monetary network.

For this reason, the goal of this thesis was to provide some technical insight about the design and functionality of Bitcoin using a different design specification than what Nakamoto provided. Instead of visualizing Bitcoin as strictly a peer-to-peer electronic cash system, the author provided a way for people to visualize Bitcoin as the means through which people can project physical power against each other through cyberspace.

Using computer theory, we can see that once society has figured out how to use computers to physically secure *financial* bits of information, that would also mean they have figured out a way to physically secure *any* bits of information, including but not limited to financial bits of information. This would suggest that Bitcoin is not merely a monetary network. Instead, Bitcoin could be more generally described as a physical security network which happens to be named after its first widely-adopted use case of securing financial bits of information. From another perspective, we gain the insight of seeing Bitcoin as a global-scale softwar technology that could revolutionize digital-age society's perception of physical confrontation. In the author's opinion, this is a more compelling description of Bitcoin's potential sociotechnical and strategic value than strictly "peer-to-peer electronic cash." *Bitpower* could be far more impactful to society than just *bitcoin*, and they could be one-in-the-same technology.

For thousands of years, people have adopted belief systems which allow ruling classes to wield asymmetric levels of abstract power and control authority over them. Populations have tried to keep themselves secure against the systemic exploitation of these abstract power hierarchies using logical constraints encoded into rules of law, but time and time again, these logical constraints have proven to be demonstrably incapable of preventing systemic exploitation and abuse of people's belief systems. No matter what logical constraints are encoded into the design of their abstract power hierarchies, people have not succeeded at sufficiently constraining belligerent people from exploiting them. Eventually, someone finds a way to exploit the population's belief system for their own personal advantage.

This is a fundamental systemic security threat for all monarchies, ministries, legislatures, parliaments, republics, presidents, prime ministers, and senators. This also happens to be the same systemic security threat shared by computer programs and computer programmers. No matter what they're called, no matter how their imaginary powers are encoded, and no matter how much people attempt to logically constrain them using carefully-designed rulesets, all belief systems which rely exclusively on imaginary sources of power create trust-based, inegalitarian, and permission-based abstract power hierarchies that are demonstrably incapable of securing people against exploitation and abuse. These systems routinely become captured by tyrannical, oppressive rulers who abuse their abstract power to benefit themselves, their friends, and their private interests at the expense of the users they ostensibly serve. Within the greater context of animals establishing intraspecies dominance hierarchies, the abstract power hierarchies imagined by sapiens routinely break down.

Abstract power hierarchies are the mechanism through which an extremely small number of systemic predators can and routinely do prey on entire populations by exploiting them through their belief systems. We have 5,000 years of recorded history to demonstrate that abstract power hierarchies are fundamentally flawed from a systemic security perspective. There seems to be no way to logically constrain people enough to stop them from systemically exploiting people.

Because software is fundamentally an abstract belief system, software recreates this same systemic security vulnerability, but at an even larger scale. Somehow, despite thousands of years' worth of empirical evidence of this systemic security flaw, populations keep allowing themselves to adopt exploitable belief systems which construct abstract power hierarchies and then entrust people not to exploit their imaginary power and control authority. So long as people remain oblivious to the fundamental systemic security flaws of their belief systems and do nothing about it, they remain vulnerable to this form of exploitation, particularly in the digital age where abstract power hierarchies are created using software.

A goal of this thesis has been to highlight this threat and illustrate how vulnerable 21st century society could become by adopting centrally-controlled computer networks running centrally-controlled software. There are people pulling the strings behind all of the computer programs running on all of the computer networks we currently use in cyberspace. Perhaps society could counteract this vulnerability and learn to physically secure itself by converting our globally-distributed power grid into a planetary-scale computer over which nobody can have complete control. To that end, proof-of-work (a.k.a. proof-of-power) protocols like Bitcoin have the potential to give people a way to countervail an emerging systemic security vulnerability that digital-age century society might not be aware of. Like many disruptive innovations to come before it, Bitcoin could fix a problem that many people don't realize they have.

No matter what it's called – warfare or revolution – physical power keeps proving itself to be an effective way to prevent centralization of control and expansion of abstract power. One needs only to open a history book to see proof of the value of physical power competition in every domain humans operate, and there's no reason to expect that cyberspace will remain an exception. If it is true that society has now figured out how to project physical power in, from, and through cyberspace, then it stands to reason that society is going to use this technology to fight future wars and revolutions in, from, and through cyberspace. The main question doesn't appear to be *if* digital-age society is going to do this; it seems to be *how* and *when* digital-age society is going to do this.

Despite the changing form of the tactics, techniques, and technologies used to produce and project both abstract and physical power, this question about how and when the next major physical conflict will break out boils down to the same question that society has repeatedly confronted for thousands of years: For how long are people going to let themselves be systemically exploited through their belief systems until they start to physically resist? If the theories presented in this thesis are valid, then the answer to this question is extraordinary: thirteen years ago, when the world started to adopt and build the Bitcoin protocol. This would imply that a third world war has already started, but people don't recognize it yet because it's a soft war fought in, from, and through cyberspace.

6.1.2 To Understand Bitcoin, Understand (Revolutionary) Warfare

"Until an independence is declared, the Continent will feel itself like a man who continues putting off some unpleasant business from day to day, yet knows it must be done, hates to set about it, wishes it over, and is continually haunted with the thoughts of its necessity."
Thomas Paine [188]

A key insight needed to understand the sociotechnical and national strategic security implications of Bitcoin is to understand the role of power projection and the complex emergent social benefits of warfare. Warfare is the means through which agrarian society physically constrains other people who claim to have abstract power and control authority over their resources. Warfare is the means through which agrarian society prevents the centralization and expansion of oppressive rulers. Warfare is how agrarian society prevents one group of people from gaining too much control authority over any valuable resource.

With an appreciation for how warfare works, it's easy to see how Bitcoin functions as a warfighting protocol, not just a peer-to-peer electronic cash system. Satoshi Nakamoto didn't just solve the Byzantine general's problem when he created Bitcoin – he didn't just find a way to get a distributed computing network to come to consensus on the legitimate state of a ledger. **Nakamoto appears to have gifted humanity with a non-kinetic (thus non-destructive and non-fratricidal) form of globally-scalable warfare, which they can use to settle policy disputes, manage resources, and establish their dominance hierarchies in a zero-trust and egalitarian way just like all animals utilize physical power to perform these same functions, starting with something that could become recognized as one of the most vital resources to emerge in the information age: bits**.

In the process of creating what appears to be the first globally-adopted softwar protocol, Nakamoto appears to have bypassed the kinetic power projection paradox and solved a global strategic nuclear stalemate. Almost exactly as Tesla predicted, agrarian society appears to be adopting a never-before-seen form of physical power projection technology where intelligent machines compete against other intelligent machines over the maximum rate of energy delivery, while humans watch eagerly from the sidelines. Tesla never lived to see the first operational computers, nor did he live to see nuclear warfare. Nevertheless, using first principles, he was able to envision a system which sounds remarkably like Bitcoin. At the same time, Bitcoin aligns perfectly to what Ford and Edison predicted in the 1930s, that monetizing electric power could represent a way to reduce the need for traditional warfare.

To the untrained eye, these theories about Bitcoin probably sound ridiculous. But when people take the time to understand primordial economics, dominance hierarchies, human metacognition, the differences between abstract and physical power, power dynamics of agrarian society, dysfunctions of abstract power hierarchies, complex emergent social benefits of warfighting, computer theory, and cyber security, then these ideas start to become more digestible. The author used first-principles and systems thinking to produce a grounded theory that Bitcoin is fundamentally an electro-cyber warfighting protocol, which happens to align to theories produced by some of last centuries' greatest minds.

The social value of warfare is to physically constrain the expansion of abstract power, thus physically decentralize people's abstract power over resources using a global-scale physical power projection competition. This is precisely the same design concept that Nakamoto encoded into the Bitcoin protocol, except Bitcoin does what Einstein claims is possible: it swaps matter for energy.

Instead of imposing severe physical costs by applying forces to masses, the Bitcoin system imposes severe physical costs by passing charges across resistors. The result is a non-kinetic, massless, software-defined form of warfare that could become the primary way nations physically secure their property (e.g. money) and policies in a way that isn't trapped in a quagmire of a potential strategic nuclear stalemate. Nations like El Salvador have already started adopting this technology to secure their property, policy, and freedom of action in cyberspace against nuclear-armed superpowers (namely the US), and other countries like Russia appear to be pivoting to do something similar. How many nations need to make these types of strategic pivots before we (the United States of America) acknowledge that this technology represents something far more strategically valuable than just peer-to-peer electronic cash?

6.1.3 Those who Criticize Bitcoin as being "Inefficient" Illustrate that they Don't Understand Bitcoin

"If you don't believe it or don't get it, I don't have the time to try to convince you, sorry."
Satoshi Nakamoto [20]

Time will tell how systems like Bitcoin will be utilized in the future. For now, all we can do is endeavor to appreciate the theoretical potential of this technology and the precedents it sets not just for cybersecurity, but for the broader field of computer theory. To the untrained eye, proof-of-work computing systems like Bitcoin look inefficient and "bad for the environment." This highlights how misunderstood this technology is, including and especially by computer scientists and software engineers who have adopted a culture of ignoring everything happening at the *bottom* of their computer tech stacks.

The reader should now understand that producing a substantial electric power deficit in the surrounding local environment is not an undesired emergent property of physical cost function protocols like Bitcoin, it's the primary value-delivered function. In simple terms, the point of Bitcoin is to be physically expensive, because physically expensive bits are a key missing ingredient to cyber security. Therefore, when people criticize Bitcoin's high energy consumption, they are ironically *validating* how well this novel design works exactly as it was intended to work.

To revisit a core concept from Chapter 3, criticizing Bitcoin for its energy consumption is like criticizing antlers for having an awkward shape and a tendency to entangle. These people clearly don't understand the value of having the capability physically constrain or impose severe physical costs on other people's computers and computer programs in, from, and through cyberspace. Hopefully, the concepts discussed in this thesis using Power Projection Theory have helped the reader gain a better appreciation for the potential complex emergent benefits of this novel capability.

In a virtual world filled with an infinitely growing number of infinitely-less-physically-costly bits of information, physical cost function protocols like Bitcoin introduce an infinitely scarce supply of infinitely-more-physically-costly bits of information. **This technology isn't "wasteful," it's extraordinary**. It appears to be a novel form of reverse-optimized computing that people don't understand because (1) they don't know computer theory, and (2) they've never seen anything like it before. If one were to actually take the time to look under the hood of this tech stack, they would see that proof-of-work protocols like Bitcoin utilize an opposite approach to ordinary computing. **This is an intentionally reverse-optimized computing system that has been explicitly designed to be *more* physically constrained and *more* thermodynamically restricted, confounding computer scientists and software engineers who don't realize they've spent the past 80 years operating under an implicit (and virtually untested) assumption that *less* physically constrained and *less* thermodynamically restricted computer networks are unquestionably better**. In a world devoted to creating the least-physically-restricted bits possible, Bitcoin is going directly against the mainstream approach to computing and endeavoring to create the most-physically-restricted bits possible. Instead of criticizing it, we should seek to understand why this could be counterintuitively beneficial. This thesis will hopefully move the ball down field in understanding the benefits of physically costly computing.

Few seem to be willing to take the time to consider what the complex emergent benefits of a larger, heavier, slower, and more energy-expensive computer would be. Enter Bitcoin, what could be the largest, heaviest, and most energy-expensive computing network ever built, directly challenging the design assumptions of traditional computing. Of course it's easier to criticize Bitcoin as "inefficient computing" than it is to take the time to re-examine the assumptions that computer and software engineers have been making for decades. But **without taking the time to seriously consider the potential benefits of reverse-optimized computing, then the critics of Bitcoin's energy usage are clearly not providing informed opinions about it. These are the unresearched opinions of people who haven't considered the idea that reverse-optimized computing could have complex**

emergent behavior that could greatly benefit a digital-age society that is clearly facing unprecedented systemic challenges.

The electric power deficit produced by computer networks running physical cost functions doubles as a power-modulated control signal that changes the state of our surrounding physical environment, thereby changing the complex emergent properties and behavior of our software. People aren't being inefficient when they run these systems, they're programming our planet as if it were a computer. They're using our globally-decentralized electric power grid as the motherboard of a planetary-scale state mechanism over which no single person or group can fully control, converting the watts it produces into (physically restricted) bits of information. By doing this, they're adding novel features and capabilities to cyberspace which weren't there before. **They're opening a portal between worlds that imports the real-world physical constraints and thermodynamic restrictions of shared objective reality into virtual reality, anchoring one reality to the other in order to make cyberspace function and behave more like physical reality.** The theoretical benefits of doing this are quite compelling, particularly as it relates to systemic security. The ideas presented in this thesis barely scratch the surface of what could be enabled by this technology in the future.

Therefore, instead of criticizing technologies like Bitcoin with accusations about how inefficient it is, the author challenges computer scientists and software engineers to take inventory of their assumptions and ask themselves, "what could be the value of having increasingly more physically restricted command actions and bits of information in the global cyber domain?" The author has provided several potential benefits to this approach, like producing real-world physical properties (e.g. scarcity, decentralization), or imposing severe physical costs on computers, computer programs, and computer programmers in, from, and through cyberspace, thus ushering in a new age of electro-cyber warfighting and power competition.

This thesis has attempted to inspire the reader to think of more answers to this challenging question. Much could be gained by introducing physical costs and restrictions into a domain that is clearly suffering from superfluous cybercrime and systemic abuse. The benefits are clear to those who understand the dynamics of physical power projection. But these benefits aren't understood by the general public, else this technology would probably not be labeled as "inefficient." Therefore, perhaps a missing ingredient for understanding Bitcoin is a novel theory about power projection that explores how power is projected in nature, human society, and cyberspace. Through the lens of Power Projection Theory, Bitcoin's potential value as something more than just money becomes easier to grasp.

6.1.4 Bitcoin Could Disrupt Human Power-Based Dominance Hierarchies at their Foundation

"Bitcoin is not an investment. It's a revolutionary act."
Dan Held [189]

Power Projection Theory can be summarized as follows: human civilization uses two types of power projection tactics to settle their property disputes and establish their dominance hierarchies: physical power and abstract power. The projection of physical power is what we call war. The projection of abstract power is what we call law. War is at least 10,000 years old (with an additional ~4 billion years of precedent set by other organisms). Written law is at least 5,000 years old.

The technologies which human society uses to form both their physical and abstract power-based dominance hierarchies are rapidly changing after the invention of computers, and it seems like humanity is struggling to understand (and reconcile) the new balance of power. Modern agrarian society has only known how to write their policies on parchment and enforce/defend them using *kinetic* power projection tactics, a.k.a. "hard" forms of warfare. They have only known how to build their new empires this way because up until ~80 years ago, non-kinetic electro-cyber power projection tactics were not an option. Humans simply haven't had the ability to create polities any other way other than through the combination of written law (i.e. abstract power projection) and kinetic warfighting (i.e. physical power projection), because they haven't had access to technology which lets them do both differently.

Until very recently, global-scale human polities (a.k.a. organized societies) have never had trans-national communities adopt policies written in computer programming languages before. They have certainly never seen how it is possible to physically enforce/secure policies using electro-cyber power projection tactics or "soft" forms of warfare like Bitcoin. This is probably because people don't normally spend their time thinking about human metacognition or the systemic differences between abstract power and physical power, much less do they think about what would happen to our societies if humanity figured out how to project both abstract and physical power in a completely new way, from a completely new domain. Therefore, if new technologies emerged which changed the way people can project both abstract and physical forms of power, it's not reasonable to expect that they would even be able to recognize what's happening. At the very least, it would likely be a confusing and misunderstood phase of history.

This thesis is an attempt to alleviate some of that confusion and misunderstanding. The key insight of Power Projection Theory can be summarized in two sentences: **technology like Bitcoin could represent the discovery of an unprecedented power projection tactic which gives people a new way to establish their dominance hierarchies and physically secure the properties and policies they value. This technology has the potential to disrupt human polities at their very foundation by changing the way our species manages one of its most precious resources: information itself, and all of the knowledge and abstract power derived from it.**

The way societies settle policy disputes could change forever, because Bitcoin offers a way for people to peacefully settle policy disputes in a zero-trust and egalitarian manner that isn't as systemically vulnerable to corrupt, abusive, self-serving people who must be trusted not to exploit their abstract power and control authority over the existing internet architecture. The way society determines who has control over resources could change forever, because Bitcoin offers a way for people to physically secure and physically compete for zero-trust and permissionless control over resources (namely digital information) in a non-lethal and egalitarian way. The way society achieves international consensus on the legitimate state of ownership and chain of custody of property could change forever, because Bitcoin offers a way for people across the world to achieve consensus on the legitimate state of ownership and chain of custody of bits of information without having to trust each other. Bitcoin's bits can represent any kind of precious information that people want it to represent, including but of course not limited to financial information.

Bitcoin could represent a "strangler pattern" replacement of our legacy internet infrastructure with a modern architecture that isn't so clearly vulnerable to systemic exploitation and abuse. Bitcoin could represent the adoption of a new internet architecture with a foundationally different state mechanism at the base layer of the tech stack. Bitcoin could represent the operating system of a planetary-scale computer that we haven't seen before and don't understand we need, one that gives people zero-trust and permissionless access to cyberspace regardless of who they are and what they want to use it for. This new version of cyberspace could be one that gives people the egalitarian right to compete for administrative control over it using their respective physical power projection capabilities. And as a complex emergent effect of that physical power projection competition, administrative control over this new planetary-scale computer can remain decentralized, thus far less vulnerable to the population-scale systemic exploitation that is increasingly plaguing the current internet architecture.

As the new internet architecture "strangles" the legacy internet architecture, the first of many services that digital-age society would likely want to transfer from the legacy internet architecture to the new architecture would be financial services. Financial bits of information are obviously valuable bits of information. People are going to want to physically secure their ability to transfer financial bits of information from one side of cyberspace to the other without it being denied or exploited. But after this financial security service is ported over to the new internet architecture, it's possible that many other services will follow, services like free speech. Insofar as humans continue to value information in the information age, they are probably going to want an internet infrastructure that allows them to physically secure their bits of information. This could transform how organized societies work, because prior to the information age, physical security of money, speech, policy, and property rights was almost exclusively performed by the state. Now it's starting to be performed (at least in part) by those willing to contribute physical power (a.k.a. watts) to enforcing/securing an open-source system known as Bitcoin.

Using Power Projection Theory, we gain a new perspective about the strategic significance of proof-of-work protocols like Bitcoin. **This technology isn't just disruptive to monetary systems; it's disruptive to how our species answers the existentially important question of managing its resources. Proof-of-work technologies like Bitcoin could change the dynamics of 10,000 years' worth of physical power competitions, consequently disrupting how societies organize themselves and shifting the global balance of power**. It's difficult to understate just how disruptive electro-cyber power projection technologies like Bitcoin can be to our current power dynamics once we open the aperture and stop looking at this technology like it's strictly "digital currency". By viewing Bitcoin as a "softwar" system rather than just a peer-to-peer electronic payment system, the strategic implications of this technology becomes clearer. **This technology could dramatically change the way humans settle disputes and compete for control over the things they value, resetting the global balance of power in ways that only full-scale kinetic world wars have done in the past**.

With the global adoption of cyberspace combined with the global adoption of an electro-cyber form of physical power competition enabled by proof-of-work technologies like Bitcoin, humanity could be at the dawn of creating a completely new type of polity that it has never seen before – a new or adjusted type of governance system which enables the formation of an organized society that resembles something on par with (or perhaps even superior to) a traditional government. With the emergence of cyberspace and a new way to project power in, from, and through cyberspace, humanity could be crossing the event horizon of a highly impactful discovery akin to the discovery of agriculture. These technologies could reshape our understanding of what the term "national defense" means by changing what "national" means, because this technology could change how abstract power hierarchies are forged in the first place.

6.2 Recommendations for Future Research

"We also think there is very high stakes game theory at play here, whereby if bitcoin adoption increases, the countries that secure some bitcoin today will be better off competitively than their peers. Therefore, even if other countries do not believe in the investment thesis or adoption of bitcoin, they will be forced to acquire some as a form of insurance."

2022 Fidelity Report [190]

6.2.1 Develop More Theories about Bitcoin that Aren't based on Financial, Monetary, or Economic Theory

The author has two recommendations for future Bitcoin research. The first recommendation is to consider doing other grounded theory research efforts like this, where people endeavor to create alternative theoretical frameworks to analyze Bitcoin's sociotechnical and national strategic implications as something other than strictly a monetary technology. This thesis was merely an attempt to produce something different after recognizing that academia has been making the same baked-in presumptions with their current analytical efforts which could possibly be contributing to so much analytical bias that it could represent a national strategic security hazard.

6.2.2 Deductively Analyze Bitcoin from the Perspective of Power Projection Theory

If people aren't interested in producing alternative theories, then a second recommended way ahead for future researchers is to consider using the concepts of Power Projection Theory presented in this thesis to derive new hypotheses about Bitcoin which can be deductively analyzed using traditional hypothesis-deductive approaches. This thesis is filled with different concepts and hypotheses that could be used for follow-on research efforts. The author tried to provide something for the research community to pick apart and validate or invalidate using their own fields of knowledge and expertise. It could be beneficial to hear other people's perspectives on these ideas. Maybe there is something more to this than the author didn't think about because of his own limited field of knowledge.

If the reader is interested in deductively analyzing concepts presented in this thesis, then the author recommends focusing on the novel computer theoretics presented in section 5.9. This section offers the richest source of candidate hypotheses for future analysis. For example, is the author's hypothesis valid that physical cost function protocols like Bitcoin could represent the operating system of a completely new type of planetary-scale computer? Does it have completely different emergent properties which no ordinary computer could replicate? What are some of the other benefits that could be gained from a reverse-optimized, planetary scale computer that makes bits of information extraordinarily expensive? Each of these questions could represent a rich research topic, particularly in fields related to computer science and cyber security.

6.3 Recommendations for Future Policy Making Efforts

"I have a foreboding of an America in my children's or grandchildren's time – when the United States is a service and information economy... when awesome technological powers are in the hands of a very few, and no one representing the public can even grasp the issues..."
Carl Sagan [191]

6.3.1 Stop Relying Exclusively on Financial, Monetary, and Economic Theorists to Influence Bitcoin Policy

For policy makers who are working on shaping public policy as it relates to Bitcoin, proof-of-work technologies, cybersecurity technologies, or national security, the author offers the following five recommendations. The first recommendation is to stop relying exclusively on financial, monetary, and economic theories to inform public policy about proof-of-work technologies like Bitcoin. Subject matter expertise in finance only makes one qualified to talk about *financial* use cases of Bitcoin, not *all* use cases of Bitcoin and especially not about proof-of-work technologies as a whole. Financial expertise is mostly irrelevant to computer theory, power projection tactics, and military strategy. Subject matter expertise in economics or monetary theory doesn't make one qualified to talk about the technical merits of a computer system arbitrarily called a *coin*, just like subject matter expertise in meteorology doesn't make one qualified to talk about the technical merits of a computer system arbitrarily called a *cloud*.

The reason why the author makes this first recommendation is because Bitcoin could have far higher balance-of-power implications related to national security as an electro-cyber power projection technology (a.k.a. softwar protocol) rather than a monetary technology. Consequently, not having people who understand Bitcoin from theoretical frameworks such as the one outlined in this thesis could be detrimental to the development of responsible public policy surrounding proof-of-work technologies. If the US policy makers are truly interested in creating responsible public policy for proof-of-work technologies, then they should be willing to consider alternative perspectives to eliminate blind spots. This thesis is an attempt to provide staffers with the background information needed to understand the "so what" of proof-of-work technologies like Bitcoin from the perspective of someone who has devoted their career to designing weapons systems for a new warfighting domain (space), understanding emerging power projection technologies, and advising senior leaders.

6.3.2 Think of Bitcoin as an Electro-Cyber Security System rather than a Monetary System

The author's second recommendation is to consider thinking about Bitcoin as first and foremost an electro-cyber security system rather than strictly a monetary system, because it could be a national strategic security hazard to mischaracterize Bitcoin as *only* a monetary system if it is indeed something more intrinsically useful than *just* a peer-to-peer electronic payment system.

Unless society decides to reverse an 80-year-old trend of becoming increasingly more reliant on their computers, then national strategic security is going to depend heavily on cyber security in the future. It's no secret that the success of future warfighting campaigns will probably hinge on our ability to keep our systems secure against cyber attacks, attacks which would include systemic exploitation of all cyber systems and cyber networks. This thesis has outlined how and why proof-of-work technologies like Bitcoin could represent a completely new type of electro-cyber power projection technology that could transform the architecture of our cyber systems (to include the base-layer physical infrastructure of the internet) to make them more physically securable. There are many applications for Bitcoin as first and foremost a cyber security protocol to protect *any* bits of information, including but not limited to *financial* bits of information. We need to open the aperture and consider these broader applications, else we could forfeit a major technological lead in cyber security to US adversaries.

Bitcoin demonstrates that proof-of-work works as a physics-based (as opposed to logic-based) cyber security system. Computer scientists have been investigating the cyber security applications of what we now call proof-of-work for thirty years – more than twice as long as they have been using proof-of-work technology for monetary applications like Bitcoin. If Bitcoin does indeed represent an operationally successful cyber security technology, then it could be of vital national strategic interest to take it seriously and to encourage its adoption. Conversely, it could be a strategic security hazard to *discourage* adoption.

If Bitcoin secures bits of digital information in cyberspace like nothing else before it, then people (to include nations) are going to want to use it to secure *all* their valuable bits of information. If the theories about the "softwar" neologism presented in this thesis are valid, then technologies like Bitcoin could represent another international strategic Schelling point – a completely new way to project physical power in a new domain that has become instrumental to the structure of digital age society. **So long as we refuse to accept this technology as anything *other* than peer-to-peer electronic cash, we are at risk of missing a much bigger strategic picture: this could be a new way of warfighting that we will have no choice but to adopt and master if we want to stay competitive in the digital age**.

Cyberspace is already changing the balance of power in modern agrarian society. Cyberspace is a battlefield which does not appear to give an upper hand to nuclear superpowers, but could theoretically give a major upper hand to electro-cyber superpowers. It stands to reason that we should at least begin to seriously consider the strategic importance of posturing this nation to be an electro-cyber superpower. As we speak, nations are actively starting to adopt Bitcoin to protect and defend their national interests. They are raising public funds to build and run hash forces. As they do this, they are validating Tesla's theories about the future of warfare. Bitcoin is demonstrating how to decrease the utility of kinetic warfare by making all nations, big or small, equally able to defend themselves in, from, and through cyberspace. We are watching Tesla's theories play out in real time, and few seem to recognize it.

Power projection tactics in the digital age have clearly changed, and we need to learn how to adapt to these new power projection tactics, else we should not expect to remain a global superpower in the digital age regardless of our kinetic power projection capabilities. Our combined military kinetic strength is practically useless in the abstract domain of cyberspace where there are no masses to displace with forces. Assuming the theories presented in this thesis are valid, this would imply that **if the US wants to be powerful in, from, and through cyberspace, they're going to need to consider stockpiling bitpower (a.k.a. bitcoin) and they're going to need to consider supporting a robust domestic hashing industry**.

It may even be worth considering the standup of organizations dedicated US kinetic power projection resources to protecting and defending our domestic hashing industry. In other words, the US might one day consider using its military resources to physically secure the mass-based infrastructure and integrity of the Bitcoin network (e.g. internet, power grid, computing facilities, chip fab resources, etc.). To that end, it's conceivable that the US may one day employ some type of combatant component dedicated to protecting and defending the nation's Bitcoin hashing infrastructure, just like we already do for other strategically important assets. This concept is illustrated in Figure 96.

Figure 96: Notional Combatant Component Dedicated to Securing Allied Hashing Industry [192]

6.3.3 Consider the Idea of Protecting Bitcoin under the Second Amendment of the US Constitution

The author's third recommendation is for US policy makers to explore the idea of protecting proof-of-work technologies like Bitcoin under the second amendment. An argument can be raised that current efforts by some policy makers to ban Bitcoin is a violation of the second amendment of the US Constitution. People have a right to physically secure what they freely choose to value. In the information age, the technologies people use to physically secure what they value should be expected to change, but the intent of the second amendment should not change. **US citizens have a right to bear arms (in whatever form those arms take, kinetic or electric) to physically secure their bits, especially if those bits double as their sovereign property**. It shouldn't matter if the both property as well as the system used to impose severe physical costs on attackers, takes the form of a computer system. In fact, it shouldn't be surprising that it would in the digital age of information.

An argument could likely be made that a non-lethal, electro-cyber security system like Bitcoin might qualify as some type of non-lethal, digital armament (again, Nakamoto himself called it an arms race). If it does, then common sense says that Americans should have a right to bear non-lethal arms to physically defend their bits of information against attack, including and especially against attacks from their own overreaching government policies (the fact that it can't cause injury is even more reason to support it, in the author's personal opinion). Thus, trying to ban Bitcoin on the grounds that it uses too much energy could be a blatant affront to American values.

6.3.4 Recognize that Proof-of-Stake is not a Viable Replacement for Proof-of-Work

The author's fourth recommendation is to recognize that proof-of-stake systems are not viable replacements to proof-of-work systems. Not recognizing this opens up a window for systemic exploitation and abuse, it also directly threatens US national strategic security. Proof-of-stake is essentially a form of sybil attack – a way for a singular organization with centralized administrative privileges to obfuscate their abstract power and masquerade as a decentralized system, allowing them to systemically exploit a weighted voting/reputation system. "Stake" is nothing more than an abstract name given to the zero-sum and inegalitarian administrative privileges that have already been assigned to an unknown group of anonymous users (probably those who awarded themselves with the majority supply of ETH before the protocol was publicly launched and several years before the protocol forked to proof-of-stake). To adopt a proof-of-stake system is to put oneself at the mercy of the users who control this so-called "stake" based on nothing more than blind faith that "stake" isn't already majority-controlled by a single group of people who could easily exploit its reputation system by dividing that "stake" across multiple addresses and masquerading as a decentralized group of people.

The fact that proof-of-stake software developers (who, in the case of Ethereum, are non-US citizens who openly admit to awarding themselves the majority supply of stake-able ETH years before forking the protocol to a proof-of-stake system) are claiming to have created a decentralized system without proof-of-work is more likely to be fraud than it is to be innovation, because we know from the basic tenets of computer theory that it's physically impossible for an object-oriented software abstraction like "stake" to be verifiably decentralized in the first place. "Stake" doesn't physically exist, so it's incontrovertibly true that it's physically impossible for the special administrative privileges which Ethereum developers arbitrarily named "stake" to be verifiably decentralized. If there's no way for people to independently validate that the special administrative privileges of a given software system are decentralized, then the system can only be decentralized in name only (DINO). For a more detailed discussion about this, see Section 5.10.

6.3.5 If we Make the Mistake of Expecting the Next World War to Look Like the Last World War, we could Lose it before we Realize that it has Already Started

"The general population doesn't know what's happening,
and it doesn't even know that it doesn't know."
Noam Chomsky [193]

History makes something explicitly clear: the discovery of new power projection tactics, techniques, and technologies changes the balance of power. The forging of new power projection tools often leads directly to the forging of new empires. This power projection game is not just a 10,000-year-old game played by human beings, it's a 4-billion-year-old game played by practically all surviving organisms. The lesson to be gained from history is quite simple: **adapt to new physical power projection tactics as they emerge, or else lose the empire**. Physical power projection has never been optional or negotiable. Physical power projection affects us the same whether we agree with it and we're sympathetic to it or not. If any organism is to survive in a world filled with predators and entropy, then newly discovered physical power projection tactics must be adopted and mastered when they emerge. It's imperative for leaders at the top of existing power hierarchies to take new power projection technologies seriously and to recognize that speed of adoption is critically important.

With this concept in mind, the author's fifth recommendation for US policy makers is to recognize the millennia-old mistake of expecting the next war to look like the last war. If the Power Projection Theory presented in this thesis is valid, then it's possible to make three important observations about Bitcoin: (1) this technology could represent a new form of electro-cyber warfighting that will empower people in new ways and disrupt existing power structures, (2) people at the top our current power structures would probably not recognize what's happening (or struggle greatly to reconcile it) because they were expecting the next war to look like the last war, and (3) the war could start and the balance of power could shift quickly and dramatically without many people realizing it's happening before it's too late.

Add these three observations together, and it's possible to draw the following conclusion: **the next major global power competition could be a "soft" or electro-cyber form of warfare fought in, from, and through cyberspace using a "softwar" protocol.** It could be some kind of revolutionary balance-of-power struggle between an existing ruled and ruling class (not necessary against governments, but against software system administrators as a whole, which could include but not be limited to governments). Additionally, this "soft" revolutionary war could happen simultaneously across the planet and therefore technically qualify as a third world war. Lastly and most importantly, this soft, non-lethal, third world war could have already started, but our leaders don't recognize it yet because they're expecting it to be a kinetic war fought in three-dimensional space rather than an electric war fought in cyberspace. This would imply that **leaders could forfeit the dominion of their empires because they're expecting the next world war to look like the last world war. Not only is it possible that a "soft" electro-cyber world war has already started, it's also possible that the resulting balance-of-power transformation is already well underway, and that existing superpowers could be behind the future power curve – all because their leaders didn't understand enough about computer theory or the profession of warfighting to recognize that a**

new, non-lethal, electro-cyber warfighting technology has emerged that dramatically changes the balance of power in digital-age society, and it's called Bitcoin.

6.4 Closing Thoughts

"If you need a machine and don't buy it, then you will ultimately find that you have paid for it and don't have it.
Henry Ford [1]

6.4.1 Bitcoin is Fundamentally about Balance-of-Power, not Money

The primary value-delivered function of Bitcoin is not about money, it's about physical empowerment and physical balance of power. Bitcoin changes the dynamics of physical power projection and takes control over a precious resource (bits of information) out of the hands of people with imaginary power who cannot be trusted. People do not appear to understand this because they do not appear to understand the dynamics of power projection. If this is true, then perhaps all that's needed to understand the "so what" of Bitcoin is a new theoretical framework from which to analyze it – one that addresses its primary value-delivered function. To that end, the author developed this grounded theory.

Living creatures have been using physical power to settle disputes, establish control authority over resources and achieve consensus on the legitimate state of ownership and chain of custody of property for 4 billion years. There has never been a viable alternative to physical power for satisfying these functions. No matter how much sapiens desperately want to *imagine* an alternative to physical power as a basis for settling our disputes and establishing our dominance hierarchies, we have never been able to replace physical power (a.k.a. watts) because watts clearly have complex emergent social benefits that can't be replicated by our imaginations.

To date, no amount of trust, faith, or written logic has been able to keep a human belief system secure against other humans. This is probably because humans are first and foremost apex predators. Give a human any kind of imaginary power, and they will be tempted to exploit it. It doesn't matter what type of stories people write or whether they're written on parchment or python. **Create any kind of belief system, and systemic predators will emerge to exploit those belief systems for their own personal advantage unless they are *physically restricted* from doing so**. This is one of many reasons why physical power projection is necessary, a lesson that "civilized" societies keep allowing themselves to forget over and over again. It appears that digital-age society has made the same mistake.

Amongst the most precious resources of modern society is information, including (but not limited to) financial information. 80 years ago, humans figured out how to digitize information and store it on intelligent machines using machine code. These intelligent machines now control most of our bits of information. This begs an important question: who controls these intelligent machines?

Those who program our computers now control one of digital-age society's most precious resources. Computer programmers are starting to recognize this, and they're deliberately designing software to give themselves extraordinary amounts of abstract power and control authority over society's bits. Not surprisingly, they're also starting to consolidate and exploit this abstract power. A new class of technocratic god-kings are emerging, and their abstract strength is derived from the software we're using, as software itself represents nothing more than an encoded belief system – another way to wield abstract power over others.

Look around, and you'll see evidence of a growing imbalance of power. The richest and most prominent among us are often the ones who control the computers which manage our financial bits of information. Behind our computer networks are programmers pulling the strings of our intelligent machines, and they have become incredibly influential. As long as we (1) continue to design systems where we give control over our bits to people wielding this imaginary source of power, and (2) do nothing to impose severe physical costs on people who can exploit that imaginary power, then we should expect to be systemically abused in, from, and through our computer networks as the abstract power of these systems becomes increasingly imbalanced. This is one of the most commonly recurring lessons of history – we have thousands of years of testimony warning us about the dangers of abstract power imbalances.

Bitcoin could fix this situation the same way physical power has always been used to end systemic exploitation and abuse. To keep people systemically secure against exploitative god-kings, simply give people the ability to impose severe physical costs on their god-kings. Like magic, the god-king's abstract power will evaporate, and the balance of power will improve for the oppressed. But how can people do this in cyberspace using nothing but their computers? How does one impose physical costs on anonymous computer programmers hiding behind the software they design? Bitcoin shows us how: convert physical power into bits so that it can be wielded in, from, and through cyberspace. Simply adopt a common protocol & infrastructure where people can apply Boolean logic to large sums of electric power drawn from our power grids. Such a simple concept; such profound implications.

By simply using large quantities of watts to represent our precious bits of information (as opposed to transistor states), the heavily consolidated and unimpeachable abstract power of digital-age god-kings can be dramatically rebalanced. The solution to the threat of systemic exploitation via cyberspace could simply be to use our globally-decentralized power grid as the underlying state mechanism of a new internet architecture. But to accomplish this, we would need to design an operating system which can convert the power grid into a planetary-scale computer over which no single person or group of people can have consolidated and unimpeachable control. Incidentally, this is exactly what Bitcoin appears to do.

Using this planetary-scale computer, we could theoretically reverse-optimize bits of information to be as expensive as possible to transmit, and then refuse to execute exploitable commands or control actions without these expensive bits of information. We can turn physical power into bits of information that prove the presence of real-world power, and then we can demand proof of that real-world power to impose a physical constraint on those who wield imaginary power. By doing so, we can impose severe physical costs on oppressors in, from, and through cyberspace and mitigate digital-age abstract power imbalances caused by software. This new power projection tactic has the potential to be quite disruptive. It has the potential to upend all existing power and dominance hierarchies and subvert national power, and the world is starting to recognize this.

For the sake of argument, let's assume a new physical power projection technology has indeed emerged in the form of Bitcoin. If the theories presented in this thesis are valid, then the US has the same options that previous empires had when innovative new power projection technologies emerged during their time. Ignoring it wasn't an option back then, and it probably still isn't an option today. Stopping adversaries from adopting it and using it against them wasn't an option back then, and it probably still isn't an option today. **Assuming the primordial dynamics of power projection still hold true, then nations can either embrace this new power projection technology and adopt it as soon as possible, or suffer from the success of those who do. Not being affected by the resulting rebalance of power isn't an option; all anyone can do at this point is maneuver to put themselves in the most favorable position possible**. And the clock is ticking.

6.4.2 Electro-Digital Antlers Would be Worth Every Watt

"Innovation is non-biological evolution. Evolution is biological innovation.
They're the same process, just different substrates..."
Robert Breedlove [194]

Aside from the national strategic implications of Bitcoin, there could be more inspiring implications. People like to condemn human society's inclination to fight wars, but maybe the problem with society is not that they fight wars to settle their disputes and secure their resources. After all, practically all living organisms do this. Instead, **maybe the problem with human society is that we have not figured out how to make our wars *soft*. That is – we haven't learned how to wage our wars in a way that avoids fratricide**. Maybe we just need to figure out a way to settle the disputes we have about our most valuable resources non-destructively. In other words, maybe humans simply need to evolve their own version of antlers – a way to compete against each other and establish dominance hierarchies using physical power competitions, but in a way that doesn't cause so much injury to their own kind.

It's unlikely that humans will grow antlers any time soon, but we could be clever enough to develop a special type of technology to perform the same function as antlers. As Tesla predicted in 1900, it could involve intelligent machines competing against each other in an energy competition, while humans watch eagerly from the sidelines. Just like Ford predicted in 1921, it could involve turning electricity into money. The same technology could accommodate both theories simultaneously.

If we could figure out a way to setup our dominance hierarchies over our valuable resources using a globally-adoptable, non-lethal, physical power projection technology, that would not just represent a technological breakthrough that could change the nature of warfighting – that would be an evolutionary breakthrough. It would represent the evolution of a technology that would (literally) empower humans to establish intraspecies dominance hierarchies and determine the right balance-of-power dynamics using non-lethal (electric) physical power rather than lethal (kinetic) physical power, all while still retaining the complex emergent benefits of physical-power-based resource control structures.

This is what I find so inspiring about Bitcoin. The idea of non-lethal warfighting is what motivated me to risk my professional career and devote my entire US national defense fellowship towards researching this technology and developing this new theoretical framework. If the theories presented in this thesis are valid, then proof-of-work technologies like Bitcoin could function as human antlers. We may have discovered how to use our globally-decentralized electric power grid as the "bone structure" we need to project power and impose severe physical costs on each other to settle our disputes, secure our resources, and establish our dominance hierarchies in a way that dramatically reduces the need to obliterate each other using traditional methods of warfare. This concept is illustrated in Figure 97.

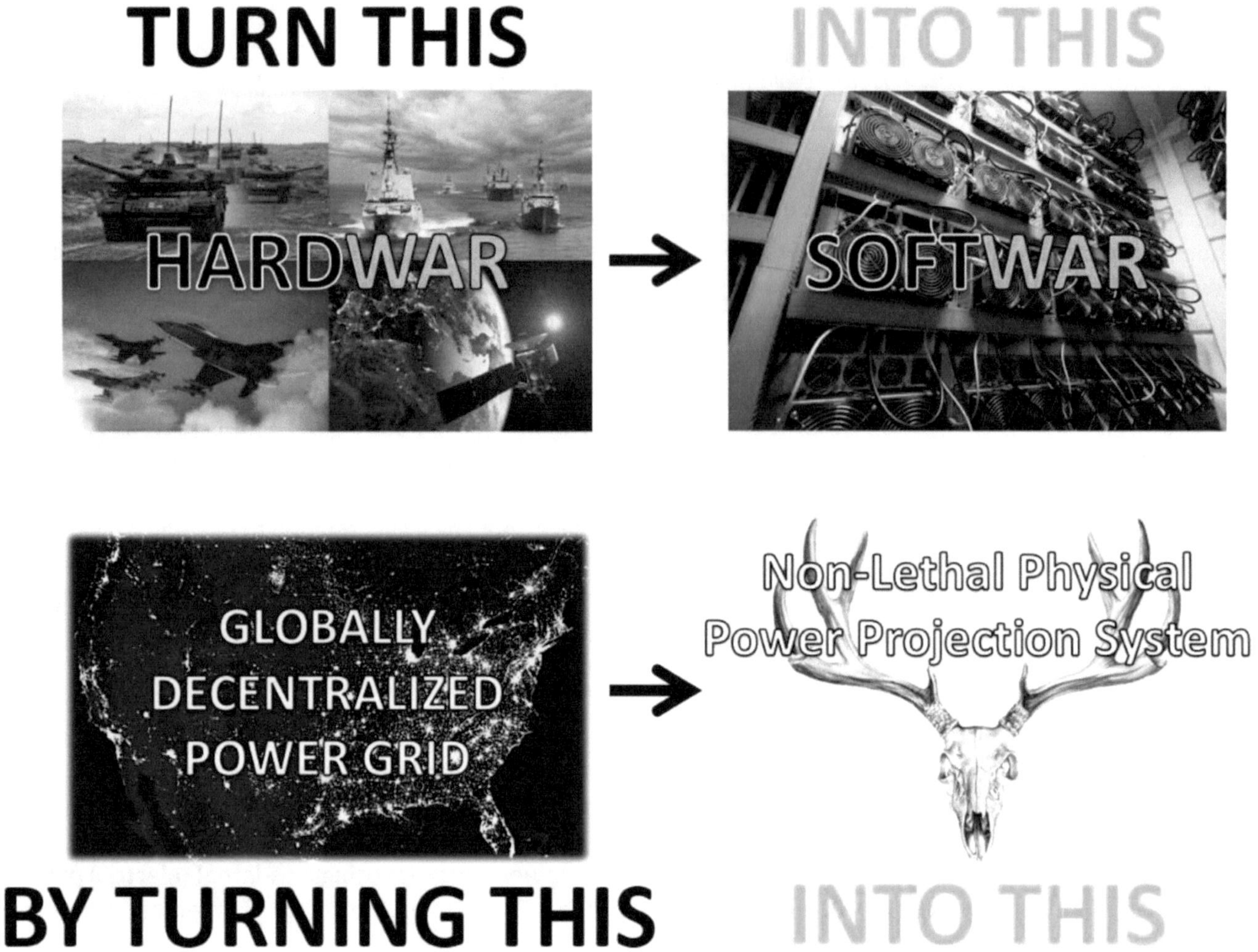

Figure 97: Illustration of Softwar Concept
[1, 2, 3, 4, 5, 149, 150, 151, 152]

The bottom line is that Bitcoin could represent the emergence of an electro-cyber power projection system and a new form of warfare that digital-age society could utilize to help mutually assure the preservation of our species while still retaining the complex sociotechnical benefits of our physical power competitions (what we otherwise call warfare). For a population entrapped by more than 10,000 years of kinetic warfighting that has now scaled their kinetic power projection capabilities to the brink of mutually assured nuclear destruction, proof-of-work technology like Bitcoin could not only function like a human version of antlers, it could also be our own personal version of multicellular cooperation – our way to escape a bounded prosperity trap that we created for ourselves.

Like archaeon did two billion years ago, proof-of-work technology could give humanity a way to overcome a fiery hellscape of kinetic warfare by giving people a mechanism to stick together and cooperate at a global scale even though we clearly can't trust each other enough to form a singular unified government with a central ruling class. Bitcoin (the system, to include its users) could function as a global-scale polity that transcends national borders and unites people across the world in a counterintuitive way: by creating a belief system where everyone agrees that no one can be trusted with abstract power. Bitcoin could allow humans to put their faith back into physical power (e.g. watts) as the preferred way to secure their resources and settle their property disputes, the way nature demands. In what would be a romantic twist of fate, the general-purpose computers we used to design the nuclear bomb could also help humanity protect itself from the threat of the nuclear bomb by utilizing software to wage *softwar*. If this theory proves to be valid, then of course Bitcoin wouldn't be a waste of energy; it would be worth every watt.

References

[1] M. Mareen, "German Main Battle Tank Leopard 2A7 (Image license owned by Jason Lowery)," Adobe Stock, [Online]. Available: https://stock.adobe.com/. [Accessed 28 December 2022].

[2] Alejandro, "Formation of NATO Military Ships in the Atlantic Ocean (Image license owned by Jason Lowery)," Adobe Stock, [Online]. Available: https://stock.adobe.com. [Accessed 28 December 2022].

[3] phaisarnwong2517, "The Fighter Jets are Taking Off for an Attack (Image license owned by Jason Lowery)," Adobe Stock, [Online]. Available: https://stock.adobe.com. [Accessed 28 December 2022].

[4] J. L. Stephens, "3D Rendering of a Satellite orbiting the Earth with Illuminated Cities at Night (Image license owned by Jason Lowery)," Adobe Stock, [Online]. Available: https://stock.adobe.com. [Accessed 28 December 2022].

[5] artiemedvedev, "Bitcoin miners in large farm (Image license owned by Jason Lowery)," Adobe Stock, [Online]. Available: https://stock.adobe.com. [Accessed 28 December 2022].

[6] M. Saylor, "Bitcoin: There is No Second Best with Michael Saylor and Greg Foss," Bitcoin Magazine, 4 November 2021. [Online]. Available: https://bitcoinmagazine.com/culture/there-is-no-second-best-with-michael-saylor.

[7] N. Tesla, "A Machine to End War," *Liberty,* 1937.

[8] N. Tesla, "The Problem of Increasing Human Energy," *The Century Magazine,* 1900.

[9] "Ford Would Replace Gold With Energy Currency and Stop Wars," New York Tribune, New York, 1921.

[10] J. Lewis, "Technological Advancement: The Choice Before Us," Center for Strategic & International Studies, 2011.

[11] "JIAAW Workplace," Brown University, [Online]. Available: https://www.brown.edu/Departments/Joukowsky_Institute/courses/13things/7687.html.

[12] W. H. Marios Philippides, The Siege and the Fall of Constantinople in 1453: Historiography, Topography, and Military Studies, Routledge, 2011.

[13] E. Gray, The Devil's Device: Robert Whitehead and the History of the Torpedo, US Naval Institute Publishing, 1991.

[14] T. Green, "The Development of Air Doctrine in the Army Air Arm 1917-1941," *Library of Congress,* 1955.

[15] L. Morton, The Decision to Use the Atomic Bomb, US Army Center of Military History, 1957.

[16] S. Baker, "Albert Einstein wrote to the US pleading with the government to build an atomic bomb 80 years ago. Here's what he said.," Business Insider, www.businessinsider.com, 2019.

[17] "National Security Archive, 35+ Years of Freedom of Information Action," George Washington University, 1946. [Online]. Available: https://nsarchive.gwu.edu/document/21885-document-10-robert-oppenheimer-president#_edn10. [Accessed 2022].

[18] "International Youth for a Better World," University of Virginia, 2018. [Online]. Available: https://batten.virginia.edu/international-youth-better-world.

[19] C. v. Clausewitz, M. Howard and P. Paret, On War (English translation of Vom Kriege), Princeton, New Jersey: Princeton University Press, 1976.

[20] P. Champagne, The Book of Satoshi: The Collected Writings of Bitcoin Creator Satoshi Namamoto, 2014.

[21] L.-M. Miranda, Artist, *Hamilton (musical).* [Art]. 2015.

[22] P. Turchin, Ultrasociety: How 10,000 Years of War Made Humans the Greatest Cooperators on Earth, Connecticut: Beresta Books, LLC, 2016.

[23] G. A. Luis Ferrer-Sadurni, "New York Enacts 2-Year Ban on Some Crypto-Mining Operations," The New York Times, 22 Novemeber 2022. [Online]. Available: https://www.nytimes.com/2022/11/22/nyregion/crypto-mining-ban-hochul.html#:~:text=New%20York%20became%20the%20first,The%20legislation%20signed%20by%20Gov..

[24] W. Broad, "Ukraine Gave Up a Giant Nuclear Arsenal 30 Years Ago. Today There Are Regrets.," The New York Times, 5 February 2022. [Online]. Available: https://www.nytimes.com/2022/02/05/science/ukraine-nuclear-weapons.html.

[25] J. Biden, "Presidential Executive Order on Ensuring Responsible Development of Digital Assets," White House, Washington DC, 2022.

[26] M. n. Cynthia Dwork, "Pricing via Processing or Combatting Junk Mail," in *CRYPTO'92: Lecture Notes in Computer Science No.740,* 1993.

[27] J. B. Ari Juels, *Client Puzzles: A Cryptographic Countermeasure Against Connection Depletion Attacks,* Bedford, MA: RSA Laboratories, 1999.

[28] A. J. Markus Jakobsson, Proofs of Work and Bread Pudding Protocols, Kluwer Academic Publishers, 1999, pp. 258-272.

[29] S. Nakamoto, *Bitcoin: A Peer-to-Peer Electronic Cash System,* https://bitcoin.org/bitcoin.pdf, 2009.

[30] C. W. Krueger, "Software Reuse," *ACM Computing Surveys,* vol. 24, no. 2, pp. 131-183, 1992.

[31] R. Frost, Artist, *The Road Not Taken.* [Art]. 1915.

[32] J. Yarow, "Paul Krugman responds to all the people throughing around his old internet quote," Business Insider, 2013. [Online]. Available: https://www.businessinsider.com/paul-krugman-responds-to-internet-quote-2013-12. [Accessed 2022].

[33] J. C. Anselm Strauss, "Grounded Theory Methodology, an Overview," in *Strategies of Qualitative Inquiry*, SAGE Publications, 1998.

[34] O. Patterson, "Overreliance on the Pseudo-Science of Economics," Harvard Kennedy School, 2015. [Online]. Available: https://www.hks.harvard.edu/centers/mrcbg/programs/growthpolicy/overreliance-pseudo-science-economics.

[35] M. Atwood, All Author, [Online]. Available: https://allauthor.com/amp/quotes/220095/.

[36] B. G. Anselm Strauss, The Discovery of Grounded Theory: Strategies for Qualitative Research, New Brunswick: Aldine Transaction, 1967.

[37] K. Sebastian, "Distinguishing Between the Types of Grounded Theory," *Journal for Social Thought,* vol. 3, no. 1, 2019.

[38] A. S. Julet Corbin, Basics of Qualitative Research: Techniques & Procedures for Develping Grounded Theory - 4th Edition, Los Angeles: SAGE Publications, Inc, 2015.

[39] J. L. A. B. Robert Breedlove, "WIM175: Bitcoin, Proof of Work, and the Future of War with Adam Back & Jason Lowery," What is Money Show, https://www.youtube.com/watch?v=37w4vxeYL8U, 2022.

[40] S. R. Maike Vollstedt, "An Introduction to Grounded Theory with a Special Focus on Axial Coding and the Coding Paradigm," 2019. [Online]. Available: https://link.springer.com/chapter/10.1007/978-3-030-15636-7_4#:~:text=Coding%20in%20grounded%20theory%20methodology,the%20evaluation%20process%20may%20begin

[41] I. Kant, "Theory without experience is intellectual play," Fordham Institute, [Online]. Available: https://fordhaminstitute.org/national/commentary/theory-without-experience-intellectual-play#:~:text=In%20Immanuel%20Kant%27s%20words%2C%20%E2%80%9CExperience,experience%20is%20mere%20intellectual%20play.%E2%80%9D.

[42] "Business Research Methodology," [Online]. Available: https://research-methodology.net/research-methods/data-collection/grounded-theory/.

[43] C. Darwin, "Letter from Charles Darwin to C.Lyell," *The Life and Letters of Charles Darwin,* vol. II, p. 108, June 1, 1860.

[44] W. R.-M. James Dale Davidson, The Soveriegn Individual, Simon & Schuster, 1999.

[45] I. Ostenberg, "Veni Vidi Vici and Caesar's Triumph," *The Classical Quarterly,* vol. 63, no. 2, 2013.

[46] B. Greene, "@alphacentauriA," Youtube, 15 December 2022. [Online]. Available: https://www.youtube.com/shorts/j-n5SB3FSH8.

[47] R. Scott, Director, *Prometheus.* [Film]. 20th Century Fox, 2012.

[48] H. Gee, A (Very) Short History of Life on Earth (ISBN 978-1-250-27665-0), New York: St. Martin's Press, 2021.

[49] O. Stone, Director, *Any Given Sunday.* [Film]. United States: The Donner's Company, Ixtlan Production, 1999.

[50] L. Megginson, "Lessons from Europe for American Business," *Soutwestern Social Science Quarterly,* vol. 36, no. 1, pp. 91-95, 1963.

[51] J. Lennon and K. Locey, "Aeon," September 2018. [Online]. Available: https://aeon.co/ideas/there-are-more-microbial-species-on-earth-than-stars-in-the-sky. [Accessed June 2022].

[52] "How Many Stars are there in the Universe?," European Space Agency, [Online]. Available: https://www.esa.int/Science_Exploration/Space_Science/Herschel/How_many_stars_are_there_in_the_Universe. [Accessed June 2022].

[53] F. Nietzche, Twilight of the Idols, 1888.

[54] D. Mills, "The origin of phagocytosis in Earth history," *Interface Focus,* vol. 10, no. 4, 2020.

[55] University of Bristol, "World's greatest mass extinction triggered switch to warm-bloodedness," 16 October 2020. [Online]. Available: https://www.bristol.ac.uk/news/2020/october/warm-blooded-mammals.html#:~:text=University%20of%20Bristol%20palaeontologist%20Professor,mass%20extinction%20of%20all%20time.. [Accessed July 2022].

[56] N. d. Tyson, "Youtube," 4 September 2022. [Online]. Available: https://www.youtube.com/shorts/uK4MrVVnRu8.

[57] C. Christensen, The Innovator's Dilemma, Cambridge: Harvard Business Review, 1997.

[58] D. Thomas, "Do not go gentle into that good night," *Bottenghe Oscure,* 1951.

[59] F. S. Fitzgerald, The Great Gatsby, 1925.

[60] Plato, Republic, Athens, 375 BCE.

[61] Y. M. B.-O. e. al, "The Biomass Distribution on Earth," *Proceedings of the National Academy of Sciences USA,* 2018.

[62] G. Lucas, Director, *Star Wars Episode 1: The Phantom Menace.* [Film]. Lucasfilm Ltd., 1999.

[63] J. Cameron, Director, *Avatar: The Way of Water.* [Film]. United States: Lightstorm Entertainment, TSG Entertainment, 2022.

[64] *Yellowstone Season 4.* [Film]. United States: Taylor Sheridan, John Linson, 2018.

[65] J. Rogan, "Forrest Galante," Joe Rogan Experience #1403, 2019.

[66] S. Aral, The Hype Machine: How Social Media Disrupts Our Elections, Our Economy, and Our Health -- and How We Must Adapt, London: Harper Collins Publishers, 2020.

[67] Y. Bar-On, R. Phillips and R. Milo, "The biomass distribution on Earth," *Proceedings of the National Academy of Sciences,* vol. 115, no. 25, p. https://www.pnas.org/doi/10.1073/pnas.1711842115, 2018.

[68] C. Foster, Being a Human, New York: Metropolitan Books, 2021.

[69] I. T, Composer, *Don't Hate the Playa.* [Sound Recording]. The Seventh Deadly Sin. 1999.

[70] D. Grossman, On Killing, Back Bay Books, 1996.

[71] G. R. R. Martin, A song of Ice and Fire, Bantam Books, 1996.

[72] A. Carrell, "Shadi Yousefian: A Restrospective," lawreviewofbooks.org, 21 October 2017. [Online]. Available: https://blog.lareviewofbooks.org/essays/shadi-yousefian-retrospective/.

[73] A. Einstein, "alberteinsteinsite.com," [Online]. Available: http://www.alberteinsteinsite.com/quotes/einsteinquotes.html.

[74] S. O'Malia, "cftc.gov," Commodity Futures Trading Commission, 19 June 2012. [Online]. Available: https://www.cftc.gov/PressRoom/SpeechesTestimony/opaomalia-16#:~:text=Albert%20Einstein%20said%2C%20%E2%80%9CThe%20true,its%20oversight%20mission%20going%20forward..

[75] L. Huica, "We suffer more in imagination than in reality - explained," Aurelius Foundation, 2022. [Online]. Available: https://aureliusfoundation.com/blog/we-suffer-more-in-imagination-than-in-reality-explained-2022-04-01/.

[76] B. Logo, "Free Brain PNG," vecteezy.com, [Online]. Available: https://www.vecteezy.com/png/1201180-brain. [Accessed 2022].

[77] Nikolae, "Sensory organs hand, nose, ear, mouth, and eye (Licensed by Jason Lowery)," Adobe Stock, [Online]. Available: https://stock.adobe.com/images/sensory-organs-hand-nose-ear-mouth-and-eye/49148505?asset_id=49148505. [Accessed 28 December 2022].

[78] S. Logo, "Free Snake Clipart," Public Domain Vectors, [Online]. Available: https://openclipart.org/detail/230794/diamondback-water-snake. [Accessed 2022].

[79] serikbaib, "ancient stone age stick (Licensed by Jason Lowery)," Adobe Stock, [Online]. Available: https://stock.adobe.com/search?filters%5Bcontent_type%3Aphoto%5D=1&filters%5Bcontent_type%3Aillustration%5D=1&filters%5Bcontent_type%3Azip_vector%5D=1&filters%5Bcontent_type%3Avideo%5D=1&filters%5Bcontent_type%3Atemplate%5D=1&filters%5Bcontent_type%3A3d%5. [Accessed 28 December 2022].

[80] E. Gamillo, "The Science Behyind Those Big Ol' Puppy-Dog Eyes," Smithsonian Magazine, 7 April 2022. [Online]. Available: https://www.smithsonianmag.com/smart-news/science-behind-those-big-ol-puppy-dog-eyes-180979877/.

[81] E. Bulwer-Lytton, Artist, *Cardinal Richelieu.* [Art]. 1839.

[82] A. W. J. L. Patrik Linderfors, "Dunbar's Number Deconstructed," The Royal Society Publishing, 5 May 2021. [Online]. Available: https://royalsocietypublishing.org/doi/10.1098/rsbl.2021.0158#:~:text=%27Dunbar%27s%20number%27%20is%20the%20notion,other%20individuals%20in%20the%20group..

[83] M. Casalenuovo, *High on Life (Video Game),* Squanch Games, 2022.

[84] B. Kirov, PLato: Quotes & Facts, 2015.

[85] Z. Snider, Director, *300.* [Film]. United States: Legendary Pictures, 2006.

[86] G. M. Hopf, Those Who Remain: A Postapocalyptic Npve;, CreateSpace Independent Publishing Platform, 2016.

[87] J. A. Tainter, The Collapse of Complex Societies, Cambridge University Press, 1990.

[88] T. Ovidiu, "Medieval Sword & Shield illustration (Licensed by Jason Lowery, #2087884927)," [Online]. Available: https://www.istockphoto.com/vector/medieval-sword-and-shield-vector-illustration-isolated-on-white-background-gm1148137302-309961466. [Accessed 28 December 2022].

[89] B. Logo, "Free Bomb Transparent PNG," toppng.com, [Online]. Available: https://toppng.com/photo/207782. [Accessed 2022].

[90] C. Logo, "Vector illustration with color crown.," dreamstime.com, [Online]. Available: https://www.dreamstime.com/vector-illustration-color-crown-isolated-clipart-image151688837. [Accessed 2022].

[91] "Logical Fallacies, Formal and Informal," [Online]. Available: https://web.archive.org/web/20111122045422/http://usabig.com/iindv/articles_stand/perm/fallacies.php.

[92] "Hypostatize vs Reify - What's the difference?," [Online]. Available: https://wikidiff.com/reify/hypostatize. [Accessed September 2022].

References

[93] A. Whitehead, Science and the Modern World, New York: Free Press, 1967.

[94] "Reification (fallacy)," psychology wiki, 2022. [Online]. Available: https://psychology.fandom.com/wiki/Reification_(fallacy).

[95] P. Verhoeven, Director, *Starship Troopers.* [Film]. United States: Tristar Pictures, Touchstone Pictures, 1997.

[96] R. Sienra, "Oculus Rift Creater Designs Frightening VR Set that Kills You for Real if You Die in a Video Game," My Modern Met, 10 November 2022. [Online]. Available: https://mymodernmet.com/oculus-vr-set-nervegear/#:~:text=Oculus%20founder%20Palmer%20Luckey%20designed,they%20die%20in%20the%20game.&text=If%20you%20die%20in%20the,honor%20of%20Sword%20Art%20Online..

[97] J. Marrs, Crossfire: The Plot that Killed Kennedy, Carroll & Graf Publishers, Inc., 1993.

[98] J. C. o. Staff, "Message from Joint Chiefs on US Capital Riot," 12 January 2021. [Online]. Available: https://news.usni.org/2021/01/12/message-from-joint-chiefs-on-u-s-capitol-riot.

[99] L. Rose, "John Lock vs Rousseau Essay," bartleby.com, [Online]. Available: https://www.bartleby.com/essay/John-Locke-Vs-Rousseau-FKESM3BZA4FP. [Accessed 2022].

[100] S. Aral, "The Hype Machine: How Social Media Disrupts Our Elections, Our Economy, and Our Health - and How We Must Adapt," Penguin Random House LLC, 2021.

[101] C. Cracknell, Director, *Persuasion.* [Film]. United States: MRC, 2022.

[102] "Countries by Government System," Genuine Impact, Lightyear, 25 10 2022. [Online]. Available: genuineimpact.substack.com.

[103] R. Ingersoll, "Did Lincoln Say, 'If You Want to Test a Man's Character, Give Him Power'?," 1883. [Online]. Available: https://www.snopes.com/fact-check/lincoln-character-power/.

[104] Y. N. Harari, Sapiens, HarperCollins Publishers, 2015.

[105] Nietzsche, Beyond Good and Evil: Prelude to a Philosophy of the Future, 1886.

[106] J. Booth, "@JeffBooth," Twitter, 22 November 2022. [Online]. Available: https://twitter.com/JeffBooth/status/1595044356180021248?s=20&t=0AvipEztzKRs6YyrJWegrA.

[107] M. T. Cicero, "Quotes and Aphorisms on Congress," www.daimon.org, [Online]. Available: https://www.daimon.org/lib/search-quotes/spider/congress-quotes.html.

[108] Y. K. Blue, Artist, *About the Future.* [Art]. 2009.

[109] L. Bergreen, Columbus: The Four Voyages 1492-1504, Penguin Books, 2012.

[110] A. Jacobsen, Operation Paperclip: The Secret Intelligence Program that Brought Nazi Scientists to America, Little, Brown and Company, 2014.

[111] G. Will, "The Doctrine of Preemtion," Imprimis, Volume 34 Issue 9, September 2005. [Online]. Available: https://imprimis.hillsdale.edu/the-doctrine-of-preemption/.

[112] J. Saunders, The History of the Mongol Conquests, University of Pennsylvania, 2001.

[113] *Pirates of the Caribbean.* [Film]. Walk Disney Pictures, 2003.

[114] "Warning Leaflets," Atomic Heritage Foundation, [Online]. Available: https://ahf.nuclearmuseum.org/ahf/key-documents/warning-leaflets/.

[115] R. Emmerich, Director, *Independence Day.* [Film]. Dean Devlin, 1996.

[116] F. Logo, "free flamethrower png," pngwing.com, [Online]. Available: https://www.pngwing.com/en/free-png-ngcka. [Accessed 2022].

[117] C. Logo, "Free Cannon Clipart," Clipart Library, [Online]. Available: http://clipart-library.com/cannons-cliparts.html.

[118] S. Logo, "Free Soldiers Clip Art," clker.com, [Online]. Available: http://www.clker.com/cliparts/v/Z/V/T/D/p/soldiers-md.png. [Accessed 2022].

[119] T. Logo, "Free Tank Profile Clip Art," openclipart.org, [Online]. Available: https://openclipart.org/detail/258912/tank-profile-illustration. [Accessed 2022].

[120] A. C. Logo, "Free US Navy Ship Silhouette," clipart-library.com, [Online]. Available: http://clipart-library.com/clip-art/us-navy-ship-silhouette-12.htm. [Accessed 2022].

[121] M. Logo, "Free Missile Clip Art," toppng.com, [Online]. Available: https://toppng.com/free-image/missile-png-PNG-free-PNG-Images_120675. [Accessed 2022].

[122] N. Logo, www.iaea.org, [Online]. Available: https://commons.wikimedia.org/wiki/File:Nuclear_symbol.svg. [Accessed 2022].

[123] D. Logo, "Free Drone Clipart," svgsilh.com, [Online]. Available: https://svgsilh.com/image/161415.html. [Accessed 2022].

[124] N. Leveson, An Introduction to System Safety Engineering, Cambridge, MA: MIT Press, 2022.

[125] A. B. William Aspray, Papers of John von Nuemann on Computing and Computer Theory, Cambridge: The MIT Press, 1986.

[126] L. Blum, "Alan Turing and the Other Theory of Computation (expanded)".

[127] A. Bromley, "The Antikythera Mechanism," *Horological Journal,* 1990.

[128] "Computer History Museum," 2008. [Online]. Available: https://www.computerhistory.org/babbage/.

[129] D. L. F. Eric G. Swedin, Computers: The Life Story of a Technology, Westport: Greenwood Press, 2005.

[130] B. Editors, "Ada Lovelace Biography," A&E Television Networks, 2 April 2014. [Online]. Available: https://www.biography.com/scholar/ada-lovelace.

[131] J. Agar, Turing and the Universal Machine: The Making of the Modern Computer, Icon Books, 2017.

[132] A. Powell, "Mark 1, Rebooted," Harvard Gazette, [Online]. Available: https://seas.harvard.edu/news/2021/07/mark-1-rebooted.

[133] B. Cohen, Howard Aiken: Portrait of a Computer Pioneer, Cambridge: The MIT Press, 2000.

[134] M. Swaine, "ENIAC Computer," Britannica, September 2022. [Online]. Available: https://www.britannica.com/technology/ENIAC. [Accessed 2022].

[135] E. R. D. H. Anthony Ralston, Encyclopedia of Computer Science, Wiley, 2003.

[136] N. Leveson, Safeware 2: Safety Engineering for a High-Tech World, 2022.

[137] AJ, "Computer," openclipart.org, [Online]. Available: https://openclipart.org/detail/17924/computer. [Accessed 2022].

[138] B. West, "Romeo & Juliet," Wikipedia Commons, [Online]. Available: https://commons.wikimedia.org/wiki/File:Attributed_to_Benjamin_West_and_studio_Romeo_and_Juliet.jpg. [Accessed 2022].

[139] S. Morgan, "Cybercrime to Cost the World $10.5 Trillion Annually by 2025," Cyber Seucirty Ventures, 13 November 202. [Online]. Available: https://cybersecurityventures.com/cybercrime-damage-costs-10-trillion-by-2025/.

[140] J. R. B. Jr., "whitehouse.gov," 12 May 2021. [Online]. Available: https://www.whitehouse.gov/briefing-room/presidential-actions/2021/05/12/executive-order-on-improving-the-nations-cybersecurity/. [Accessed 2022].

[141] N. Leveson and J. Thomas, STPA Handbook, 2018.

[142] D. Parnas, "Software Aspects of Strategic Defense Systems," *Communications of the ACM,* vol. 28, no. 12, pp. 1326-1335, 1985.

[143] J. Shore, The Sachertort Algorithm and Other Antidotes to Computer Anxiety, Penguin Books, 1986.

[144] T. Wachowskis, Director, *The Matrix.* [Film]. United States: Warner Bros., 1999.

[145] "Imagining the Internet: A History & Forecast," Elon University, [Online]. Available: https://www.elon.edu/u/imagining/time-capsule/150-years/back-1830-1860/. [Accessed 2022].

[146] N. Carlson, "Well, These New Zuckerberg IMs Won't Help Facebook's Privacy Problems," Business Insider, 13 May 2010. [Online]. Available: https://www.businessinsider.com/well-these-new-zuckerberg-ims-wont-help-facebooks-privacy-problems-2010-5?utm_source=reddit.com.

[147] P. Niquette, "niquette.com," [Online]. Available: http://www.niquette.com/books/softword/part0.htm. [Accessed 2022].

[148] W. Gibson, Neuromancer, Ace, 1984.

[149] valerybrozhinsky, "Abstract Futuristic Cyberspace with binary Code (Image license owned by Jason Lowery)," Adobe Stock, [Online]. Available: https://stock.adobe.com. [Accessed 28 December 2022].

[150] kmiragaya, "Tablet with cuneiform writing of the ancient Sumerian or Assyri (Image license owned by Jason Lowery)," Adobe Stock, [Online]. Available: https://stock.adobe.com. [Accessed 28 December 2022].

[151] Roundicons.com, "Flags Icons (Image license owned by Jason Lowery)," Adobe Stock, [Online]. Available: https://stock.adobe.com. [Accessed 28 December 2022].

[152] S. Logos, "Vectorstock Image 29580405 (License #45279500 owned by Jason Lowery)," Vector Stock, [Online]. Available: https://www.vectorstock.com/royalty-free-vector/social-media-logo-icon-set-on-circle-button-vector-29580405. [Accessed 28 December 2022].

[153] L. Alden, "Twitter @LynAldenContact," 24 October 2022. [Online]. Available: https://twitter.com/LynAldenContact/status/1584714423130017793?s=20&t=DGJNi0k3oEVPMWvSp_Znkg.

[154] "PC Monitor Clip Art," illustoon, [Online]. Available: https://illustoon.com/?id=2772.

[155] L. Logo, "Free Leg Iron Clipart," wannapik.com, [Online]. Available: https://www.wannapik.com/vectors/2770. [Accessed 2022].

[156] C. Logo, "Free Metal Chain Clipart," publicdomainvectors.org, [Online]. Available: https://publicdomainvectors.org/en/free-clipart/Metal-chain-clip-art/83531.html. [Accessed 2022].

[157] A. Back, "Hashcash - A Denial of Service Counter-Measure," 2002. [Online]. Available: http://www.hashcash.org/hashcash.pdf.

[158] "Free Sparkle Effect PNG Image With Transparent Background," toppng.com, [Online]. Available: https://toppng.com/free-image/sparkle-effect-png-PNG-free-PNG-Images_105161. [Accessed 2022].

[159] "power lines electric power free image," pixabay.com, [Online]. Available: https://pixabay.com/illustrations/power-lines-electric-power-hot-4582650/.

[160] Aquir, "Approved red grudge round vitabe rubber stamp (License #2087884682)," iStock by Getty Images, [Online]. Available: https://www.istockphoto.com/vector/approved-red-grunge-round-vintage-rubber-stamp-gm652331292-118446189?irgwc=1&cid=IS&utm_medium=affiliate&utm_source=FFF%20Web%20Media%20Inc.&clickid=wcfSc1UDXxyNWKazf7X%3ArVZwUkAzR-3NRRnYxg0&utm_term=ADP&utm_campaign=&ut. [Accessed 2022].

[161] J. S. Nye, "Soft Power," *Foreign Policy,* vol. 80, pp. 153-171, 1990.

[162] M. Saylor, "@saylor," Twitter, 18 September 2020. [Online]. Available: https://twitter.com/saylor/status/1307029562321231873?s=20&t=V8Juco2oQ_sdHWYZ3RJKXg.

[163] Anatolir, "Adobe Stock #318583143 (Licensed by Jason Lowery)," Adobe Stock, [Online]. Available: https://stock.adobe.com/images/yellow-notebook-icon-flat-illustration-of-yellow-notebook-vector-icon-for-web-design/318583143?as_campaign=ftmigration2&as_channel=dpcft&as_campclass=brand&as_source=ft_web&as_camptype=acquisition&as_audience=users&as_content. [Accessed 28 December 2022].

[164] M. Saylor, *The Atlas Society Gala Speech,* 2022.

[165] T. Paine, 23 December 1776. [Online]. Available: https://www.ushistory.org/paine/crisis/c-01.htm.

[166] M. Irving, "Record-breaking chip can transpit entire internet's traffic per second," newatlas.com, 2022.

[167] "Antikythera," Berggruen Institute, [Online]. Available: https://www.berggruen.org/work/antikythera-2/.

[168] "wikipedia.com," NeXTcube, 20 April 2015. [Online]. Available: https://en.wikipedia.org/wiki/Motherboard#/media/File:NeXTcube_motherboard.jpg. [Accessed 1 December 2022].

[169] "nasa.gov," [Online]. Available: https://www.nasa.gov/topics/earth/images/index.html. [Accessed 1 December 2022].

[170] "Earth & Ethernet," [Online]. Available: https://vpn-services.bestreviews.net/socks5-and-vpn/. [Accessed 1 December 2022].

[171] I. Logo, "Free Internet Network Clipart," pixabay.com, [Online]. Available: https://pixabay.com/vectors/internet-network-scheme-154450/. [Accessed 2022].

[172] M. Saylor, "@saylor," Twitter, 28 April 2022. [Online]. Available: https://twitter.com/saylor/status/1519664335047598080?s=20&t=fKn6N4d6dMIH8ms5lh44ZQ.

[173] B. Logo, "Free Bitcoin Logo Clipart," Wikipedia Commons, [Online]. Available: https://commons.wikimedia.org/wiki/File:Bitcoin_logo.svg. [Accessed 2022].

[174] Sensvector, "Cryptocurrency in cyberspace (Licensed by Jason Lowery)," Adobe Stock, [Online]. Available: https://stock.adobe.com/search?filters%5Bcontent_type%3Aphoto%5D=1&filters%5Bcontent_type%3Aillustration%5D=1&filters%5Bcontent_type%3Azip_vector%5D=1&filters%5Bcontent_type%3Avideo%5D=1&filters%5Bcontent_type%3Atemplate%5D=1&filters%5Bcontent_type%3A3d%5. [Accessed 28 December 2022].

[175] H. S. Wolk, "The Genius of George Kenney," *Air and Space Forces Magazine,* 2002.

[176] B. Seidl, "Transistors in Different Housings," Wikipedia Commons, [Online]. Available: https://commons.wikimedia.org/wiki/File:Transistors-white.jpg. [Accessed 2022].

[177] Ю. Кондратьева, "ASIC bitcoin miner and ASIC mining farm (Licensed by Jason Lowery)," Adobe Stock, [Online]. Available: https://stock.adobe.com/images/asic-bitcoin-miner-and-asic-mining-farm-bitcoin-mining-cryptocurrency-mining-equipment-and-hardware-flat-vector-illustration-isolated-on-white/454558824?prev_url=detail. [Accessed 28 December 2022].

[178] Bitboy, "Bitcoin Logo," [Online]. Available: https://commons.wikimedia.org/wiki/File:Bitcoin_logo.svg#/media/File:Bitcoin.svg.

[179] "Ethereum Logo," cryptologos.cc, [Online]. Available: https://cryptologos.cc/ethereum.

[180] V. Buterin, "Ethereum Foundation," Youtube, 14 September 2022. [Online]. Available: https://www.youtube.com/watch?v=Nx-jYgI0QVI.

[181] "MEV Watch," [Online]. Available: https://www.mevwatch.info/. [Accessed 12 January 2022].

[182] "Total Bitcoin Hash Rate," blockchain.com, 2022. [Online]. Available: https://www.blockchain.com/explorer/charts/hash-rate.

[183] B. A.G., "Twitter," [Online]. Available: https://twitter.com/GetBittr/status/1593573892060332032?s=20&t=iEsHUtj6T2pxe_HX3vLUig.

[184] Atlantic Council, "The Implications of a MADCOM World - Three Scenarios for the Future," JSTOR, 1 September 2017. [Online]. Available: https://www.jstor.org/stable/resrep03728.5?mag=to-fight-fake-news-broaden-your-social-circle#metadata_info_tab_contents.

[185] "Gabriel's Horn," Soul of Mathematics, 21 March 2021. [Online]. Available: https://soulofmathematics.com/index.php/gabriels-horn/.

[186] H. Finney, "RPOW - Reusable Proofs of Work," https://nakamotoinstitute.org/finney/rpow/, 2004.

[187] V. Putin, "Public Comments, 2019 Russian Energy Week International Forum".

[188] T. Paine, *Common Sense (Pamphlet),* 1776.

[189] D. Held, "@danheld," Twitter, 8 January 2023. [Online]. Available: https://twitter.com/danheld/status/1612158532324835328?s=20&t=rC7Y8_-CgrHztTMMk2bfog.

[190] M. Fox, "Bitcoin adoption for countries is a "high stakes' game," Business Insider, 13 Jan 2022. [Online]. Available: https://markets.businessinsider.com/news/currencies/bitcoin-adoption-countries-high-stakes-game-theory-central-banks-crypto-2022-1. [Accessed 2022].

[191] C. Sagan, The Demon-Haunted World: Science as a Candle in the Dark, 1995.

[192] F. viewpoints, "Investing in Bitcoin: What to Consider," Fidelity, 2022. [Online]. Available: https://www.fidelity.com/learning-center/trading-investing/investing-in-bitcoin.

[193] N. Chomsky, "Noam Chomsky Quotes from the Father of Linguistics," Everyday Power, [Online]. Available: https://everydaypower.com/noam-chomsky-quotes/.

References

[194] R. Breedlove, "Robert Breedlove on Jason Lowery's 'Power Project' through Bitcoin PoW Thesis," The Your Life! Your Terms! Show, 2022. [Online]. Available: https://fountain.fm/clip/srY2zNY9bKI22057vd6x.

[195] J. Vanuga, Artist, *Snarling Gray Wolf.* [Art]. https://www.jeffvanuga.com/.

[196] R. Breedlove, Interviewee, [Interview].

[197] A. Lincoln, Writer, *Cooper Union Speech.* [Performance]. 1860.

[198] T. Draxe, Writer, *Bibliotheca Scholastica Instructissima.* [Performance]. 1618.

[199] S. F. S. T. T. A. A. D. S. Tobias Andermann, "The past and future human impact on mammalian diversity," *Science Advances,* vol. 6, no. 36, 2020.

[200] H. Ritchie, "Wild mammals have declined by 85% since the rise of humans, but there is a possible future where they flourish," Our World Data, ourworldindata.org, 2021.

[201] J. Kosinski, Director, *Oblivion.* [Film]. United States: Relativity Media, 2013.

[202] W. Shakespeare, Writer, *Julius Caesar.* [Performance]. 1599.

[203] T. E. o. Encyclopaedia, Britannica, Encyclopedia Britannica, 2018.

[204] G. M. Hopf, Those Who Remain: A Postapocalyptic Novel, CreateSpace, 2016.

[205] E. Johansson, Artist, *A Marvel of Engineering.* [Art]. www.erikjo.com, 2022.

[206] C. Nolan, Director, *The Dark Knight Rises.* [Film]. United States: Warner Bros., 2012.

[207] "coinyuppie.com," 21 September 2022. [Online]. Available: https://coinyuppie.com/bitcoin-prequel-3-crypto-wars-in-the-90s/.

[208] E. Rice, "The Second Amendment and the Struggle Over Cryptography," *Hastings Science & Technology Law Journal,* vol. 9, 2017.

[209] H. Ford, *Clay Worker magazine vol. 85-86,* p. 230, 1926.

[210] W. B. Wriston, Bites, Bytes, and Balance Sheets: The New Economic Rules of Engagement in a Wireless World, Hoover Institution Press, 2007.

[211] T. Local, "Nobel Descendant slams Economics Prize," 28 September 2007. [Online]. Available: https://web.archive.org/web/20071014012248/http://www.thelocal.se/2173/20050928/. [Accessed 2022].

[212] "Darwin Corresondence Project," University of Cambridge, [Online]. Available: https://www.darwinproject.ac.uk/people/about-darwin/six-things-darwin-never-said/evolution-misquotation#:~:text=Megginson%20wrote%20in%201963%3A,in%20which%20it%20finds%20itself..

[213] A. More, V for Vendetta, Vertigo, 2008.

[214] A. C. B. S. Prabhupad, Bhagavad Gita, 2nd Century CE.

[215] "power lines electric power," pixabay.com, 2022. [Online]. Available: https://pixabay.com/illustrations/power-lines-electric-power-hot-4582650/.

[216] Aquir, "Approved red grunge round vintage rubber stamp (License #)," iStock, 2022. [Online]. Available: https://www.istockphoto.com/vector/approved-red-grunge-round-vintage-rubber-stamp-gm652331292-118446189?irgwc=1&cid=IS&utm_medium=affiliate&utm_source=FFF%20Web%20Media%20Inc.&clickid=wcfSc1UDXxyNWKazf7X%3ArVZwUkAzR-3NRRnYxg0&utm_term=ADP&utm_campaign=&ut.

[217] J. Lowery, "@JasonPLowery," Twitter, 1 November 2021. [Online]. Available: https://twitter.com/JasonPLowery/status/1455217564641075207?s=20&t=VUsTmelIU1MtERDQ_P3sCA.

[218] Q. Logo, "Free Drone Clipart," www.pngitem.com, [Online]. Available: https://www.pngitem.com/middle/TwRTTT_clipart-drone-clipart-png-transparent-png/. [Accessed 2022].

[219] D. Logo, "Free White-Tailed Deer Silhouette," favpng.com, [Online]. Available: https://favpng.com/png_view/deer-white-tailed-deer-silhouette-clip-art-png/M3ydkq0N. [Accessed 2022].

[220] L. Fridman, "@lexfridman," Twitter, 12 October 2022. [Online]. Available: https://twitter.com/lexfridman/status/1580265560314871808?s=20&t=mL-c2loP9_TdSTlGFY9Ljw.

[221] J. Renard, "Jules Renard Quotes," quote.org, [Online]. Available: https://quote.org/quote/failure-is-not-our-only-punishment-for-298256. [Accessed 2022].

[222] S. Ruocco, "@sergio_GS," Twitter, 26 December 2022. [Online]. Available: https://twitter.com/Sergio_GS/status/1607600404405293057?s=20&t=nnFt1U_VIMKh6MUrxe2lGg.

Made in United States
Troutdale, OR
03/31/2025

30040361R00204